Power Electronics Semiconductor Switches

Power Electronics Semiconductor Switches

R.S. Ramshaw

Department of Electrical and Computer Engineering
University of Waterloo
Ontario
Canada

CHAPMAN & HALL
London · Glasgow · New York · Tokyo · Melbourne · Madras

Published by Chapman & Hall, 2–6 Boundary Row, London SE1 8HN

Chapman & Hall, 2–6 Boundary Row, London SE1 8HN, UK

Blackie Academic & Professional, Wester Cleddens Road, Bishopbriggs, Glasgow G64 2NZ, UK

Chapman & Hall Inc., 29 West 35th Street, New York NY10001, USA

Chapman & Hall Japan, Thomson Publishing Japan, Hirakawacho Nemoto Building, 6F, 1–7–11 Hirakawa-cho, Chiyoda-ku, Tokyo 102, Japan

Chapman & Hall Australia, Thomas Nelson Australia, 102 Dodds Street, South Melbourne, Victoria 3205, Australia

Chapman & Hall India, R. Seshadri, 32 Second Main Road, CIT East, Madras 600 035, India

---

First edition 1993

Printed in Great Britain by St Edmundsbury Press Ltd, Bury St Edmunds, Suffolk

ISBN 0 412 28870 2

A catalogue record for this book is available from the British Library

Library of Congress Cataloging-in-Publication data available

∞ Printed on permanent acid-free text paper, manufactured in accordance with the proposed ANSI/NISO Z 39.48–199X and ANSI Z 39.48–1984

CONTENTS

PREFACE

Power electronics plays a significant role in the application of electric power wherever there is a need to change voltage, current or frequency from the standard values that are available. At the heart of power electronics is a fast acting switch that is used, not just to isolate the source from the load, but to modulate the power that reaches the load.

This is an introductory text about semiconductor switches and their uses to modulate power to a form best suited to a load. Today the subject of power electronics has expanded to include not only power semiconductor devices, but also to incorporate converter topologies, analysis and simulation, control and estimation techniques and control hardware together with the attendant software. Each of these is a topic in its own right and is left for further study following this text.

In the first part of the text there is a description of how power is converted from ac to dc, ac to ac, dc to dc and dc to ac forms in the terms of the periodic opening and closing of a single switch or a group of switches that are ideal. That is, the basics of power conditioning are generalized.

In the second part of the text, systems of supply, converter and load are described. The equations that govern the behaviour of those converters with ideal switches are formulated to provide the fundamentals of the analysis of power conditioning in basic circuits. The result is twofold. The basic responses of currents voltages and average power are determined for the input and output of the converter, starting with the simplest load and progressing to complex passive loads. Further, the performance factors that describe the goodness of power conversion are applied to each type of converter. This description is broadly based with simple analytical treatments of simple systems to show in first principles what is possible to be done with more complex switching systems.

With this background to the use of semiconductor switches, the important switches are treated separately in the third and main part of the text. The purpose is to help the reader make a choice of the right switch for a particular application and to help in the use of the switches in circuits. For each switch there is a description of operation in general terms of a *pn* junction and charge flow under the influence of electric fields. The physics of operation is omitted, because the text is aimed at those who may have the desire to use switches in a converter. Circuit operation and switch control are the important objectives of this text.

No switch is perfect. Accordingly, some attempt is made to accentuate the performance characteristics of the switch in terms of steady-state and transient operation. From this, the main forms of protection become evident.

Design techniques for circuits and systems that incorporate switches are left to manufacturers' manuals. The analytical descriptions, that are given in this text, are mainly to show the principles of switch and circuit behaviour and to provide a base so that manufacturers' data sheets and manuals can be used to the best effect. The best effect usually means minimum losses and waveforms as close to the ideal as possible.

This book is intended as a text for readers who wish to become users of semiconductor switches and who wish to understand basics of power electronics and switch applications. It is for this reader that there are many worked examples and many more problems for practice. Hands-on experience quickly shows the fragility and limitations of power electronics circuits, just like solving problems enables rapid learning of principles and an acknowledgement of the limitations of our modelling and analytical tools. Many of the problems make use of chopper circuits to exemplify a switch's performance, because choppers provide extreme characteristics that are relatively easy to analyze.

Although this book sets out to be a first course in the subject of power electronics, it does rely on the readers' familiarity with a fundamental electrical engineering course in circuit theory and its attendant mathematics. The level of analysis has increased in recent years. The knowledge of mathematics that is required has not changed much, but the complexity of the switched-circuit analysis has been extended by the availability of mathematical tools for computer aided analysis. There are software packages (MATLAB, MATHCAD etc.) available to take away the drudgery of solving a transcendental equation and to solve problems of Fourier analysis, of complex integrals or simultaneous transcendental equations that might have been left alone before.

As this book is an introductory text, the bibliography given at the end of each chapter is an invitation for further reading.

Acknowledgement and gratitude are due to Elizabeth Nicklin who has taken the author's scribblings and sketches and transformed them into the typeset text and figures set before you. The University of Auckland kindly provided the facilities for the preparation of the text. At the University of Auckland James Thompson was kind enough to solve many of the MATLAB problems and Grant Spencer gave his time generously to proof-read the final draft. Thanks are due to Philips Export B.V. and Harris Semiconductor for giving permission to reproduce their data sheets.

Raymond Ramshaw
Waterloo.

CHAPTER 1

POWER CONDITIONING

1.1. INTRODUCTION

Power electronics is an established technology that bridges the power industry with its need for fast controllers, and the semiconductor industry with its attempt to produce devices with greater power handling capabilities. In essence, what power electronics does is to condition the power from a supply to suit the needs of the load. The main element of power electronics is the semiconductor switch. A description of semiconductor switches and how they are used is the subject of this book.

In this chapter the concepts of power conditioning are laid out. The concepts are based on switching techniques that are idealized. These are expanded in Chapter 2 to provide a base for the analysis of circuits with converters that employ switches. The following chapters describe the real switches that are used in the converters.

1.2. POWER ELECTRONICS

Power electronics emerged in the late 1950s when the silicon controlled rectifier (SCR) was developed. The SCR became known as the thyristor. This thyristor was a semiconductor switch that could control power of the order of kilowatts. It could switch on and off faster than any electromechanical switch and its voltage and current ratings were higher than any transistor. Manufacturing techniques of thyristors improved and with these the ratings of the switches increased. Today, power electronics has the capability of controlling blocks of power of the order of hundreds of megawatts. Over a period of 35 years this is a great advance.

As with all switching devices, the thyristor has limitations. To fill some of the shortfalls of thyristors, other semiconductor switches have been developed for use in the power industry. They include the triac, the gate-turn-off thyristor (GTO), the bipolar transistor, the power MOSFET and the insulated-gate bipolar transistor (IGBT), together with some emerging variations that are discussed in the following chapters. Even these do not cover adequately all the range of possible applications. However, there is continual development of semiconductor switches and combinations of switches in an attempt to satisfy the almost insatiable need to control blocks of power at higher rates of switching for greater efficiency and better quality conditioning. We need to know the characteristics of those switches that are available to us in order to choose the best switch for any application.

No matter what detailed form the control of power actually takes, the system configuration is usually the same. Figure 1.1 illustrates a general system configuration in block diagram form.

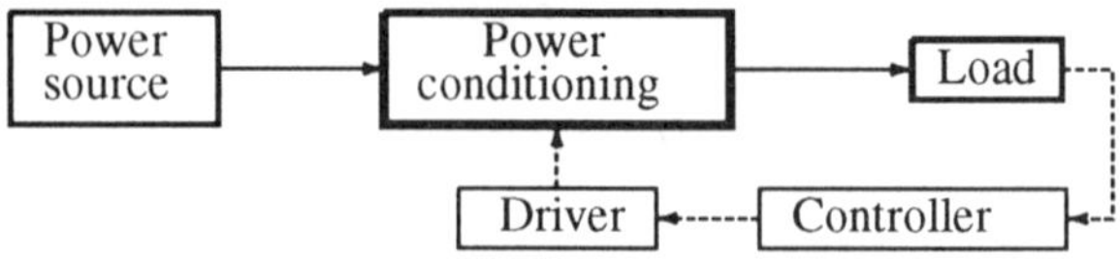

Fig. 1.1 Power conditioning.

The power-source unit is usually a constant voltage, fixed frequency, ac supply. In some cases it is a constant voltage dc supply. Occasionally, as in aircraft, the source could be a non-constant voltage, variable frequency, ac supply. Then, there is the load unit, which converts the energy from the supply into a form to do useful work. The loads can be many and varied, from lighting to heating and from mechanical drives to dc links in high voltage ac power systems. Because not all loads match all sources, power conditioning must be performed in many cases. The power conditioning is accomplished by a switch or group of switches. Since the switches must be turned on and off, there must be driver circuits for these functions. Also, since the switches must be turned on and off in a special sequence, there is a need for a controller unit. Together, the power conditioning switches, the driver and the controller units constitute the power electronics component of the system.

1.3. POWER MODULATION

Power conditioning is known also as power modulation or power conversion. This conditioning may take a specialized form such as the regulation of voltage, or the conversion of alternating current to direct current, or the conversion of direct current to alternating current.

There is no one configuration of the power-electronics system to satisfy all types of conditioning. This becomes evident as a number of systems is considered. Indeed, the conversion of one given type of supply (usually specified by voltage and frequency) into any other type of supply has to be designed on its own merits. Seven simple types are described in this section to exemplify the concepts that are involved. These types are shown in Table 1.1. The first three types convert an ac or dc fixed-voltage source into an adjustable-voltage dc supply. Types 4 and 5 create an adjustable ac voltage from a fixed voltage ac source. The most versatile types are 6 and 7. They allow the conversion of a fixed voltage source into an adjustable voltage, adjustable frequency supply.

The power-electronics switches that are used to create the power conditioning are allowed to be ideal. Ideal, in the sense of switching, means that it takes no time at all to turn on or to turn off the switch. While the switch is on it offers no impedance to current and while it is off it offers infinite impedance to current. There is no need for switch protection. During the on-state, there is no limit to the current and during the off-state the switch can withstand any value of voltage or rate of change of voltage without changing state or being destroyed. The use of a perfect switch allows the investigation of the principles of power conditioning without the need to compensate for the shortcomings of real switches. Chapters 3 to 10 are reserved for the true characteristics of switches.

1.3.1. DC Supplies to a Load

Table 1.1 indicates that there are at least three possibilities (types 1, 2 and 3) to obtain an adjustable dc voltage by power electronics. In practice there are many complex circuits. The reader needs only view the different dc power supplies on the market to confirm this. However, the principles are not so different from those generic forms that follow.

TABLE 1.1 Power conditioning

	Type	Description
1	ac ⟩ to ⟩ dc	Integral-cycle control
2	ac ⟩ to ⟩ dc	Point-on-wave switching
3	dc ⟩ to ⟩ dc	Chopper control
4	ac ⟩ to ⟩ ac	Integral-cycle control
5	ac ⟩ to ⟩ ac	Point-on-wave switching
6	ac ⟩ to ⟩ ac	Cycloconverter
7	dc ⟩ to ⟩ ac	Inverter

AC source to adjustable dc supply (type l).

Type-1 conversion is defined as *integral-cycle* control. Figure 1.2a depicts a circuit that allows the ac source to be rectified in such a controlled manner that the average dc voltage across the load has an adjustable value. The ac source has a voltage v_s whose waveform is sinusoidal, as shown in Fig. 1.2b.

The four ideal switches *Swa*, *Swb*, *Swc* and *Swd* are formed into a bridge configuration. This allows the use of the switches as a rectifier to obtain a dc voltage across the load. However, the switches can interrupt at the end of any half cycle of the source voltage (hence the name, integral-cycle control), so that any number of half cycles of positive voltage can appear across the load and any number of half cycles can be blocked.

Figure 1.2c shows the instantaneous voltage across the load with an arbitrary block of one half cycle every three half cycles of the source voltage. The average voltage V_{av} across the load is shown also. This method requires the switching, on or off, to take place at the voltage-zero crossing. From the figure the sequence of switching is as follows.

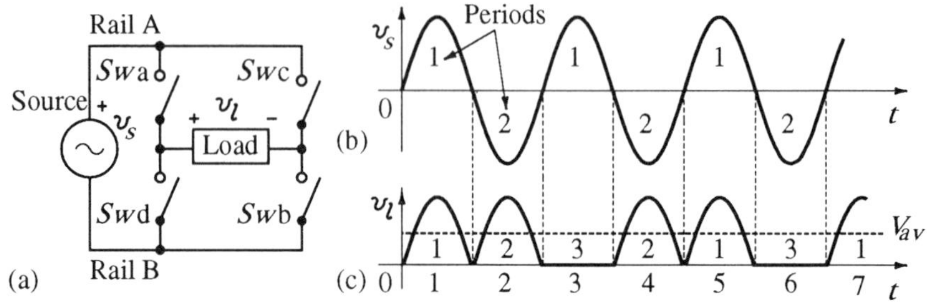

Fig. 1.2 Integral cycle control, ac to dc conversion.
(a) Circuit diagram, (b) source voltage, (c) load voltage.

1. Period 1: switches *Swa* and *Swb* on, switches *Swc* and *Swd* off.
2. Period 2: switches *Swa* and *Swb* off, switches *Swc* and *Swd* on.
3. Period 3: switches *Swa*, *Swb*, *Swc* and *Swd* off.
4. Period 2: switches *Swa* and *Swb* off, switches *Swc* and *Swd* on.
5. Period 1: switches *Swa* and *Swb* on, switches *Swc* and *Swd* off.
6. Period 3: switches *Swa*, *Swb*, *Swc* and *Swd* off.
7. Repeat 1.

For the load voltage v_l to have the polarity as shown in Fig. 1.2a, rail A must be positive while switches *Swa* and *Swb* are closed and rail B must be positive while switches *Swc* and *Swd* are closed. The average voltage V_{av} across the load is reduced by blocking more half cycles. What would be the sequence of switching if it were required that the polarity of the load voltage change, and if the magnitude of the average load voltage were to be one half of the maximum average value?

This method of conversion seems good. The method is simple, and, because complete half cycles are switched, no EMI (electromagnetic interference) is generated and the power factor on the ac side is high. However, this method is not common, and the reason could be that there are easier ways to accomplish the same results.

A disadvantage of this method is that the frequency of the source must be high to stop mechanical oscillation about the mean speed if the load is a motor. A period of no source current would tend to make a motor slow down. The reinstatement of a source current would tend to make the motor increase speed. The value of the ensuing speed oscillation depends on the motor inertia, but would decrease with an increase in the source frequency.

AC source to adjustable dc supply (type 2).

Type-2 conversion of an ac source to a dc supply is classified as *point-on-wave control* in Table 1.1. It is also known as ac-line control or phase control. The ac source is not only rectified, but the point in every half cycle that the voltage v_l appears across the load is delayed by an angle α that is called the *trigger angle*, or *firing angle*, or *delay angle*.

Figure 1.3a shows the ac source, the dc load and four switches *Swa*, *Swb*, *Swc* and *Swd* in a bridge configuration. Switching *Swa* and *Swb* on while rail A is positive and switching *Swc* and *Swd* on while rail B is positive rectifies the source voltage whose waveform is shown in Fig. 1.3b. At every half cycle of the voltage waveform there is a delay α before switches *Swa* and *Swb* (or *Swc* and *Swd*) are turned on. The delay angle α can be adjusted from zero to π radians. However, it is shown in Fig. 1.3c that *Swa* and *Swb* (or *Swc* and *Swd*) are turned off at the end of the half cycle. This demonstrates the principles. For a particular load voltage the firing angle α is fixed.

The sequence of switching is as follows.

1. Period 1: switches *Swc* and *Swd* are off.
   At instant $1^{'}$ (α), *Swa* and *Swb* are switched on.
2. At instant $1^{''}$ (π), *Swa* and *Swb* are switched off.
3. Period 2: switches *Swa* and *Swb* are off.
   At instant $2^{'}$ (α), *Swc* and *Swd* are switched on.
   At instant $2^{''}$ (π), *Swc* and *Swd* are switched off.
4. Repeat 1.

The maximum average voltage across the load is obtained for $\alpha = 0$. A reduction of voltage is achieved by increasing the value of the firing angle α, until, at $\alpha = \pi$, the switches never turn on and the load voltage is zero.

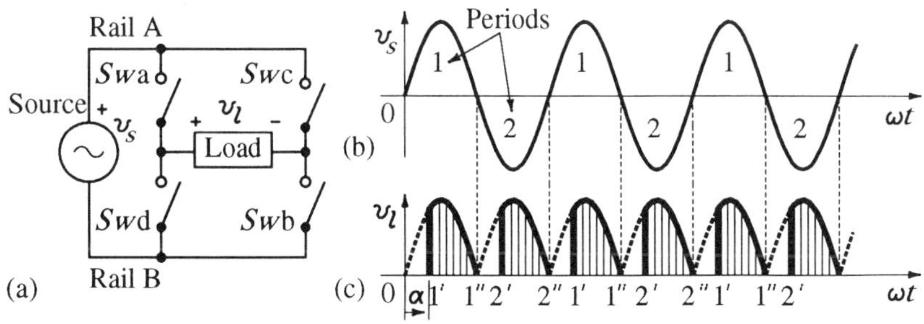

Fig. 1.3 Point-on-wave control, ac to dc conversion.
(a) Circuit diagram, (b) source voltage, (c) load voltage.

This form of switching has simple control. Therefore, the method is common. The disadvantages are that this switching mode produces a poor power factor on the ac side, and that there is EMI generated by the nonsinusoidal waveform at the load.

DC source to adjustable dc supply (type 3).
Type-3 form of conversion is classified as a *chopper* in Table 1.1. The source voltage v_s is an unvarying dc voltage V_s as shown in Fig. 1.4a and 1.4b. By turning the switch *Swa* on and off in a particular sequence, illustrated in Fig. 1.4c, the average voltage V_{av} across the load can be adjusted from zero to V_s.

The switching sequence is as follows.

1. Switch *Swa* is turned on for a time t_{ON}. $v_l = V_s$ during this time.
2. Switch *Swa* is turned off for a time t_{OFF}. $v_l = 0$ during this time.
3. Repeat 1.

By inspection of Fig. 1.4c the average voltage V_{av} across the load is given by

$$V_{av} = V_s \frac{t_{ON}}{T} = V_s \frac{t_{ON}}{t_{ON} + t_{OFF}} \tag{1.3.1}$$

where $t_{ON}/T \triangleq m$ = duty cycle (or mark-space ratio).

Any of t_{ON}, t_{OFF} or T can be made a variable in order to adjust the value of V_{av}. The most common mode is to fix the period T of switching and make the on-time t_{ON} the independent variable. This method is called *pulse-width modulation* (PWM) and is used often in both ac and dc applications.

1.3.2. AC Supplies to a Load

Table 1.1 indicates that there are at least four types of conversion that provide an ac supply to a load. The sources can be either ac or dc supplies. Each type utilizes a different technique of switching and has its own special areas of application. The techniques will be described in the following subsections and the applications will be described in Chapter 2.

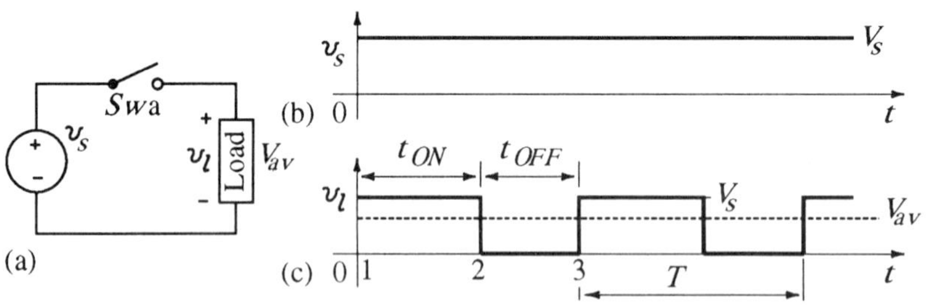

Fig. 1.4 Chopper, dc to dc conversion.
(a) Circuit diagram, (b) source voltage, (c) load voltage.

AC source to adjustable ac supply (type 4).
Type-4 conversion is classified as *integral-cycle control* in Table 1.1. This method is similar to type 1 except that rectification is excluded. Frequency is not adjustable.

Figure 1.5a illustrates the circuit comprising the source, the load and the power electronics (switch *Swa*) that produces the conversion. Figure 1.5b shows the waveform of the ac source voltage v_s and Fig. 1.5c shows the waveform of the load voltage for the particular switching sequence that follows.

1. Period 1: rail A positive, switch *Swa* on for a full half-cycle.
2. Period 2: rail B positive, switch *Swa* on for a full half-cycle.
3. Period 3: rail A positive, switch *Swa* off for a full half-cycle.
4. Period 2: rail B positive, switch *Swa* on for a full half-cycle.
5. Period 1: rail A positive, switch *Swa* on for a full half-cycle.
6. Period 3: rail B positive, switch *Swa* off for a full half-cycle.
7. Repeat 1.

The maximum load voltage is obtained if switch *Swa* is on continuously. A reduction of effective load voltage is achieved by increasing the number of blocked half-cycles (period 3). In order that the load response to the blocked half-cycles is minimal instantaneously but acceptable on average, the frequency of the supply has to be high.

The switch driver circuit and controller are complex, so that this type of conversion is rarely found in practice.

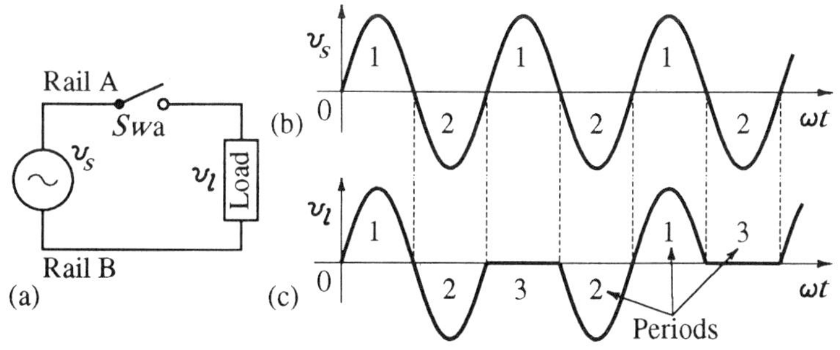

Fig. 1.5 Integral cycle control, ac to ac conversion.
(a) Circuit diagram, (b) source voltage, (c) load voltage.

AC source to adjustable ac supply (type 5).
The type-5 conversion in Table 1.1 is one of the simplest forms of converting an ac source of fixed voltage to an ac supply of adjustable voltage (adjustable in the sense that the rms value can be changed). The frequency at the load is always the same as the frequency at the source.

The *point-on-wave* (ac-line control or phase control) technique is employed in the same way as in type-2 conversion, but no rectification is required. Figure 1.6 depicts the circuit, the source-voltage waveform and the load-voltage waveform. The switching sequence can be followed by inspecting Fig. 1.6b and 1.6c. The sequence is as follows.

1. Period 1: rail A positive, instant $1'$ (α), switch *Swa* turned on.
2. Switch *Swa* turned off at instant $1''$.
3. Period 2: rail B positive, instant $2'$ (α), switch *Swa* turned on.
4. Switch *Swa* turned off at instant $2''$.
5. Repeat 1.

The voltage is reduced by increasing the delay angle α. If $\alpha = 0$, then the load voltage is a maximum ($V_{l\,rms} = V_{s\,rms}$). At $\alpha = \pi$, the load voltage is a minimum ($V_{l\,rms} = 0$). Hence, the load voltage can be controlled to be any value between zero and the source-voltage value.

AC source to adjustable ac supply (type 6).
According to Table 1.1 a type-6 converter is called a *cycloconverter.* It utilizes an ac source. By rectification and point-on-wave control a supply of adjustable voltage and frequency can be generated. Figure 1.7 is an aid to the principle of the cycloconverter. The circuit is shown in Fig. 1.7a. The switch arrangement is a bridge configuration, that is supplied by an ac voltage v_s, whose waveform is shown in Fig. 1.7b, and that supplies a complex load voltage v_l, whose waveform is shown in Fig. 1.7c.

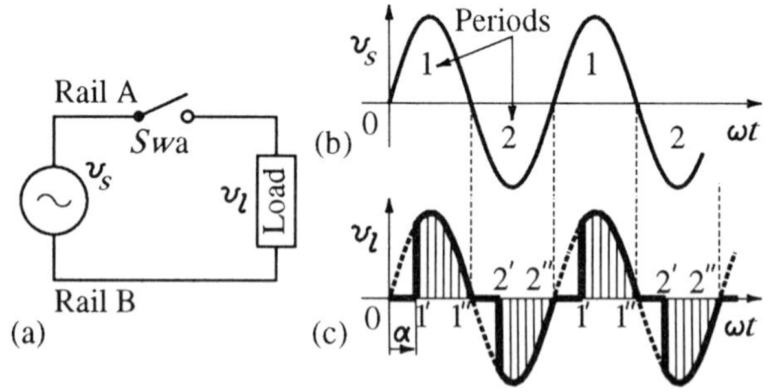

Fig. 1.6 Point-on-wave control, ac to ac conversion.
(a) Circuit diagram, (b) source voltage, (c) load voltage.

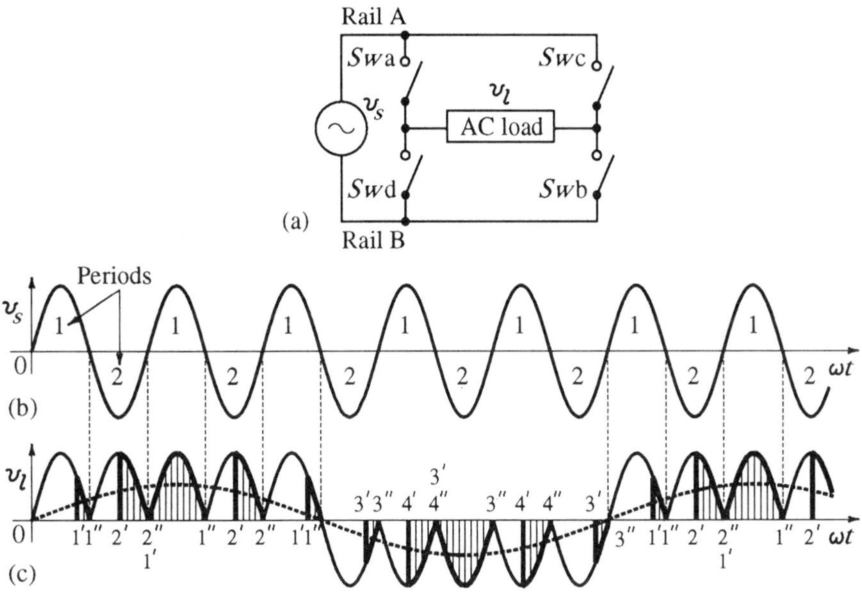

Fig. 1.7 Cycloconverter, ac to ac conversion.
(a) Circuit diagram, (b) source voltage, (c) load voltage.

The power conditioning for the load can be split into two parts, the frequency and the magnitude of the voltage. With reference to Fig. 1.7c the switching sequence of switches *Swa*, *Swb*, *Swc* and *Swd* is such as to rectify the supply, so that 5 half cycles of the supply provide a positive polarity at the load. Then 5 half cycles of the supply provide a negative polarity at the load. These 10 half cycles of the supply provide 2 half cycles for the load, one positive and one negative, to constitute one cycle of an ac supply to the load. Consequently, the load frequency is obtained by the rectification mode. Voltage adjustment is then governed by point-on-wave modulation. This control of the firing angle α can also improve the wave shape. Angle α is shown to vary from a large value at the beginning of the half cycle, to zero at the middle of the half cycle, and back to a large value at the end. In this way, the wave can look more like the required sine wave that is shown as a broken line in the figure.

This particular switching sequence is as follows.

1. Period 1: rail A positive, *Swc* and *Swd* off, *Swa* and *Swb* on at $1'$.
2. Period 1: *Swa*, *Swb*, *Swc* and *Swd* off at $1''$.
3. Period 2: rail B positive, *Swa* and *Swb* off, *Swc* and *Swd* on at $2'$.
4. Period 2: *Swa*, *Swb*, *Swc* and *Swd* off at $2''$.

Repeat for a total of 5 half cycles of the source voltage.

.

11. Period 2: rail A positive, *Swa* and *Swb* off, *Swc* and *Swd* on at $3'$.
12. Period 2: *Swa*, *Swb*, *Swc* and *Swd* off at $3''$.
13. Period 1: rail B positive, *Swc* and *Swd* off, *Swa* and *Swb* on at $4'$.
14. Period 1: *Swa*, *Swb*, *Swc* and *Swd* off at $4'$.

.

Repeat for a total of 5 half cycles of the source voltage.

.

This sequence is repeated to provide a continuous output of the required frequency and voltage.

It can be seen that the method is complex. The frequency of the output is lower than the source frequency and the harmonic content of the output voltage waveform is high. There is an advantage, however. It is that this method is a direct ac to ac conversion. No dc links are required. Aircraft use the cycloconverter because it is light weight. The example used here takes a fixed-frequency, fixed-voltage supply and describes the principle of producing a lower frequency and an adjustable voltage. In an aircraft the generators coupled to the engines produce varying high frequency, varying voltage at the terminals because the speed is varying. The cycloconverter switching for this application is to produce a constant-voltage, constant-frequency supply.

DC source to adjustable ac supply (type 7).

A converter that modulates a dc power source so that a load can be supplied with ac voltage is named an *inverter*. One such configuration is shown in Fig. 1.8a. It comprises a bridge. The input to the bridge is a constant-voltage source, depicted in Fig. 1.8b and the output is a square-wave ac voltage, as shown in Fig. 1.8c.

From an inspection of the figure the switching sequence is

1. Period 1: switches *Swc* and *Swd* are off, switches *Swa* and *Swb* are on.
2. Period 2: switches *Swa* and *Swb* are off, switches *Swc* and *Swd* are on.
3. Repeat 1.

It is sometimes required to have both an adjustable voltage and an adjustable frequency at the output of the inverter. Since the on and off sequencing of the switches can be programmed for almost any pattern[1], it is possible to produce any analogue waveform very closely by this digital means. Figure 1.9 illustrates how the output voltage and frequency can be changed independently. The switching sequence is as follows.

1. Switches *Swa*, *Swb*, *Swc* and *Swd* are off.
2. Period 1: switches *Swc* and *Swd* are off, switches *Swa* and *Swb* are on.
3. Switches *Swa*, *Swb*, *Swc* and *Swd* are off.
4. Period 2: switches *Swa* and *Swb* are off, switches *Swc* and *Swd* are on.
5. Repeat 1.

[1] Without special precautions it is not wise to have switches *Swa* and *Swd* or *Swc* and *Swb* on together.

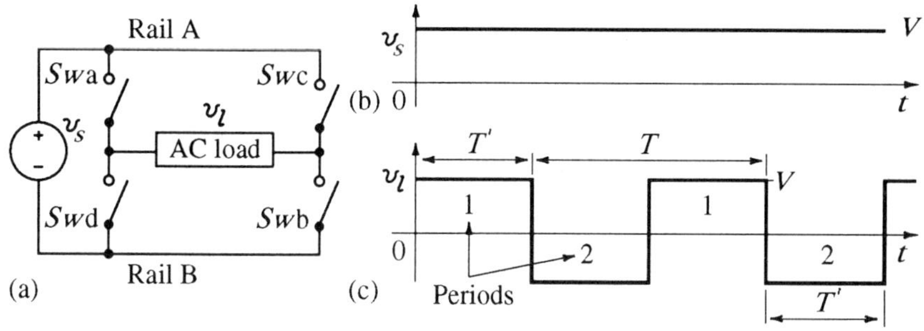

Fig. 1.8 Inverter.
(a) Circuit diagram, (b) source voltage, (c) load voltage.

The ac voltage at the load is a function of the area under the voltage curve. So the adjustment of time T', that the pairs of switches are on, alters the voltage. The frequency is controlled by the period T of switching ($f = 1/T$). If the output voltage is specified (T' is fixed), what is the maximum frequency of the ac output[2]?

Although the method described here is simple in principle, it can be complex in practice. The waveforms in Figs 1.8c and 1.9 have a high harmonic content. For many applications this is inefficient and undesirable. A pure sinewave is the ideal requirement. To approach this requirement many pulse-width-modulation algorithms have been devised. In this case the rectangular voltage waveforms of periods 1 and 2 in Fig. 1.9 are divided into many pulses of different widths by means of an increased frequency of switching. Figure 1.10 shows a general PWM pattern.

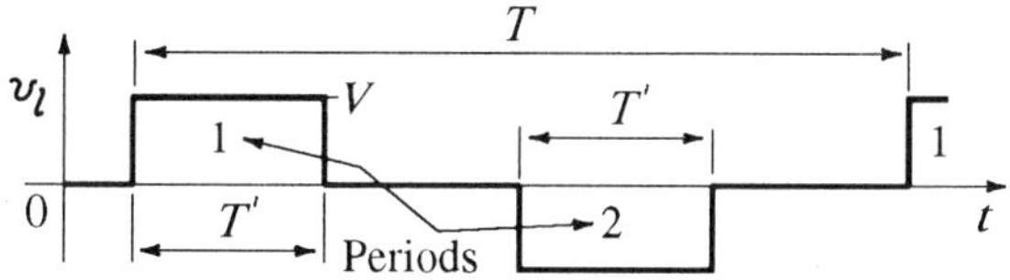

Fig. 1.9 Adjustable voltage and frequency.

[2] The maximum frequency is $1/(2T')$.

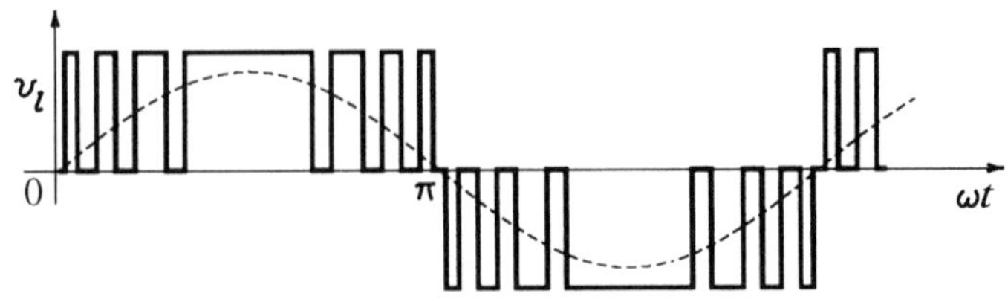

Fig. 1.10 Pulse-width modulation (PWM) pattern.

1.4. WAVEFORM DISTORTION

Any method of digital switching that is designed to simulate an analogue signal will fall short of the ideal requirement by some extent.

In the analysis of the performance of loads with *chopped* voltage waveforms it is necessary to determine average or effective (rms) values of voltage, current and power. Further, a knowledge of the harmonic content of a waveform is often needed, if losses are to be calculated or if comparisons of waveforms are to be made. This section acts as a brief review of these subjects, and the problems at the end of this chapter provide exercises as a reminder. Applications of these calculations are used throughout the book.

1.4.1. Average Values

Determination of the average value of the voltage across a dc load is important. If the instantaneous value of the voltage $v(t)$ is periodic with a period T, then the average voltage is given by

$$V_{av} = \frac{1}{T}\int_0^T v(t)\,dt\,. \tag{1.4.1}$$

It can be convenient to transform time t to radian θ measure, so that

$$V_{av} = \frac{1}{2\pi}\int_0^{2\pi} v(\theta)\,d\theta \tag{1.4.2}$$

where $\theta = \omega t$, $\omega = 2\pi f$ and $f = 1/T$.

EXAMPLE 1.1

Consider a full-wave rectified sine wave, as shown in Fig. 1.3c. Find a general expression to describe the average load voltage at any firing angle α.

Solution

Assume $v_s = \hat{V}\sin\omega t$.

Therefore, $V_{l\,av} = \frac{1}{\pi}\int_{\alpha}^{\pi} v_s\, d(\omega t) = \frac{1}{\pi}\int_{\alpha}^{\pi} \hat{V}\sin\omega t\, d(\omega t)$.

That is, $V_{l\,av} = \frac{\hat{V}}{\pi}(1 + \cos\alpha)$.

The average value ranges from $0.637\hat{V}$ down to zero, but it is a nonlinear function of α.

1.4.2. RMS Values

The rms (root mean square) value of a current that varies periodically with time is the *effective* value that is equivalent to a constant dc current in terms of heating. That is, the periodic current and direct current produce the same average power in an element. If the instantaneous value of the current $i(t)$ is periodic with a period T, then the rms current is given by

$$I_{rms} = \left[\frac{1}{T}\int_0^T i^2(t)\,dt\right]^{1/2}. \qquad (1.4.3)$$

In radian measure

$$I_{rms} = \left[\frac{1}{2\pi}\int_0^{2\pi} i^2(\theta)\,d\theta\right]^{1/2}. \qquad (1.4.4)$$

The rms value is useful to determine the value of average power in ac loads. It is also useful if the load current is direct current but the voltage and current are not constant values.

EXAMPLE 1.2

Consider an inverter circuit as depicted in Fig. 1.8. Let the load be a resistor of value R. For the output as shown, determine the rms value of the load current $i_l(t)$.

Solution

The current through the load is $i_l(t) = v_l(t)/R$.

Therefore, $I_{l\,rms} = \left[\frac{1}{T/2}\int_0^{T/2} (V/R)^2 dt\right]^{1/2}$.

That is, $I_{l\,rms} = V/R$.

This is a special case in which the rms value is the same as the average value.

1.4.3. Form Factor

The term *form factor* is applied to ac waveforms. It provides an indication of both how much distortion a waveform has from a pure sinewave and how much loss can be expected. The definition is

$$\text{form factor} \triangleq \frac{\text{rms value}}{\text{average value}} . \tag{1.4.5}$$

In ac terms, the average value implied here is half-wave average, because the average value of a periodic function with symmetry is zero.

Any periodic voltage waveform given by the Fourier series

$$\begin{aligned} v_l(t) = V_{dc} &+ \hat{V}_1 \sin(\omega t - \theta_1) + \hat{V}_2 \sin(2\omega t - \theta_2) \\ &+ \hat{V}_3 \sin(3\omega t - \theta_3) + \cdots \end{aligned} \tag{1.4.6}$$

has an rms value given by

$$V_{l\,rms} = (V_{dc}^2 + V_1^2 + V_2^2 + V_3^2 + \cdots)^{1/2} \tag{1.4.7}$$

where $V_1 = \hat{V}_1/\sqrt{2}$ is the rms value of the fundamental component of the waveform, V_{dc} is the dc component and the other terms represent the various harmonics.

For a required fundamental rms value of voltage, any dc level or harmonic will increase the overall rms value. The fundamental component does useful work usually and all other components produce losses[3]. For a pure sinewave the form factor is 1.11. Therefore, any deviation of the form factor from the value of 1.11 is a general indication of waveform distortion and load losses.

1.4.4. Harmonics

Power conditioning with semiconductor switches inevitably introduces harmonics into the ac system. Further, in an analysis of a system in switch-mode operation, it can be difficult to determine the response to a nonsinusoidal, but periodic forcing function. However, the Fourier method permits a solution. The periodic function is expressed as a series of sinusoidal functions (harmonics), the response to each harmonic can be determined, and, if the system is linear, the complete response is determined by superposition. Accordingly, the initial problem is to find the magnitudes and phases of the harmonics of a waveform.

A Fourier series representing a periodic waveform can be written in trigonometric form as

$$v(t) = \frac{1}{2}a_0 + a_1 \cos\omega t + a_2 \cos 2\omega t + a_3 \cos 3\omega t + \cdots +$$

[3] This does not apply to heating applications.

$$+ b_1 \sin\omega t + b_2 \sin 2\omega t + b_3 \sin 3\omega t + \cdots . \tag{1.4.8}$$

The coefficients a_n (where $n = 0, 1, 2, 3 \cdots$) are evaluated from the integration

$$a_n = \frac{1}{\pi} \int_0^{2\pi} v(t) \cos n\omega t \, d(\omega t) . \tag{1.4.9}$$

The coefficients b_n (where $n = 1, 2, 3 \cdots$) are evaluated from the integration

$$b_n = \frac{1}{\pi} \int_0^{2\pi} v(t) \sin n\omega t \, d(\omega t) . \tag{1.4.10}$$

EXAMPLE 1.3

Consider the output waveform of the inverter shown in Fig. 1.8. Determine the Fourier series of this square wave.

Solution

The singular functions are
$v_l(t) = V$ for $0 < \omega t < \pi$ and $v_l(t) = -V$ for $\pi < \omega t < 2\pi$.
The average value of the waveform is zero by inspection. So, $a_0 = 0$ and $V_{dc} = 0$. Because of the position of the origin the waveform is an odd function. That means that $a_n = 0$, proved by evaluating eq. (1.4.9).

For the sine terms eq. (1.4.10) is evaluated in two parts.

$$b_n = \frac{1}{\pi} \int_0^{\pi} V \sin n\omega t \, d(\omega t) + \frac{1}{\pi} \int_{\pi}^{2\pi} (-V) \sin n\omega t \, d(\omega t) .$$

That is, $b_n = \dfrac{2V}{\pi n}(1 - \cos n\pi)$.

Consequently, the harmonic series is

$$v_l(t) = \frac{4V}{\pi} \sin\omega t + \frac{4V}{3\pi} \sin 3\omega t + \frac{4V}{5\pi} \sin 5\omega t + \cdots .$$

The rms value of the fundamental component of voltage is $0.9V$ $(4V/\sqrt{2}\pi)$, whereas the rms value of the waveform is V (see EXAMPLE 1.2). If the load responded usefully to the fundamental component only and if the equivalent load to all harmonics were a resistance R, then the efficiency would be 81%. That is, the losses due to the nonsinusoidal nature of the forcing function v_l would be 19% of the input power to the load.

1.4.5. Total Harmonic Distortion (THD)

There are many applications of power conditioning where it is necessary to synthesize a sine wave from discrete pulses. Figure 1.8c shows the simplest case and possibly the worst case. There are many algorithms for switching power electronics inverters to obtain close approximations to a sine wave. In order to pick the best, there must be a measure of goodness. Besides cost, there is harmonic content of the waveform.

If a requirement of an inverter output were that there were to be no third and fifth harmonics, for example, then an inspection of a Fourier analysis would suffice. However, if the overall harmonic content were of interest, because of losses in the load, then there is a better measure. It is total harmonic distortion (THD). Since the rms value of a waveform is given by

$$V_{rms}^2 = V_{dc}^2 + V_1^2 + V_2^2 + \cdots + V_r^2 + \cdots \tag{1.4.11}$$

where V_r is the rms value of the rth harmonic, and since the fundamental rms value V_1 is a suitable reference, the total harmonic distortion can be defined by

$$(\text{THD})^2 \triangleq V_{dc}^2 + V_2^2 + V_3^2 + \cdots + V_r^2 + \cdots . \tag{1.4.12}$$

That is,

$$(\text{THD})^2 = V_{rms}^2 - V_1^2 . \tag{1.4.13}$$

As a per unit quantity

$$\text{pu THD} = \left(\frac{V_{rms}^2 - V_1^2}{V_1^2} \right)^{1/2} \tag{1.4.14}$$

so that the deviation of the waveform from the pure sine wave is given by how much the THD digresses from zero.

1.4.6. Average Power

The instantaneous value p of power in a load is the product of the instantaneous value of voltage v across the load and the instantaneous value of the current i. That is,

$$p = v\,i . \tag{1.4.15}$$

This value of power has significance if the voltage and current have constant dc values. If either the voltage or the current is not constant, then the important value is the average power.

In switching circuits that modulate the power from a source to a load the voltages and currents are periodic. Thus, the average power P in a load is defined by

$$P \triangleq \frac{1}{T}\int_0^T p\,dt = \frac{1}{T}\int_0^T vi\,dt \tag{1.4.16}$$

where T is the period of the waveform.

If the nature of the load is known and if one of the variables, voltage or current, is known, the average power can be calculated. The variables may be constants, sinusoids, rectangular waves, or chopped sine waves, so in some cases the calculation may be complex. However, if power is to be modulated, it is necessary to determine the value of that entity which is being controlled.

EXAMPLE 1.4

The voltage $v = \hat{V}_s \sin\omega t$ across a load invokes the current response $i = \hat{I}\sin(\omega t - \theta)$. Determine the average power absorbed by the load.

Solution

The average power P in the load is

$$P = \frac{1}{2\pi}\int_0^{2\pi} vi \, d\omega t = \frac{1}{2\pi}\int_0^{2\pi} \hat{V}_s \hat{I} \sin\omega t \sin(\omega t - \theta) \, d\omega t .$$

So, $$P = \frac{\hat{V}_s \hat{I}}{2 \times 2\pi}\int_0^{2\pi} (\cos\theta - \cos(2\omega t - \theta)) \, d\omega t = \frac{\hat{V}_s \hat{I}}{2}\cos\theta .$$

Since the rms values of sinewaves are the amplitudes divided by $\sqrt{2}$,

$$P = V_s I \cos\theta$$

where V_s and I are rms values and $\cos\theta$ is known as the power factor.

EXAMPLE 1.5

Consider a full-wave rectified sine wave, as shown in Fig. 1.3c. Find a general expression to describe the average power in a load that is purely resistive for any firing angle α.

Solution

In a load of resistance R, the load current is $i = v_l/R$, and the load voltage v_l has the absolute value of the supply voltage $v_s = \hat{V}_s \sin\omega t$ over the interval $\pi - \alpha$ every half cycle. Otherwise the load voltage is zero.
The average power absorbed by the load is

$$P = \frac{1}{\pi}\int_\alpha^\pi v_l i \, d\omega t = \frac{1}{\pi R}\int_\alpha^\pi \hat{V}_s^2 \sin^2\omega t \, d\omega t = \frac{V_{l\,rms}^2}{R} = I_{rms}^2 R .$$

So, the average power is proportional to the square of the rms value of the load voltage or current. In terms of α,

$$P = \frac{\hat{V}_s^2}{2\pi R}\left[\omega t - \frac{\sin 2\omega t}{2}\right]_\alpha^\pi = \frac{\hat{V}_s^2}{2\pi R}\left[\pi - \alpha + \frac{\sin 2\alpha}{2}\right].$$

The average power can be adjusted from zero to $\hat{V}_s^2 / 2R$ with α control.

1.5. POWER SEMICONDUCTOR SWITCHES

In this chapter we have discussed how the on and off action of a switch can be used to modulate power from any kind of supply to suit any kind of load. The type of switch was not specified. It was an ideal switch that dissipated no power. It could be turned on in no time with a signal of zero energy. While on, the switch offered zero impedance to the conduction of any current. It could be turned off in no time with a signal of zero energy. While off, the switch offered an infinite impedance to the conduction of current. That is, no current was conducted. Further, the switch would block any voltage, and any rate of change of voltage would not alter the state of being off. The use of the ideal switch is helpful for system analysis.

In practice, there is no ideal switch. The closest we can come to the ideal power switch is the group of semiconductor switches that utilize silicon in different forms. The base of all high-power silicon switches is the *pn* junction, so this will be described briefly here. This description will form the base of all the practical switches discussed in Chapters 3 through 10.

Silicon is a semiconductor element from group IV of the Periodic Table. That is, it has four electrons in its outer shell. Its resistivity is high, but not as high as an insulator. The resistivity can be decreased considerably by adding small concentrations of certain doping materials to the silicon.

Elements from group V of the Periodic Table have five electrons in the outer orbit of the atoms. Examples are antimony, arsenic and phosphorus. If the silicon is doped with a small percentage of one of these elements[4], there is covalent sharing of electrons in the outer orbit. This means there are free electrons available for conduction, and the resistivity is low, but it is not as low as a conductor. Silicon that is doped to have an excess of electrons is called an *n*-type material because of the negative charge. The electrons in an *n*-type semiconductor are called majority carriers. Any positive charge carriers in an *n*-type material are called minority carriers.

Elements from group III of the Periodic Table have 3 electrons in the outer orbit of each atom. Examples of these are aluminium, boron and gallium. If an intrinsic silicon crystal is doped with one of these elements, then, within the covalent sharing, there will be some atoms that will have one electron deficient. This deficiency of a negative charge is equivalent to a positive charge, called a *hole*. Semiconductors that have excess holes for conduction are called *p*-type

---

[4] The proportion of impurity is about 1 part in 10 million.

materials (*p* for *positive*). The majority carriers in *p*-type silicon are holes, whereas any mobile electrons that are present in the same material are called minority carriers. These *p*-type semiconductors have lower resistivities than the pure intrinsic silicon but they have higher resistivities than conductors[5].

Silicon is processed in the form of a very pure single crystal, sliced and cut to form dies. The doping is achieved by diffusion processes to get different concentrations of *p*-type and *n*-type regions on the silicon die. There are alloying and epitaxial growth techniques too. Materials with n^- regions and p^- regions are lightly doped, and n^+ regions and p^+ regions are heavily doped.

A semiconductor wafer that is processed to have adjoining *p* and *n* layers is called a *pn* junction and has special electric characteristics. This *pn* junction is the basic building block in any power semiconductor switch and in itself has the property of being an uncontrolled, unidirectional switch. The *pn* junction is on and will conduct current, if the *p* region is positively biased from an external voltage source. The *pn* junction is off and will block the conduction of current if the *n* region is positively biased from an external voltage source. There is no control over this ON/OFF action.

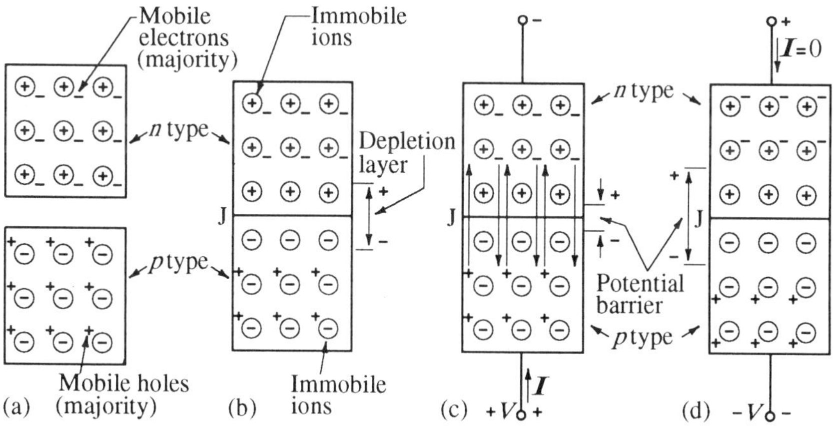

Fig. 1.11 The *pn* junction.
(a) *p* and *n* types, (b) unbiased, (c) forward biased, (d) reverse biased.

[5] The resistivity of doped silicon has a value similar to the carbon that is used in the manufacture of resistors and potentiometers.

Figure 1.11a depicts two doped materials. One is an *n*-type material with donor impurity atoms, that shows the majority carriers (mobile electrons) to be free and uniformly distributed, and shows the immobile positively charged atoms to be encircled. The other is a *p*-type material with acceptor impurity atoms, that shows the majority carriers (mobile holes) to be free and uniformly distributed, and shows the immobile negatively charged atoms to be encircled.

If both *n*-type and *p*-type regions are formed on the same silicon wafer, a junction J is formed between the two regions. This is shown in Fig. 1.11b, and this configuration of the semiconductor is called a *pn junction.* During the formation of the junction there is a dynamic reaction between the majority carriers of both regions. In each region the majority charges experience the repulsive forces of like charges within the same vicinity. Consequently, these charges diffuse towards regions of lower concentration. Mobile electrons and holes diffuse across the junction where they cancel by recombination[6] with opposite charges. That is, a hole that is a majority carrier in the *p* region diffuses across the junction, becomes a minority carrier, recombines with a majority carrier electron of the *n* region and so depletes the region near the junction of a mobile carrier. In general, there becomes a depletion of mobile carriers on each side of the junction as shown in Fig. 1.11b. What is left near the junction is a region of immobile charged atoms that is called a *space charge.* This space charge is distributed in such a way that the depletion layer can be considered to be an equivalent capacitance, called the *depletion-layer capacitance,* whose value is of the order of picofarads. Also, this space charge creates an electric field gradient that allows the depletion layer to form a potential barrier[7]. The potential barrier repels further majority carriers migrating across the junction and, in this neutral form, a steady condition is reached in the *pn* junction.

Figure 1.11c illustrates the *pn* junction as a switch in the on-state. An external source of voltage V is applied across the terminals of the *pn* junction such that the *p* region is positive and the *n* region is negative. The applied electric field E is in such a direction to reduce the potential barrier and narrow the depletion layer. Majority-carrier holes are attracted to the negative terminal and majority-carrier electrons are attracted to the positive terminal. More majority carriers with enough kinetic energy from the applied field can cross the junction. (Note that force $F = qE$ and energy $W = Fdu/dt = 1/2mu^2$.)[8] The drift of these carriers of charge q under the influence of the E field constitutes the current I. There is also a contribution by minority carriers created by thermal ionization and flowing in the opposite direction. The magnitude of the current I increases exponentially with the increase of the applied voltage V until the depletion layer vanishes completely. At this point the voltage drop across the device is about 1 V, the current I assumes a value determined by the circuit voltage and impedance, and the device, as a

[6] Capture of a free electron in an empty covalent bond is an example of recombination, which includes the trapping of carriers by impurities in the crystal.

[7] The potential difference across the depletion layer is about 0.6 volts.

[8] Mass is m and speed is u.

switch, is on.

Figure 1.11d indicates a voltage V that creates a reverse bias to the *pn* junction. That is, the *p* side of the junction is negative and the *n* side of the junction is positive. The electric field due to the voltage, forces majority-carrier electrons in the *n* region away from the junction and more towards their electrode, and forces majority-carrier holes in the *p* region away from the junction and more towards their electrode. This action widens the depletion layer and increases the space charge to suppress the diffusion of majority carriers across the junction. So, there is no current and the device, as a switch, has assumed the blocking condition of the bias voltage and is in the off-state. Ideally, the current is zero for the condition of reverse bias, but in practice there is a very small reverse leakage current. Thermal ionization in the semiconductor material causes the breaking of bonds and the creation of minority carriers. These carriers are free to be accelerated by the applied electric field to some average velocity. This flow constitutes the leakage current.

The *pn* junction, as an uncontrolled switch that allows unidirectional current conduction, has a limited application. However, it is the building block for the manufacture of power semiconductor switches such as those thyristors and transistors that are described in this book.

The thyristor is a controlled semiconductor switch that can control greater blocks of power than any other silicon device. It is chosen arbitrarily to represent the switches in the ideal converter circuits that are described in Chapter 2. Therefore, it is worth characterizing the ideal thyristor here.

The thyristor is known also as a silicon controlled rectifier, which describes its function admirably. As shown in Fig. 1.12a it comprises a number of *pn* junctions within the *pnpn*-layer construction. There are three terminals, anode A, cathode K and gate G.

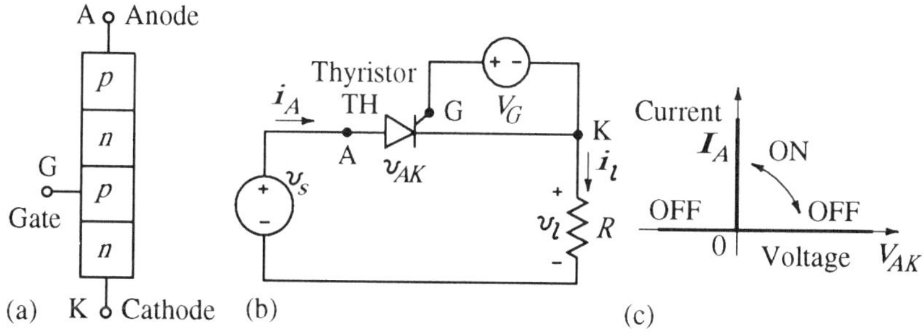

Fig. 1.12 The thyristor semiconductor switch.
(a) Junctions, (b) circuit diagram, (c) ideal characteristics.

Figure 1.12b illustrates a circuit with the thyristor TH connected between a power source of voltage v_s and a power load of resistance R. The purpose of the switch is to modulate how much power is absorbed by the load. This modulation is achieved, in part, by the signal source V_G. Let the thyristor be off and let the signal source V_G be zero. The switch will remain off no matter what polarity of voltage v_{AK} is applied to the thyristor terminals A and K. This is shown as the OFF characteristic in Fig. 1.12c. Consequently, the thyristor does not conduct current between the main terminals A and K and $i_A = 0$. This is verified by an inspection of Fig. 1.12a. There is always a *pn* junction blocking, no matter what polarity of voltage is applied to the anode.

In order to turn on the thyristor a positive signal V_G must be applied to the gate terminal G, and the anode A must be made positive with respect to the cathode K by means of the source voltage v_s. Under these conditions, the switch is on. No impedance is offered to the conduction of current, so $i_A = v_s/R$. Once on, the thyristor remains on even if the gate signal is removed. This is shown as the ON characteristic in Fig. 1.12c. In the on-state $v_{AK} = 0$ ideally.

If the thyristor is switched on, it remains on to conduct the current i_A until that current is brought to a zero value. As soon as $i_A = 0$ the thyristor turns off and remains off until both V_{GK} and V_{AK} are positive again.

Just like the *pn*-junction diode, the thyristor allows only unidirectional current to be conducted. The current i_A cannot be negative ideally. That is, the thyristor blocks all reverse bias voltages (negative V_{AK}).

This description of the ideal thyristor switch allows us to analyze switched circuits in a relatively simple manner. Chapter 2 concerns such circuits.

1.6. APPLICATIONS

Power electronics finds application wherever a load needs controlled power modulation. It is found in the control of light dimming and the control of steel mills. It is found in the control of domestic appliances and the control of HV dc links in power systems.

In terms of the two variables, installed power of semiconductor equipment and operating frequency of the conditioned supply, the product of the two is a function that is almost constant. At low frequency (0 to 120 Hz), the high power has applications covering induction heating, motor drives and uninterruptible power supplies (UPS), whereas, at the higher frequencies (1 kHz and above), the relatively low installed power capacity covers applications such as fluorescent lighting inverters and ultrasonic induction heating.

The control of electric drives may involve the glamour areas of robotics, computer disk drives, gyros and space applications or it may be associated with more conventional loads like fans, compressors, pumps and hoists. Whatever the mechanical load, the power semiconductor converter lends itself to the control of adjustable speed drives, which may be dc or ac drives.

DC drives require voltage control. This is provided by controlled rectifiers (type 5) or by choppers (type 3), depending upon what kind of source is available. The dc drive is expensive and the rectifier is inexpensive.

AC drives require voltage and frequency control. Inverters (type 7) and sometimes cycloconverters (type 6) accomplish this task. The ac drive is inexpensive and the inverter is expensive.

The choice of which type of drive is better depends on cost and versatility. In general the dc drive is found useful in many applications. However ac drives are becoming prominent as inverter technology continues to improve.

Whatever load a power-electronics system of switches is to control, there is usually one semiconductor switch that is best for the application. It may be best because it can switch the fastest, or it can block the highest voltage, or it has minimum losses, or it costs the least. A choice of switch can be made. This book describes the characteristics of semiconductor switches in order to aid the choice of switch as well as how to use it.

1.7. SUMMARY

Power conditioners utilizing power-semiconductor switches are called converters. By discrete-switching sequences any output voltage can be approximated. In this way a supply of one kind can be converted to a supply of any other kind. Particular switching patterns can produce a dc voltage of adjustable magnitude or an ac voltage of adjustable amplitude and frequency from either an ac or a dc source.

The principles of switch-mode power conditioning give rise to distorted output waveforms. It becomes necessary in an analysis to determine the performance of the power conditioner and load. This is facilitated by being able to calculate the rms and average values and the harmonic content of the output waveforms.

In practice, converters comprise switches that are silicon devices. The *pn* junction forms the building block of the increasing variety of power semiconductor switches that are available.

1.8. PROBLEMS

Section 1.4

1.1 Calculate the rms, average, half-wave average values and form factors of the following voltage waveforms.

(**a**) A sine wave $v(t) = 100\sin 377t$.

(**b**) A full-wave rectified sine wave, where the first half cycle is described by $v(t) = 100\sin 377t$.

(**c**) A half-wave rectified sine wave, where the first half cycle is described by $v(t) = 100\sin 377t$.

(**d**) A phase-controlled sine wave $v(t) = 100\sin 377t$, where the delay angle $\alpha = 90$ degrees for both positive and negative half cycles. See Fig. 1.6c and assume the switch opens as the voltage crosses zero.

(**e**) A sine wave $v(t) = 100\sin 377t$ with ac integral-cycle control giving 3 cycles on, 2 cycles off, 3 cycles on, etc.

(f) A rectangular wave with both positive and negative half cycles having an amplitude of 100 V and a frequency 60 Hz.
(g) A triangular wave, with both positive and negative half cycles having an amplitude of 100 V and a frequency of 60 Hz.

1.2 Calculate the harmonics of a fully rectified sine wave whose first half cycle is described by $v(t) = \hat{V} \sin\omega t$. If the load is a 1 ohm resistor, determine the average load power as a sum of harmonic powers.

1.3 Calculate the harmonics for a half-wave rectified sine wave.

1.4 Calculate the harmonics of the current waveform in the form of a phase-controlled, half-wave rectified sine wave, if the delay angle $\alpha = 90$ degrees, and if the load is resistive.

1.5 Estimate the total harmonic distortion of the square waveform shown in Fig. 1.8c.

1.6 Figure 1.3 depicts ac-dc conversion with point-on-wave control. If the ac source voltage is 120 V and the load is resistive with a value 1.2 Ω, calculate the average power delivered by the source for a delay angle $\alpha = \pi/2$ radians.

1.7 Consider a dc-dc converter, as shown in Fig. 1.4. If the dc source voltage is 1000 V and the load is resistive with a value 10 Ω, calculate the average power in the load for a chopper duty ratio $m = 0.6$.

1.8 Figure 1.6 illustrates an ac-ac voltage converter with point-on-wave control. If the ac source is 240 V and the load is resistive with a value 5.76 Ω, calculate the average power absorbed by the load for a delay angle $\alpha = \pi/6$ radians.

1.9 An inverter is shown in Fig. 1.8 and Fig. 1.9. If the dc source voltage is 200 V, if the load is resistive with a value of 2 Ω and if the switches are controlled to give pulses of load current for an interval $T' = 0.1$ ms every 0.2 ms ($T = 0.4$ ms), determine the average power absorbed by the load.

1.10 Consider the integral-cycle ac-dc converter shown in Fig. 1.2. For an ac supply of 100 V at 2 kHz and a resistive load of 0.1 Ω determine the average power modulated if the switching frequency is 500 Hz and the on time for each cycle is 0.25 ms.

CHAPTER 2

SWITCHES IN CIRCUITS

2.1. INTRODUCTION

This chapter is concerned with the description and analysis of circuits that employ switches to modulate power in the load. In a circuit a configuration of switches that modulates or conditions power is called a converter. Thyristors are the switches that are shown in the figures of converters throughout this chapter. Any other appropriate semiconductor switch could be substituted, since, in the most part, the switch is treated as being ideal in this chapter and the analysis is, therefore, the same. The following chapters describe the different kinds of switches, the details of their characteristics and their areas of application.

There are applications of switches associated with power supplies, high voltage dc transmission, induction heating, light dimmers and battery chargers, to name a few. Each of these has many particular variations from alarms and appliances, through the alphabet, to voltage regulators and welding. There are so many cases that it is impossible to cover every particular circuit. Instead, we will concentrate on some general circuits that are commonly found and we will highlight the concepts and the method of analysis. In an analysis, what are usually required to be found are the load voltage, current and power, and the ratings of the switches.

Figure 1.4a illustrates a chopper circuit that employs an ideal switch *Sw* to modulate power from a dc source to a load. In Fig. 2.1a the same chopper circuit has a particular switch *TH* that replaces the general switch *Sw*. The two switches are treated in the same way. They can be turned on and turned off at will. While they are on they offer no impedance to current. While they are off they block voltage and there is no conduction of current. The particular switch *TH* is called a thyristor, which is a semiconductor switch sometimes known as a silicon controlled rectifier. Details of this switch are to be found in section 1.5 and Chapter 5.

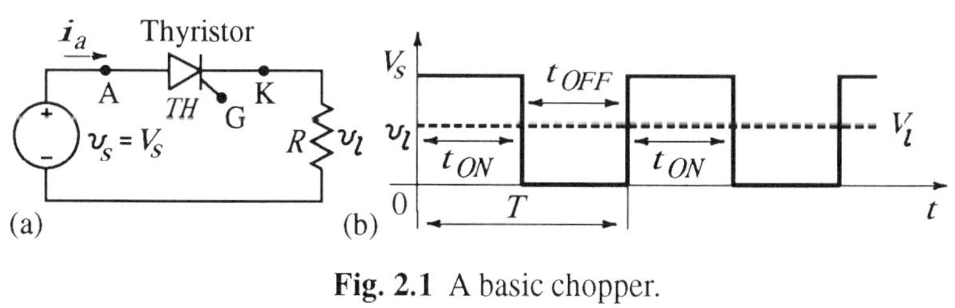

Fig. 2.1 A basic chopper.
(a) Circuit diagram, (b) output voltage waveform.

For this chapter, the thyristor *TH* is considered an ideal switch that can conduct current in only one direction. In Fig. 2.1 the thyristor *TH* is shown to have three electrodes, A (anode), K (cathode) and G (gate). If the anode has a positive voltage with respect to the cathode, and if a small positive signal is applied to the gate, the thyristor switch turns on. Current is conducted in the direction from anode to cathode. The thyristor remains on until the gate signal is removed and the anode current is brought to zero. At this point the thyristor is off. An ideal thyristor blocks voltage that is positive at the cathode with respect to the anode so that there can be no current directed from the cathode to the anode in the device.

Real switches are not ideal, as just described. They have leakage currents when they are off and they have voltage drops across them when they are on. It takes a finite time to go through the processes of turning on and off. These imperfections of the switch must be taken into account at some stage of the design and implementation of a circuit. However, in the analysis of the circuit, the performance of the switch is, most often, good enough to consider the switch to be ideal.

2.2. DC TO DC CONVERSION

The purpose of any power modulator, that is connected between a dc source and a dc load, is to modulate the power by adjusting the output voltage or the output current. The conventional way is to have voltage control at the output. If the modulator is almost lossless it can be considered to be a dc transformer. The action is to switch the modulator on and off in a predetermined fashion that can be called chopping. This kind of dc-dc converter is called a *switched-mode power supply*.

The general action of choppers is described in section 1.3.1 of Chapter 1. The basic circuit diagram is shown in Fig. 2.1a. The input voltage v_s from the source is a constant, V_s. The thyristor *TH* (or any appropriate switch) is turned on for a time t_{ON} and turned off for a time t_{OFF}. If this is repeated with the same period T, the average voltage V_l across the load is given by

$$V_l = mV_s \tag{2.2.1}$$

where m is the duty ratio, $m = t_{ON}/T$, and $T = t_{ON} + t_{OFF}$. The variables are t_{ON}, t_{OFF} and T. It is usual to have a fixed frequency ($f = 1/T$) of switching that depends on the type of switch and the application, and it is usual to control the on time t_{ON}. The duty ratio m can be controlled from zero to unity so that the average output voltage V_l can have any value from 0 to V_s in this case. This method of control is called *Pulse-width Modulation* (PWM) and is frequently incorporated in special purpose chips and in single-chip computers.

A shortcoming of the basic chopper circuit is that, literally, it chops the supply voltage into discrete pulses. This suggests the use of a capacitor C across the load to smooth the output voltage. The problem with a switch and a capacitor is that the capacitor looks like a short circuit initially. Therefore, an inductance has to be added to limit the change of current. The problem with switches and inductors is that a large voltage ($L\,di/dt$) can be induced in the inductor if the switch is turned off quickly. Therefore, a free-wheeling diode has to be added to limit the change of voltage. Not only has the chopper circuit become more complex, but there are three different configurations of switch, capacitor, inductor and diode. Each configuration has a different characteristic and the name of the three are buck converter, boost converter and buck/boost converter. Two of these converters are described in the following sections.

2.2.1. Buck Converter

A buck converter is a dc-dc converter, whose output voltage V_l is less than or equal to the supply voltage V_s. Figure 2.2a depicts the configuration of a switch, a capacitor, an inductor and a diode that satisfies this characteristic. We want to find the load voltage V_l in terms of the supply voltage V_s and the duty cycle *m*. It is assumed that conditions are in the steady state and the chopping action is periodic.

It is assumed that the load voltage V_l is maintained reasonably constant by the capacitor *C*. The action of switching the thyristor on and off in sequence gives rise to the node *X* voltage *v* waveform as shown in Fig. 2.2b. While the thyristor *TH* is on, the voltage v_L across the inductance *L* is

$$v_L = V_s - V_l = L\frac{di}{dt} \approx \text{constant}. \tag{2.2.2}$$

Thus, while *TH* is on, the inductor current *i* is

$$i = \frac{V_s - V_l}{L}t + I_{\min}. \tag{2.2.3}$$

While the switch *TH* is off, the voltage drop across the inductance *L* is

$$v_L = -V_l = L\frac{di}{dt} \approx \text{constant}. \tag{2.2.4}$$

Thus, while *TH* is off, the inductor current *i* is

$$i = -\frac{V_l}{L}t + I_{\max}. \tag{2.2.5}$$

Here, time *t* is measured from the instant that *TH* is turned off. The linear rise and decay of current *i* is depicted in Fig. 2.2c. In the steady state $I_{\min}$ has the same value after each period *T*. Consequently, the average V_L value of the inductor voltage is

$$V_L = \frac{1}{T}\int_0^T v_L\,dt = \frac{1}{T}\int_0^T L\frac{di}{dt}\,dt = \frac{L}{T}\int_{I_{\min}}^{I_{\min}} di = 0. \tag{2.2.6}$$

Hence, the constraint on the current is that the average inductor voltage is zero. Figure 2.2d shows the waveform of the instantaneous value of the inductor voltage $v_L = L\,di/dt$. Because the average voltage is zero,

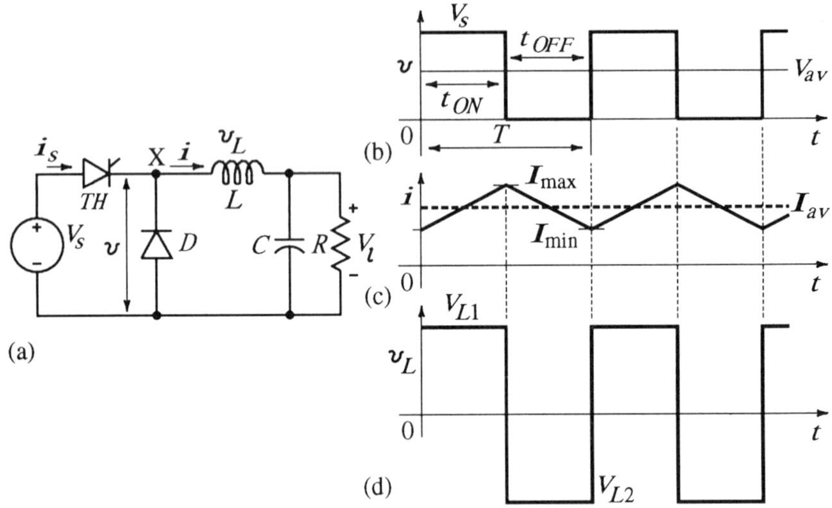

Fig. 2.2 A buck converter.
(a) Circuit diagram, (b) voltage at node X,
(c) current waveform, (d) voltage across inductance.

$$V_{L1} \times \frac{t_{ON}}{T} + V_{L2} \times \frac{t_{OFF}}{T} = 0 \tag{2.2.7}$$

or,

$$mV_{L1} + (1 - m)V_{L2} = 0. \tag{2.2.8}$$

Since,

$$V_{L1} = V_s - V_l \text{ and } V_{L2} = - V_l \tag{2.2.9}$$

then,

$$V_l = mV_s = V_{av}. \tag{2.2.10}$$

With this particular circuit configuration, the output voltage V_l across the load can be adjusted from zero to the value of the source voltage V_s; that is, the output voltage is equal to or less than the input voltage. In the next section we will analyze the circuit configuration that provides an output voltage that is greater than the input voltage.

EXAMPLE 2.1

A buck converter operates from a voltage source of 100 V. The thyristor is switched at 1000 pulses per second and the maximum power to the load is 10 kW. If it can be assumed that the converter efficiency is 100%, calculate the value of the inductance such that its current value reaches zero only once per cycle, if the power to the load is 5 kW.

Solution

We can refer to Fig. 2.2c for the shape of the current waveform. If the current i reaches zero once each cycle, the current is just continuous and $I_{min} = 0$. The output power is V_l^2/R and the average input power from the source is mV_sI_{av} (prove this). These powers are equal.

The inductor average current $I_{av} = I_{max}/2$ for the condition specified.

The voltage across the inductance is $V_{L1} = LI_{max}/t_{ON}$,

or, $I_{max} = \dfrac{V_{L1}}{L}\dfrac{m}{f} = \dfrac{V_s - V_l}{L}\dfrac{m}{f}$.

Eliminating the current using the power equality, the inductance value is

$L = \dfrac{R(1-m)}{2f}$.

Maximum power in the load is when $m = 1$, $V_l = V_s = 100\,\text{V}$ and $R = V_l^2/P = 100^2/10^4 = 1\,\Omega$. So, for 5 kW dissipated in the load

$m = \dfrac{1}{V_s}\sqrt{PR} = \dfrac{1}{100}\sqrt{5000} = 0.707$

and $L = \dfrac{R(1-m)}{2f} = \dfrac{1 \times (1 - 0.707)}{2 \times 1000} = 0.147\,\text{mH}$.

2.2.2. Boost Converter

As the name implies the boost converter enables the load voltage to be greater than the supply voltage. Figure 2.3a shows the configuration of this type of dc-dc converter. In concept the switch causes a discontinuity in the waveform of the current, which in turn tends to boost the voltage ($v_L = L\,di/dt$).

The operation of this circuit involves the periodic switching of the thyristor *TH*. In the steady-state case the required voltage V_l across the load is maintained sensibly constant by means of the capacitor *C*. While the thyristor is on, the constant supply voltage V_s is applied across the inductor *L*, since the switch is considered to be ideal. The linear rise in source current (shown in Fig. 2.3b) is given by the equation

$$v_L = V_s = V_{L1} = L\,di_s/dt\,. \tag{2.2.11}$$

During the charging time t_{ON}, the diode *D* prevents the capacitor from discharging through the thyristor *TH*, so that any discharge of the capacitor is into the load resistor *R* to maintain V_l constant. The voltage v_L across the inductor is $v_L = V_{L1} = V_s$.

While the thyristor is off, the voltage across the inductor is given by

$$v_L = V_s - V_l = V_{L2} = L\,\frac{di_s}{dt}\,. \tag{2.2.12}$$

If the voltage across the load is greater than the voltage across the supply, energy

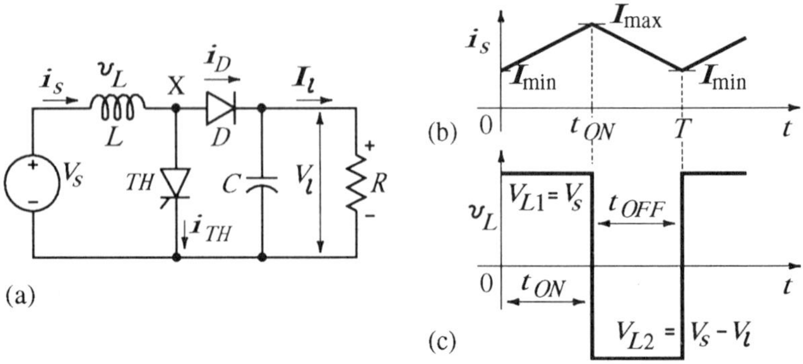

Fig. 2.3 A boost converter.
(a) Circuit diagram, (b) and (c) waveforms.

is being delivered to the load from the stored energy in the inductor as well as from the supply. The voltage v_L across the inductor is $V_s - V_l$ and is constant (V_{L2}). This means that di_s/dt is negative and constant (from eq. (2.2.12)), so the current i_s is reducing linearly from I_{max} to I_{min} as shown in Fig. 2.3b.

We want to find the value of the output voltage in terms of the input voltage V_s and the duty cycle m (t_{ON}/T). The average voltage V_L across the inductor over the period T is

$$V_L = \frac{1}{T}\int_0^T v_L \, dt = \frac{1}{T}\int_0^T L\frac{di_s}{dt}\, dt = \frac{L}{T}\int_{I_{min}}^{I_{min}} di_s = 0. \tag{2.2.13}$$

If the average inductor voltage is zero, the area under the curve (Fig. 2.3c) over the period T must be zero. That is,

$$V_{L1}t_{ON} + V_{L2}t_{OFF} = 0. \tag{2.2.14}$$

If we divide each side by T and substitute for the values of V_{L1} and V_{L2} we obtain

$$V_s m + (V_s - V_l)(1 - m) = 0 \tag{2.2.15}$$

or, finally, the output voltage V_l is

$$V_l = V_s/(1 - m). \tag{2.2.16}$$

Since $0 \le m \le 1$, then $V_l \ge V_s$. There are practical limitations on how much greater the load voltage can be. Limitations are the power and the associated current ratings of the components. Another practical limitation is stability.

EXAMPLE 2.2

An ideal boost converter (see Fig. 2.3) modulates power from a 100-V dc source to a resistive load of $3\,\Omega$. If the average current rating of the diode is 100 A, and the frequency of modulation is 1 kHz, determine **(a)** the maximum average power absorbed by the load, **(b)** the associated duty cycle m of the thyristor switch TH and **(c)** the average current rating of the thyristor.

Solution

Refer to Fig. 2.3.

(a) The average current in the capacitor is zero for steady-state switching. Therefore, the average current I_l in the load is the same as the average current in the diode. That is, $I_l = 100$ A. Since the capacitor maintains the load voltage V_l virtually constant, the average power P absorbed by the load is

$$P = \frac{1}{T}\int_0^T v_l i_l \, dt = \frac{V_l}{T}\int_0^T i_l \, dt = V_l I_l = I_l^2 R = 100^2 \times 3 = 30 \times 10^3 \text{ W}.$$

(b) The average voltage V_l across the load is given by
$V_l = P/I_l = 30 \times 10^3 / 100 = 300$ V.
The average voltage V_l across the load is also given by
$V_l = V_s/(1-m)$. See eq. (2.2.16).
Therefore, the switching duty cycle m of the thyristor is
$m = 1 - V_s/V_l = 1 - 100/300 = 0.667$.

(c) The converter is considered to be 100% efficient, so the power input is equal to the power output. That is, $P = V_s I_s$, where I_s is the average source current (V_s is constant). Therefore, the average source current is
$I_s = P/V_s = 30 \times 10^3 / 100 = 300$ A.
The instantaneous current i_{TH} in the thyristor is given by Kirchhoff's current law at the node X. That is,
$i_{TH} = i_s - i_D$.
In terms of average values
$I_{TH} = I_s - I_D = I_s - I_l$.
Therefore, $I_{TH} = 300 - 100 = 200$ A.
The current rating of the thyristor would be greater than 200 A to accommodate the maximum possible average power plus a safety factor for a reduction of load resistance and any transient that occurs from time to time.
It is important to calculate the semiconductor device ratings before constructing a converter. A thyristor that could conduct the source current in this case would be over-rated.

2.3. AC TO DC CONVERSION

AC to dc conversion takes place if the power source is an ac supply and if the application is a dc load such as battery charging, dc motors, electrolysis, electromagnets, entertainment equipment and dc power supplies for computers, etc. An important application is to provide a dc link between an ac source of one frequency and an ac load of a different frequency.

The general principles of power conditioning for ac-dc conversion are described in section 1.3.1 of Chapter 1. Point-on-wave control is of main concern here. However, this method of control is proved to produce a low power factor. Accordingly, other methods of control must be investigated.

The ac-dc converter may be required to have a fixed dc voltage output, so that diodes would give the rectification. However, controlled rectification with the use of thyristors or other controlled semiconductor switches would comprise the converter for an adjustable dc voltage output. For a single-phase or a three-phase supply, the conversion can be a half-wave converter, a bridge half-controlled converter or a full-bridge converter, depending on the load power. Passive loads will be considered in this chapter.

Analysis of any converter has two aspects. One aspect is how the load voltage and power are affected by the control parameters (usually the delay angle α). The second aspect is the effect of the load and control parameters on factors of the ac side of the converter such as harmonics and power factor. During the description of operation of some of the converters, attention will be given to these aspects of power quality.

In any analysis in this chapter, the elements of all converter circuits are assumed to be ideal. This is reasonable for supply voltages greater than 100 V.

2.3.1. Converter Performance and Operation Modes

Figure 2.4 depicts a general single-phase converter. The currents and voltages are shown as instantaneous values. In general, the dc load voltage v_l and current i_l will not be constant and may be discontinuous.

The input voltage v_s is usually sinusoidal. The semiconductor switches of the converter often utilize the delay angle α to control the magnitude of the average output voltage, so that the input current i_s is usually discontinuous and non-sinusoidal. Associated with these voltages and currents there are performance parameters that describe the goodness of the converter.

Input current as a Fourier series.

The Fourier series was introduced in section 1.4.4. For the general converter configuration in Fig. 2.4, if the input current i_s waveshape can be described, then so too can the Fourier series, which has the general form

$$i_s(t) = I_{dc} + \sum_{n=1,2\cdots}^{\infty} (a_n \cos n\omega t + b_n \sin n\omega t) \tag{2.3.1}$$

where the dc component (average value) is given by

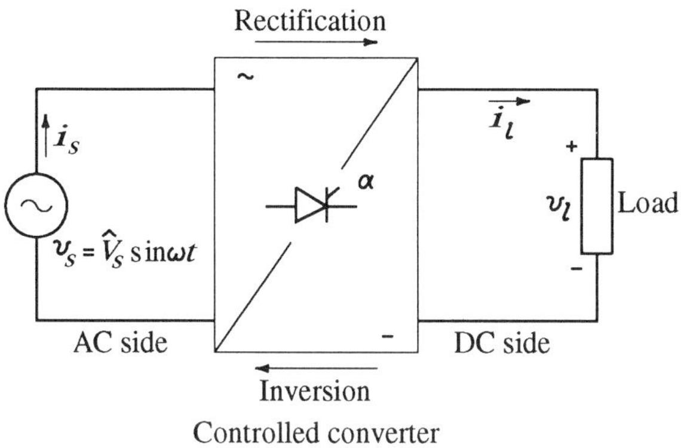

Fig. 2.4 A single-phase converter general schematic.

$$I_{dc} = \frac{1}{2\pi} \int_{\alpha}^{2\pi+\alpha} i_s(t)\, d\omega t \tag{2.3.2}$$

and the coefficients a_n and b_n are expressed by

$$a_n = \frac{1}{\pi} \int_{\alpha}^{2\pi+\alpha} i_s(t) \cos n\omega t\, d\omega t = \mathrm{f}(\alpha) \tag{2.3.3}$$

and

$$b_n = \frac{1}{\pi} \int_{\alpha}^{2\pi+\alpha} i_s(t) \sin n\omega t\, d\omega t = \mathrm{f}(\alpha)\,. \tag{2.3.4}$$

Usually there is symmetry, so $I_{dc} = 0$, and $i_s(t)$ can be expressed as

$$i_s(t) = \sum_{n=1,2\cdots}^{\infty} \sqrt{2}\, I_{sn} \sin(n\omega t + \theta_n) \tag{2.3.5}$$

where the displacement angle θ_n of the nth harmonic current is given by

$$\theta_n = \tan^{-1} a_n/b_n = \mathrm{f}(n,\alpha) \tag{2.3.6}$$

and the rms value of the nth harmonic input current is

$$I_{sn} = \frac{1}{\sqrt{2}}(a_n^2 + b_n^2)^{1/2}\,. \tag{2.3.7}$$

The rms value of the input current is then

$$I_s = (I_{s1}^2 + I_{s2}^2 + I_{s3}^2 + I_{s4}^2 + \cdots)^{1/2}\,. \tag{2.3.8}$$

With this information it is possible to determine harmonic, displacement and power factors relating to the input side of the converter.

Input total harmonic distortion (THD).

Total harmonic distortion was introduced in section 1.4.5. Equation (2.3.8) gives an expression for the rms value of the ac line current. The ideal value of the input current would comprise only a fundamental component. There would be no harmonics. Since harmonics are present, the input current waveform is distorted. The distortion component DC is formed from eq. (2.3.8) to be

$$\mathrm{DC} = (I_s^2 - I_{s1}^2)^{1/2} = \left[\sum_{n=2}^{\infty} I_{sn}^2 \right]^{1/2} . \tag{2.3.9}$$

As a measure of the harmonic content of the converter input current i_s, *total harmonic distortion* THD is defined by the ratio of the harmonic component to the fundamental component. That is, the normalized distortion component is

$$\mathrm{THD} \triangleq \frac{(I_{s\,rms}^2 - I_{s\,1\,rms}^2)^{1/2}}{I_{s\,1\,rms}} . \tag{2.3.10}$$

The ideal value of this harmonic factor would be zero.

Input power factor PF and displacement power factor DPF.

The power factor PF of a converter-fed load, in terms of the ac supply quantities, is defined by

$$\mathrm{PF} \triangleq \frac{\text{average power}}{V_{s\,rms} I_{s\,rms}} . \tag{2.3.11}$$

With pure sinewave excitation the power factor is the cosine of the phase angle between the voltage and current. However, with a controlled converter, the supply current i_s is not sinusoidal, so we must use the definition above.

In most cases the supply voltage is sinusoidal; that is, the only voltage harmonic is the fundamental component. Accordingly, there is no power associated with the harmonic components of the currents, and the average power P from the supply is associated with the fundamental quantities $V_{s\,1\,rms}$ and $I_{s\,1\,rms}$ in the form

$$P = V_{s\,1\,rms} I_{s\,1\,rms} \cos\theta_1 \tag{2.3.12}$$

where θ_1 is the phase-angle difference between the fundamental components of the voltage and current. From the definition of power factor

$$\mathrm{PF} = \frac{V_{s\,1\,rms} I_{s\,1\,rms} \cos\theta_1}{V_{s\,rms} I_{s\,rms}} = \frac{I_{s\,1\,rms}}{I_{s\,rms}} \cos\theta_1 . \tag{2.3.13}$$

By definition, the *input distortion factor* is defined as the ratio of the rms fundamental component of current and the total rms value of the current. That is,

$$\text{input distortion factor} \triangleq \frac{I_{s\,1\,rms}}{I_{s\,rms}} . \tag{2.3.14}$$

Also, $$\text{input displacement power factor DPF} \triangleq \cos\theta_1 . \tag{2.3.15}$$

If the input current i_s were sinusoidal (for $\alpha = 0$), the power factor PF would be equal to the *displacement power factor* DPF and the input distortion factor would be unity. Ideally, the power factor is unity, but, since $I_{s\,rms} > I_{s\,1\,rms}$, the power factor is less than the ideal case whenever controlled rectification is employed.

Output form factor FF.

On the output side of a converter, shown in Fig. 2.4, the load voltage v_l would be a constant dc value ideally. The converter provides an actual output voltage that is a rectified partial sinewave. As a measure of the *goodness* of the shape of the output voltage we have the *form factor* FF, which is defined as the ratio of the rms value of the output voltage and the average value of the output voltage. That is,

$$\mathrm{FF} \triangleq \frac{V_{l\,rms}}{V_{l\,av}}. \tag{2.3.16}$$

The ideal value of the form factor FF is unity. That is, $V_{l\,rms} = V_{l\,av}$. This occurs if the output voltage is a constant dc value.

Output ripple factor RF.

The converter output voltage v_l can be considered to consist of two components, the average value $V_{l\,av}$ and a ripple (ac component), whose rms value is $V_{l\,ac}$. The rms value $V_{l\,rms}$ of the output voltage is given by

$$V_{l\,rms}^2 = V_{l\,av}^2 + V_{l\,ac}^2. \tag{2.3.17}$$

From this, the ac component of voltage is

$$V_{l\,ac} = (V_{l\,rms}^2 - V_{l\,av}^2)^{1/2}. \tag{2.3.18}$$

The *ripple factor* RF of the output voltage v_l is defined as the ratio of the rms ac component to the dc component. That is,

$$\mathrm{RF} \triangleq \frac{V_{l\,ac}}{V_{l\,av}} = \left[\frac{V_{l\,rms}^2}{V_{l\,av}^2} - 1\right]^{1/2}. \tag{2.3.19}$$

Ideally, the value of RF is zero. In terms of the form factor FF, the ripple factor is

$$\mathrm{RF} = (\mathrm{FF}^2 - 1)^{1/2}. \tag{2.3.20}$$

The ripple factor of the output current i_l can be defined in the same way as the ripple factor of the output voltage.

Rectification ratio.

The *rectification ratio* RR is defined by the expression

$$\mathrm{RR} \triangleq \frac{V_{l\,av} I_{l\,av}}{V_{l\,rms} I_{l\,rms}}. \tag{2.3.21}$$

This is known also as the *converter efficiency,* but this can be confusing.

Ideally, the output voltage v_l and current i_l are unvarying, so that $V_{l\,rms} = V_{l\,av}$. In this case the efficiency is a maximum with RR = 1. Efficiency of a non-ideal converter involves losses and has the conventional definition.

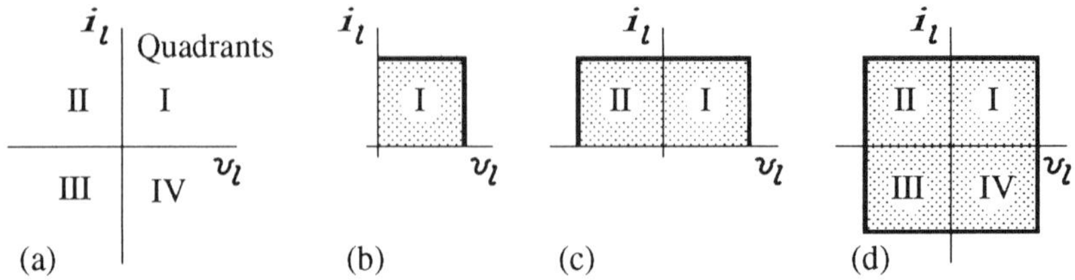

Fig. 2.5 Quadrant operation.
(a) Quadrants, (b) one-quadrant operation,
(c) two-quadrant operation, (d) four-quadrant operation.

Converter quadrant operation.
A general converter is illustrated in Fig. 2.4. There is a number of converter configurations. Each can produce a different characteristic. One form of characterization is defined by quadrant operation, as depicted in Fig. 2.5. Figure 2.5a shows the coordinate system, the four quadrants I, II, III, and IV and the axis variables, load voltage v_l and current i_l.

One quadrant operation, as depicted in Fig. 2.5b, indicates that the converter allows only one polarity of load voltage v_l and current i_l and this is the normal rectifier mode of the converter. *Two quadrant operation* indicates greater versatility. The converter can rectify with positive polarities of v_l and i_l, but it can have the condition of positive current i_l and a reverse voltage v_l that is defined as the inverter mode (quadrant II). This is shown in Fig. 2.5c.

Figure 2.5d indicates that, with *four quadrant operation*, any combination of polarity of load voltage v_l and current i_l is possible. To implement this operation requires two converters (called a dual converter) with one connected in reverse to the other. In this way, rectification and inversion are obtained from one converter, as in quadrants I and II. Quadrants III and IV represent a second converter operation that rectifies with reverse polarity of voltage v_l and current i_l (quadrant IV) compared with the first converter, and that inverts with opposite polarity of voltage v_l and the same polarity of current i_l (quadrant III).

In the sections that follow different converter configurations are described. For each converter the mode of quadrant operation will be pointed out.

2.3.2. Single-phase Half-wave Converter

Figure 2.6a shows a circuit diagram, in which the ac supply of voltage v_s is rectified by the thyristor switch *TH* and the power in the resistive load is modulated by controlling the delay angle α.

The operation of the circuit can be described with respect to the waveforms in Fig. 2.6b. The sinusoidal supply voltage v_s acts as a reference. Initially, the switch *TH* is off, so there is no current in the circuit. At some angle $\omega t = \alpha$ the switch *TH* is turned on and is immediately equivalent to a short circuit. The current response is $i_A = v_s/R$. When the current falls naturally to zero, the switch turns off, blocks all voltage that is positive at the cathode K (reverse bias) and stays off until the anode A is positive and a gate G turn-on signal is applied again (at $\omega t = 2\pi + \alpha$). This is repeated to provide a steady periodic waveform.

What is important is the power in the load as a function of the supply voltage v_s, the value of the load resistance R and the delay angle α of the thyristor *TH*. Also of importance is the average current in the load because this is related to the rating of the thyristor. Figure 2.6b shows voltage and current waveforms. The load current falls to zero in phase with the voltage since the load is resistive. Therefore, the thyristor commutates (switches off) at zero voltage points.

In the following calculation the thyristor is assumed to be ideal. The average current $I_{l\,av}$ in the circuit for any trigger delay angle α is

$$I_{l\,av} = \frac{1}{2\pi}\int_{\alpha}^{\pi} i_l \, d\omega t = \frac{1}{2\pi}\int_{\alpha}^{\pi} \frac{v_s}{R} \, d\omega t = \frac{1}{2\pi}\int_{\alpha}^{\pi} \frac{\sqrt{2}\,V_s}{R} \sin\omega t \, d\omega t$$

$$= \frac{\sqrt{2}\,V_s}{2\pi R}[-\cos\omega t]_{\alpha}^{\pi} = \frac{V_s}{\sqrt{2}\,\pi R}(1 + \cos\alpha). \qquad (2.3.22)$$

The maximum value of the average current occurs if $\alpha = 0$, that is, the maximum value is $\sqrt{2}\,V_s/\pi R$. From this value the current rating of the thyristor is determined. Ideally, the current rating is $\sqrt{2}\,V_s/\pi R$, but in reality a tolerance must be allowed to protect the thyristor against overloads.

The average power P in the load for any trigger angle α is given by

$$P = \frac{1}{2\pi}\int_{0}^{2\pi} v_l i_l \, d\omega t = \frac{1}{2\pi}\int_{\alpha}^{\pi} \frac{v_l^2}{R} \, d\omega t = \frac{V_s^2}{\pi R}\int_{\alpha}^{\pi} \sin^2\omega t \, d\omega t$$

$$= \frac{V_s^2}{2\pi R}\left[\omega t - \frac{\sin 2\omega t}{2}\right]_{\alpha}^{\pi} = \frac{V_s^2}{2\pi R}\left(\pi - \alpha + \frac{\sin 2\alpha}{2}\right). \qquad (2.3.23)$$

The average power can be varied from zero to $V_s^2/2R$ with α control.

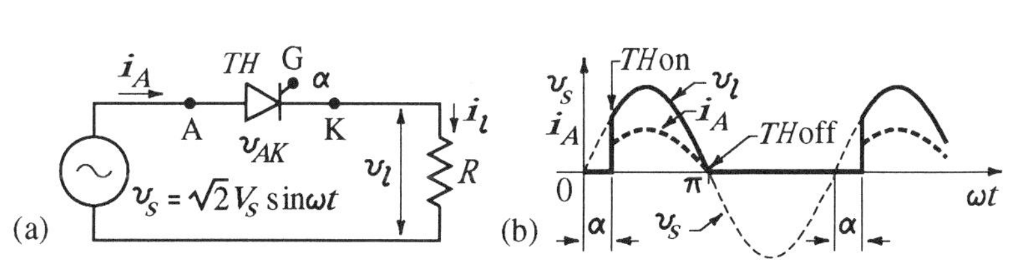

Fig. 2.6 Controlled rectification.
(a) Circuit diagram, (b) waveforms.

EXAMPLE 2.3

A supply delivers a rectangular-wave ac voltage of amplitude 200 V at 100 Hz. A thyristor modulates the power to a load resistor of value 2 ohms. Determine the rms value of the current and the average load power for a trigger angle $\alpha = \pi/2$ radians. What are the current and voltage ratings of the thyristor?

Solution

The rms value of the load current is

$$I_{l\,rms} = \left[\frac{1}{2\pi}\int_{\alpha}^{\pi}\frac{V_s^2}{R^2}\,d\omega t\right]^{1/2} = \frac{V_s}{R}\left[\frac{1}{2\pi}(\pi-\alpha)\right]^{1/2} = 50\text{ A for } \alpha = \pi/2 \text{ radians.}$$

The average load power P for $\alpha = \pi/2$ radians is
$P = I_{l\,rms}^2 R = 50^2 \times 2 = 5 \times 10^3$ W.
The maximum average current $I_{l\,av}$ (when $\alpha = 0$) is

$$I_{l\,av} = \frac{1}{2\pi}\int_0^{\pi}\frac{V_s}{R}d\omega t = \frac{V_s}{2R} = 50\text{ A.}$$

Therefore, the current rating for the thyristor would be some value greater than 50 A. To block a voltage of 200 V with some safety, the thyristor voltage rating might be 300 V.

Inductive load. It is common for the load of a converter to have an inductive component L in addition to resistance R, as shown in Fig. 2.7a. This added element will cause a delay in the rise and fall of current i_l, as shown in Fig. 2.7b. Consequently, the thyristor does not turn off until the current extinguishes at angle β. The load voltage v_l is negative from π to β and this reduces the average load voltage across the load.

This negative portion of the voltage across the load changes the character of the converter by allowing two-quadrant operation. See Fig. 2.7c.

For any firing angle α, as shown in Fig.2.7b, while the current i_l is rising in value, the voltage v_l across the load is positive. The energy from the supply is absorbed by the resistance R and stored in the inductance L. After the current i_l starts to fall in value, the voltage across the inductor changes polarity (di_l/dt is negative) and the voltage v_l across the load becomes negative. Some energy that was stored in the inductance L is returned to the ac supply and the rest is absorbed by the resistance R.

In Fig. 2.7d, it is shown that the thyristor is turned on late ($\alpha > \pi/2$) and it is indicated that the current comes to zero late ($\beta > 3\pi/2$). At this point the thyristor extinguishes. Is this feasible in this circuit?

The way Fig. 2.7b is drawn, it can be seen by inspection that the interval that the instantaneous load voltage v_l is positive is greater than the time it is negative. Therefore, the average value of the load voltage $V_{l\,av}$ is positive and the average value of the load current $I_{l\,av}$ is positive. By definition this is the rectification

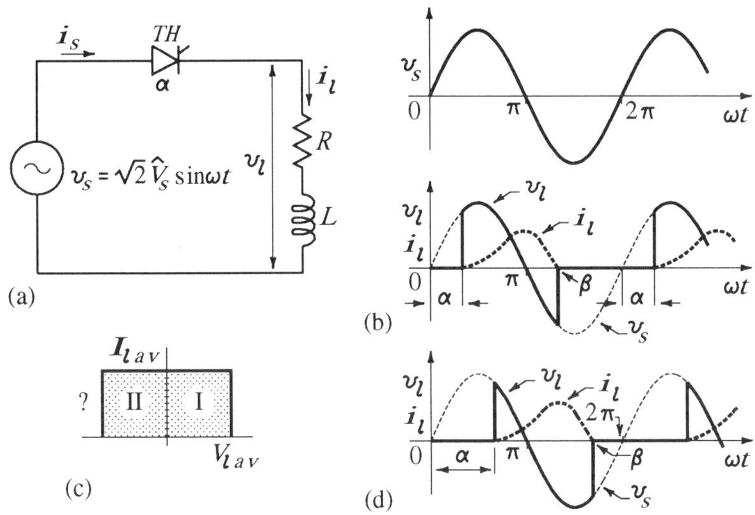

Fig. 2.7 Single-phase half-wave conversion.
(a) Circuit diagram, (b) rectifier waveforms,
(c) two-quadrant operation, (d) inversion waveforms.

mode of operation, quadrant I in Fig. 2.7c.

The way Fig. 2.7d is drawn, it can be seen by inspection that the average load voltage $V_{l\,av}$ has a negative value and the average load current has a positive value. By definition this is the inversion mode of operation, quadrant II in Fig. 2.7c. For a half-wave rectifier this does not appear possible from an intuitive point of view. Instantaneously, it is acceptable to have an exchange of energy that alternates from source to load and back again, but it is not acceptable in average terms.

The condition for a negative value of the average load voltage can be seen from Fig. 2.7d to be

$$\beta - \pi > \pi - \alpha. \tag{2.3.24}$$

If the load is purely resistive then the thyristor turns off at $\beta = \pi$. No matter what value of α is used to turn on the thyristor, the condition of eq. (2.3.24) cannot be met, so there is no inversion. If the load is purely inductive, there are no losses; what energy is absorbed by the inductor from the ac supply is delivered back again. The average value of voltage across an inductor that is excited with a periodic, steady-state function is zero, because the current starts at zero and ends at zero. There is symmetry of current response and, therefore, symmetry of instantaneous load voltage. If the thyristor is turned on at $\alpha = 0$, it won't turn off until $\beta = 2\pi$. If the thyristor is turned on at $\alpha = \pi/2$, it won't turn off until $\beta = 3\pi/2$. For any value of α, by symmetry $\beta = 2\pi - \alpha$. This indicates that the condition of inversion defined by eq. (2.3.24) is not met.

Finally, if the load comprises both resistance R and inductance L, the load current i_l will begin to rise at $\omega t = \alpha$ and will fall to zero at some angle $\omega t = \beta$. The angle β must be at some value between the two extremes of π for resistance and $2\pi - \alpha$ for inductance. Therefore, the condition in eq. (2.3.24) can never be met. There can be no inversion based on average values. This type of converter can only operate in the rectification mode.

The circuit depicted in the diagram of Fig. 2.7a is simple in concept but it is not simple to analyze. The average load voltage $V_{l\,av}$ is given by

$$V_{l\,av} = \frac{1}{2\pi}\int_{\alpha}^{\beta} \sqrt{2}\, V_s \sin\omega t \, d\omega t = \frac{V_s}{\sqrt{2}\,\pi}(\cos\alpha - \cos\beta)\,. \tag{2.3.25}$$

The unknown in this equation is the thyristor extinction angle β. To find this angle involves the numerical solution of a transcendental equation, which is the solution of the current response from

$$v_s = Ri_l + L\frac{di_l}{dt} \tag{2.3.26}$$

over the region of positive i_s (from α to β). Once β has been found, the rms value of the load current $I_{l\,rms}$ can be calculated and this result leads to the determination of the average power $(I_{l\,rms}^2 R)$ absorbed by the load as a function of the firing angle α.

EXAMPLE 2.4

Consider the single-phase, half-wave converter, depicted in Fig. 2.7. The power supply is a 115-V, 60-Hz source and the load has a resistance R of 8 Ω and an inductance L of 0.02 H. If the thyristor firing angle is $\alpha = \pi/3$ radians, what are the average values of the load voltage and current?

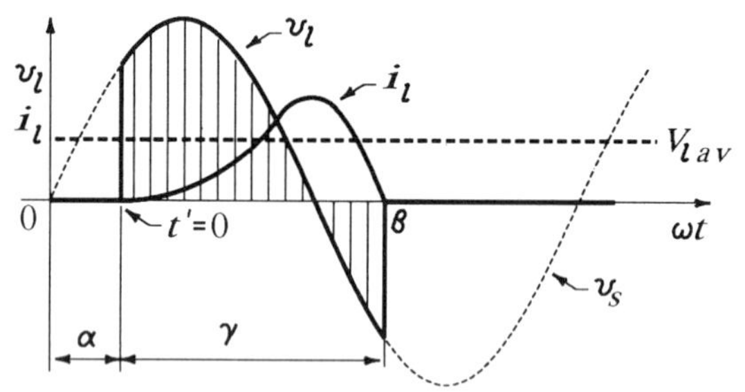

Fig. EX2.4

Solution

The solution of this problem may be obtained using the Laplace method, but, because of the simplicity of the circuit, it is probably quicker to use the classical approach. That is, the solution of the differential equation has a particular integral and a complementary function to create the steady-state component and the transient component of the current.

The forms of the load voltage and current are shown in Fig. EX2.4. It is assumed that the thyristor extinction angle β is such that the current is discontinuous. This circuit would require an inductance of infinite value for the load current to be continuous.

If the load current i_l has the steady-state component i_{ss} and the transient component i_{trans}, then the solution has the form

$i_l(t) = i_{ss} + i_{trans}$

as long as the thyristor is in the on-state. From ac theory, the steady-state component is

$$i_{ss} = \frac{\hat{V}_s}{[R^2 + \omega^2 L^2]^{1/2}} \sin(\omega t - \theta), \text{ where } \tan\theta = \omega L / R.$$

Upon substitution of the given values $i_{ss} = 14.792 \sin(\omega t - 0.7558)$ A.

The transient component of the current is $i_{trans} = I_o \, e^{-Rt/L}$.

At $\omega t = \alpha$, the load current is zero, so upon substitution of values $0 = 14.792 \sin(\alpha - 0.7558) + I_o \, e^{-400\alpha/\omega}$.

For $\alpha = \pi/3$ rad and $\omega = 377$ rad/s, $I_o = -12.91$ A.

Between α and β, that is, over the interval γ of thyristor conduction

$i_l(t) = 14.792 \sin(\omega t - 0.7558) - 12.91 \, e^{-400t}$ A.

From this equation, the load current i_l is zero at $t' = 0$ and $t' = 10.3$ ms.[1]

This indicates that the thyristor conducts for an amount $\omega t' = \gamma = 2.836$ radians and the extinction angle is $\beta = 3.883$ radians. $t' = t - \alpha/\omega$.

The average load voltage can be calculated from the source voltage $v_s = \hat{V}_s \sin\omega t$ between the limits of thyristor turn-on and turn-off, that is, between the limits of α and β (where $\beta = \alpha + \gamma$). So,

$$V_{l\,av} = \frac{1}{2\pi} \int_{\alpha}^{\beta} \hat{V}_s \sin\omega t \, d\omega t = \frac{\sqrt{2} \times 115}{2\pi} \int_{\pi/3}^{\pi/3 + 2.836} \sin\omega t \, d\omega t = 31.55 \text{ V}.$$

Since the average voltage across the inductor L is zero, then the average voltage across the resistor R has the same value as $V_{l\,av}$.

Consequently, for a firing angle of $\alpha = \pi/3$ radians, the average load current is $I_{l\,av} = V_{l\,av}/R = 31.55/8 = 3.94$ A.

We have seen that, if the converter load is inductive, there is a negative instantaneous voltage across the load, because the thyristor does not commutate until the load current drops to zero. A freewheeling diode across the load prevents the load voltage going negative and causes the thyristor to commutate as the supply

[1] This latter solution is obtained from a trial and error iterative method or by using a math software package.

voltage falls to zero. This is because the diode provides the path for the load current, so the thyristor current drops to zero. At zero source current i_s the thyristor turns off. Figure 2.8a shows the circuit diagram.

While the source voltage provides a positive voltage at the thyristor anode A, the switch *TH* blocks until the firing pulse is applied at $\omega t = \alpha$. Then the thyristor turns on, so the current i_l in the load rises until the source voltage falls to $v_s = 0$. This is the same as in the circuit illustrated in Fig. 2.7, because, over the positive half cycle from $\omega t = \alpha$ to $\omega t = \pi$, the diode is reverse biased and $i_D = 0$.

As soon as the source voltage v_s goes negative, the diode *D* becomes forward biased and turns on. In the on-state the diode had no impedance ideally, so it offers a short circuit to the load. This means that the load current i_l, sustained by the energy in the inductance *L*, has a closed path formed by the diode.

In the circuit loop formed by the source, the diode and the thyristor, as the source voltage v_s goes negative, the diode is forward biased and turns on immediately. This means that the source voltage v_s is applied directly across the thyristor, so the voltage v_{AK} tends to go negative. If there is no inductance in this loop to maintain current i_s, the reverse bias voltage turns the thyristor off, and the current $i_s = 0$ and voltage $v_{AK} = -v_s$. Kirchhoff's current law at the node K provides the information that $i_D = i_l$.

During the interval from $\omega t = \pi$ to the time that the thyristor is fired again ($\omega t = 2\pi + \alpha$), the thyristor blocks and the load current i_l decays exponentially to zero. For this condition $i_D = i_l = 0$. This is shown to occur at $\omega t = \beta$ in Fig. 2.8b. When the thyristor is turned on again, the source voltage v_s will be positive at the anode A and positive at the cathode K. Thus, the voltage source v_s reverse biases the diode, and this puts the diode in the off-state (open circuit). The cycle now repeats with the conditions $i_l = i_s$, $v_l = v_s$ over the interval $(2\pi + \alpha) \leq \omega t \leq 3\pi$.

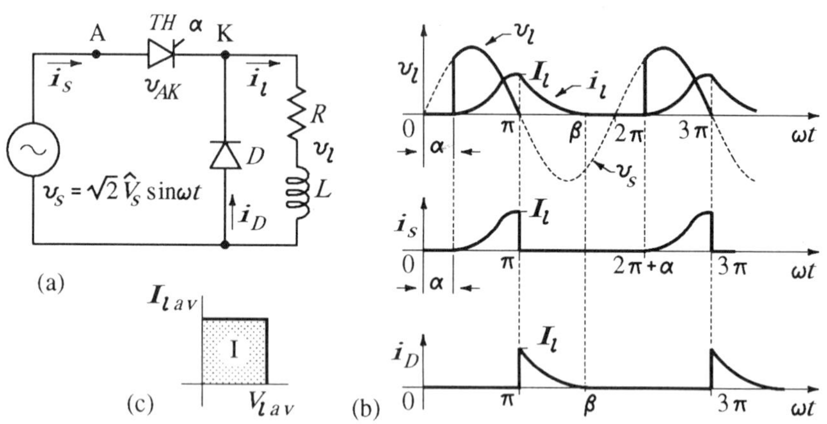

Fig. 2.8 Single-phase half-wave rectification.
(a) Circuit with diode, (b) waveforms, (c) single-quadrant operation.

The load voltage cannot go negative because of the diode, so the converter can only rectify. This is depicted in Fig. 2.8c as single-quadrant operation. The average voltage across the load is greater for the same firing angle, if the freewheeling diode[2] is connected across the load. The source current i_s is the same as the load current only while the thyristor is on. As soon as the source voltage v_s goes negative, the diode D turns on and keeps the load voltage v_l equal to zero. In this way, most of the energy that is stored in the inductance L is transferred to the load resistance R. Some of the energy is dissipated in the diode, if the diode is not ideal. The instantaneous voltage v_l across the load is given by

$$v_l = Ri_l + L\frac{di_l}{dt}. \tag{2.3.27}$$

Since the average voltage across the inductor is zero for periodic steady-state changes of current, in terms of average values the average load voltage is given by

$$V_{l\,av} = RI_{l\,av} \tag{2.3.28}$$

where,

$$V_{l\,av} = \frac{1}{2\pi}\int_{\alpha}^{\beta} v_s \, d\omega t = \frac{V_s}{\sqrt{2}\,\pi}(1+\cos\alpha). \tag{2.3.29}$$

Equations (2.3.28) and (2.3.29) help to determine the value of the average load power P only for highly inductive loads. In such cases it can be assumed that the load current is virtually unvarying[3], $V_{l\,av}$ is calculated from eq. (2.3.29), $I_{l\,av}$ follows from eq. (2.3.28) and the power in the load is $P \approx V_{l\,av} I_{l\,av} = I_{l\,av}^2 R$.

For circuits that do not have highly inductive loads, the current response $i_l(t)$ has to be determined from eq. (2.3.27) between the limits α and π and then from

$$i_l = I_l e^{-tR/L} \tag{2.3.30}$$

between π and β. (The extinction angle β approximately equals $4 \times \omega L/R$, measured from $\omega t = \pi$.) The next step is to calculate the rms value of the load current $I_{l\,rms}$, and the average power P in the load is $I_{l\,rms}^2 R$.

EXAMPLE 2.5

Consider the single-phase, half-wave converter depicted in Fig. 2.8. The power supply is a 115-V, 60-Hz source. The load has a resistance R of 8 Ω and an inductance L of 0.02 H. If the thyristor firing angle is $\alpha = \pi/3$ radians, what are the average values of the load voltage and current?

---

[2] This diode is often referred to as the *commutating* diode because it causes the thyristor to turn off just as the supply voltage goes negative.

[3] If the load current is virtually unvarying the current has the same rms and average values.

Solution

This problem differs from EXAMPLE 2.4 in that a commutating diode is employed across the load. While the thyristor is on, the load current is under the influence of the source voltage v_s. After the thyristor commutates at $\omega t = \pi$ the load current decays exponentially to zero under the influence of L and R. Figure EX2.5 shows the waveforms.

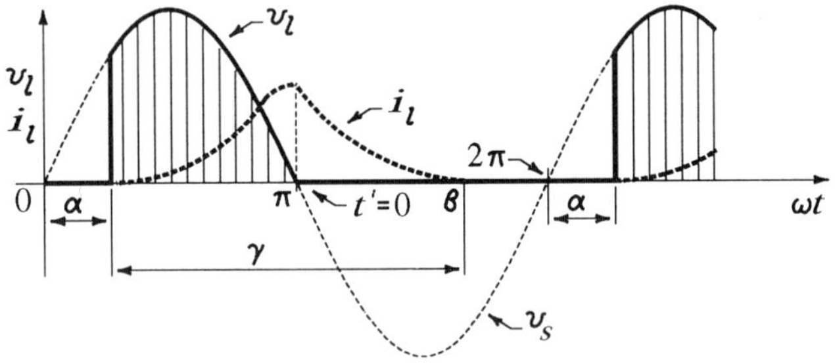

Fig. EX2.5 Load voltage and current waveforms.

The average load voltage is

$$V_{l\,av} = \frac{1}{2\pi}\int_{\alpha}^{\pi} \hat{V}_s \sin\omega t \, d\omega t = \frac{\hat{V}_s}{2\pi}(1+\cos\alpha) = \frac{\sqrt{2}\times 115}{2\pi}(1+0.5) = 38.83\text{V}.$$

Over the range $\alpha \le \omega t \le \pi$ the load current can be described by $i_l(t) = i_{ss} + i_{trans}$.

The steady-state component i_{ss} is

$$i_{ss} = \frac{\hat{V}_s}{[R^2 + \omega^2 L^2]^{1/2}} \sin(\omega t - \theta), \text{ where } \theta = \tan^{-1}\frac{\omega L}{R}.$$

Upon substitution $\theta = \tan^{-1} 2\pi \times 60 \times 0.02/8 = 0.756$ rads and

$$i_{ss} = \frac{\sqrt{2}\times 115}{[8^2 + (120\pi)^2 \times 0.002^2]^{1/2}} \sin(\omega t - 0.756) = 14.79 \sin(\omega t - 0.756).$$

The transient component i_{trans} is $i_{trans} = I_o\, e^{-Rt/L}$.

At $\omega t = \alpha$, $i_l(t) = 0$.

So, $-I_o\, e^{-\alpha R/\omega L} = 14.79 \sin(\alpha - 0.756) = 4.247$.

That is, $I_o = 4.247 \times e^{8\pi/(3\times 2\pi\times 60\times 0.02)} = -12.90$ A.

For $\alpha = \pi/3$, and for the interval that the supply voltage v_s appears across the load,

$i_l(t) = 14.792 \sin(\omega t - 0.7556) - 12.90\, e^{-400t}$ A.

At $\omega t = \pi$ the load current is

$i_l(\pi/\omega) = 14.792 \sin(\pi - 0.756) - 12.90\, e^{-400\pi/120\pi} = 9.687$ A.

Therefore, while the freewheeling diode is conducting

$i_l = 9.687\, e^{-Rt'/L} = 9.687\, e^{-400t'}$ A,

where $t' = 0$ at the time the thyristor ceases to conduct. That is $t' = t - \pi/\omega$. This current will decay to zero in about four time constants, that is, in about 0.01 s. This time is equivalent to 3.77 radians (216°). Thus, the total load current duration from the time the thyristor is fired at $\omega t = \alpha$ to the time the current is zero

again is
$\gamma = (\pi - \alpha) + 3.77 = 5.86$ radians.
The average value of the load current is

$I_{l\,av} = \frac{1}{2\pi}\int_{\alpha}^{\beta} i_l \, d\omega t.$ That is,

$$I_{l\,av} = \frac{1}{2\pi}\int_{\pi/3}^{\pi} 14.792\sin(\omega t - 0.756)\,d\omega t - \frac{1}{2\pi}\int_{\pi/3}^{\pi} 12.9\,e^{-400t}\,d\omega t$$
$$+ \frac{1}{2\pi}\int_{0}^{3.77} 9.687\,e^{-400t'}\,d\omega t' = 4.85\text{ A}.$$

This is the same as $V_{l\,av}/R$.
There is an increase in the average values of voltage and current for the same firing angle α when the diode D is added to the circuit.

2.3.3. Single-phase Bridge Converter

Figure 2.9 shows four configurations[4] of a single-phase full-wave converter. For full control the bridge of Fig. 2.9a consists of four thyristors. While rail A is positive, thyristors $TH1$ and $TH2$ can be triggered at angle α. While rail B is positive thyristors $TH3$ and $TH4$ can be triggered at angle α. If the load current i_l is continuous and in steady state, the load voltage waveform looks like that in Fig. 2.10a. The current intervals are depicted in Fig. 2.10b. If the load current were ripple free, the intervals would be the waveshapes.

While thyristors $TH1$ and $TH2$ of the fully-controlled bridge are positively biased, gate pulses will turn them on. They will conduct the load current i_l until both the thyristors $TH3$ and $TH4$ are forward biased and receive gate pulses. At this point thyristors $TH3$ and $TH4$ turn on, thyristors $TH1$ and $TH2$ become reverse biased and turn off as the load current is transferred from the outgoing thyristors to the incoming thyristors. The average load voltage $V_{l\,av}$ is

$$V_{l\,av} = \frac{1}{\pi}\int_{\alpha}^{\pi+\alpha} v_l\,d\omega t = \frac{\hat{V}_s}{\pi}\int_{\alpha}^{\pi+\alpha} \sin\omega t\,d\omega t = \frac{\hat{V}_s}{\pi}[-\cos\omega t]_{\alpha}^{\pi+\alpha}.$$

Therefore,

$$V_{l\,av} = \frac{2}{\pi}\hat{V}_s\cos\alpha. \tag{2.3.31}$$

This equation holds only if the current is continuous. An inspection of eq. (2.3.31) indicates that the single-phase full-wave bridge with this particular configuration has two quadrant operation. For a range of trigger angle $0 \le \alpha \le \pi/2$ the average load voltage $V_{l\,av}$ is positive, varying from $2\hat{V}_s/\pi$ to 0 volts. This is

[4] Those converters that employ uncontrolled devices (that is, diodes) are called half-controlled or asymmetric converters.

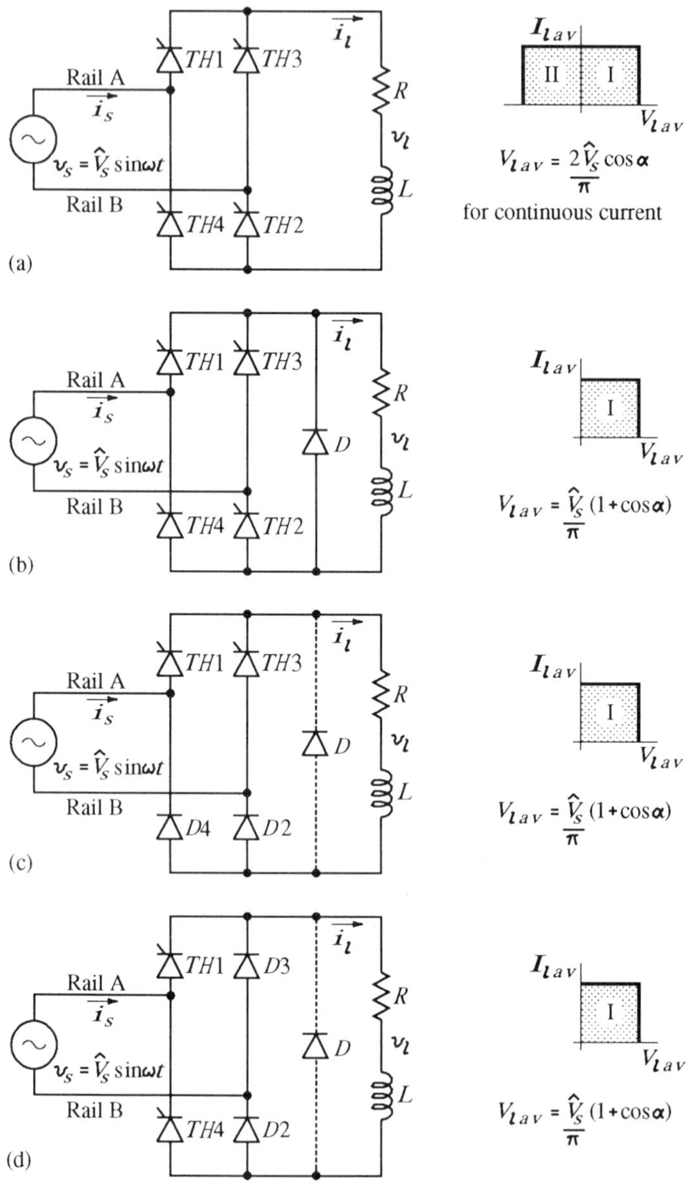

Fig. 2.9 Single-phase full-wave converters. (a) Inversion, (b) rectification, (c) and (d) half-controlled rectification.

(a)

(b)

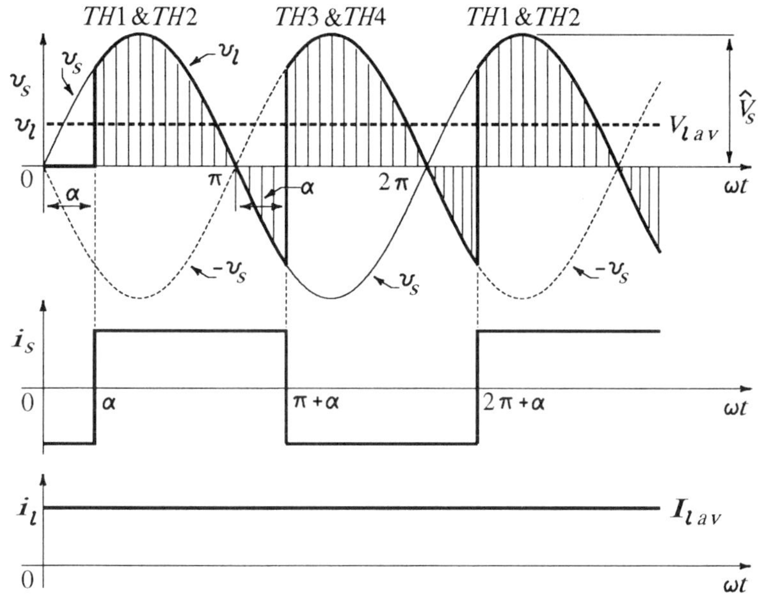

Fig. 2.10 Waveforms for two-quadrant operation.
(a) Voltage waveforms, (b) current intervals.

quadrant I mode of rectification. However, over the range of trigger angle $\pi/2 \leq \alpha \leq \pi$, $\cos\alpha$ is negative, so the average load voltage $V_{l\,av}$ is negative, varying from 0 to $-2\hat{V}_s/\pi$ volts. This is quadrant II mode of inversion.

EXAMPLE 2.6

Consider a fully-controlled, single-phase, full-wave, bridge converter as depicted in Fig. 2.9a. The supply is 120 V at 60 Hz. The load comprises a resistor of 8 Ω and an inductor whose value is high enough to assume that the load current is continuous and virtually ripple free. If the firing angle α of the thyristors is $\pi/6$ radians, calculate **(a)** the output voltage form factor FF, **(b)** the output voltage ripple factor RF, **(c)** the rectification ratio RR, **(d)** the expression for the input current as a Fourier series, **(e)** the input harmonic distortion THD, **(f)** the input power factor PF, **(g)** the input displacement power factor DPF and **(h)** the input distortion factor.

Solution

The performance factors of a converter are described in section 2.3.1.

(a) The output form factor FF is the ratio of the rms to the average load voltage. From Fig. 2.10 the rms load voltage $V_{l\,rms}$ can be expressed as

$$V_{l\,rms} = \left[\frac{1}{\pi} \int_{\alpha}^{\pi+\alpha} v_l^2 \, d\omega t \right]^{1/2} = \left[\frac{\hat{V}_s^2}{\pi} \int_{\alpha}^{\pi+\alpha} \sin^2 \omega t \, d\omega t \right]^{1/2} .$$

That is, $V_{l\,rms} = \hat{V}_s / \sqrt{2} = V_s$.

The rms value of the load voltage is constant, independent of the firing angle α, and equal to the rms value of the supply voltage V_s, if the load current is continuous. The average value of the load voltage $V_{l\,av}$ is given by eq. (2.3.31) to be

$V_{l\,av} = (2\hat{V}_s / \pi)\cos\alpha$.

For $\hat{V}_s = \sqrt{2}\,V_s = \sqrt{2} \times 120 = 170$ V and $\alpha = \pi/6$ radians,

$V_{l\,av} = (2 \times 170/\pi)\cos\pi/6 = 93.7$ V.

Consequently, the form factor $\mathrm{FF} = V_{l\,rms}/V_{l\,av} = 120/93.7 = 1.28$.

(b) The ripple factor RF is a measure of how much more the output voltage is greater than the dc component. If $V_{l\,ac}$ is the ac component of the output voltage, then the ripple factor is expressed in eq. (2.3.19) to be

$$\mathrm{RF} = \frac{V_{l\,ac}}{V_{l\,av}} = \frac{(V_{l\,rms}^2 - V_{l\,av}^2)^{1/2}}{V_{l\,av}} = \left[\frac{V_{l\,rms}^2}{V_{l\,av}^2} - 1 \right]^{1/2} = (\mathrm{FF}^2 - 1)^{1/2}.$$

Therefore, $\mathrm{RF} = (1.28^2 - 1)^{1/2} = 0.8$.

The ripple factor for the output current is zero.

(c) From eq. (2.3.21), the rectification ratio RR is

$$\mathrm{RR} = \frac{V_{l\,av} I_{l\,av}}{V_{l\,rms} I_{l\,rms}} = \frac{V_{l\,av}}{V_{l\,rms}} = \frac{93.7}{120} = 0.78 .$$

This value is high because the output current is filtered by the inductance.

(d) While thyristors $TH1$ and $TH2$ are on, the source current i_s is equal to the constant load current i_l and, while thyristors $TH3$ and $TH4$ are on, the source current i_s has the same magnitude but of opposite polarity. That is, the waveshape of the source current is rectangular.

By inspection of Fig. 2.10b the dc component I_{dc} of the Fourier series for the input current i_s is zero, and by symmetry there are no even harmonics.

Using eqs (2.3.3) through (2.3.7), the coefficients are

$$a_n = \frac{1}{\pi} \int_{\alpha}^{2\pi+\alpha} I_{l\,av} \cos n\omega t \, d\omega t$$

$$\text{or,} \quad a_n = \frac{1}{\pi} \int_{\alpha}^{\pi+\alpha} I_{l\,av} \cos n\omega t \, d\omega t - \frac{1}{\pi} \int_{\pi+\alpha}^{2\pi+\alpha} I_{l\,av} \cos n\omega t \, d\omega t.$$

That is, $a_n = -\dfrac{4}{n\pi} I_{l\,av} \sin n\alpha$ for $n = 1,3,5 \cdots$.

Similarly, $b_n = \dfrac{4}{n\pi} I_{l\,av} \cos n\alpha$ for $n = 1,3,5, \cdots$

and $\theta_n = \tan^{-1} a_n / b_n = -n\alpha$.

These are important results because they are functions of the angle α.

It follows that $i_s(t) = \displaystyle\sum_{n=1,3,5\cdots}^{\infty} \frac{4}{n\pi} I_{l\,av} \sin(n\omega t - n\alpha)$.

$\alpha = \pi/6$ and $I_{l\,av} = V_{l\,av}/R = 93.7/8 = 11.7$ A, so the Fourier series is
$i_s(t) = 14.9\sin(\omega t - \pi/6) + 4.97\sin(3\omega t - \pi/2) + 2.98\sin(5\omega t - 5\pi/6) + \cdots$.

(e) The total harmonic distortion THD is equal to $(I^2_{s\,rms} - I^2_{s\,1\,rms})^{1/2}/I_{s\,1\,rms}$ and is the ratio of the harmonic component of current to the fundamental component. The fundamental component of current is
$i_{s\,1} = 14.9\sin(\omega t - \pi/6)$, so $I_{s\,1\,rms} = 14.9/\sqrt{2} = 10.54$ A.
The rms value of the input current $I_{s\,rms}$ is equal to the value of the unvarying load current $I_{l\,av}$. That is, $I_{s\,rms} = 11.7$ A.
Thus, THD $= (11.7^2 - 10.54^2)^{1/2}/10.54 = 0.4824$ (or 48.24%).
The harmonic content of a rectangular wave is high, mainly because of the third harmonic component.

(f) The ratio of average power to volt-amperes is the definition of the input power factor PF. In this case of unvarying load current

$$\text{PF} = \frac{P}{V_s I_{s\,rms}} = \frac{V_{l\,av} I_{l\,av}}{V_s I_{s\,rms}} = \frac{V_{l\,av}}{V_s} = \frac{2\sqrt{2}}{\pi}\cos\alpha = 0.78\,.$$

The power factor is a function of the firing angle α, and has a maximum possible value of 0.9. The power factor can have very low values. This is not good.

(g) The input displacement power factor DPF is defined as $\cos\theta_1$, where θ_1 is the angle between the fundamental components of voltage and current. The voltage V_s has been taken as reference, so, from the Fourier series of the current,
DPF $= \cos\theta_1 = \cos\alpha = \cos\pi/6 = 0.866$.
Note that the displacement power factor DPF varies with the delay angle α.

(h) The input distortion factor is defined in eq. (2.3.14) as the ratio of the fundamental to total component of rms current, $I_{s\,1\,rms}/I_{s\,rms}$. In this case of ripple-free output current, the distortion factor is $I_{s\,1\,rms}/I_{l\,av} = 10.54/11.7 = 0.9$.

Single-phase full-wave bridge (single quadrant).

If a diode is connected across the load of a single-phase, full-wave, fully-controlled converter, as in Fig. 2.9b, the converter is a rectifier and operates only in the one quadrant mode. This is because the freewheeling diode D prevents a reverse voltage across the load. Consequently, as the supply voltage starts to reverse polarity (rail A goes negative, for example), diode D conducts the load current, the conducting thyristors $TH\,1$ and $TH\,2$ are reverse biased and, since the load current has been transferred to the diode, no source current is delivered, and thyristors $TH\,1$ and $TH\,2$ commutate.

The load voltage v_l has a waveform like that shown in Fig. 2.11. Its average value $V_{l\,av}$ is

$$V_{l\,av} = \frac{1}{\pi}\int_{\alpha}^{\pi} v_l\, d\omega t = \frac{\hat{V}_s}{\pi}\int_{\alpha}^{\pi}\sin\omega t\, d\omega t = \frac{\hat{V}_s}{\pi}[-\cos\omega t]_{\alpha}^{\pi}.$$

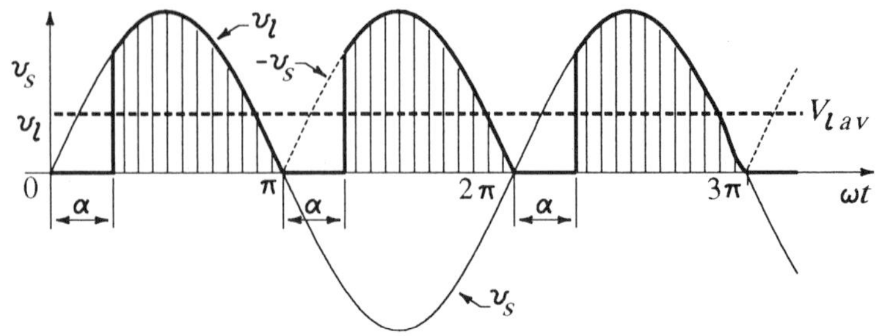

Fig. 2.11 Voltage waveforms for one-quadrant operation.

Therefore,

$$V_{l\,av} = \frac{\hat{V}_s}{\pi}(1 + \cos\alpha)\,. \tag{2.3.32}$$

The value of the average voltage ranges from $2\hat{V}_s/\pi$ to 0 as the thyristor delay angle α increases from 0 to π radians. This voltage is never negative.

The diode D aids commutation of the converter, but it does create the disadvantage of producing a discontinuous supply current with its attendant reduction of performance factors.

Single-phase full-wave half-controlled converters.

Figures 2.9c and 2.9d show converter configurations, in which only two thyristors are employed in the bridge. The other two thyristors have been replaced by diodes. Because of the added diodes, the circuit will operate in the one quadrant fashion with or without the freewheeling diode D. These configurations are called half-controlled converters. Less thyristors mean less expense.

Full-wave rectification is obtained by firing each thyristor alternately in each half cycle and the diodes freewheel to prevent the voltage becoming negative across the load. For example, if thyristor $TH1$ is forward biased and fired at angle α, it will conduct the load current i_l. The diode $D2$ would provide the return path. As the supply voltage v_s reverses polarity, $D2$ becomes reverse biased (see Fig. 2.9c) and commutates, and the source current falls to zero. If diode D is not present, $TH1$ and $D4$ provide the freewheeling action to maintain the load current. When thyristor $TH3$ of Fig. 2.9c is turned on, thyristor $TH1$ is reverse biased and turns off, while $TH3$ and $D4$ conduct the load current (and source current). The reader may like to go through a similar argument about the operation of the half-controlled converter illustrated in Fig. 2.9d. The average voltage $V_{l\,av}$ across the load is

$$V_{l\,av} = \frac{\hat{V}_s}{\pi}(1 + \cos\alpha)\,. \qquad (2.3.33)$$

For both half-controlled converters the average load voltage $V_{l\,av}$ is the same as that given by eq. (2.3.32). That is, the voltage is the same as a fully-controlled converter that utilizes a freewheeling diode. If a freewheeling diode D is connected across the load of a half-controlled rectifier, the diode conducts the load current while the applied load voltage V_l is zero and, literally, takes the load off the bridge components. This allows the thyristors to commutate earlier.

Other ac-dc converter configurations.
There is a number of other forms of controlled rectifiers that can give equal or better performance than those basic converters already described in this section.

There is the single-phase, full-wave, centre-tapped converter, illustrated in Fig. 2.12. The circuit diagram of a three-phase, half-wave converter is depicted in Fig. 2.13. Figure 2.14 shows the circuit diagrams of three-phase, full-wave converters. The reader may care to confirm the expressions of the average load voltage.

Pulse-width modulation techniques can be applied to ac-dc converter systems. These techniques do not improve the power factor greatly but the harmonic distribution can be changed.

2.4. DC TO AC CONVERSION

Converters that modulate power from a dc source to provide ac power to a load are called *inverters*. The two parameters of importance in inverters are the output voltage and the output frequency. There are direct ac-ac converters that are called cycloconverters and there are ac-ac converters that have a dc link. The latter type converts ac power to dc power and then converts the dc power to the required kind of ac power. The previous section dealt with ac-dc conversion. This section is concerned with a few examples of the dc-ac conversion, and is an expansion of the basic principles described in section 1.3.2.

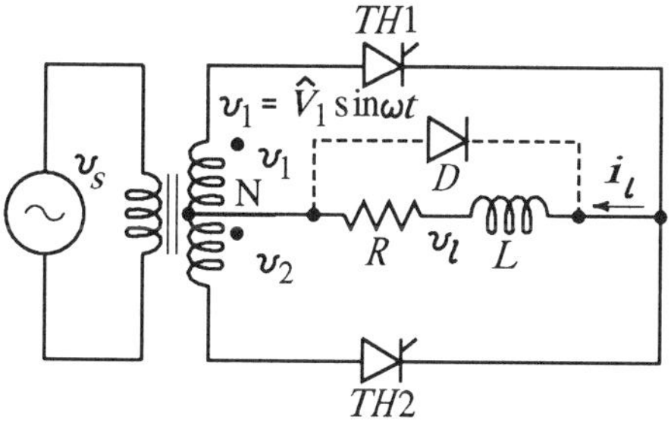

Without D and with continuous current i_l

$$V_{l\,av} = \frac{2\,\hat{V}_1 \cos\alpha}{\pi}.$$

With D $V_{l\,av} = \dfrac{\hat{V}_1\,(1 + \cos\alpha)}{\pi}$.

Fig. 2.12 Centre-tapped converter.

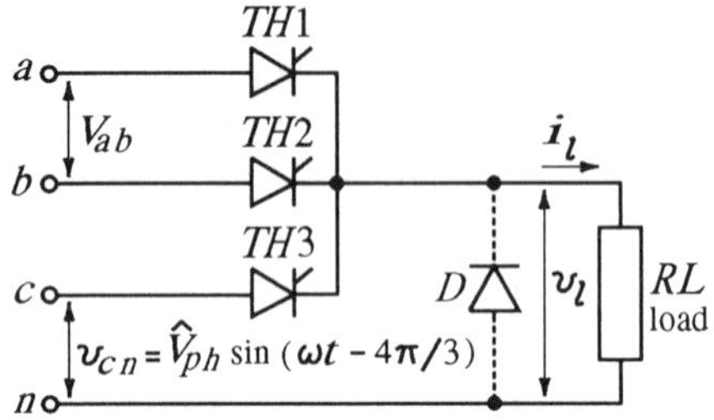

With R load, or D

$$V_{l\,av} = \frac{\sqrt{3}\,\hat{V}_{ab}}{2\pi}\left[1 + \cos(\alpha + \pi/6)\right],$$

range $\pi/6 \leqslant \alpha \leqslant 2\pi/3$ or $\pi/3 \leqslant \omega t \leqslant 5\pi/6$.

$$V_{l\,av} = \frac{3}{2\pi}\hat{V}_{ab}\cos\alpha$$

range $0 \leqslant \alpha \leqslant \pi/6$ or $\pi/6 \leqslant \omega t \leqslant \pi/3$.

RL load (no diode)

$$V_{l\,av} = \frac{3}{2\pi}\hat{V}_{ab}\cos\alpha,$$

range $0 \leqslant \alpha \leqslant 2\pi/3$ or $\pi/6 \leqslant \omega t \leqslant 5\pi/6$.

Fig. 2.13 Three-phase, half-wave converter.

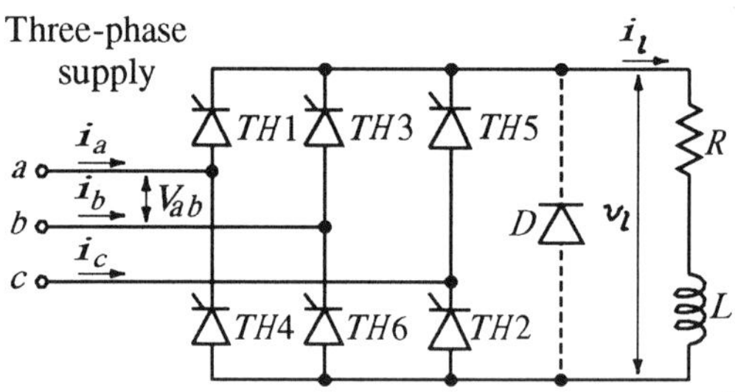

(a)

RL load (no diode)

$$V_{l\,av} = \frac{3}{\pi}\hat{V}_{ab}\cos\alpha.$$

Continuous current

R or RL and D load

range $0 \leqslant \alpha \leqslant \pi/3$,

$$V_{l\,av} = \frac{3}{\pi}\hat{V}_{ab}\cos\alpha.$$

Continuous current

Range $\pi/3 \leqslant \alpha \leqslant 2\pi/3$

$$V_{l\,av} = \frac{3}{\pi}\hat{V}_{ab}\left[1 + \cos(\pi/3 + \alpha)\right].$$

Discontinuous current

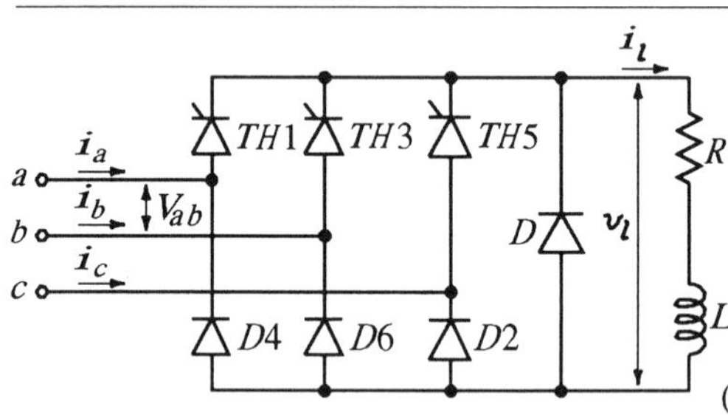

(b)

Range $0 \leqslant \alpha \leqslant 2\pi/3$,

$$V_{l\,av} = \frac{3}{2\pi}\hat{V}_{ab}(1 + \cos\alpha).$$

Fig. 2.14 Three-phase converters.
(a) Fully-controlled converter, (b) half-controlled converter.

The switch-mode inverter is the most common inverter found in service. Power semiconductor switches, in different configurations, chop the dc supply waveform so that the load experiences rectangular waves that periodically change polarity to give alternating voltage. Concern is given to the quality of power absorbed by the load. In most cases a sinewave of a single frequency is desired to minimize losses produced by other harmonics. In switch-mode inverters there are always some harmonics because the output waveform is synthesized digitally from the rectangular waveforms. Techniques to produce a near-sinewave are called pulse-amplitude modulation (PAM) or pulse-width modulation (PWM) methods.

There are two types of switch-mode inverters. They are the voltage-type and the current-type inverters. As its name implies, the voltage-type inverter is fed from a constant voltage source. The voltage chopping to create changing intervals of constant voltage at the load causes the current to change. This type is the one most commonly used. The current-type inverter supplies a constant source current that is not interrupted. The pattern of switching changes the voltage across the load in this case. Current-type inverters are used in some large ac motor drives.

If it is a premium to have an inverter with no harmonics, then, instead of switched-mode inverters, resonant inverters can be used. Resonance of energy between a capacitor and an inductor at the required output frequency will provide good sinewave outputs.

We will investigate the quality of an inverter output in the same way we analyzed the ac side of an ac-dc converter. The same performance factors will be used. See section 2.3.1.

In the following figures of inverters, thyristors are shown to be the semiconductor switches. Any appropriate switch can be used in practice. The inverter-system analysis is general, in that a switch can be turned on at any time it is positively biased, it can be turned off at any time and has unidirectional current when it conducts. How it does this is of no concern in this chapter. These details are left to the following chapters.

The inverters have a similar configuration to those of the rectifying converters. If we replace ac sources with loads and replace loads with dc sources in the converters illustrated in section 2.3, Figs 2.9, 2.12 and 2.14 we arrive at inverter circuits with voltage sources. Figure 2.15 depicts four inverters for the conversion of direct voltage to alternating voltage.

Current-source inverters are derived by taking the dual of the voltage-source inverters. Current sources replace voltage sources. Current-filter inductors replace voltage-filter capacitors, and series diodes (to block reverse voltages at the switches) replace inverse-parallel diodes (to conduct reverse current that a switch would block).

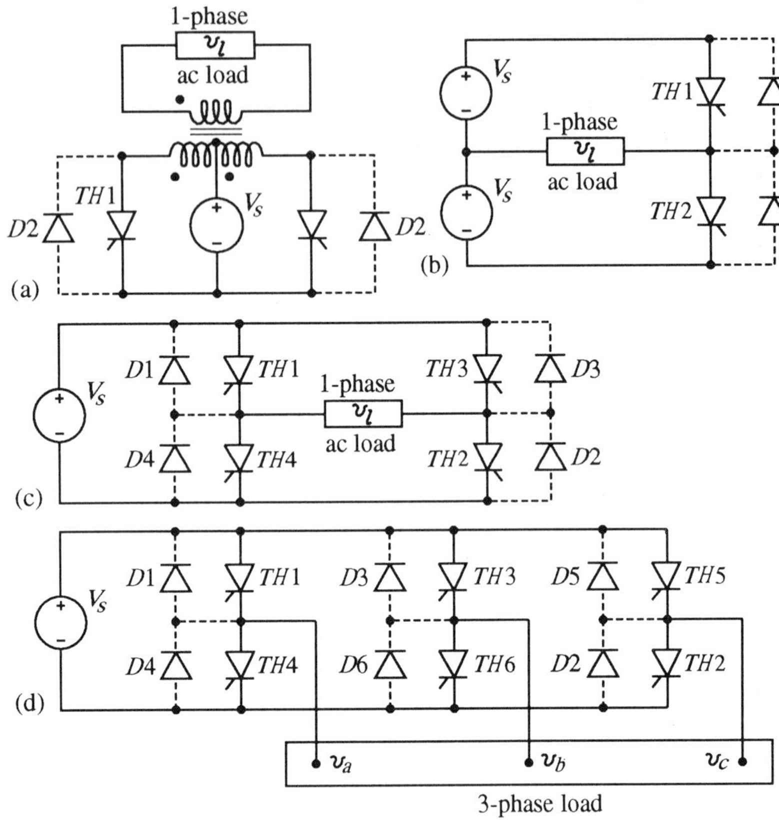

Fig. 2.15 Voltage-source inverters.
(a) Centre-tapped transformer, (b) centre-tapped source (half bridge),
(c) single-phase bridge, (d) three-phase bridge.

2.4.1. Centre-tapped Source Inverter

Figure 2.16a depicts a single-phase inverter with a centre-tapped supply and a resistive load. If a centre-tapped supply is not available, an alternative arrangement is shown in Fig. 2.16b. The capacitors, which form the voltage divider, would have large values in order to minimize the voltage change during current conduction. This inverter is also known as a single-phase, half-bridge inverter.

The overall circuit can be seen as two choppers with a common load R. One chopper circuit comprises the elements V_{s1}, $TH1$ and R. This chopper provides an adjustable positive voltage v_l at the load. The second chopper comprises the elements V_{s2}, $TH2$ and R. This chopper provides an adjustable negative voltage v_l at the load.

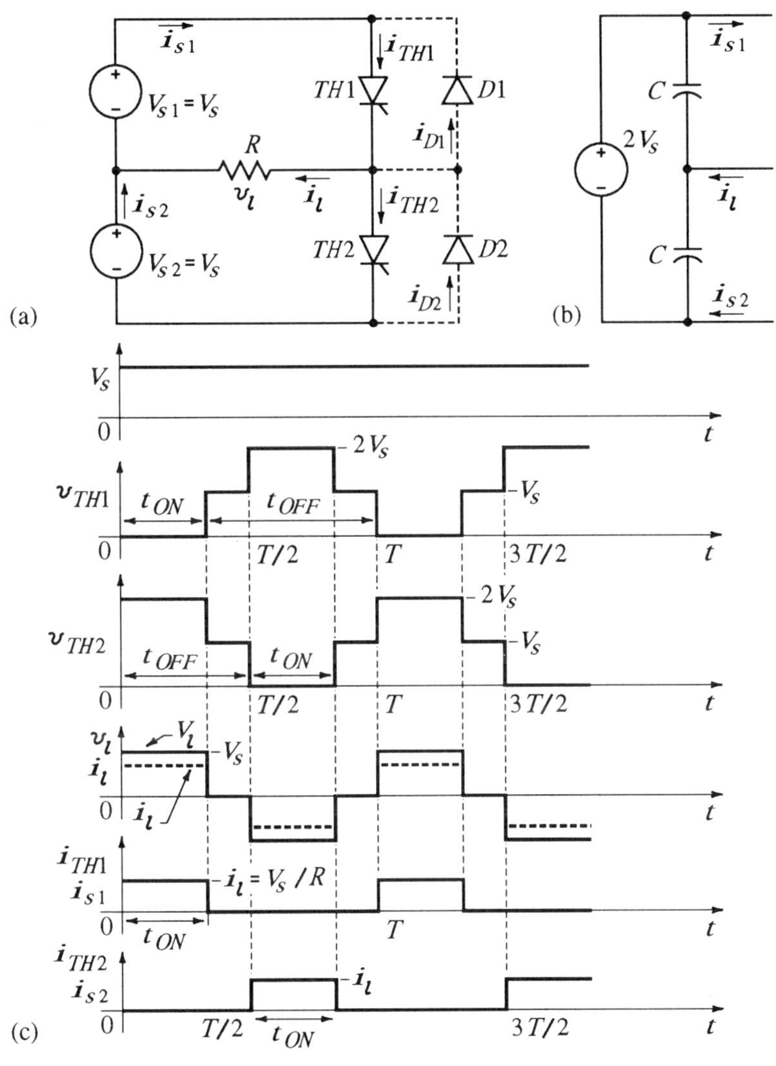

Fig. 2.16 Centre-tapped source inverter with resistive load. (a) Circuit diagram, (b) alternative source, (c) waveforms.

If each chopper has its switch turned on for an interval t_{ON} over a period T, but out of phase by 180° (equivalent to $T/2$ seconds) from each other, the waveforms are like those shown in Fig. 2.16c. Note that the thyristors must withstand a forward voltage that is twice the supply voltage V_s, and the average current in the thyristor is half of the average half-wave load current.

Both thyristors are off initially. Thyristor $TH1$ is turned on at time $t = 0$, is allowed to conduct the load current for an interval t_{ON}, and is then turned off. At time $t = T/2$ (half the switching period $T = 1/f$, where f is the switching frequency of each thyristor), while thyristor $TH1$ remains off, thyristor $TH2$ is turned on to conduct the reverse load current. Thyristor $TH2$ remains on for an interval t_{ON} (for symmetrical output waveforms) and is then turned off. Thyristors $TH1$ and $TH2$ remain off for the rest of the period T. At time $t = T$, the switching cycle is repeated by turning thyristor $TH1$ on for another interval t_{ON}. The output voltage v_l has an alternating rectangular waveform of magnitude V_s. Since the load is resistive, the output current i_l has an alternating rectangular waveform in phase with the voltage v_l and of magnitude V_s/R.

If the load has any inductance L, it is important to provide a path for the load current when any one of the thyristors is switched off. Otherwise, a thyristor that turns off will interrupt the load current in a very short time and a voltage spike can be generated (Ldi_l/dt). This can cause damage to semiconductor devices. A simple way to remove the stored energy ($Li_l^2/2$) from the inductance, after a controlled switch has been turned off, is to use diodes that are connected in reverse parallel to the thyristors, as shown in Fig. 2.16a. As soon as thyristor $TH1$ has been turned off, load inductance will tend to maintain current i_l in the load, directed from right to left. Together with the source V_{s2} and load R, the diode $D2$ provides a path for the current, and the inductive energy is absorbed by the source and load resistance (and by the diode if it is nonideal) during the interval $t_{ON} < t < T/2$. Similarly, diode $D1$ provides a path for the load current when thyristor $TH2$ is turned off.

The purpose of the inverter is to provide modulated ac power to the load from a dc supply. There are two factors associated with the ac power. One factor is the frequency of alternation and the other factor is the voltage across the load. The frequency of the alternating voltage is controlled by the switching rate of the thyristors. If the periodic time of the switching of the thyristors is $t = T$ seconds, the frequency f of the alternating voltage (or current) is

$$f = \frac{1}{T} \text{ Hz.} \tag{2.4.1}$$

In an ac circuit, rms values of voltage and current are of concern, because they are equivalent to the heating values of constant dc currents and voltages.

In some applications it is important that the load voltage waveform is sinusoidal. The load voltage waveform in Fig. 2.16 is far from sinusoidal. There is a fundamental sinewave component at frequency f, but there are also substantial harmonic components at frequencies $3f$, $5f$, $7f$, etc., (called the third harmonic, fifth harmonic, seventh harmonic, etc., respectively). Some of the harmonics can be filtered out using LC components. At low frequencies this can be expensive because the components have large values and large sizes. Another way is to shape the output waveform to something that is close to the ideal. There are two ways to shape a waveform. One way is to use a switching technique that is called pulse width modulation PWM, whereby a large number of short rectangular pulses are created to closely simulate a sinewave. The other way is to use a resonating link, and by switching at voltage and current zeroes very good sinewaves are produced.

Here, we will investigate the basic switching philosophy and show by example how improvements can be made. This calls for an analysis of the generated waveforms to determine the goodness or quality of the inverter output. We can make use of the performance factors that were used in section 2.3.1 for ac-dc converters. These concern the power output, the power factor, the displacement power factor and the total harmonic distortion. Since it is often required to reduce or eliminate certain harmonics, a harmonic analysis is important. It shows the various harmonic levels with respect to the fundamental component which is the reference. A useful term is the *harmonic factor* HF that is defined for each harmonic as

$$\mathrm{HF} = \frac{V_{l\,n}}{V_{l1}} \tag{2.4.2}$$

where V_{l1} is the value of the fundamental component and $V_{l\,n}$ is the value of the nth harmonic component. So, HF is a normalized or per unit value with the fundamental component being the base value. The aim is to keep the values of HF low.

The half-wave average of the load voltage is

$$V_{l\,av} = \frac{1}{T/2}\int_0^{T/2} v_l\,dt = \frac{1}{T/2}\int_0^{t_{ON}} V_s\,dt = 2V_s\frac{t_{ON}}{T}. \tag{2.4.3}$$

We can define the duty cycle m in the same way as for choppers. That is,

$$m = \frac{t_{ON}}{T}, \quad \text{where} \quad 0 \le m \le 0.5. \tag{2.4.4}$$

So, the average voltage over a half cycle becomes

$$V_{l\,av} = 2mV_s. \tag{2.4.5}$$

In terms of rms values, the load voltage is

$$V_{l\,rms} = \left[\frac{1}{T/2}\int_0^{T/2} v_l^2\,dt\right]^{1/2} = \left[\frac{1}{T/2}\int_0^{t_{ON}} V_s^2\,dt\right]^{1/2} = \sqrt{2m}\,V_s. \tag{2.4.6}$$

Consequently, the magnitude of the output voltage depends on the value of the input voltage V_s and the switching duty cycle m; the minimum value is zero and the maximum value is V_s for both average and rms values.

The average power P absorbed by a resistive load is

$$P = \frac{1}{T}\int_0^T v_l i_l\,dt = \frac{1}{T/2}\int_0^{t_{ON}} \frac{V_s^2}{R}dt = 2m\frac{V_s^2}{R} = \frac{V_{l\,rms}^2}{R} = I_{l\,rms}^2 R\,. \qquad (2.4.7)$$

The average power P_s delivered by the sources is

$$P_s = \frac{1}{T}\int_0^T (V_{s1}i_{s1} + V_{s2}i_{s2})dt = \frac{V_s}{T}\int_0^{t_{ON}} i_l\,dt + \frac{V_s}{T}\int_{T/2}^{T/2+t_{ON}} i_l\,dt\,. \qquad (2.4.8)$$

So,

$$P_s = V_s I_{l\,av} = V_s \times \frac{2mV_s}{R} = \frac{2mV_s^2}{R} = \frac{V_{l\,rms}^2}{R}\,. \qquad (2.4.9)$$

In an ideal system, as portrayed here, it is expected that the average power output P would be equal to the average power P_s from the sources.

EXAMPLE 2.7

A centre-tapped source inverter, as depicted in Fig. 2.16, modulates power from a 200 V dc source to a purely resistive load whose value is $R = 2\,\Omega$. If the switches have a duty cycle $m = 0.4$, determine **(a)** the average power absorbed by the load, **(b)** the total harmonic distortion THD of the voltage waveform, **(c)** the harmonic factor HF of the third harmonic of voltage and **(d)** the voltage and current ratings of the switches. **(e)** Plot the harmonic factors for any m.

Solution

(a) From the load-voltage waveform in Fig. 2.16c the average power P is

$$P = \frac{1}{T/2}\int_0^{t_{ON}} \frac{v_l^2}{R}dt = \frac{2}{RT}\int_0^{t_{ON}} V_s^2 dt = \frac{2mV_s^2}{R} = \frac{2\times 0.4}{2}\times 200^2 = 16\times 10^3\ \text{W}.$$

(b) The total harmonic distortion THD $= \left(\dfrac{V_{l\,rms}^2}{V_{l1}^2} - 1\right)^{1/2}$.

$V_{l\,rms}^2 = P \times R = 16\times 10^3 \times 2 = 32\times 10^3$. Therefore, $V_{l\,rms} = 178.9$ V.

The fundamental value of the rms load voltage is obtained from a harmonic analysis. The coefficients of the Fourier series are obtained as follows. From an inspection of the load-voltage waveform in Fig. 2.16c there is no dc component and there are no even harmonics. From eq. (2.3.3), the coefficient a_1 for the fundamental component is

$$a_1 = \frac{1}{T/2}\int_0^T v_l \cos\omega t\,dt = \frac{2V_s}{\omega T}[\sin\omega t]_0^{t_{ON}} - [\sin\omega t]_{T/2}^{T/2+t_{ON}} = \frac{2V_s}{\pi}\sin 2m\pi\,.$$

From the eq. (2.3.4), the other coefficient is b_1

$$b_1 = \frac{2}{T}\int_0^T v_l \sin\omega t\,dt = \frac{2V_s}{\omega T}[-\cos\omega t]_0^{t_{ON}} + \frac{2V_s}{\omega t}[\cos\omega t]_{T/2}^{T/2+t_{ON}} = \frac{2V_s}{\pi}(1-\cos 2m\pi)\,.$$

$$V_{l1} = (a_1^2 + b_1^2)^{1/2}/\sqrt{2} = \frac{\sqrt{2}\,V_s}{\pi}(\sin^2 2m\pi + (1-\cos 2m\pi)^2)^{1/2}\,.$$

That is, $V_{l1} = \dfrac{2\sqrt{2}\,V_s}{\pi}\sin m\pi\,.$

For $V_s = 200$ V and $m = 0.4$,

$$V_{l1} = \frac{2\sqrt{2}}{\pi} \times 200 \times \sin 0.4\pi = 171.2 \text{ V}.$$

$$\text{THD} = \left(\frac{V_{l\,rms}^2}{V_{l1}^2} - 1 \right)^{1/2} = \left(\frac{178.9^2}{171.2^2} - 1 \right)^{1/2} = 0.32 \ (32\%).$$

In general, for this type of inverter

$V_{l\,rms} = \sqrt{2m}\, V_s$ and $V_{l1} = 2\sqrt{2}\, V_s(\sin m\pi)/\pi$.

$$\text{Therefore, THD} = \left(\frac{m\pi^2}{4\sin^2 m\pi} - 1 \right)^{1/2}.$$

THD depends only on m for single rectangular waveforms.

(**c**) It can be noted that in general for this inverter with a resistive load the harmonic coefficients are

$$a_n = \frac{2V_s}{n\pi}\sin 2nm\pi \text{ and } b_n = \frac{2V_s}{n\pi}(1 - \cos 2nm\pi).$$

This gives the rms value of the nth harmonic to be

$$V_{ln} = \frac{2\sqrt{2}\,V_s}{n\pi}\sin nm\pi.$$

For this example the third harmonic is

$$V_{l3} = \frac{2\sqrt{2} \times 200}{3\pi}\sin(3 \times 0.4 \times \pi) = -35.3 \text{ V}.$$

The harmonic factor for the third harmonic voltage is

$$\text{HF} = \frac{V_{l3}}{V_{l1}} = \frac{35.3}{171.2} = 0.21 \text{ per unit } (21\%).$$

In general, the harmonic factor for the nth harmonic voltage is

$$\text{HF} = \frac{V_{ln}}{V_{l1}} = \frac{\sin nm\pi}{n \sin m\pi}.$$

(**d**) For $m = 0.5$ there is maximum possible conduction time for each thyristor. The load current i_l is

$i_l = v_l / R = 200/2 = 100$ A continuous for $m = 0.5$.

The average thyristor current for this condition is

$$I_{TH1\,av} = \frac{1}{T}\int_0^T i_{TH1}\,dt = \frac{1}{T}\int_0^{t_{ON}} i_l dt = \frac{1}{T}\int_0^{T/2} 100\,dt = 50 \text{ A}.$$

Therefore, the current rating of the thyristor must be greater than 50 A.

From the waveforms in Fig. 2.16c, it can be seen that both thyristors must withstand a forward bias voltage of $2V_s$ while they are off. Therefore, the voltage rating of each thyristor must be greater than 400 V.

(**e**) From the values of V_{ln} the harmonic factors V_{ln}/V_{l1} can be plotted against m. This is shown in Fig. EX2.7.

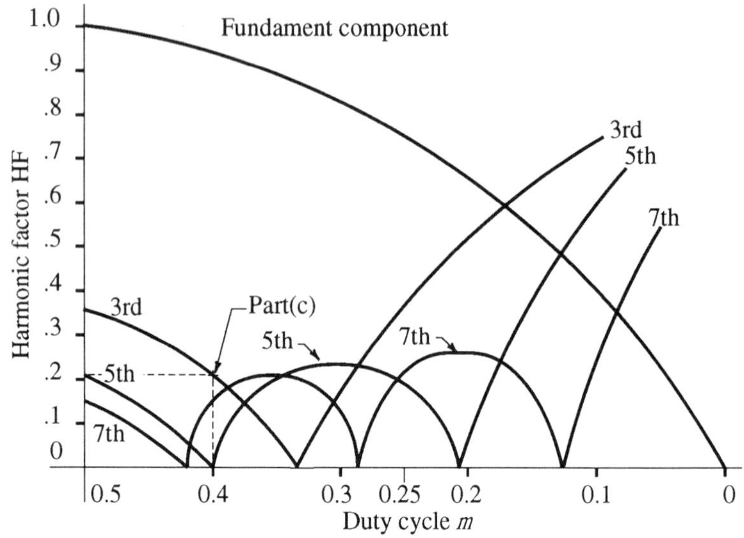

Fig. EX2.7

Sinewave by resonance.
There is a large percentage of harmonic distortion in the output voltage and current waveforms of the inverter configuration, shown in Fig. 2.16. However, nearly pure sinewaves can be produced by resonant circuits. Most loads have inductance. If capacitance is added to the load, as illustrated in Fig. 2.17a, the result is a resonant circuit.

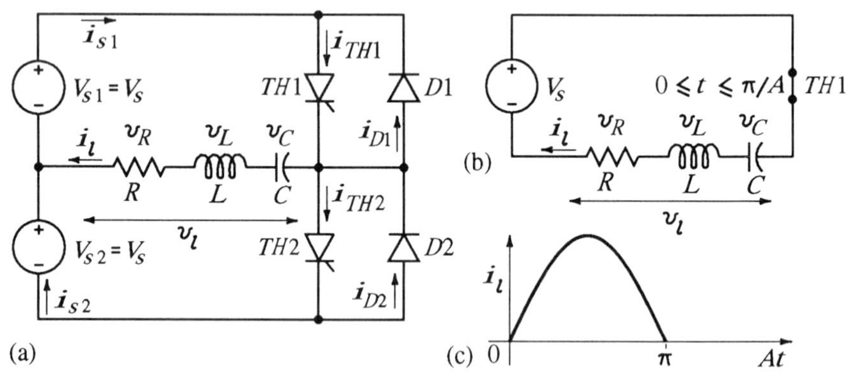

Fig. 2.17 Sinewave by resonance.
(a) Half-bridge with *RLC* load, (b) equivalent circuit, (c) current response.

Consider this resonant circuit for the interval that switch $TH1$ is on and switch $TH2$ is off. Figure 2.17b depicts the equivalent circuit and Fig. 2.17c shows the load current response while i_l is positive. While switch $TH1$ is on, the equation governing the behaviour of the response is given by Kirchhoff's voltage law

$$V_s = v_l = v_R + v_L + v_C \,. \tag{2.4.10}$$

That is,

$$V_s = Ri_l + L\frac{di_l}{dt} + \frac{1}{C}\int i_l dt \,. \tag{2.4.11}$$

The solution[5] of this differential equation has the form

$$i_l = \frac{V_s}{L}\frac{1}{A}e^{-Rt/2L}\sin At \tag{2.4.12}$$

where, for zero initial conditions, $A^2 = 1/(LC) - R^2/(4L^2)$, and $A = \omega_o$ is the natural angular frequency of resonance.

The load is allowed to conduct the first half cycle of this quasi-sinewave of current before the switch $TH1$ is turned off. Most semiconductor switches will turn off if a controlling signal is applied to or removed from the device. In this case the thyristor has the facility that, if the current in it reduces to zero, the thyristor turns off naturally. This serves the purpose of this inverter. As the load current i_l goes through zero at the end of the first half cycle, thyristor $TH1$ turns off.

If, at this point, thyristor $TH2$ is turned on, the RLC load sees a supply voltage V_s of the reverse polarity. The current response has the same form as that portrayed in Fig. 2.17c and eq. (2.4.12) except that the direction of current has reversed and the initial conditions alter the peak value. At the end of this second half cycle of current, thyristor $TH2$ will turn off naturally.

By switching the thyristors on at a frequency equal to the natural resonant frequency, the result is that the load current is very close, in form, to a sinewave. The load is usually fixed, so that the frequency of the output can be controlled by the value of the capacitance C.

EXAMPLE 2.8

Consider the half-bridge inverter that is illustrated in Fig. 2.17a. The dc source has a voltage $V_s = 200$ V. The load has a resistance $R = 1\,\Omega$ and an inductance $L = 1$ mH. The frequency of the inverter is to be 1 kHz. Determine **(a)** the necessary value of the capacitance C so that the inverter can operate satisfactorily with a quasi-sinewave of current and **(b)** the total harmonic distortion THD of the load current waveform for thyristor $TH1$ being switched for the first time (that is, zero initial conditions).

---

[5] More details for the solution of eq. (2.4.11) can be found in section 5.6.1.

Solution

(a) The resonant frequency is to be the operating frequency of the inverter for a load current response that is close to a sinewave. That is, from eq. (2.4.12),

$$f = \frac{\omega_o}{2\pi} = \frac{A}{2\pi} = \frac{1}{2\pi}\left[\frac{1}{LC} - \frac{R^2}{4L^2}\right]^{1/2}$$

or, $C = \dfrac{L}{4\pi^2 f^2 L^2 + R^2/4} = \dfrac{10^{-3}}{4\pi^2 + 1/4} = 25.2\times10^{-6}$ F.

Note that the value of R has little effect on the value of C.

(b) From eq. (2.4.12), the first half cycle of the current response is given by

$i_l = \dfrac{V_s}{L}\dfrac{1}{A}e^{-Rt/2L}\sin At$.

For $V_s = 200$ V, $L = 1$ mH, $A = \omega_o = 2\pi f = 6283$ rad/s and $R = 1\ \Omega$,

$i_l = 31.83e^{-0.07958\omega_o t}\sin\omega_o t$ A.

The rms value of the load current is given by

$$I_{l\,rms}^2 = \frac{1}{\pi}\int_0^\pi i_l^2\, d\omega_o t = \frac{1}{\pi}\int_0^\pi (31.83e^{-0.07958\omega_o t}\sin\omega_o t)^2 d\omega_o t .$$

From a math software package $I_{l\,rms}^2 = \dfrac{1244.6}{\pi}$, so, $I_{l\,rms} = 19.904$ A.

The load current has no dc component and has no even harmonics. From Fourier analysis, the coefficient a_1 of the fundamental component is given by eq. (2.3.3) to be

$$a_1 = \frac{2}{\pi}\int_0^\pi i_l\cos\omega_o t\, d\omega_o t = \frac{2}{\pi}\int_0^\pi 31.83e^{-0.07958\omega_o t}\sin\omega_o t\cos\omega_o t\, d\omega_o t .$$

So, $a_1 = \dfrac{2}{\pi}\times 1.7627 = 1.1222$.

The coefficient b_1 is $b_1 = \dfrac{2}{\pi}\displaystyle\int_0^\pi 31.83e^{-0.07958\omega_o t}\sin^2\omega_o t\, d\omega_o t$.

So, $b_1 = \dfrac{2}{\pi}\times 44.1514 = 28.108$.

The rms value of the fundamental component of the current waveform is

$$I_{l1} = \frac{1}{\sqrt{2}}(a_1^2 + b_1^2)^{1/2} = \frac{1}{\sqrt{2}}(1.1222^2 + 28.108^2)^{1/2} = 19.89 \text{ A}.$$

The total harmonic distortion THD $= \left(\dfrac{I_{l\,rms}^2}{I_{l1}^2} - 1\right)^{1/2}$.

So, THD $= \left(\dfrac{19.9^2}{18.89^2} - 1\right)^{1/2} = 0.03$ (3%).

In the above example the resonant circuit has reduced the harmonic distortion to 3%. This is small enough to assume that the load current in such an inverter can be considered to be sinusoidal with negligible harmonics. This does ease the analysis.

Up to this stage only a transient analysis has been performed on the inverter circuit. For the first few cycles of operation the current sinusoid will have an increasing amplitude while the initial conditions of voltage across the capacitor change. Eventually there will be a balance between the energy stored in the capacitor at the time of switching and the amount of energy absorbed by the resistance R. During each transient cycle the current waveform will have the same quasi-sinusoidal shape. Accordingly, for steady conditions, we can say that the current waveform is such that we can consider the fundamental component and ignore the harmonics.

In the following example a steady-state analysis, based on the results of EXAMPLE 2.8, is performed on an inverter.

EXAMPLE 2.9

A dc ac converter, shown in Fig. 2.17a, is used to modulate power from a source whose voltage is $V_s = 200$ V to a load whose resistance is $R = 1\,\Omega$, whose inductance is $L = 1$ mH and whose capacitance is $C = 25.2\,\mu$F. The frequency of thyristor switching is 1 kHz. For steady-state operation, determine **(a)** the average power delivered to the load and **(b)** the current ratings of the thyristors and the diodes.

Solution

(a) $\omega = 2\pi f = 2\pi \times 1000 = 6283$ rad/s.

$\omega L = 1\,\Omega$, $1/\omega C = 1\,\Omega$, so the circuit is in resonance with the natural frequency equal to the applied frequency. This means that the sinusoidal current response in the load is in phase with the rectangular-wave voltage at the output. Considering the voltage wave to be an odd function, the Fourier coefficient b_1 is

$$b_1 = \frac{1}{\pi}\int_0^{2\pi} v_l \sin\omega t\, d\omega t = \frac{2}{\pi}\int_0^{\pi} V_s \sin\omega t\, d\omega t = \frac{4V_s}{\pi}.$$

Thus, the rms value of the fundamental component of voltage is

$$V_{l1} = \frac{b_1}{\sqrt{2}} = \frac{4V_s}{\sqrt{2}\,\pi} = \frac{4\times 200}{\sqrt{2}\,\pi} = 180\text{ V}.$$

The fundamental value of the load impedance is

$$Z_1 = (R^2 + (\omega L - \frac{1}{\omega C})^2)^{1/2} = R = 1\,\Omega.$$

Since the load current is approximately a sine wave, its rms current is

$I_{l\,rms} = V_{l1}/Z_1 = 180/1 = 180$ A.

Therefore, the average power P absorbed by the load is

$P = I_{l\,rms}^2 R = 180^2 \times 1 = 32.4 \times 10^3$ W.

(b) Since the load current is in phase with the load voltage wave,

$i_l = \sqrt{2} \times 180 \sin\omega t$ and the current is zero at the switching times of the thyristors. Hence, the diodes have no current to conduct.

The half-wave average of the load current is

$$I_{l\,av} = \frac{1}{\pi}\int_0^{\pi} \hat{I}_l \sin\omega t\, d\omega t = 2\frac{\hat{I}_l}{\pi} = \frac{2\sqrt{2}\times 180}{\pi} = 162\text{ A}.$$

This current is shared by the two thyristors, so the current rating of each thyristor must be greater than 81 A.

2.4.2. Single-phase Bridge Inverter

Resistive load. Figure 2.18a depicts a single-phase inverter with a dc source voltage V_s, a resistive load R and a bridge configuration of switches.

The overall circuit comprises two subsidiary chopper circuits with a common dc source and a common load. One chopper circuit consists of the elements V_s, $TH1$, $TH2$ and R and this provides an adjustable positive voltage v_l across the load. Switches $TH1$ and $TH2$ are usually on together. The second chopper circuit consists of the elements V_s, $TH3$, $TH4$ and R and this allows an adjustable negative voltage v_l to be applied across the load. The choppers are never on together because a short-circuited voltage supply cannot be tolerated.

If each chopper has its pair of switches turned on alternately for an interval t_{ON} (where $t_{ON} \le T/2$) over each period T, the circuit voltage and current waveforms are like those that are shown in Fig. 2.18b. In this inverter the thyristors must withstand a forward voltage that is equal to the value of the source voltage V_s, and the average current in each thyristor is half of the average half-wave load current. This indicates that the bridge inverter of Fig. 2.18 can handle twice the power of the half-bridge inverter of Fig. 2.16, if all the thyristors have the same ratings.

Operation of this inverter is best described by assuming all four thyristors are off initially. Thyristors $TH1$ and $TH2$ are switched on at time $t = 0$, are allowed to conduct the load current i_l for an interval t_{ON}, and are then switched off. With all thyristors off, each thyristor blocks half the supply voltage V_s.

At time $t = T/2$ (half the switching period $T = 1/f$, where f is the switching frequency), while thyristors $TH1$ and $TH2$ remain off, thyristors $TH3$ and $TH4$ are switched on to conduct the reverse load current. During the interval that thyristors $TH3$ and $TH4$ conduct, both thyristors $TH1$ and $TH2$ must block the full supply voltage V_s. Thyristors $TH3$ and $TH4$ are allowed to remain on for an interval t_{ON}, in order to provide a symmetrical alternating current at the output, and then they are switched off for the rest of the period T. At time $t = T$ the cycle of switching is repeated. The output voltage v_l has an alternating rectangular waveform of magnitude V_s. Since the load is resistive, the output current i_l has an alternating rectangular waveform, that is in phase with the load voltage waveform. The magnitude of the current waveform is $i_l = V_s/R$.

A bridge configuration of diodes accompanies the inverter bridge of unidirectional switches, as illustrated in Fig. 2.18a. The diodes accommodate inductive load currents at the time that controlled switches are turned off. In this way, high voltages ($v = L di_l/dt$) are not inflicted on sensitive semiconductor devices. Diodes $D3$ and $D4$ furnish a current path together with the source and inductive load if thyristors $TH1$ and $TH2$ are switched off. Similarly, the load current has a path provided by diodes $D1$ and $D2$ at the times that thyristors $TH3$ and $TH4$ are switched off.

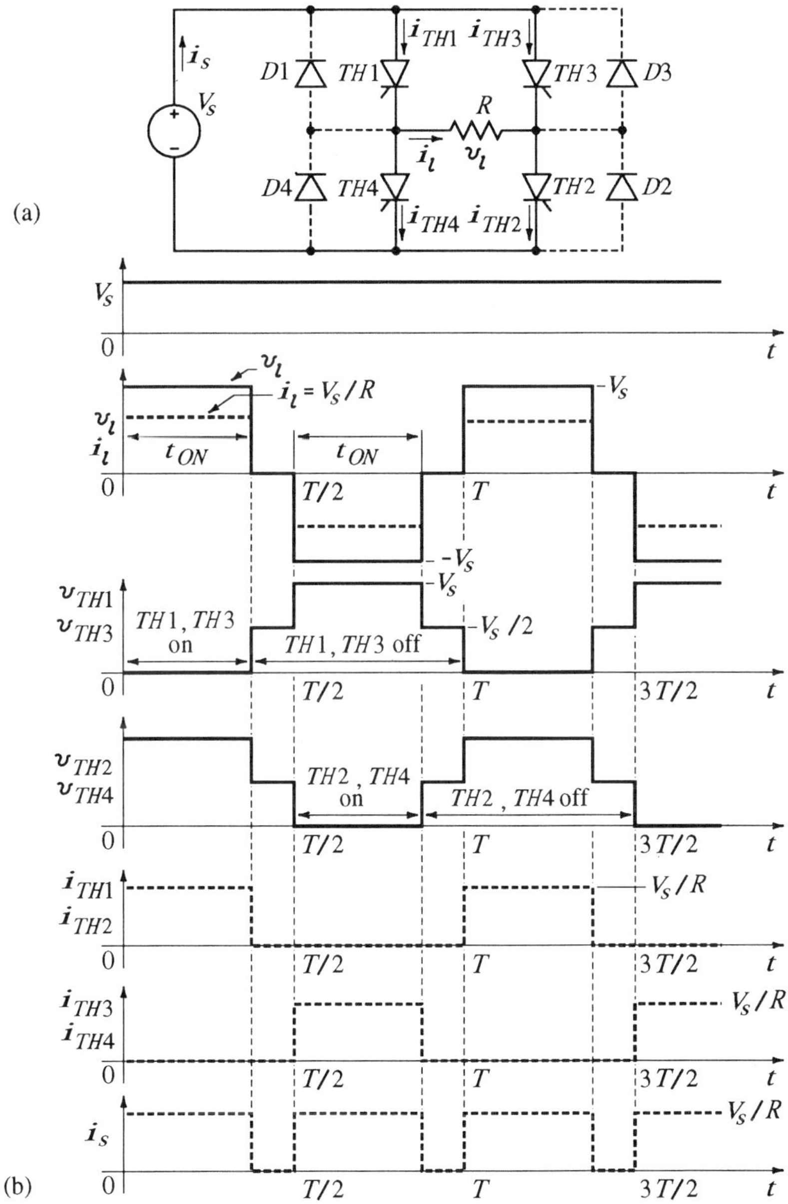

Fig. 2.18 Single-phase bridge inverter.
(a) Circuit diagram, (b) waveforms.

An inspection of Fig. 2.16c and Fig. 2.18b shows that the load voltage and current waveforms of the half-bridge and full-bridge inverters are identical. Therefore, the analysis of the performance of the inverter output, carried out in section 2.4.1, does not have to be repeated. The performance equations for the bridge inverter of Fig. 2.18 will be summarized by reference to the previous section.

If we use the definition of duty cycle m as

$$m = \frac{t_{ON}}{T}, \quad \text{where } 0 \le m \le 0.5 \tag{2.4.13}$$

then we can use the expression in eq. (2.4.6) to state that the rms value of the load voltage is

$$V_{l\,rms} = \left[\frac{1}{T/2} \int_0^{T/2} v_l^2 \, dt \right]^{1/2} = \sqrt{2m}\, V_s \,. \tag{2.4.14}$$

The average power absorbed by the load is, from eq. (2.4.7),

$$P = \frac{1}{T} \int_0^T v_l i_l \, dt = 2m \frac{V_s^2}{R} \,. \tag{2.4.15}$$

From EXAMPLE 2.7, for any duty cycle m, the load voltage has a fundamental rms value

$$V_{l1} = \frac{2\sqrt{2}}{\pi} V_s \sin m\pi \tag{2.4.16}$$

and the total harmonic distortion is

$$\text{THD} = \left(\frac{V_{l\,rms}^2}{V_{l1}^2} - 1 \right)^{1/2} = \left(\frac{m\pi^2}{4\sin^2 m\pi} - 1 \right)^{1/2} . \tag{2.4.17}$$

The rms value of the nth harmonic load voltage is

$$V_{ln} = \frac{2\sqrt{2}}{n\pi} V_s \sin nm\pi \,. \tag{2.4.18}$$

EXAMPLE 2.10

Consider the single-phase bridge inverter, illustrated in Fig. 2.18a. The dc-source voltage is $V_s = 600$ V and the load has a resistance of $R = 20\,\Omega$. If the inverter is to operate at 500 Hz with an rms load voltage of 500 V, find **(a)** the average power absorbed by the load, **(b)** the average source current, **(c)** the average current in each thyristor, **(d)** the thyristor on-time each period, **(e)** the total harmonic distortion THD, **(f)** the instantaneous value of the load voltage up to the ninth harmonic and **(g)** the harmonic factor HF of the fifth harmonic of voltage.

Solution

(a) Refer to Fig. 2.18. The average power P absorbed by the load is
$P = V_{l\,rms}^2/R = 500^2/20 = 12.5\times10^3$ W.

(b) Since the system is ideal, the source power $P_s = P$.
Also, $P_s = V_s I_{s\,av}$, so the average source current is
$I_{s\,av} = P_s/V_s = 12.5\times10^3/600 = 20.83$ A.

(c) From Fig. 2.18b, it is seen that the average thyristor current is half the average source current, if the load is only resistive.
So, $I_{TH\,av} = I_{s\,av}/2 = 20.83/2 = 10.42$ A.

(d) The rms load voltage in terms of the duty cycle is given by eq. (2.4.14).
$V_{l\,rms} = \sqrt{2m}\,V_s$. For $V_{l\,rms} = 500$ V and $V_s = 600$ V, this gives $m = 0.347$.
The thyristor on-time t_{ON} is given by
$t_{ON} = mT = m/f = 0.347/500 = 0.694\times10^{-3}$ s.

(e) From eq. (2.4.18), the rms value of voltage for the nth harmonic is

$$V_{ln} = \frac{2\sqrt{2}}{n\pi}V_s \sin n m\pi .$$

The fundamental component ($n = 1$) for $V_s = 600$ V and $m = 0.347$ is

$$V_{l1} = \frac{2\sqrt{2}}{\pi}\times 600\sin(0.347\pi) = 479 \text{ V}.$$

$$\text{Therefore, THD} = \left(\frac{V_{l\,rms}^2}{V_{l1}^2} - 1\right)^{1/2} = \left(\frac{500^2}{479^2} - 1\right)^{1/2} = 0.30\ (30\%).$$

(f) From eq. (2.3.3) and eq. (2.3.4) the Fourier series coefficients for rectangular waves are

$$a_n = \frac{2V_s}{n\pi}\sin 2nm\pi \ \text{ and } \ b_n = \frac{2V_s}{n\pi}(1-\cos 2nm\pi).$$

For n = 1, 3, 5, 7, 9, $V_s = 600$ V and $m = 0.347$,
$a_1 = 313.2$, $a_3 = 32.44$, $a_5 = -76.06$, $a_7 = 23.54$, $a_9 = 29.6$,
$b_1 = 600.6$, $b_3 = 4.2$, $b_5 = 68.7$, $b_7 = 5.34$, $b_9 = 12.06$.
$\hat{V}_{ln} = (a_n^2 + b_n^2)^{1/2}$, so
$\hat{V}_{l1} = 677.4$ V, $\hat{V}_{l3} = 32.7$ V, $\hat{V}_{l5} = 102.5$ V, $\hat{V}_{l7} = 24.1$ V and $\hat{V}_{l9} = 32$ V.
Also, $\tan\theta_n = a_n/b_n$, so
$\theta_1 = 0.48$ rad, $\theta_3 = 1.44$ rad, $\theta_5 = -0.84$ rad, $\theta_7 = 1.35$ rad, $\theta_9 = 1.18$ rad.
This gives the instantaneous value of the load voltage,

$$v_l = 677\sin(\omega t + 0.48) + 32.7\sin(3\omega t + 1.44) + 102.5\sin(5\omega t - 0.84) + 24.1\sin(7\omega t + 1.35) + 32\sin(9\omega t + 1.18) + \cdots .$$

For this value of m the 5th harmonic has a large value.

(g) The harmonic factor $\text{HF} = \dfrac{\hat{V}_{l5}}{\hat{V}_{l1}} = \dfrac{102.5}{677} = 0.15\ (15\%)$.

Inductive load. The loads of inverters can range between the extremes of pure resistance R and pure inductance L for any ratio of L/R and can also include capacitance. Capacitance usually improves the quality of the output current waveform of an inverter. This was demonstrated in section 2.4.1. We should investigate the effect of a purely inductive load.

EXAMPLE 2.11

A single-phase bridge inverter is depicted in Fig. EX2.11a. The dc source voltage is $V_s = 600$ V and the load is an electromagnet whose resistance can be neglected and whose inductance is $L = 10$ mH. In steady-state operation, the inverter frequency is 200 Hz and the switching duty cycle is $m = 0.5$. **(a)** Sketch the voltage and current waveforms. Determine **(b)** the thyristor and diode current ratings and **(c)** the average power delivered by the dc supply.

Solution

(a) With $m = 0.5$ the voltage across the load is a full rectangular wave. At any time the voltage across the load is $v_l = \pm V_s$. Also, the voltage across the load is

$v_l = L\dfrac{di_l}{dt}$, slope $\dfrac{di_l}{dt} = \dfrac{v_l}{L} = \pm\dfrac{V_s}{L} = \pm\text{constant}$.

While $v_l = +V_s$, the load current slope is constant and positive. While $v_l = -V_s$, the load current slope is constant and negative. An equivalent circuit is shown in Fig. EX2.11b, in which the source is an alternating voltage of rectangular form and the load is the inductance L. The current response is shown.

The average current in an ac circuit is zero for steady conditions.

$$I_{l\,av} = \frac{1}{T}\int_0^T i_l dt = \frac{1}{T}\int_0^{T/2} i_l(t)dt + \frac{1}{T}\int_0^{T/2} i_l(t')dt', \text{ where } t' = t - T/2.$$

For $0 \le t \le T/2$ and $T/2 \le t \le T$, or $0 \le t' \le T/2$,

$i_l(t) = \dfrac{V_s}{L}t + I_1$ and $I_l(t') = -\dfrac{V_s}{L}t' + I_2$.

For $I_{l\,av} = 0$, then integration shows $I_1 = -I_2$.

But, $I_2 - I_1 = \dfrac{V_s}{L}\times\dfrac{T}{2}$. Therefore, $I_2 = -I_1 = \dfrac{V_s}{L}\times\dfrac{T}{4} = I$.

With these constraints the load current waveform is redrawn in Fig. EX2.11c.

While the load current i_l is positive and the load voltage is positive, thyristors $TH1$ and $TH2$ conduct the load current ($T/4 \le t \le T/2$). During this interval energy from the source is being stored in the inductor. At $t = T/2$, thyristors $TH1$ and $TH2$ are turned off and $TH3$ and $TH4$ are turned on. The diodes $D3$ and $D4$ must conduct the load current i_l while it is positive ($T/2 \le t \le 3T/4$). During this interval stored energy in the inductor ($Li_l^2/2$) is being returned to the source. The load current tends to reverse under the influence of $-V_s$ and thyristors $TH3$ and $TH4$ conduct the load current until $t = T$, at which point they are switched off. The load current i_l is transferred to diodes $D1$ and $D2$ until the current comes to zero again. The cycle is then repeated.

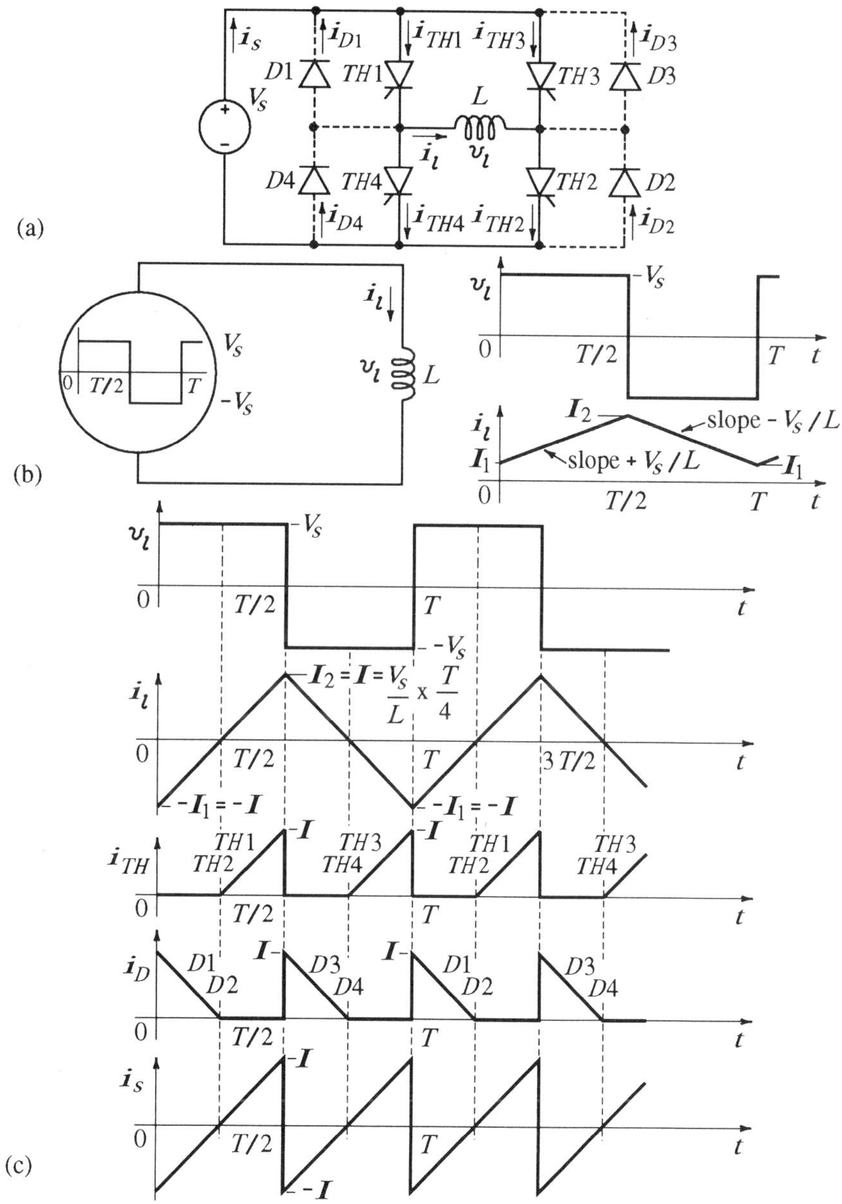
(a)
i_s
i_{D1}
i_{TH1}
i_{TH3}
i_{D3}
D1
TH1
TH3
D3
V_s
L
i_l
v_l
D4
TH4
TH2
D2
i_{D4}
i_{TH4}
i_{TH2}
i_{D2}
(b)
V_s
$-V_s$
0
T/2
T
i_l
v_l
L
v_l
V_s
t
i_l
I_2
slope $- V_s / L$
I_1
slope $+ V_s / L$
I_1
(c)
v_l
V_s
$-V_s$
i_l
$I_2 = I = \frac{V_s}{L} \times \frac{T}{4}$
3T/2
$-I_1 = -I$
$-I_1 = -I$
i_{TH}
TH1
TH2
TH3
TH4
I
i_D
D1
D2
D3
D4
i_s
$-I$

Fig. EX2.11

The source current i_s is negative and equal to the diode currents, while they conduct the load current i_l. The source current i_s is positive and equal to the thyristor currents while they conduct the load current. Figure EX2.11c shows all the current waveforms.

(b) The peak current of any of the devices is

$$I = \frac{V_s}{L} \times \frac{T}{4} = \frac{600}{10 \times 10^{-3}} \times \frac{1}{200 \times 4} = 75 \text{ A}.$$

The average current in all thyristors and diodes is the same.

$$I_{TH\,av} = I_{D\,av} = \frac{1}{2} I \frac{T}{4} \times \frac{1}{T} = \frac{75}{8} = 9.4 \text{ A}.$$

(c) By inspection of Fig. EX2.11c the average source current is zero. Consequently the average power delivered by the source ($V_s I_{s\,av}$) is zero. This is to be expected, since there is no resistance in the circuit to dissipate power. In steady ac operation the inductor absorbs power for half the cycle and returns it to the source during the other half cycle.

RL load. The most common load of an inverter consists of inductance as well as resistance.

It is important that we know how to control the frequency, the voltage and the power output of the inverter. In order to calculate the average power absorbed by the load resistance, we have to determine the current response to the switching arrangement. This response allows us to find the current ratings of the inverter semiconductors also. Let us find an expression for the load current.

Figure 2.19a illustrates a single-phase bridge inverter with a series *RL* load. We will assume a frequency f of thyristor switching, so that the period T of the voltage and current waveforms is $T = 1/f$. In addition, we assume the condition for maximum average power. This occurs if the on-time t_{ON} of the switches is $t_{ON} = T/2$. This gives the maximum voltage v_l at the load, as shown in Fig. 2.19b.

The current response i_l to a rectangular waveform of voltage in an *RL* circuit has an exponential form. Figure 2.19b shows the general nature of the current response. While a positive V_s is applied to the load, current i_l is a rising exponential time function. At the next half cycle, while a negative V_s is applied to the load, current i_l is a falling exponential function.

The load current i_l is shown to start at a value $i_l = -I_1$ at $t = 0$ and to rise to $i_l = I_2$ at $t = T/2$, the end of the first half cycle. In steady-state operation, during the second half cycle, the load current must fall from a value $i_l = I_2$ at $t = T/2$ to a value $i_l = -I_1$, at $t = T$, so that i_l can start at $-I_1$ at the begining of the next cycle.

In the steady state, with an applied alternating voltage $\pm V_s$, the average load current must be zero since there can be no dc level. For a zero average value, the positive and negative peak values of current must be the same. That is, at $t = nT/2$ (for $n = 0,1,2,3 \cdots$)

(a)

(b)

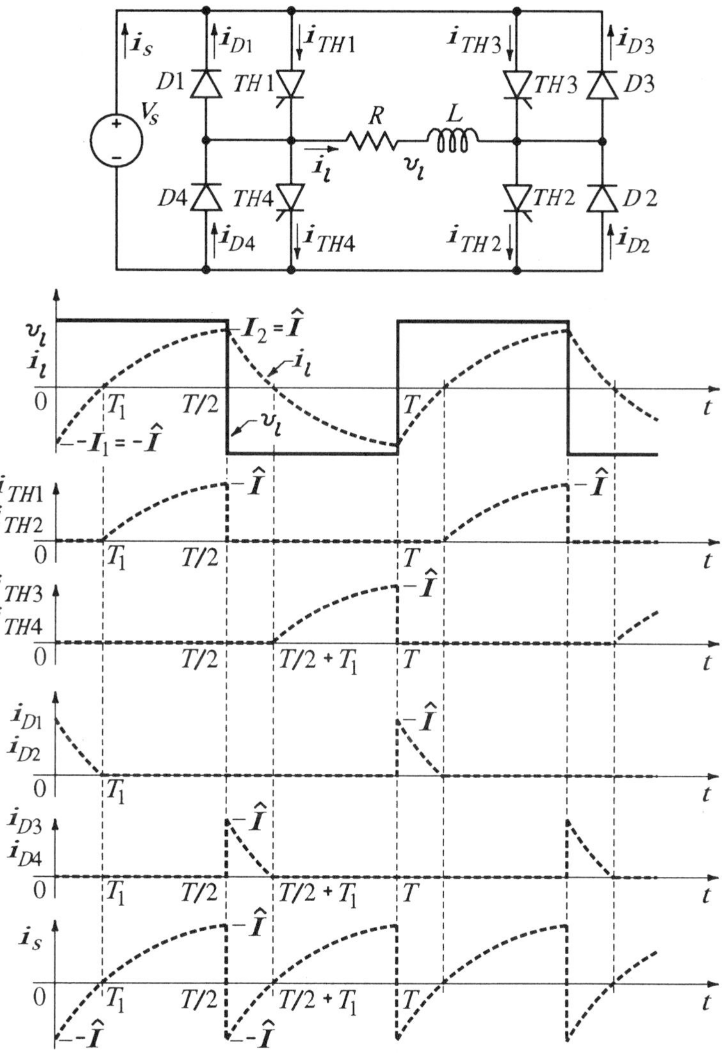

Fig. 2.19 Bridge inverter with *RL* load.
(a) Circuit diagram, (b) waveforms.

$$I_1 = I_2 \triangleq \hat{I}\,. \tag{2.4.19}$$

From this we can determine the current sharing by the switches, controlled and uncontrolled. This facilitates the analysis to find an expression for the power that is modulated by the inverter.

During the intervals that the load voltage v_l and the load current i_l are both positive, thyristors $TH1$ and $TH2$ conduct the current i_l. While the load voltage v_l is negative and the load current i_l is positive, diodes $D3$ and $D4$ conduct the current i_l. During the intervals that the load voltage v_l and the load current i_l are negative, thyristors $TH3$ and $TH4$ conduct the current i_l. For the last part of the cycle, while the load voltage v_l is positive and the load current i_l is negative, the diodes $D1$ and $D2$ conduct the current i_l. Finally, the source current i_s is positive and the source is delivering power if the thyristors conduct current, and the source current is negative and the source is absorbing power if the diodes conduct current. Figure 2.19b shows the waveforms.

For the conditions of steady state and maximum average power, the current response i_l is obtained from

$$\pm V_s = Ri_l + L\frac{di_l}{dt}. \tag{2.4.20}$$

Over the range $0 \le t \le T/2$ for $+V_s$, $-t/\tau = \ln(i_l - I_o) + \mathrm{k}$, where $I_o = V_s/R$, $\tau = L/R$ and k is a constant.
Since $i_l = -\hat{I}$ at $t = 0$,

$$i_l(t) = I_o - (I_o + \hat{I})e^{-t/\tau}. \tag{2.4.21}$$

At $t = T/2$, $i_l = \hat{I}$, so we can find the peak current $\hat{I}$ from

$$\hat{I} = I_o \frac{1 - e^{-T/2\tau}}{1 + e^{-T/2\tau}} = I_o \tanh\left(\frac{T}{4\tau}\right). \tag{2.4.22}$$

This is an important equation. If the time constant τ of the circuit and the period of alternation are such that $5\tau \le T/2$, then the peak value of current $i_l = \hat{I}$ will be close to the steady value $i_l = I_o = V_s/R$.

For the interval $T/2 \le t \le T$, or $0 \le t' \le T/2$, where $t' = t - T/2$,

$$i_l(t') = -I_o + (I_o + \hat{I})e^{-t'/\tau}. \tag{2.4.23}$$

In order to find the current sharing among the thyristors and diodes we must find the time $t = T_1$ at which the load current i_l is zero. The solution of eq. (2.4.21) for $i_l = 0$ at $t = T_1$ provides the value

$$T_1 = \tau \ln(1 + \hat{I}/I_o). \tag{2.4.24}$$

From this information we obtain the device ratings.

The average current in any thyristor is

$$I_{TH\,av} = \frac{1}{T}\int_{T_1}^{T/2} i_l(t)dt = \frac{1}{T}\left[I_o(T/2 - T_1) + \tau(I_o + \hat{I})(e^{-T/2\tau} - e^{-T_1/\tau})\right]. \tag{2.4.25}$$

The average current in any diode is

$$I_{D\,av} = \frac{1}{T}\int_0^{T_1} i_l(t)dt = \frac{1}{T}\left[-I_o T_1 - \tau(I_o + \hat{I})(e^{-T_1/\tau} - 1)\right]. \tag{2.4.26}$$

The average source current is

$$I_{s\,av} = \frac{1}{T/2}\int_0^{T/2} i_l(t)dt = 2(I_{TH\,av} - I_{D\,av})$$

$$= \frac{2}{T}\left[I_o \frac{T}{2} + \tau(I_o + \hat{I})(e^{-T/2\tau} - 1)\right]. \tag{2.4.27}$$

The average power P delivered by the source and absorbed by the load is

$$P = V_s I_{s\,av}\,. \tag{2.4.28}$$

In order to determine the performance factors of the inverter output a harmonic analysis must be undertaken.

An adjustable voltage can be obtained at the output of a bridge inverter by adjusting the on-time of the switches within the half period of oscillation. That is, the duty cycle m can be a control variable. The system performance can be determined in terms of the parameter m, if a new analysis is carried out. That pleasure is left for the reader to enjoy.

EXAMPLE 2.12

Consider the single-phase bridge inverter, depicted in Fig. 2.19a. The dc source voltage is $V_s = 600$ V and the load has a resistance $R = 20\,\Omega$ in series with an inductance $L = 10$ mH. The inverter operates at 500 Hz and delivers maximum power to the load in the steady state. Determine **(a)** the peak current from the dc source, **(b)** the current ratings of the switches and **(c)** the average power controlled by the inverter.

Solution

Refer to the inverter waveforms shown in Fig. 2.19b. Since maximum power is to be delivered to the load, the switches operate to apply a full rectangular voltage waveform (duty cycle $m = 0.5$) across the load.

(a) Current $I_o \triangleq V_s/R = 600/20 = 30$ A.
The load time constant $\tau = L/R = 10^{-2}/20 = 0.5\times10^{-3}$ s.
The half period is $T/2 = 1/(2f) = 1/(2\times500) = 10^{-3}$ s.
The source current i_s has the same waveform as the load current over the range $0 \le t \le T/2$. From eq. (2.4.21), the source current is
$i_s(t) = I_o - (I_o + \hat{I})e^{-t/\tau}$.
At $t = T/2$, $i_s(T/2) = \hat{I}$. Hence, the peak source current is

$$\hat{I} = I_o \frac{1 - e^{-T/2\tau}}{1 + e^{-T/2\tau}} = I_o \tanh\left(\frac{T}{4\tau}\right) = 30\tanh\left(\frac{2\times10^{-3}}{4\times0.5\times10^{-3}}\right) = 22.85\text{ A}.$$

(b) The load current is zero at $t = T_1$. That is, $i_l(t = T_1) = 0 = I_o - (I_o + \hat{I})e^{-T_1/\tau}$.
So, $T_1 = \tau \ln(1 + \hat{I}/I_o) = 0.5 \times 10^{-3} \ln(1 + 22.85/30) = 0.283 \times 10^{-3}$ s.
The average current in each thyristor is

$$I_{TH\,av} = \frac{1}{T}\int_{T_1}^{T/2} i_l(t)dt = \frac{1}{T}\left[I_o(T/2 - T_1) + \tau(I_o + \hat{I})(e^{-T/2\tau} - e^{-T_1/\tau})\right].$$

$$I_{TH\,av} = \frac{10^3}{2}[30 \times 0.717 \times 10^{-3} + 0.5 \times 10^{-3} \times 52.85(0.135 - 0.568)] = 5.04 \text{ A}.$$

The current rating of each thyristor must be greater than 5 A.
The average current in each diode is

$$I_{D\,av} = \frac{1}{T}\int_{0}^{T_1} i_l(t)dt = \frac{1}{T}\left[-I_o T_1 - \tau(I_o + \hat{I})(e^{-T_1/\tau} - 1)\right].$$

$$I_{D\,av} = \frac{10^3}{2}[-30 \times 0.283 \times 10^{-3} - 0.5 \times 10^{-3} \times 52.85(0.568 - 1)] = 1.47 \text{ A}.$$

The current rating of each diode must be greater than 1.5 A.

(c) The average source current is
$I_{s\,av} = 2(I_{TH\,av} - I_{D\,av}) = 2 \times (5.04 - 1.46) = 7.15$ A.
The average power delivered to the load is
$P = V_s I_{s\,av} = 600 \times 7.15 = 4.29 \times 10^3$ W.

2.4.3. Three-phase Inverters

We took a fleeting glance at three-phase conversion from ac sources to dc loads in Fig. 2.14. The same switch configuration can be used for the conversion from dc sources to ac loads.

There is a need for ac power at levels higher than that which single-phase bridge converters can provide. Three-phase bridge converters can satisfy the need. Induction motors, above the rating of a few kilowatts, have three-phase windings. If such motors are to have their speed adjusted, three-phase bridge inverters must be able to have an output that has both adjustable voltage and adjustable frequency.

In single-phase bridge inverters the output frequency can be changed by altering the rate of switching of the semiconductor devices in sequence. The voltage at the output can be adjusted by duty cycle control. In three-phase bridge inverters the output frequency can be changed in the same way as for single-phase bridges, but duty cycle control cannot be employed for adjustable voltage. Fortunately, there is a number of alternatives that can be employed. These alternatives are mainly concerned with pulse-width modulation (PWM), that is the technique for synthesizing the shape of the output voltage to closely resemble a sinewave. Other methods involve multi-connected, out-of-phase inverters. We will make a brief study of three-phase inverters. Of interest is the power output and the quality of the output waveforms.

Figure 2.15d depicts a dc source of voltage V_s and a three-phase load with voltages v_a, v_b, and v_c. The power is modulated by three half bridges of switches, also called legs or poles, that comprise the three-phase bridge inverter. There are six controlled switches (thyristors for higher powers and transistors for lower powers) to shape the output waveform. There are six uncontrolled switches (diodes) to provide a path for inductive currents whenever the controlled switches are turned off. The twelve switches form two bridges connected in inverse parallel. Thyristors are shown as the controlled switches in the figures, but any switch with controlled turn-on and turn-off can be substituted.

Resistive load. Figure 2.20a shows the ideal circuit diagram of a three-phase bridge inverter with a resistive wye-connected load. The ideal output of the inverter would be a set of sinusoidal voltages, as shown in Fig. 2.20b. Do you know why the ideal waveshape is a sinusoid[6]?

We will settle for a simple switching sequence in order to obtain a three-phase set of output voltages without being concerned about the waveshape. From the sinusoids illustrated in Fig. 2.20b, we can derive a sequence that each switch in the bridge is turned on and off. Note the labelling of switch numbers in Fig. 2.20a. It is done in such a way as to make it easy to remember the overall switching pattern.

The phase voltage v_{an} is positive during the interval 0 to π radians. In order to implement this requirement in practice, thyristor $TH\,1$ must be on and thyristor $TH\,4$ must be off. At the same time, either thyristor $TH\,6$ or thyristor $TH\,2$ must be on to complete the path to the supply and allow conduction in phase a. Thyristor $TH\,1$ is shown to have a conduction interval from 0 to π radians in Fig. 2.20c. For cyclic repetition, thyristor $TH\,1$ must be conducting again for the interval 2π to 3π radians because v_{an} is positive once more.

The phase voltage v_{an} in Fig. 2.20b is negative over the interval π to 2π radians. This can be accomplished by having thyristor $TH\,4$ on and thyristor $TH\,1$ off. Consequently, thyristor $TH\,4$ is shown to have a conduction interval from π to 2π radians in Fig. 2.20c for a negative v_{an}.

If we repeat this argument for all positive and negative voltages of phases b and c, we arrive at a conduction pattern for all the thyristors covering all intervals of time. In terms of each individual thyristor, it is on for π radians and off for π radians each period. This is called 180° switching.

An inspection of Fig. 2.20c indicates that, after every $\pi/3$ rad interval, there is a change of switching pattern as follows.

- From 0 to $\pi/3$ rad, $TH\,5$, $TH\,6$, $TH\,1$ are on. All others are off.
- From $\pi/3$ to $2\pi/3$ rad, $TH\,6$, $TH\,1$, $TH\,2$ are on. All others are off.
- From $2\pi/3$ to π rad, $TH\,1$, $TH\,2$, $TH\,3$ are on. All others are off.

[6] Think about the current-voltage relations for linear R, L and C elements and think about the value of the instantaneous power in a three-phase system.

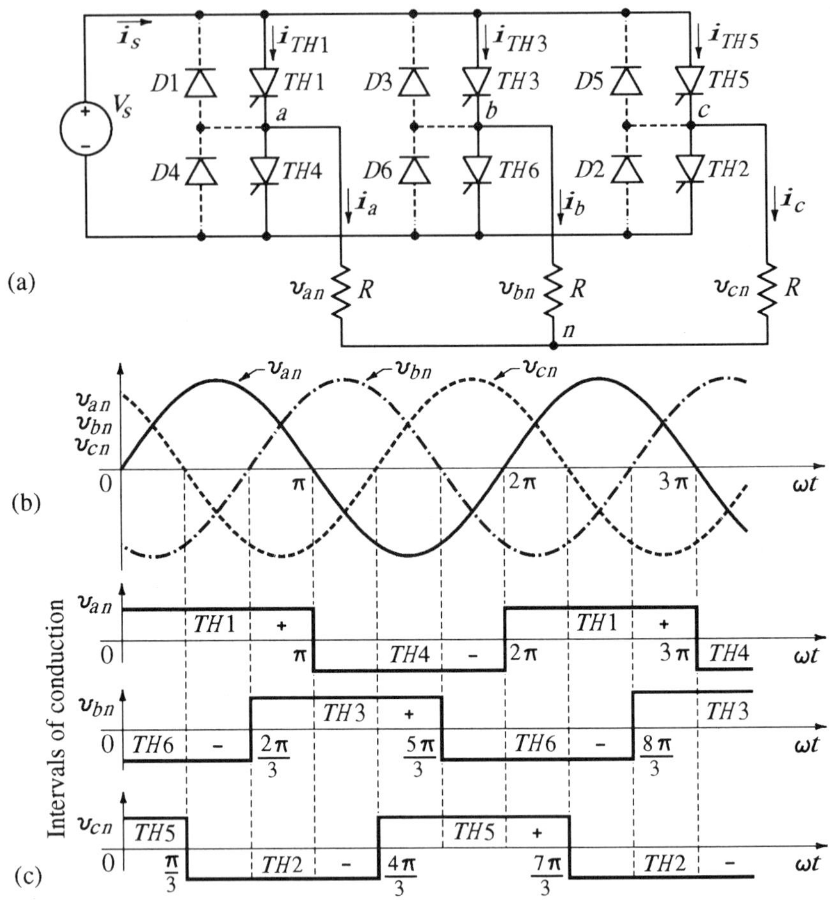

Fig. 2.20 Three-phase bridge inverter.
(a) Circuit diagram, (b) sinewaves, (c) conduction intervals.

- From π to $4\pi/3$ rad, $TH2, TH3, TH4$ are on. All others are off.
- From $4\pi/3$ to $5\pi/3$ rad, $TH3, TH4, TH5$ are on. All others are off.
- From $5\pi/3$ to 2π rad, $TH4, TH5, TH6$ are on. All others are off.
- From 2π to $7\pi/3$ rad, $TH5, TH6, TH1$ are on, and the cycle repeats.

There are six steps of switching to create a cyclic three-phase pattern, 1,2,3, 2,3,4, 3,4,5, 4,5,6, 5,6,1 and 6,1,2. This sequence of six steps, that gives rise to the name *six-pulse inverter,* is easy to remember. It can be checked that when three switches are on at any one time to satisfy the three-phase load requirements, there is always a complete path for conduction between the dc source and the ac load. For a voltage source inverter like this, there should never be two switches on together in any one leg. Otherwise, there would be a short circuit across the supply.

The waveshapes of the three-phase voltages can be obtained once the switching pattern has been established. It is assumed that each phase load has the same value of resistance R. Figure 2.21a shows the phase a, b and c connections to the dc source for the interval 0 to $\pi/3$ radians, while thyristors $TH5$, $TH6$ and $TH1$ are on. From this equivalent circuit the voltages and currents associated with each phase of the load can be determined. The values of voltages are shown.

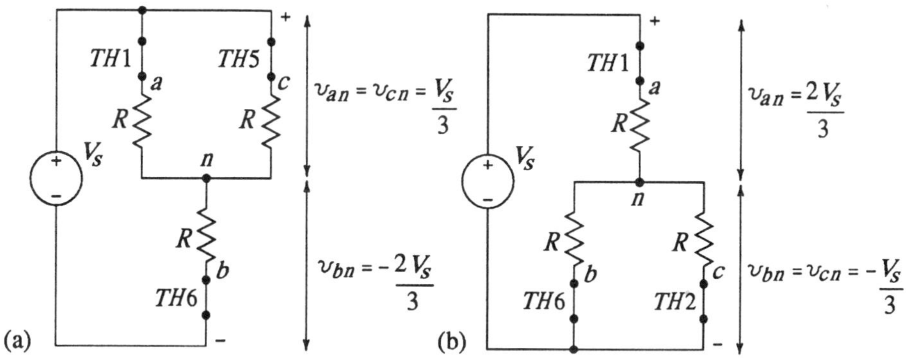

Fig. 2.21 Equivalent circuits.
(a) $TH5$, $TH6$, $TH1$ on, (b) $TH6$, $TH1$, $TH2$ on.

For the interval $\pi/3$ to $2\pi/3$ radians, thyristors $TH6$, $TH1$ and $TH2$ are on, so the equivalent circuit has a diagram like that shown in Fig. 2.21b. The arrangement of the two parallel resistances of phases b and c in series with the resistance of phase a gives an unequal voltage division, as shown.

From similar equivalent circuits for each switching pattern, the complete voltage pattern for each phase voltage can be built up. A summary of the results is as follows.

- From 0 to $\pi/3$, $v_{an} = V_s/3$, $v_{bn} = -2V_s/3$, $v_{cn} = V_s/3$.
- From $\pi/3$ to $2\pi/3$ rad, $v_{an} = 2V_s/3$, $v_{bn} = -V_s/3$, $v_{cn} = -V_s/3$.
- From $2\pi/3$ to π rad, $v_{an} = V_s/3$, $v_{bn} = V_s/3$, $v_{cn} = -2V_s/3$.
- From π to $4\pi/3$ rad, $v_{an} = -V_s/3$, $v_{bn} = 2V_s/3$, $v_{cn} = -V_s/3$.
- From $4\pi/3$ to $5\pi/3$ rad, $v_{an} = -2V_s/3$, $v_{bn} = V_s/3$, $v_{cn} = V_s/3$.
- From $5\pi/3$ to 2π rad, $v_{an} = -V_s/3$, $v_{bn} = -V_s/3$, $v_{cn} = 2V_s/3$.

The waveforms of the phase voltages are shown in Fig. 2.22a. They are not exactly like the sinewaves that are desired, but they do have some resemblance.

The line voltages v_{ab}, v_{bc} and v_{ca} of the three-phase load are defined by

$$v_{ab} = v_{an} - v_{bn}, \quad v_{bc} = v_{bn} - v_{cn} \quad \text{and} \quad v_{ca} = v_{cn} - v_{an}. \tag{2.4.29}$$

From these equations and the waveforms in Fig. 2.22a the waveforms of the line voltages can be constructed. For example, for the interval 0 to $\pi/3$ radians the

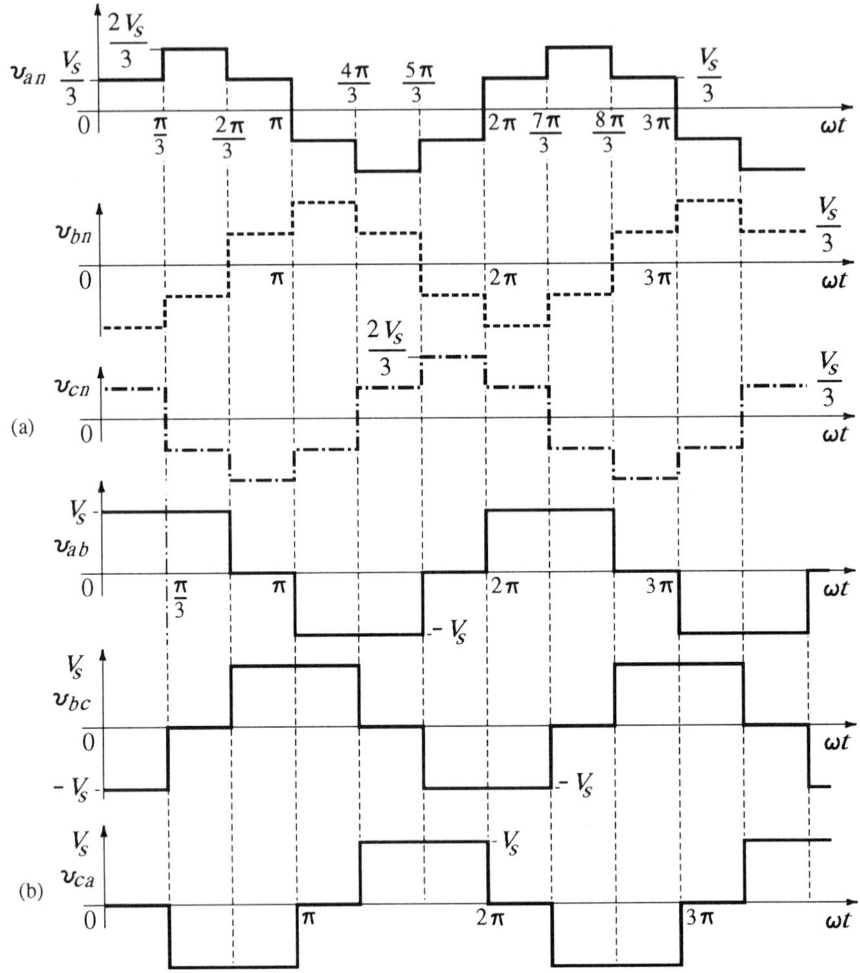

Fig. 2.22 Output voltage waveforms.
(a) Phase voltages, (b) line voltages.

line voltage v_{ab} is

$$v_{ab} = v_{an} - v_{bn} = V_s/3 - 2V_s/3 = V_s \, . \tag{2.4.30}$$

All the line voltage waveforms are shown in Fig. 2.22b.

For a purely resistive load, the current waveforms are in phase with the phase voltages for a three-phase wye connection. For these fairly simple and regular waveforms, average power in the load and waveform quality factors can be calculated readily.

EXAMPLE 2.13

Consider the three-phase bridge inverter, depicted in Fig. 2.20a. The dc source voltage is $V_s = 600$ V. The balanced load is wye connected and has only resistance of value $R = 20$ ohms per phase. The inverter operates at 100 Hz in the 6-pulse mode. Determine **(a)** the source current i_s as a function of time, **(b)** the average power delivered to the three-phase load, **(c)** the rms value of the phase voltage across the load, **(d)** the rms value of the line voltage across the load, **(e)** the value of the third harmonic in the line voltage waveform of the load and **(f)** the total harmonic distortion THD of the line voltage of the load.

Solution

(a) Since the system is balanced and there is no inductance, the diodes are inoperative. From Fig. 2.21 it can be deduced that, at all times, there are two parallel resistances in series with a resistance to form a load for the source. Thus, the source sees a load resistance R_l of $R/2 + R = 3R/2 = 3\times 20/2 = 30\,\Omega$.
The source current $i_s = V_s / R_l = 600/30 = 20$ A. This is a constant.

(b) The average power P delivered by the source to the load is
$P = V_s I_{s\,av} = 600\times 20 = 12\times 10^3$ W.
Therefore, the average power per phase P_{ph} in the load is
$P_{ph} = P/3 = 12\times 10^3/3 = 4\times 10^3$ W per phase.

(c) The average power P_{ph} across each phase of the load is given by
$P_{ph} = I^2_{a\,rms} R = V^2_{an\,rms}/R$. (See Fig. 2.20a for currents i_a, i_b, i_c.)
Therefore, the phase voltage across the load is
$V_{an\,rms} = V_{bn\,rms} = V_{cn\,rms} = \sqrt{P_{ph}\times R} = \sqrt{4\times 10^3 \times 20} = 283$ V.

(d) The line voltage is given by $V_{ab\,rms} = \sqrt{3}\,V_{an} = \sqrt{3}\times 283 = 490$ V.
If the reader is not confident about using $\sqrt{3}$ in a balanced, three-phase system calculation, we can go back to first principles. From Fig. 2.22b and the definition of rms value,

$$V_{ab\,rms} = \left[\frac{1}{\pi}\int_0^{\pi} v_{ab}^2\,d\omega t\right]^{1/2} = \left[\frac{V_s^2}{\pi}\int_0^{2\pi/3} d\omega t\right]^{1/2} = V_s\frac{\sqrt{2}}{\sqrt{3}} = 490\text{ V}.$$

(e) An inspection of the line voltage waveforms in Fig. 2.22b, shows that there is a 60° dwell (no voltage) in the half period of 180°. Therefore, there is no third harmonic in the line voltage.
If the reader is not confident about this conclusion, we can use a Fourier analysis on the line voltage waveform in Fig. 2.22b. Consider the waveform of v_{ab} and make a new origin at $\omega t = \pi/3$ so that the waveform can be described as an even function. The Fourier coefficient a_n is calculated from

$$a_n - \frac{1}{\pi}\int_{-\pi}^{\pi} v_{ab}\cos n\,\omega t = \frac{4}{\pi}\int_0^{\pi/2} v_{ab}\cos n\omega t\,d\omega t\,.$$

So, $a_n = \frac{4V_s}{n\pi}[\sin n\omega t]_0^{\pi/3} = \frac{4V_s}{n\pi}\sin\frac{n\pi}{3}$.

For the third harmonic, $n = 3$. So, $a_3 = 0$.

Since $b_n = 0$ for all n, there is no third harmonic.

(f) Total harmonic distortion THD $= \left(\frac{V_{l\,rms}^2}{V_{l1}^2} - 1\right)^{1/2}$.

$V_{l\,rms} = V_{ab\,rms} = 490$ V from part (d). From part (e)

$$V_{l1} = V_{ab1} = \frac{a_1}{\sqrt{2}} = \frac{4V_s}{\sqrt{2}\pi}\sin\frac{\pi}{3} = \frac{4\times 600}{\sqrt{2}\pi}\times\frac{\sqrt{3}}{2} = 468 \text{ V}.$$

Therefore, THD $= \left(\frac{490^2}{468^2} - 1\right)^{1/2} = 0.31$ (31%).

Six-pulse, 120° mode. The analysis of the six-pulse, three-phase inverter has been based on each thyristor having a turn-on and a turn-off interval equal to one half of the period of alternation. This was called the 180° conduction mode.

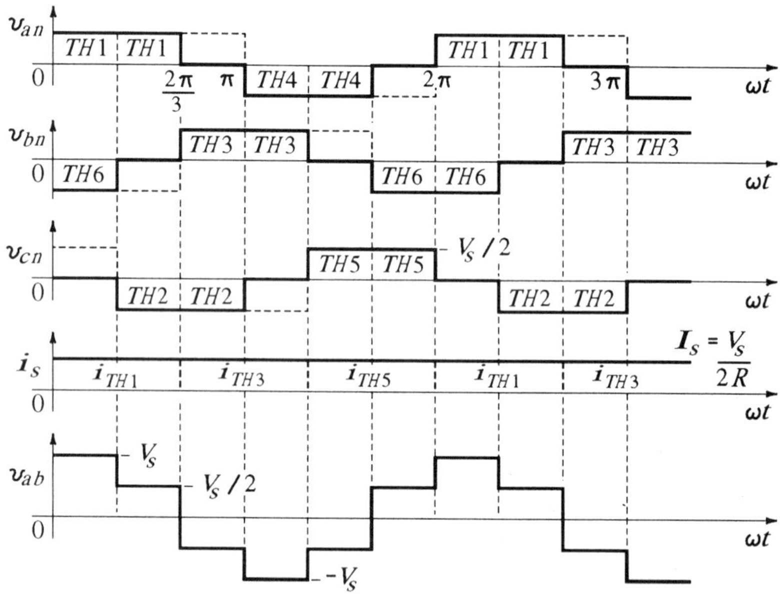

Fig. 2.23 Waveforms for the 120°, 6-pulse inverter.

There is another mode. It is the 120° mode of switching, for which each thyristor has an on-time that is two thirds of the 180° mode. This provides a simple means of reducing the load voltage without changing the frequency. Figure 2.23 shows how the 120° mode of switching affects the voltage and current waveforms for a resistive load. There are only two thyristors on at any one time, and the diodes are inoperative.

The switching sequence is $TH6, TH1$, $TH1, TH2$, $TH2, TH3$, $TH3, TH4$, $TH4, TH5$, $TH5, TH6$ in any one cycle. This gives six pulses per cycle, the voltage across each phase is $V_s/2$ for 120° of the half cycle and zero for the remainder of the half cycle. While the phase voltage is finite, the phase current is $V_s/(2R)$ for a resistive load. This is the same as the source current I_s, since there is always one of the odd numbered thyristors and one of the even numbered thyristors on at any one time.

The power delivered by the source is

$$P = V_s I_s = \frac{V_s^2}{2R} \tag{2.4.31}$$

and is constant. This is 75% of the power that can be delivered in the 180° mode.

The value of line voltage for each pulse interval of $\pi/3$ radians can be found from an inspection of the phase-voltage waveforms in Fig. 2.23. The line voltage is shown for $v_{ab} = v_{an} - v_{bn}$. For the interval $0 \le \omega t \le \pi/3$, thyristors $TH1$, $TH6$ are on, so, $v_{ab} = +V_s$. For the interval $\pi/3 \le \omega t \le 2\pi/3$, thyristors $TH1$, $TH2$ are on. From Fig. 2.20a, node a is at V_s and node c is at ground (assumed), so node b is at the potential of node n, which, in a balanced, symmetrical system with ideal switches, is at $V_s/2$. So, $v_{ab} = V_s/2$. For the interval $2\pi/3 \le \omega t \le \pi$, thyristors $TH3$, $TH2$ are on, node b is at V_s, node c is at 0 V and node a is at the potential of node n. Hence, $v_{ab} = -V_s/2$. The rest of the waveform is deduced in the same way.

The rms value of the phase voltage is

$$V_{ph\,rms} = \left[\frac{1}{\pi}\int_0^{\pi} v_{an}^2\, d\omega t\right]^{1/2} = \left[\frac{1}{\pi}\int_0^{2\pi/3} \frac{V_s^2}{4}\, d\omega t\right]^{1/2} = \frac{V_s}{2}\frac{\sqrt{2}}{\sqrt{3}}. \tag{2.4.32}$$

Therefore, the line voltage is

$$V_{l\,rms} = \sqrt{3} \times V_{ph\,rms} = \frac{V_s}{\sqrt{2}}. \tag{2.4.33}$$

These values are 86.6% of those obtained in the 180° mode of switching.

The quality of the ac waveforms for a 120°, 6-pulse inverter is left as an exercise for the interested reader. The analysis has the same form as that described in EXAMPLE 2.13 and problem 2.33.

2.5. SUMMARY

This chapter focuses on the description of the performance of elementary, ideal systems that incorporate switches for power conditioning. Only a few converters were chosen to indicate the method of conversion and the means to arrive at a measure of the quality of conversion.

Choppers provide a way to convert a dc supply of fixed voltage to a dc supply of adjustable voltage. The controlling factor in dc-dc conversion is the duty cycle m. By altering the on-time of a switch in an LC circuit, the output voltage can be controlled to be less than or greater than the supply voltage.

Point-on-wave control (phase control) is the way in which ac-dc conversion takes place. The ac waveform is rectified and chopped by switches to allow dc pulses to appear at the load. The chopping is the means to obtain an adjustable dc voltage. In this particular section switches were limited to the delay angle α control. The switches were allowed to turn-off at voltage zero or current zero. The cases of more versatile switching (at any advanced angle of extinction β) is discussed in the chapters on particular switches. Accordingly, the controlling factor that determines the dc voltage output is the delay angle α.

The method of point-on-wave control is standard. The configuration of the switching circuit for rectification depends on how much power is to be modulated. For low power a single-phase supply and bridge configuration are employed. For high power a three-phase supply and the appropriate bridge of switches are used.

The quality of the output dc waveforms of an ac-dc converter suffer in the conversion. The waveform of current on the ac side is also degraded. Performance factors are used to show how the quality changes with load, the type of converter and the controlling parameter α.

Inverters for dc-ac conversion have the same kind of switch configurations as the controlled rectifier circuits. Most often, the aim is to have a sinewave output. The more sophisticated the switching pattern is the closer to a sinewave one gets. In this section only the simple forms of switching are described. In this way, by using the same performance factors derived from the Fourier series as were used in ac-dc conversion, it is possible to make a relatively straightforward prediction of how the system behaves. What has been emphasized has been the average power absorbed by the load, the voltage, the frequency and the attendant output harmonics. All of this is accomplished with simple analysis.

All the systems of switching, that demonstrate power conditioning, have been kept ideal and basic to emphasize the principles of operation and to predict behavior using uncomplicated analysis. In service, converters are found to be much more complex than described in this text. The bibliography at the end of this chapter is an invitation for further reading at the more complex level of the control of converters with passive loads.

Up to this point, ideal generic switches have been used in the circuits, although thyristors have been drawn in the diagrams. Even though power semiconductor switches give a good performance they are never perfect. It takes time to turn them on and off. They dissipate energy during the switching action and during conduction. Some switches block forward and reverse bias voltages, some only block forward bias voltage. There are some switches that can turn on more quickly than others, so that they can operate at higher frequencies. Other switches can handle very large amounts of power. We need to know how to choose the right kind of semiconductor switch for any kind of application and we need to know how to protect each semiconductor device that best fits uninterrupted service in an economic manner. The following chapters treat each semiconductor switch individually and concern the performance of the device from an elementary analytical approach.

2.6. PROBLEMS

Note. All elements of the systems in the following problems can be considered to be ideal unless otherwise stated.

Section 2.2

2.1 Consider the chopper circuit shown in Fig. 2.1. The ideal switch *TH* modulates power from a dc source of 100 V to a resistive load of 1Ω. If the duty cycle of the chopper is $m = 0.6$ and the switching frequency is 1 kHz, determine **(a)** the average current in the switch and **(b)** the average power absorbed by the load.

2.2 A buck converter (see Fig. 2.2) modulates power from a 100-V dc source to a resistive load of 2 Ω whose voltage is essentially ripple free. The inductance of the chopper has a value of 0.4 mH. If the switching frequency is $f = 1000$ Hz and if the duty cycle is $m = 0.7$, find **(a)** the average power absorbed by the load, **(b)** the minimum and maximum values of the inductor current and **(c)** the average current in the converter switch.

2.3 An ideal boost converter (see Fig. 2.3) modulates power from a 100-V dc source to a resistive load. The converter switching frequency is 1000 Hz and the duty cycle m must be less than 0.7. The source is rated at 10 kW. Calculate the value of the converter inductance so that the source current is just continuous, if the power to the load is 5 kW. Assume that the value of the converter capacitance is high enough that the load voltage is virtually constant.

Section 2.3

2.4 A switched-mode ac-dc converter causes a 120-V, 60-Hz source to supply a current whose instantaneous value is

$$i = 50\sin(377t - \pi/3) + 25\sin(1131t - 0.2)\ \text{A},$$

where the voltage is taken as reference. Determine **(a)** the input displacement power factor DPF, **(b)** the average power P delivered by the supply, **(c)** the rms value of the supply current I_{rms}, **(d)** the total harmonic distortion THD of the current waveform and **(e)** the power factor PF of the ac circuit.

2.5 An ac supply of voltage 240 V at 50 Hz is fully rectified such that the current in a resistive load of 10 Ω has a waveform comprising a periodic series of positive half waves of sinusoidal shape. **(a)** Describe the load current waveform. Determine **(b)** the load form factor FF, **(c)** the ripple factor RF of the load current and **(d)** the rectification ratio RR.

2.6 A single thyristor modulates power from a 240-V, single-phase source to a resistive load of 2Ω. See Fig. 2.6. If the trigger delay angle of the thyristor switching is $\alpha = \pi/2$ radians, **(a)** sketch v_s, v_{AK}, v_l and i_A and determine **(b)** the average value of the load voltage, **(c)** the average value of the load current, **(d)** the rms value of the load voltage, **(e)** the average power P absorbed by the load, **(f)** the rectification ratio RR of the converter, **(g)** the form factor FF of the load current and **(h)** the ripple factor RF of the output voltage. **(i)** Determine the voltage and current ratings of the thyristor.

2.7 A single thyristor modulates power from a single-phase 115-V, 10-Hz source to a purely inductive load of 10 mH. **(a)** Determine the value of the thyristor firing angle α if the peak energy delivered to the load is to be 750 J. **(b)** What are the voltage and current ratings of the thyristor?

2.8 Consider the half-wave converter, shown in Fig. 2.6. The load is a battery that can be modelled by an ideal constant voltage source of 48 V in series with a resistance whose value is 0.8 Ω. The ac source has a voltage of 115 V at 60 Hz. If the firing angle of the thyristor is set at $\alpha = 2\pi/3$ radians, find **(a)** the time it takes to charge the battery 10 Ah and **(b)** the average power dissipated as heat.

2.9 A supply delivers a rectangular wave ac voltage of amplitude 200 V at 100 Hz. A thyristor modulates the power from this source to a load whose resistance is $R = 2\,\Omega$ and whose inductance is $L = 1$ mH. See Fig. 2.7. For a thyristor gate pulse at a firing angle $\alpha = \pi/2$ radians, determine **(a)** the instantaneous value of the load current, **(b)** the extinction angle β of the thyristor and **(c)** the average power absorbed by the load. **(d)** What are the voltage and current ratings of the thyristor?

2.10 Consider the half-wave converter, illustrated in Fig. 2.7. If the source has a sinusoidal voltage and the thyristor is ideal, determine the instantaneous value of the load current for any firing angle α.

2.11 A single thyristor modulates power from a single-phase 120-V, 60-Hz source to a load, across which there is connected a freewheeling diode. See Fig. 2.8. The load resistance is $R = 0.8\,\Omega$ and the load inductance is $L = 1$ mH. For a thyristor firing angle $\alpha = \pi/3$ radians, determine **(a)** if the load current is continuous, **(b)** the instantaneous value of the load current, **(c)** the average value of the load current, **(d)** the load current at $\omega t = \pi$ radians, **(e)** the average power dissipated in the load resistance during the freewheeling intervals, **(f)** the total average power delivered by the source and **(g)** the rms value of the load current. **(h)** What are the voltage and current ratings of the thyristor?

2.12 A single-phase, full-wave ac-dc converter, as depicted in Fig. 2.9a, modulates power from a 120-V, 60-Hz source to a purely resistive load, $R = 2\,\Omega$. If the firing angle α of the thyristor is $\pi/6$ radian, determine **(a)** the output form factor FF, **(b)** the output voltage ripple factor RF, **(c)** the rectification ratio RR, **(d)** the average output power, **(e)** the input harmonic distortion THD, **(f)** the input power factor PF, **(g)** the input displacement power factor DPF and **(h)** the voltage and current ratings of the thyristors.

2.13 A single-phase, full-wave ac-dc converter, as illustrated in Fig. 2.9a, modulates power from an ac source to an *RL* load. Assume that the value of the load inductance is high enough that the load current can be considered continuous and virtually ripple free. In terms of the thyristor firing angle α, determine **(a)** the output voltage form factor FF, **(b)** the value of the load current, **(c)** the output voltage ripple factor RF, **(d)** the rectification ratio RR, **(e)** the average output power, **(f)** the input harmonic distortion THD, **(g)** the input power factor PF, **(h)** the input displacement power factor DPF and **(i)** sketch the voltage and current waveforms.

2.14 Consider the ac-dc converter shown in Fig. 2.9a. The ac supply is 120 V at 60 Hz and the load resistance is $R = 2\,\Omega$. It can be assumed that the value of the load inductance is high enough ($L \approx \infty$) that the load current is virtually constant. If the firing angle α of the thyristor is $\pi/6$ radian, determine **(a)** the output form factor FF, **(b)** the output voltage ripple factor RF, **(c)** the rectification ratio RR, **(d)** the average output power, **(e)** the input harmonic distortion THD, **(f)** the input power factor PF, **(g)** the input displacement power factor DPF and **(h)** the current and voltage ratings of each thyristor.

2.15 A single-phase, full-wave ac-dc converter, depicted in Fig. 2.9a, modulates power from a 240-V, 50-Hz source to a load whose resistance is $R = 2\,\Omega$ and whose inductance is $L = 6$ mH. **(a)** Over what range of the thyristor firing angle α is the load current discontinuous? **(b)** What is the average power absorbed by the load if the load current is just continuous and **(c)** for this condition what is the input power factor PF?

2.16 Consider the circuit diagram of the single-phase, full-wave ac-dc converter, shown in Fig. 2.9b, with a freewheeling diode across the load. For any thyristor trigger angle α, **(a)** sketch the waveforms of v_s, v_l, i_l, i_s and i_D for discontinuous and continuous load current. Determine **(b)** the load voltage form factor FF, **(c)** the load voltage ripple factor RF, **(d)** the instantaneous value of the load current i_l, **(e)** expressions for the average input power and load power and **(f)** the input power factor PF.

2.17 A single-phase, full-wave, ac-dc converter, illustrated in Fig. 2.9b, has a freewheeling diode across the load. The power is to be modulated from a 240-V, 50-Hz source to a load whose resistance is $R = 2\,\Omega$ and whose inductance is $L = 2$ mH. If the load current is just continuous determine the values of **(a)** the thyristor firing angle α, **(b)** the output voltage form factor FF, **(c)** the output voltage ripple factor RF, **(d)** the converter rectification ratio RR, **(e)** the average power absorbed by the load and **(f)** the input power factor PF. **(g)** What is the current rating of the diode?

2.18 Consider the ac-dc converter system shown in Fig.2.9b. The source is 240 V at 50 Hz. The load resistance is $R = 2\,\Omega$ and the load inductance is $L = 2\,\text{mH}$. For a thyristor firing angle $\alpha = 2\pi/3$ radians, **(a)** sketch the waveforms of v_s, v_l, i_l, i_D and i_s. Determine **(b)** the output voltage form factor FF, **(c)** the output voltage ripple factor RF, **(d)** the converter rectification ratio RR, **(e)** the average power absorbed by the load and **(f)** the input power factor PF.

2.19 A single-phase, full-wave, ac-dc converter, depicted in Fig. 2.9b, has a freewheeling diode across the load. The power is to be modulated from a 240-V, 50-Hz source to a load whose resistance is $R = 2\,\Omega$ and whose inductance is $L = 2\,\text{mH}$. For a thyristor firing angle $\alpha = \pi/3$ radians, **(a)** sketch the waveforms of the variables v_s, v_l, i_l, i_D and i_s. Determine **(b)** expressions for the instantaneous value of the load current i_l, **(c)** the load current excursion ($\Delta I = i_{l\,\max} - i_{l\,\min}$), **(d)** the converter rectification ratio RR, **(e)** the load voltage and current form factors FF, **(f)** the load voltage and current ripple factors RF, **(g)** the average power absorbed by the load and **(h)** the input power factor PF.

2.20 Consider the converter described in problem 2.19. **(a)** Calculate the maximum power that can be absorbed by the load in steady state and determine **(b)** the current ratings of the thyristors. For the condition of maximum power, find **(c)** the load current excursion, **(d)** the converter rectification ration RR, **(e)** the load voltage and current form factors FF, **(f)** the load voltage and current ripple factors RF and **(g)** the input power factor PF.

2.21 A single-phase, full-wave, ac-dc converter is illustrated in Fig. 2.9b. There is a diode connected across the load. The power is modulated from a 240-V, 50-Hz source to an *RL* load, whose resistance is $R = 2\,\Omega$ and whose inductance is high enough to consider the load current to have a ripple factor with a value of zero. For a thyristor firing angle $\alpha = \pi/3$ radians, **(a)** sketch the waveforms of the variables v_s, v_l, i_l, i_D and i_s. Determine **(b)** the converter rectification ratio RR, **(c)** the load voltage form factor FF, **(d)** the load voltage ripple factor RF, **(e)** the average power absorbed by the load, **(f)** the input power factor PF, **(g)** the total harmonic distortion THD of the input current and **(h)** the input displacement power factor DPF. **(i)** Estimate the current ratings of the thyristors and the diode.

2.22 A single-phase, full-wave, half-controlled ac-dc converter is depicted in Fig. 2.9d. Consider there to be no freewheeling diode connected across the load. The power is modulated from a 240-V, 50-Hz source to an *RL* load, whose resistance is $R = 2\,\Omega$ and whose inductance is high enough to consider the load current to have a ripple factor $\text{RF} = 0$. For a thyristor firing angle $\alpha = \pi/2$ radians, **(a)** sketch the waveforms of the variables v_s, v_l, i_l, i_s and the currents in the thyristors and diodes. Determine **(b)** the converter rectification ratio RR, **(c)** the load voltage form factor FF, **(d)** the load voltage ripple factor RF, **(e)** the average power absorbed by the load, **(f)** the input power factor PF, **(g)** the total harmonic distortion THD of the input current and **(h)** the input displacement power factor DPF. **(i)** Estimate the current ratings of the thyristors and the diodes in the bridge.

2.23 Repeat problem 2.22 for the case where thyristor *TH*4 is replaced by a diode *D*4 to form an asymmetrical bridge. Refer to Fig. 2.9d.

Section 2.4

2.24 Describe the operation of the single-phase, centre-tapped transformer inverter, illustrated in Fig. 2.15a.

2.25 Consider the half-bridge inverter illustrated in Fig. 2.16a. The dc source has a voltage $V_s = 300\,\text{V}$ and the load has a resistance $R = 18\,\Omega$. If the average power delivered by the source to the load is 3 kW, determine **(a)** the duty cycle m of switching, **(b)** the total harmonic distortion THD of the voltage waveform, **(c)** the harmonic factor HF of the fifth harmonic of voltage and **(d)** the voltage and current ratings of the switches.

2.26 A dc-ac converter, shown in Fig. 2.17a, is used to modulate power from a source whose voltage is $V_s = 100\,\text{V}$ to a load whose resistance is $R = 1\,\Omega$, whose inductance is $L = 1\,\text{mH}$ and whose capacitance is $C = 25.2\,\mu\text{F}$. The frequency of thyristor switching is 930 Hz. For steady-state operation, if the load current harmonics can be neglected, sketch **(a)** the waveforms v_l, i_l, i_{TH1}, i_{TH2}, i_{D1}, i_{D2}, i_{s1} and i_{s2}, and determine **(b)** the average power delivered to the load and **(c)** the current ratings of the thyristors and diodes.

2.27 Consider the single-phase, bridge inverter depicted in Fig. 2.18. The dc source has a voltage $V_s = 600\,\text{V}$ and the load has a resistance $R = 18\,\Omega$. If the average power delivered to the load is 3 kW, determine **(a)** the duty cycle m of inverter switching, **(b)** the average current conducted by the thyristors, **(c)** the total harmonic distortion THD and **(d)** the harmonic factor HF of the ninth-harmonic component of voltage.

2.28 A single-phase, bridge inverter has a dc voltage source $V_s = 600\,\text{V}$ and a resistive load $R = 10\,\Omega$. For the maximum power transfer condition, find **(a)** the instantaneous value of the load current as a Fourier series, **(b)** the total harmonic distortion THD, **(c)** the harmonic factor HF of the ninth-harmonic component of current and **(d)** the thyristor voltage and current ratings.

2.29 A single-phase bridge inverter, as depicted in Fig. 2.19, has a dc source voltage that is $V_s = 600\,\text{V}$ and a load that has a resistance $R = 20\,\Omega$ in series with an inductance $L = 2\,\text{mH}$. The inverter operates at 500 Hz and delivers maximum power to the load in the steady state. Determine **(a)** the peak current from the dc source, **(b)** the current ratings of the switches, **(c)** the average power absorbed by the load and **(d)** the rms value of the load current.

2.30 A single-phase bridge inverter, illustrated in Fig. 2.19, has a dc source voltage $V_s = 600\,\text{V}$ and a load whose resistance, $R = 20\,\Omega$, is in series with an inductance, $L = 10\,\text{mH}$. The inverter operates at 500 Hz and delivers maximum power to the load in steady state. Determine **(a)** the harmonic load voltages up to the ninth, **(b)** the harmonic load impedances up to the ninth, (**c**) the instantaneous value of the load current up to the ninth harmonic, (**d**) the total harmonic distortion THD of the load current and (**e**) the harmonic factor HF of the ninth-harmonic component of load current.

2.31 A single-phase bridge inverter has a dc source voltage $V_s = 600\,\text{V}$ and has a load that has resistance, $R = 20\,\Omega$, in parallel with inductance, $L = 5\,\text{mH}$. If the inverter is to operate at 500 Hz with maximum average power to the load, determine (**a**) the peak value of the source current, (**b**) the average values of current in the switches, (**c**) the average power delivered from the source to the load and (**d**) the total harmonic distortion THD of the load current.

2.32 In EXAMPLE 2.7 the rms value of the load voltage was determined in terms of the duty cycle m for a rectangular waveform. If the rectangular waveform is split up into two symmetrical pulses in the same half cycle, find (**a**) the value of the nth-harmonic component of the waveform in terms of the dc supply voltage V_s and the duty cycle m. (**b**) If $m = 0.4$, calculate the ratio of the third-harmonic voltages for single-pulse and double-pulse waveforms.

2.33 A three-phase bridge inverter has a dc source voltage $V_s = 600\,\text{V}$ and has a purely resistive load of value $R = 20$ ohms per phase connected in wye. The inverter operates at 100 Hz in the six-pulse mode with each thyristor on for half of one period (180°). See Fig. 2.20. Determine (**a**) the instantaneous value of the load current as a Fourier series, (**b**) the harmonic factors for the seventh and ninth harmonics of the load current and (**c**) the average current in the thyristors.

2.34 A three-phase bridge inverter has a dc source voltage $V_s = 600\,\text{V}$ and has a purely inductive load of value $L = 6.67\,\text{mH}$ per phase connected in wye. The inverter operates in the steady state at 200 Hz in the 180°, six-pulse mode. See Fig. 2.20a. (**a**) Sketch the current waveforms i_a, i_{TH1}, i_{TH3}, i_{TH5}, i_{D1}, i_{D3}, i_{D5} and i_s. (**b**) Find the peak value of the source current and (**c**) determine the average currents in the thyristors and diodes.

2.35 A three-phase bridge inverter has a dc source whose voltage is V_s. See Fig. 2.20a. The three-phase load is balanced, has resistance R and inductance L in series, and is wye connected. Sketch the currents i_l, i_{TH1}, i_{D1} and i_s in relation to the phase voltage v_{an}, if the inverter is operated in the 180°, 6-pulse mode and is in the steady state.

2.36 A three-phase bridge inverter has a dc source voltage $V_s = 600\,\text{V}$ and has a purely resistive load of value $R = 20$ ohms per phase connected in wye. The inverter operates at 100 Hz in a 120°, six-pulse mode (each switch is on for two thirds of one half period each cycle). See Fig. 2.23. Determine (**a**) the value of the source current i_s (see Fig. 2.20), (**b**) the average power delivered to the three-phase load, (**c**) the rms values of the phase voltage across the load, (**d**) the rms value of the line voltage across the load, (**e**) the value of the third harmonic in the phase voltage waveform, (**f**) the total harmonic distortion THD of the phase voltage, (**g**) the instantaneous value of a line current as a Fourier series, (**h**) the harmonic factors of the seventh and ninth harmonics of the load current and **(i)** the average current in the thyristors. Compare the results obtained for this problem with results obtained in EXAMPLE 2.13 and problem 2.33 for a 180°, 6-pulse inverter.

2.7. BIBLIOGRAPHY

Bose, B.K. (ED). *Modern Power Electronics - Evolution, Technology and Applications.* New York: IEEE Press, 1992.

Bird, B.M., and K.G.King. *An Introduction to Power Electronics.* New York: John Wiley & Sons Inc., 1983.

Csaki, F., I. Hermann, I. Ipsits, A. Karpati, and P. Magyar. *Power Electronics.* Budapest: Akademiai Kiado, 1979.

Fisher, Marvin J. *Power Electronics.* Boston, Mass.: PWS-Kent Publishing Co., 1991.

Hoft, R.G. *Semiconductor Power Electronics.* New York: Van Nostrand, 1986.

Kassakian, J.G., M.F. Schlecht, and G.C. Verghese. *Principles of Power Electronics.* Reading, Mass.: Addison-Wesley Publishing Co.Inc., 1991.

Lander, Cyril W. *Power Electronics.* 2nd ED. McGraw-Hill Book Co., 1987.

Laster, Clay. *Thyristor Theory and Application.* Blue Ridge Summit, PA: Tab Books Inc., 1986.

Lee, R.W. *Power Converter Handbook - Theory, Design and Application.* Peterborough: Canadian General Electric, 1979.

Mohan, N., T.M. Underland, and P. Robbins. *Power Electronics.* New York: John Wiley & Sons,Inc., 1989.

Rashid, Muhammad Harunur. *Power Electronics.* Englewood Cliffs, N.J.: Prentice Hall,Inc., 1988.

Sen, P.C. *Principles of Electric Machines and Power Electronics.* New York: John Wiley & Sons,Inc., 1989.

Shepherd, W. *Thyristor Control of AC Circuits.* London: Bradford University Press, 1976.

Shepherd, W., and P. Zand. *Energy Flow and Power Factor in Non-sinusoidal Circuits.* Cambridge: Cambridge University Press, 1979.

Tarter, R.E. *Principles of Solid State Power Conversion.* Howard W. Sams, 1985.

Thorborg, Kjeld. *Power Electronics.* Hemel Hempstead: Prentice-Hall International (UK) Ltd., 1988.

Williams, B.W. *Power Electronics, Devices, Drivers, and Applications.* New York: John Wiley & Sons, 1987.

Wood, P. *Switching Power Converters.* New York: Van Nostrand, 1982.

CHAPTER 3

THE DIODE

3.1. INTRODUCTION

The diode is a switch, but it is an uncontrolled switch. No signals can turn the diode on or off. Its action depends entirely on the voltages and currents of the circuit to which the diode is connected. It is described here in some detail because its structure is the foundation of all the devices that follow in this text, and it has important applications.

This semiconductor device, usually silicon, is more often than not a *pn* junction. That branch of a circuit with a diode is constrained to have current in only one direction. The diode blocks current in the reverse direction. Therefore, the diode is an uncontrolled rectifier and is also called a *diode rectifier.* In power applications diodes are used principally to rectify, that is, to convert alternating current to direct current. However, a diode is used also to allow current freewheeling. That is, if the supply to an inductive load is interrupted, a diode across the load provides a path for the inductive current and prevents high voltages ($L\,di/dt$) damaging sensitive components of the circuit.

Figure 3.1 illustrates the depiction of the diode. The *pn* junction structure is illustrated in Fig. 3.1a, showing the anode A terminal connected to the *p* region and the cathode K terminal connected to the *n* region. Figure 3.1b shows the circuit symbol D for the diode and the unidirectional current i_D for the case of the voltage at the anode being positive with respect to the cathode ($v_D = v_{AK} > 0$). This is forward bias. The symbol of the diode is shaped somewhat like an arrow. It indicates the only possible direction of conventional current in the element. There is the addition of the line at the cathode to represent the blocking action of current in the reverse direction. Thus, the diode rectifier is a two-layer, two terminal, silicon device that acts as an uncontrolled switch.

A typical application of a diode is shown in the simple circuit of Fig. 3.1c. Here, the diode allows only direct current in the resistive load even though the supply is an alternating current source. During the positive half cycle of the supply voltage v_s, the anode A of the diode is positive with respect to the cathode K. This forward bias turns the diode on, the current i_D is limited only by the load resistance R ($i_D \approx v_s/R$) and there is a small voltage drop v_D of about one volt across the diode. During the negative half cycle of the supply voltage, the cathode of the diode is positive with respect to the anode. This reverse bias maintains the diode in the off-state, which means that $i_D = 0$ (except for a small leakage current) and $v_D = -v_s$.

Like the thyristor, the diode is current controlled. That is, once either switch has been turned on, it will remain on until the main current in the anode-to-cathode path falls to zero. At zero current the turn-off process commences. In a

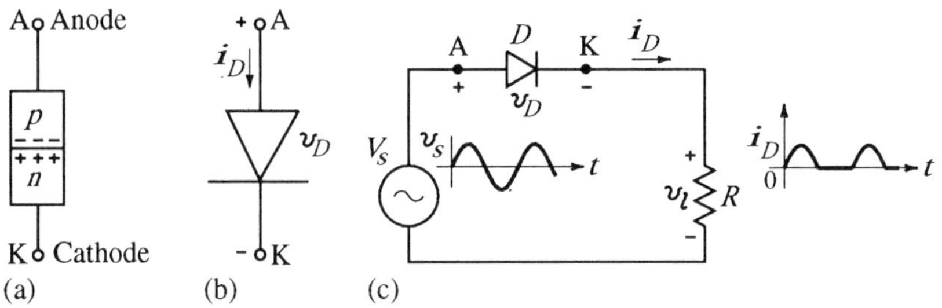

Fig. 3.1 Diode rectifier.
(a) *pn* junction, (b) circuit symbol, (c) simple circuit.

rectifier circuit, like that shown in Fig. 3.1c, if the load contains inductance the current lags the voltage. Accordingly, the source can reverse its voltage polarity and the diode remains on ($v_D \approx 0$, $v_l \approx v_s < 0$) until all the energy stored in the inductance is absorbed by the supply and the load resistance. At this point the diode turns off ($i_D = 0$, $v_l = 0$ and $v_s = v_D$). The switching times (the times to change state from on to off and from off to on) are of the order of microseconds. In general power applications, these switching times can be considered to be no time at all and are ignored in circuit analysis. However, account of these times is taken, if device power dissipation is being considered for heat sink design, for example.

The techniques of diode manufacture have advanced to a level that single devices can have voltage withstand ratings up to 5000 V and current ratings up to 3000 A. At lower ratings it is possible for specially designed diodes to switch at frequencies up to 50 kHz. Further, for still lower ratings in circuits that operate at low voltages there is the Schottky diode, a metal oxide device. This diode has a voltage drop in the on-state that is about half the value of the voltage drop of the *pn* junction diode. There is a classification of fast and slow recovery diodes.

In their own right, diodes are used for the uncontrolled rectification of alternating current to direct current in the power range from watts to megawatts. Because diodes can be connected in series and parallel, limits are set by the load rather than the converter. Power diodes are also important in complementary functions such as freewheeling action in inductive circuits that incorporate controlled semiconductor switches to modulate as well as convert power from one type to another. A further application is in snubber circuits for switch protection.

The rest of this chapter describes the action and application of the diode.

3.2. DIODE STRUCTURE

A general description of the structure and action of a silicon *pn* junction is given in section 1.5 of Chapter 1, because it is the fundamental building block of semiconductor switches. However, the *pn* junction is a switch in its own right. It forms a diode rectifier by its naturally uncontrolled switching characteristics.

The name *diode* implies *two* electr*odes* (or two terminals), anode A and cathode K, and the name *rectifier* signifies one direction only for current. As described in sections 1.5 and 3.1, if the diode switch is off initially, and then, if the circuit causes its anode to go positive with respect to its cathode ($v_D > 0$), the diode turns on. Conversely, if the switch is on and the circuit attempts to make the anode current i_D reverse direction, the diode turns off so that $v_D \leq 0$, that is, the anode is no longer at a positive voltage with respect to the cathode.

In practice, the *pn* junction of a power diode has a structure similar to that shown in Fig. 3.2. It is called the *pin* structure (*p* and *n* from the *pn* junction and the *i* signifying a near-intrinsic middle layer). It is given this physical form in order to widen the depletion layer and increase the reverse voltage withstand capability. However, the circuit symbol remains unchanged. The heavily doped n^+ region is the substrate on which the lightly doped (almost intrinsic) n^- drift region[1] is grown as a very thin layer to keep the resistance low (low on-state forward voltage drop) and yet to maintain most of the voltage in reverse bias. The heavily doped p^+ region is formed by diffusion to complete the junction before metalization that creates the contacts. This structure is then packaged.

3.3. DIODE I-V CHARACTERISTICS

Figure 3.3a depicts a simple circuit diagram of a source voltage V_s, a diode D that rectifies the current I_D, and a load R that absorbs the power delivered by the source.

In order to determine the steady-state characteristics of the diode, the source voltage V_s is adjusted and the diode voltage V_D is recorded along with the circuit current I_D. The plot of I_D versus V_D represents the I-V characteristic of the diode. Figure 3.3b illustrates the typical shape for a *pn* junction diode. The discussion that follows is an expansion of the one already made in section 1.5.

3.3.1. Forward Bias

Refer to Fig.3.3 and consider that the initial conditions represent the diode in the off-state. The voltage source is zero ($V_s = 0$) and the circuit current is zero ($I_D = 0$). For these conditions the space-charge potential barrier at the diode junction is about 0.6 V.

[1] The n^- region is called the drift region because it is principally the charge flow by drift due to the electric field of the forward bias, rather than diffusion, that gives rise to current in this region.

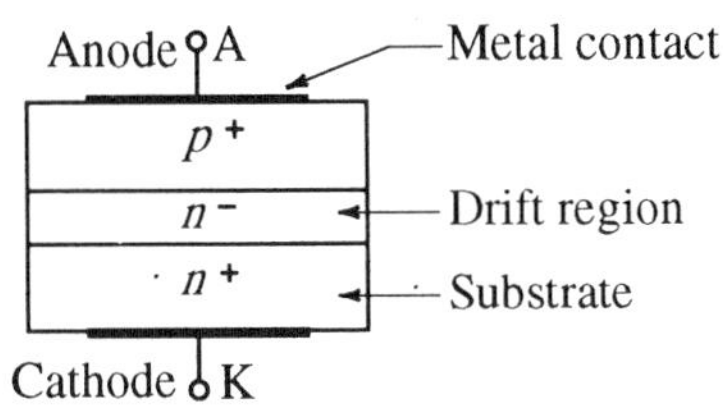

Fig. 3.2 *PIN* diode structure.

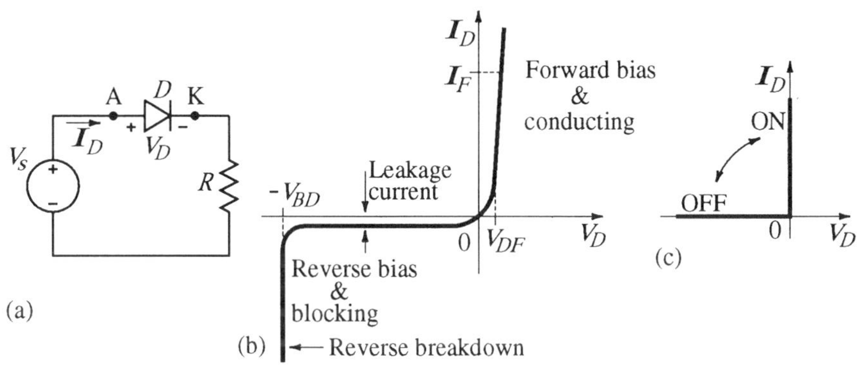

Fig. 3.3 Diode I-V characteristics.
(a) Test circuit, (b) practical case, (c) ideal case.

If the source voltage V_s is set to a small value such that the anode A is positive with respect to the cathode K, the diode is forward biased such that the p side of the junction is at a higher potential than the n side of the junction. Associated with the voltage V_{AK} there is an electric field between the metallized contacts of the electrodes. The influence of the electric field is to narrow the depletion layer and reduce the potential barrier. At the same time the electric field accelerates the majority carriers to an average velocity that allows diffusion across the reduced potential barrier. The result is a small forward current I_D and the diode is in the on-state.

As the source voltage V_s is slowly increased, the circuit current I_D increases with the diode voltage drop V_D in an exponential fashion until the potential barrier is reduced to zero. At this point $V_{AK} = V_D = V_{DF} \approx 0.7\text{V}$. Any further increase in the value of the source voltage V_s allows the possibility of a large forward current I_D, that depends mainly on the load resistance R and the source voltage V_s. The *pin* structure of the diode gives the on-state voltage drop V_D a linear characteristic once the potential barrier has been reduced to zero. The characteristic has the form shown in Fig. 3.3b and described by the equation

$$V_D = V_{DF} + R_D I_D \tag{3.3.1}$$

where R_D is the equivalent resistance of the diode.

As an example, typical values of a practical device might be $V_{DF} \approx 0.7\text{V}$ and $R_D \approx 1.0\text{m}\Omega$. In the steady state, R_D is virtually a constant, although it does reduce as the junction temperature increases. At $I_D = 100\text{A}$, the voltage drop across the diode would be $V_D = 0.8\text{V}$, whereas at the full-load value $I_D = I_F = 400\text{A}$, the voltage drop increases to $V_D = 1.1\text{V}$. In relation to source voltages greater that 100 V, this change of voltage drop is almost negligible.

While in the steady on-state there are conduction losses in the diode. The power dissipated is $P_D = V_D I_D$. Once the diode is on it remains in the on-state until the current I_D is reduced to zero by means of the voltages and currents in the rest of the circuit.

3.3.2. Reverse Bias

Again, consider the diode to be in the off-state. With reference to Fig. 3.3a, this means that $V_s = V_{AK} = V_D = 0, I_D = 0$ and the junction potential barrier is 0.6 V.

If the source voltage V_s is set to a value $-V_s$ such that the cathode K is positive with respect to the anode A, the diode is reverse biased such that the *n* side of the junction is at a higher potential than the *p* side of the junction. An electric field is created between the metallized contacts of the electrodes by the voltage difference. This electric field draws majority carriers away from the junction to widen the depletion layer and to increase the potential barrier. That is, the electric field draws electrons from the *n* region towards the cathode of positive polarity and draws holes from the *p* region towards the anode of negative polarity. Consequently, diffusion across the junction by majority carriers is suppressed and there is no current by majority carrier flow.

There is a small reverse current while the diode is reverse biased. There are always some minority carriers that are created by thermal ionization of the silicon atoms in each region. These minority carriers are accelerated across the junction, assisted by the electric field, and constitute a small leakage current in the reverse direction to that of the on-state current. The value of the leakage current is microamperes for diodes of low current ratings and is milliamperes for diodes of high current ratings. The leakage current does not vary much with the reverse voltage $-V_D$, but it does vary exponentially with temperature. The I-V characteristic for reverse bias is shown in Fig. 3.3b.

The value of the reverse voltage applied across a diode can be high but is limited. In practice it must be kept below the breakdown value $V_D = -V_{BD}$. If the reverse bias is too high, the electric field accelerates the minority carriers to such a velocity that, on collision with the lattice imperfections, the kinetic energy imparted to the lattice imperfections creates electron-hole pairs. These charges are accelerated by the same field and produce more collisions and more electron-hole pairs that contribute to the leakage current. This process, that is called impact ionization, is multiplicative. If the electric field is above a value of about 3×10^5 V/cm ($V_D = -V_{BD}$) the leakage current becomes high, due to the above process, and avalanche breakdown is said to have occurred. With the rapid increase of reverse current there is no reduction in the reverse voltage, so that the power dissipatión in the diode ($P_D = V_D I_D$) can become high enough to be thermally damaging.

3.3.3. Ideal Diode

How one considers the diode depends on the aspect of investigation. With respect to Fig. 3.3a, if the steady source voltage V_s is less than 20 V the detailed characteristics shown in Fig. 3.3b are of concern. If the diode heat sink is being designed, the practical characteristics are of concern. Otherwise, it is usual to consider all the elements in the circuit diagram of Fig. 3.3a to be ideal, and that includes the diode.

Consider the ideal case. Upon application of a forward bias to the diode, the potential difference between the anode and cathode creates an electric field. The electric field acts on the mobile charges[2] to accelerate them to some average drift velocity. This constitutes the circuit current I_D. The diode is on. Between the terminals A and K the diode looks like a short circuit. It has zero resistance to current conduction so there is no voltage drop ($V_D = 0$). The value of the current I_D depends entirely upon the source voltage and the other circuit parameters. Once the diode conducts, it is on until the forward bias is removed and the current is reduced to zero.

Upon the application of a reverse bias to the diode, the electric field forces the majority carriers away from the junction. The depletion layer widens, blocks current and allows the junction to withstand the applied voltage. Ideally, there are no minority carriers. Therefore, the diode is off. Between the terminals A and K, the diode looks like an open circuit, there is no current, and the voltage V_{AK} depends on the other circuit elements. Once the diode is off, it remains off until a forward bias is reapplied.

Figure 3.3c illustrates the ideal characteristics of the uncontrolled diode switch. While on, the diode has no voltage drop across it. While off, the diode conducts no current. In the case of a general circuit analysis to determine the steady-state or dynamic performance of the load, the use of the ideal characteristics of the diode is satisfactory as long as the source voltage V_s is greater than about 100 V. For low voltages or for detailed studies of the protection of the circuit elements, the ideal characteristics of the diode are less than satisfactory.

---

[2] The Lorentz force equation states that the force F of electric origin acting on a charge q in an electric field E is $F = q E$

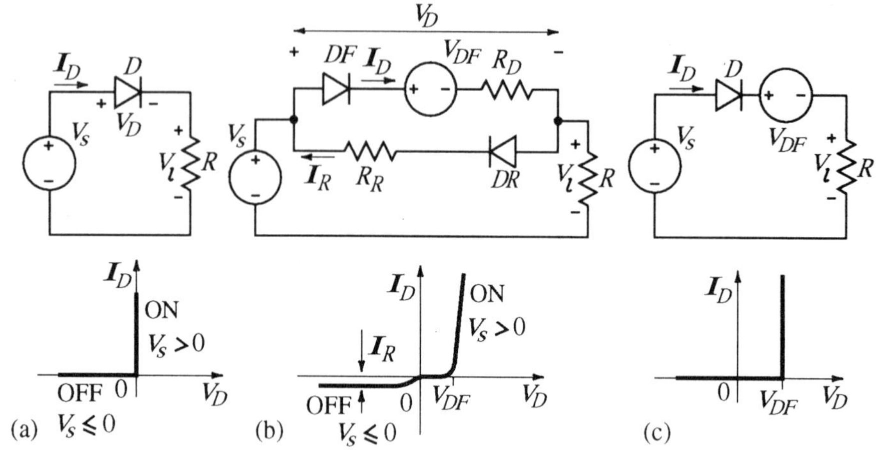

Fig. 3.4 Diode models.
(a) Ideal representation, (b) practical equivalent circuit, (c) simplified circuit.

3.4. DIODE MODELS

Since the diode's *pn* junction is the building block of semiconductor switches, it is not possible to model the junction by a diode except in the ideal case, where the *pn* junction looks like a short circuit while it is in the on-state and looks like an open circuit while it is in the off-state.

The ideal model is represented by the diode symbol in Fig. 3.4a. All the elements are considered ideal in the circuit of this figure. If the source voltage is $V_s > 0$, the circuit response is that the diode voltage drop V_D, the voltage across the load V_l and the circuit current I_D are

$V_D = 0$, $V_l = V_s$ and $I_D = V_l/R$.

The other condition is that the source voltage is $V_s \leq 0$. For this, the circuit response is

$V_D = -V_s$, $V_l = 0$ and $I_D = 0$.

The I-V characteristics for these circuit responses are shown to lie on the axes in Fig. 3.4a.

We will confine the modelling of a practical diode to equivalent circuits that comprise ideal diodes and circuit elements to approximate the characteristics shown in Fig. 3.3b. The circuit diagram in Fig. 3.4b shows the source voltage V_s and the load resistance R. The ideal elements that comprise the rest of the circuit represent an approximate equivalent circuit of the diode for both forward and reverse bias. For forward bias, with the supply voltage $V_s > 0$, the real diode has its *p* region positive and its *n* region negative. The ideal diode *DR* blocks current so $I_R = 0$. However, the ideal diode *DF* is in the on-state and offers no resistance to the forward current I_D. The forward characteristic of the real diode is

$$V_D = V_{DF} + R_D I_D . \tag{3.4.1}$$

For any forward current I_D, the voltage drop V_D can be calculated, if the diode parameters V_{DF} and R_D have been determined.

For reverse bias, with the supply voltage $V_s < 0$, the ideal diode *DF* blocks in order to satisfy the requirement that the forward current I_D is zero. However, the ideal diode *DR* is in the on-state, because it is forward biased; so it offers no resistance to the reverse leakage current I_R of the real diode. A high-value resistance R_R models the practical reverse leakage of the diode. Leakage current increases with temperature, so the use of R_R may become important at high temperatures.

Further approximations can be made to the practical equivalent circuit. For example, if $R_D \approx 0$ and $R_R \approx \infty$, a simpler circuit is derived. This is depicted in Fig. 3.4c. The only parameter is V_{DF}, the forward voltage drop. If this value cannot be obtained from the diode data sheet, it can be assumed to be about 1V.

Which of these steady-state models is used depends on the kind of analysis that is required. If the source voltage is high ($V_s > 100$ V), if the switching frequency is low ($f \le 60$ Hz) and if the load responses (V_l, I_D) are required, then the ideal model is satisfactory. For medium voltages ($V_s < 100$ V), for low switching frequencies and for load analysis, the approximate equivalent circuit, shown in Fig. 3.4c, can be used. The equivalent circuit shown in Fig. 3.4b is reserved normally for cases of low supply voltages ($V_s < 20$ V) and low frequency switching, if load analysis or the determination of diode conduction losses for thermal considerations is required.

EXAMPLE 3.1

While being reverse biased by a source of 1000 V the reverse leakage current of a diode is measured to be 1 mA at a junction temperature of 200°C. The load is a 10 Ω resistor. Calculate the parameters of the diode in reverse bias and the power dissipated by the diode.

Solution

The equivalent circuit model shown in Fig. 3.3b can be used.
$V_s = -1000$ V, $I_D = -1.0$ mA, $R = 10\,\Omega$. *DF* is an open circuit. *DR* is a short circuit.
The input or driving point resistance R_i of the circuit under these conditions is

$$R_i = \frac{V_s}{I_D} = \frac{-1000}{-10^{-3}} = 1.0\,\mathrm{M\Omega}.$$

Also, $R_i = R + R_R$.
Consequently the diode parameter $R_R = R_i - R \approx 1.0\mathrm{M\Omega}$ at 200°C.
The power dissipated in the diode is $P_D = I_D^2 R_R = 10^{-6} \times 10^6 = 1$ W.
This value is small compared with the conduction loss.
In the forward direction the conduction loss is
$P_D = V_D I_D \approx V_D \times V_s / R \approx 1 \times 1000/10 = 100$ W.

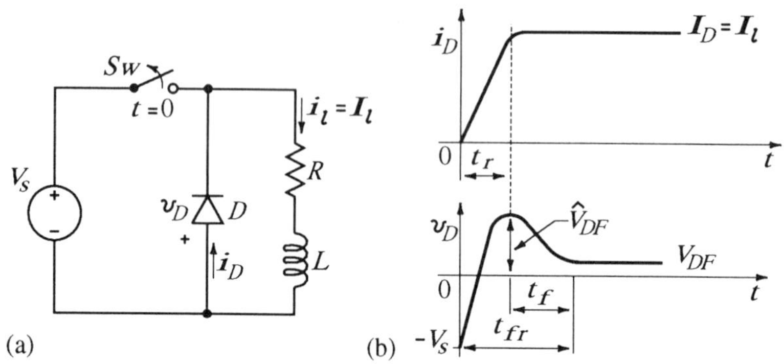

Fig. 3.5 Turn-on characteristics.
(a) Circuit, (b) waveforms.

3.5. DIODE TURN-ON

The diode is an uncontrolled switch. Consider that it is off, and that a forward-bias voltage is applied across the anode and cathode. The diode turns on. However, the diode is a charge device and this means that it takes time to change states from off to on. Conduction occurs because the forward bias causes the depletion layer to disappear. Since the p side of the junction is positive with respect to the n side, the resulting electric field can accelerate majority carriers that constitute the forward current. When all majority carriers crossing over the junction contribute to conduction, the diode is turned on.

The turn-on transient can be explained with the help of Fig. 3.5. The circuit diagram in Fig. 3.5a indicates that the load current $i_l = I_l$ is constant. This is a reasonable assumption, if the load time constant L/R is long compared with the time of turn-on t_{fr} (forward recovery). For the time $t < 0$, the switch Sw is closed. Steady conditions prevail and the diode D is reverse biased at $-V_s$. It is in the off-state, and $i_D = 0$.

At $t = 0$, the switch Sw is opened. The diode becomes forward biased, provides a path for the load current in R and L, so that the diode current i_D rises to I_l after a short time t_r (rise time) and the diode voltage drop falls to its steady value after a further time t_f (fall time). This is shown in Fig. 3.5b. The diode turn-on time is the time t_{fr}, that comprises $t_r + t_f$. It takes this time t_{fr} for charge to change from one equilibrium state (off) to the other (on).

From the time $t = 0$ to the time $t = t_r$, the tendency for the inductance L to maintain its current I_l constant as the switch opens reduces the voltage across the diode from $-V_s$ and forces the current in it to rise at some rate di_D/dt. During this time the depletion-layer space charge is discharged from its reverse-bias level to its steady thermal equilibrium level. As the diode becomes forward biased, majority carriers are accelerated across the junction. While the current is rising and the depletion layer is becoming narrow, there is no conductivity modulation. That is, the resistance of the semiconductor is relatively high. This means that the voltage drop $i_D R_D$ is high. To this voltage drop is added the $L_s di_D/dt$ voltage drop due to the stray inductance of the device and leads. The total drop v_D reaches a peak forward value $\hat{V}_{DF}$ that may be from 5 to 20 V, a value much greater than the steady value $V_{DF} \approx 1$ V. The time t_r for the voltage to reach $\hat{V}_{DF}$ is usually about 0.1 μs.

At a time $t > t_r$, the current i_D becomes constant at I_l, so there is no $L_s di_D/dt$ component to the voltage drop v_D. Further, conductivity modulation takes place because the growth of excess carriers in the semiconductor is accompanied by a reduction of resistance, especially in the n^- drift region. Consequently, the $i_D R_D$ voltage drop reduces. In the equilibrium state, that may take a time $t_f \approx 1$ μs, with a uniform distribution of excess carriers, the voltage drop v_D is at its minimum steady-state value V_{DF}.

During the turn-on interval t_r, the current is not uniformly distributed, so the current density can be high enough in some parts to cause hot spots and possible failure. Accordingly, the rate of rise of current di_D/dt should be limited until the conduction spreads uniformly and the current density decreases.

Associated with the high voltage $\hat{V}_{DF}$ at turn-on, there is high current, so there is extra power dissipation that is not evident from the steady-state model. The turn-on time is about 1 μs. This means that the power dissipation is not significant compared with the steady on-state conduction losses unless the switching frequency f is high ($f > 50$ kHz). For high switching frequencies, diodes are manufactured to have shorter carrier life times to reduce the interval t_r of the forward recovery time at turn-on. Some device design compromise has to be made, because reduced lifetimes mean that the on-state voltage drop is increased and this causes higher conduction losses.

3.6. DIODE TURN-OFF

If the diode is conducting and if a reverse bias voltage is applied to its terminals, the diode turns off as soon as the forward current is reduced to zero. Once off, the diode continues to block further conduction until a forward bias is reapplied. This is the action of an ideal diode and the transition between the on-state and the off-state at zero current takes no time.

In practice, it takes time to turn the diode off from the conducting state to the blocking state. That time ranges from tens of nanoseconds to a few microseconds and depends on how the device was manufactured. We can describe the action of turn-off by using the information in section 1.5 on the *pn* junction and by using the circuit illustrated in Fig. 3.6a. Except for the diode, the circuit elements of this simple chopper are considered to be ideal. Switching *Sw* at a regular frequency, the source of constant voltage V_s maintains a constant current I_l in the *RL* load, because it is assumed that the load time constant L/R is long compared with the period of the switching.

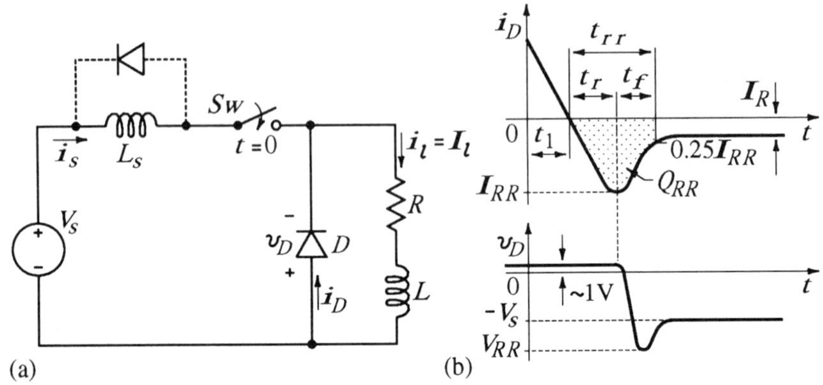

Fig. 3.6 Diode turn-off.
(a) Chopper circuit, (b) waveforms.

This is a good test circuit because it puts the semiconductor device under high stress with respect to voltage and current. While the switch *Sw* is closed, the load is being charged and the diode should be reverse biased. While the switch *Sw* is open, the diode *D* provides a freewheeling path for the load current I_l. The inductance L_s is included for practical reasons and may be the lumped source inductance and snubber inductance, that should have a freewheeling diode to suppress high voltages when the switch is opened. Let us consider that steady conditions prevail. At the time $t=0^-$ the switch *Sw* is open, the load current is $i_l=I_l$, the diode current is $i_D=I_D=I_l$, and the voltage drop v_D across the diode is small (about 1 V).

The important concern is what happens after the switch is closed at $t=0$. Figure 3.6b depicts the waveforms of the diode current i_D and voltage v_D. At $t=0^-$ there was the excess charge carrier distribution of conduction in the diode. This distribution cannot change instantaneously so at $t=0^+$ the diode still looks like a virtual short circuit, with $v_D \approx 1$ V. Kirchhoff's current law provides us with the relation

$$i_D = I_l - i_s \tag{3.6.1}$$

and Kirchhoff's voltage law yields

$$V_s = L_s \frac{di_s}{dt} = L_s \frac{d}{dt}(I_l - i_D) = -L_s \frac{di_D}{dt} \,. \tag{3.6.2}$$

Accordingly, the diode current changes at the rate

$$\frac{di_D}{dt} = -\frac{V_s}{L_s} = \text{constant.} \tag{3.6.3}$$

This means that it takes a time $t_1 = L_s I_l / V_s$ seconds for the diode current to fall to zero.

At the time $t = t_1$ the current i_D is zero, but up to this point the majority carriers have been crossing the junction to become minority carriers, so the *pn* junction cannot assume a blocking condition until these carriers have been removed. At zero current, the diode is still a short circuit to the source voltage. Equations (3.6.1) to (3.6.3) still apply and the current i_D is forced to go negative at the same rate while i_s rises above I_l at the same rate. The diode voltage v_D changes little while the excess carriers remain. The diode reverse current rises over a time t_r during which the excess charge carriers are swept out of the region.

At the end of the interval t_r the reverse current i_D can have risen to a substantial value I_{RR} (peak **R**everse **R**ecovery current), but, by this time, sufficient carriers have been swept out and recombined that current cannot be supported. Therefore, over a **f**all-time interval t_f the diode current i_D reduces to almost zero very rapidly while the remaining excess carriers are swept out or recombined.

It is during the interval t_f that the potential barrier begins to increase both to block the reverse bias voltage applied by the source voltage as i_D reduces, and to suppress the diffusion of majority carriers because the excess carrier density at the junction is zero. The reverse voltage creates the electric field that allows the depletion layer to acquire space charge and widen. That is, the electric field causes electrons in the *n* region to be forced away from the junction towards the cathode and causes holes to be forced away from the junction towards the anode. The blocking voltage v_D can rise above the voltage V_s transiently because of the additional voltage $L_s di_D/dt$ as i_D falls to zero over the time t_f.

The sum of the intervals $t_r + t_f = t_{rr}$ is known as the **r**everse **r**ecovery time, and is usually less than 1 microsecond. This time is also known as the storage time because it is the time that is taken to sweep out the excess charge Q_{RR} from the silicon by the reverse current. Q_{RR} is a function of $I_D = I_l$, di_D/dt and the junction temperature. It has an effect on the reverse recovery current I_{RR} and the reverse recovery time t_{rr}, so it is usually quoted in the data sheets. The fall time t_f can be influenced by the design of the diode. It would seem reasonable to make it short to decrease the turn-off time, but the process is expensive. The bulk of the silicon can be doped with gold or platinum to reduce carrier lifetimes and hence to reduce t_f. The advantage is an increased frequency of switching. There are two disadvantages associated with this gain in performance. One is an increased on-state voltage drop and the other is an increased voltage recovery overshoot V_{RR}, that is caused by the increased $L_s di_s/dt$ as i_D falls more quickly. The measure of the speed of recovery is the ratio $t_f/t_r = S$, where S is referred to as the *softness factor* or the *snappiness*. Softness ($S \approx 1$) signifies a slow recovery, whereas snappiness ($S \ll 1$) indicates an abrupt reduction of current to near zero together with a high voltage spike. Since there is a voltage drop associated with current during the diode turn-off process, there is power dissipation which can be as high as

50mW for very high voltages and high rates of change of current.

With sweep out and recombination accomplished, there are no majority carriers for conduction. However, thermal ionization produces minority carriers in both the *n* regions. The electric field, that is in existence because of the reverse bias voltage, causes a flow of these minority carriers across the junction. The electrons in the *p* region have a force on them ($\mathbf{F}_e = q\mathbf{E}$) attracting them to the cathode and the holes in the *n* region are attracted equally to the anode. This drift constitutes a small current I_R that is known as the **R**everse leakage current. It exists as long as the reverse bias voltage is applied.

The reverse-bias voltage across the diode must be limited below the breakdown value V_{BD}. If the reverse bias is increased too much, the minority carriers can be accelerated to such a high kinetic energy, that, upon collision with an impurity, the bonding of another minority carrier is broken. This new minority carrier is also accelerated and upon collision frees yet another. This *impact ionization* is multiplicative with the result that there is *avalanche breakdown*.[3] That is, there is a sharp increase in current without a decrease in voltage. For power diodes, care is taken to avoid this because the power dissipation would cause failure.

EXAMPLE 3.2

Consider the chopper circuit illustrated in Fig. 3.6. The switch *Sw* modulates power from the dc source of voltage $V_s = 600$ V to the *RL* load such that the current is $i_l = I_l = 100$ A and without ripple. The source inductance is $L_s = 30\,\mu$H. The diode chosen has the given specifications 1000 V, 100 A at 150°C, a reverse recovery time of $t_{rr} = 1\,\mu$s for this circuit application and a softness factor $S = 0.6$. Find the charge storage Q_{RR} at the beginning of each turn-off action and the peak reverse current I_{RR} during turn-off. Estimate the reverse voltage spike at turn-off.

Solution

Under steady conditions the switch *Sw* is closed at $t = 0$ and the current i_D in the diode falls linearly to the negative value I_{RR} under the influence of the inductance L_s. Let the current waveshape under the abscissa axis in Fig. 3.6b be approximated to a triangle. The area enclosed is equivalent to Q_{RR}, the stored charge swept out by the reverse current. Therefore,

$$Q_{RR} = I_{RR} \times t_{rr}/2 \text{ and } I_{RR} = t_r \times \frac{di_D}{dt} = \frac{t_{rr}}{1+S} \times \frac{di_D}{dt}, \text{ where } S = \frac{t_f}{t_r} = 0.6.$$

While the diode is turning off $V_s = -L_s di_D/dt$.

So, $di_D/dt = -600/(30 \times 10^{-6}) = -20$ A/μs.

$$\text{Hence, } I_{RR} = \frac{t_{rr}}{1+S} \times \frac{di_D}{dt} = \frac{1 \times 10^{-6}}{1.6} \times 20 \times 10^6 = 12.5 \text{ A}.$$

This is an overshoot of 12.5%.

$Q_{RR} = I_{RR} \times t_{rr}/2 = 12.5 \times 1 \times 10^{-6}/2 = 6.25\,\mu$C.

At the point the reverse current reaches its maximum value the diode blocks and

---

[3] This is also known as the Zener effect. In some cases this is used to maintain a constant voltage.

its voltage is
$V_{RR} \approx V_s + L_s di_D/dt \approx V_s + L_s I_{RR}/t_f$.
So, $V_{RR} = 600 + 30\times10^{-6} \times 12.5/0.375\times10^{-6} = 1600$ V.
This is high. The source inductance L_s is the controlling factor.

3.7. DIODE PROTECTION

The diode has no low-power gate circuits like the thyristor, so it is less fragile. Nonetheless, since the power diode is most often a *pn* junction, thought has to be given to protection against overcurrent, overvoltage and transients.

3.7.1. Overcurrent

Diodes have steady-state current ratings as average values, rms values, and pulse values. These values are associated with the maximum steady temperature of operation produced by the conduction losses in the device. Because of unknown contingencies such as transient circuit faults it is advisable to operate below the rated value.

Often a diode is a subsidiary element in a controlled switching circuit. For such a case, the protection of the controlled switch covers the diode overcurrent protection and nothing more need be done. If the diode is the only semiconductor in the circuit, then special fuses are used to interrupt the diode current before it rises to a value that the temperature increases to a dangerous level. Care has to be taken to match the i^2t ratings of the fuse and the diode.

The maximum current in a diode depends on the device power dissipation that sets the junction temperature to its maximum operating value. How much power can be dissipated depends on how quickly the thermal energy, that is generated by the current and the equivalent resistance, can be transferred to the ambient. This is no different from the thermal considerations of any semiconductor switch, and becomes a compromise between the choice of diode and the choice of heatsink. What has to be taken into account is the junction operating temperature T_J, the diode's on-state voltage drop V_D, which is a function of the forward current I_D, the diode's (junction to case) thermal resistance $R_{\theta JC}$ and the characteristics of the heatsink and the bonding between the diode and the heatsink. More is written on this subject in Chapter 5, The Thyristor.

EXAMPLE 3.3

A diode has ratings of 1000 V, 200 A and the maximum junction temperature is not to exceed 180°C. The data sheet provides the information that the on-state voltage drop is 1.8 V at rated current and the thermal resistance, junction to case, is $R_{\theta JC} = 0.08$°C/W. For a given heatsink the thermal resistance, sink to ambient, is $R_{\theta SA} = 0.2$°C/W and it is known that for this arrangement the thermal resistance, case to sink, $R_{\theta CS}$ can be kept below 0.03°C/W. Estimate the maximum direct current that the diode can tolerate safely if the ambient temperature is maintained at 30°C.

Solution

Since we do not have the data sheets with curves of voltage drop, current and power dissipation, we have to make an estimation. If the current is to be near the full-load value, we assume that the on-state voltage drop is close to 1.8 V. The steady power dissipation P_D from the diode to the ambient is given by the temperature difference divided by the total thermal resistance.
In general, $P_D = \Delta T / R_\theta$
where P_D is the power dissipated, ΔT is the temperature difference between two points and R_θ is the thermal resistance between the same two points.
This is like Ohm's law ($I = V/R$) for thermodynamics.
In this case, $P_D = V_D I_D = (T_J - T_A) / (R_{\theta JC} + R_{\theta CS} + R_{\theta SA})$
where T_J is the diode temperature at its junction and T_A is the ambient temperature.
So, $I_D = (T_J - T_A) / [(R_{\theta JC} + R_{\theta CS} + R_{\theta SA}) V_D]$.
That is, $I_D = (180 - 30) / [(0.08 + 0.03 + 0.2) \times 1.8] = 233\text{A}$.
The estimate is reasonable, since the maximum value of the current for the given conditions is close to the specified, full-load value, and the voltage drop will be close to 1.8 V.

3.7.2. Overvoltage

In the forward direction, with the anode positive, voltage poses no problem for the diode. It turns on if it is positively biased. Now, the voltage across the diode is low and, if there is a problem, it is one relating to the resulting current at turn-on. Once the diode turns on, the voltage, that turned it on, will appear across another element.

Overvoltage for a diode that is reverse biased is a problem to be avoided at the design stage of the circuit. An overvoltage in this case signifies that the breakover voltage has been exceeded. Avalanche breakdown will result so that a large current could be conducted by the diode while it had a large voltage drop across it. Once this happens it is likely that the device will be destroyed thermally by the high power dissipation at the junction. It is usual to choose a diode voltage rating 20 percent higher than the circuit will impose on the diode during normal operating conditions. Then, protection circuitry can be added, generally in the form of snubber circuits, to suppress transients that can give rise to higher voltages.

3.7.3. Transients

To demonstrate what transients exist in a diode circuit and how they are suppressed, use can be made of the chopper circuit diagram in Fig. 3.7a. This has been referred to in the turn-on and turn-off sections 3.5 and 3.6 respectively.

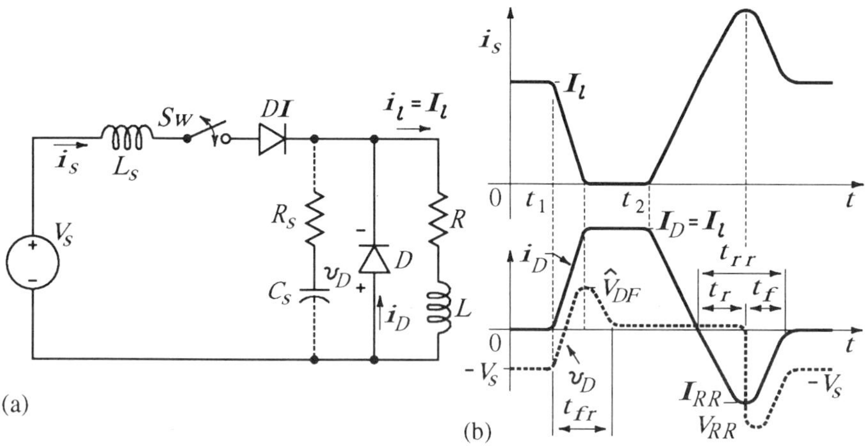

Fig. 3.7 Transients in a chopper.
(a) Circuit diagram, (b) waveforms.

All the elements in the circuit can be treated as ideal except the freewheeling diode D, until it is convenient to relax this constraint. The source is a constant voltage V_s. Initially, L_s is treated as stray inductance. The diode *DI* isolates the source from the load in reverse bias. The load is assumed to be sufficiently inductive and the switching frequency is assumed to be sufficiently high that the load current is ripple free. This isolates the effects of the switching transients and makes them clearer to separate mathematically. The freewheeling diode D, that provides a path for the load current while the chopper switch *Sw* is open, is treated in practical terms. The switch is treated as quasi-ideal for convenience. Elements R_s and C_s are not connected at this stage.

In steady-state operation, the switch *Sw* is opened at time $t = t_1$ and closed at time $t = t_2$. At $t = t_1$ the diode begins to change state from off to on, the current i_D rises from zero to I_l, during which time the voltage v_D rises (up to 20 V) due to the resistance of the junction and the stray inductance of the diode branch. See Fig. 3.7b. This transient interval t_{fr} causes substantial power dissipation in the diode because both the voltage and current of the device are high at the same time. If the frequency of switching is high this power dissipation has to be accommodated in the heatsink design. Otherwise, the diode might overheat.

At time $t = t_2$ the switch *Sw* is closed and the diode starts the transition from the on-state to the off-state. The rate of decrease of current i_D is governed by the source voltage V_s and stray inductance L_s until most of the carriers of charge storage are swept out of the *pn* junction. There is a negative recovery current up to I_{RR} during the sweep-out. This current is reflected in the source current i_s that diode *DI* conducts. Here, there are two dangers. The rate of rise of source current i_s is the same as the rate of decay of the diode current. A practical diode *DI* does not turn on with uniform carrier distribution instantly, so the rate of rise of current must be limited to keep the localized current densities within bounds of thermally acceptable values. A high rate of decay of diode D current means that the value of the peak reverse recovery current I_{RR} will be high. A high value of I_{RR} indicates a high power dissipation in the diode D during turn-off, especially in fast turn-off devices where the reverse recovery voltage V_{RR} during charge recombination can also be high. The way to overcome both dangers by limiting the rate of change of currents is to add inductance to L_s (series *snubber*). A series snubber inductance L_s reduces the di_s/dt and possible turn-on hotspots in diode DI. It also reduces di_D/dt and as a result the peak value of the reverse recovery current is reduced (see EXAMPLE 3.2).

Adding a series snubber element L_s creates a need for further protection. At the time the current in the diode DI reaches its peak value, $I_l + I_{RR}$, diode D turns off. This turn-off can be abrupt if t_f is short and the softness factor S is low. Consequently, the current in the diode DI has to reduce from its peak value to its steady value I_l with the same abruptness. During this time t_f the energy in the inductance L_s has to reduce. All this gives rise to an additional voltage $L_s di_s/dt$ that appears in reverse across the diode D. The total peak voltage V_{RR} across the diode D can be greater that the avalanche breakdown value, with resulting damage. To prevent this, protection is afforded by a freewheeling diode across the inductance L_s, if it is a snubber inductance. If the inductance L_s is stray inductance, a freewheeling diode cannot be connected across something that is distributed. Instead, a parallel snubber[4] capacitor C_s can be used to absorb the inductive energy and keep the diode reverse voltage within safe bounds. Some resistance R_s is added in series with the capacitor in order to damp any oscillations between capacitance and inductance in the circuit.

EXAMPLE 3.4

Consider the chopper circuit shown in Fig. 3.7a. Its dc voltage is $V_s = 500$V, and its load resistance of 0.1 Ω is to absorb an average power of 16kW with a virtually ripple-free current. The data sheet for the diode D provides the following information for these particular operating conditions.

Ratings 600 V, 200 A; reverse recovery time $t_{rr} = 1$ µs at $di_D/dt = -50$ A/µs with a softness factor $S = 0.5$.

Determine a suitable value for a snubber capacitor C_s in order to prevent the reverse-recovery voltage across the diode D being greater than 600 V.

Solution

Refer to Fig. 3.7 for the notation used in the solution. The capacitor C_s must absorb the excess energy stored in the stray inductance L_s of the circuit while the diode D is undergoing the turn-off process, that is, while the diode DI current falls

[4] Such storage elements as L and C are connected in the circuit to snub, or reduce the rates of change of current and voltage.

from $(I_l + I_{RR})$ to I_l in time t_f.
Accordingly, $C_s V_{RR}^2/2 = L_s[(I_l + I_{RR})^2 - I_l^2]/2$.
The inductance L_s is calculated from $V_s = L_s di_s/dt - v_D$, which is applicable during the diode D turn-off interval.
Since $i_s + i_D = I_l$ and $v_D \approx 0$,
$L_s = -V_s/(di_D/dt) = 500/(50\times 10^6) = 10\,\mu$H.
The load current is determined from $P_l = I_l^2 R$.
That is, $I_l = (P_l/R)^{1/2} = (16\times 10^3/0.1)^{1/2} = 400$ A.
From the triangular turn-off waveshape for the current of the diode D,
$I_{RR} = t_r \times di_D/dt = t_{rr}\times(di_D/dt)/(1+S)$, where $S = t_f/t_r = 0.5$.
Consequently, the absolute value of the reverse recovery current is
$I_{RR} = 1\times 10^{-6}\times 50\times 10^6/1.5 = 33.3$ A.
Upon substitution in the energy equation,
$0.5\times C_s \times 600^2 = 0.5\times 10\times 10^{-6}\,[(400+33.3)^2 - 400^2]$ and $C_s = 0.77\,\mu$F.

If there were no snubber capacitor, over the interval t_f the reverse recovery voltage V_{RR} could reach a value given by
$V_{RR} = V_s - L_s di_s/dt \approx V_s + L_s I_{RR}/t_f = V_s + L_s I_{RR}(1+S)/(St_{rr})$.
So, $V_{RR} = 500 + 10\times 10^{-6}\times 75\times 1.5/(0.5\times 10^{-6}) = 2750$ V.
This could cause the diode D to fail.

Note that the freewheeling diode has a current rating that is a fraction of the load rating.

3.8. DIODE RATINGS APPLICATIONS AND ANALYSIS

The applications of power diodes are where forward current is to be unimpeded, where reverse current is to be blocked and where there is to be no controlled modulation of power.

Principally, the main application of diodes is for rectifiers, the uncontrolled conversion of ac to dc power at a fixed voltage. Another is for freewheeling, the creation of a path for inductive current to prevent high induced load voltages at times that switches are opened in the circuit. Freewheeling in chopper circuits was used in the previous examples. The performance of rectification will be discussed briefly in this section.

For high-voltage, high-power applications, the *pn* junction diode is used exclusively. At reverse voltages less than 100V, and currents less than 300A, another kind of diode is better. It is the Schottky barrier diode, which has an *n*-type silicon-to-metal contact that forms the potential barrier. The reasons why it is better are that the on-state voltage drop is about 0.6V, and turn-off is much faster because there are no minority carriers to be recombined. The frequency of switching can be high, but it must be noted that the reverse leakage current is high in comparison with *pn* junctions. Accordingly, Schottky diodes are used in low-voltage rectifier circuits where improvements in rectifier performance are important, as in dc power supplies.

3.8.1. Ratings

Typical ratings of a Schottky diode are as follows.

Peak reverse voltage, 40 V;
maximum average forward current, $I_F = 15$ A;
peak operating junction temperature, $T_J = 150$°C;
thermal resistance, junction to ambient, $R_{\theta JA} = 5$°C/W;
maximum on-state voltage, $V_D = 0.5$ V;
maximum reverse leakage current, $I_R = 70$ mA.

A general classification divides the ratings of *pn*-junction diodes into two groups. One is for general purpose and the other is a fast recovery group. The general-purpose diode has a reverse recovery time greater that 5 μs (and as high as 50 μs) that makes it suitable for applications where the switching frequency is less than 1kHz such as line-commutated converters for dc drives. The fast-recovery diode has a reverse recovery time less than 5 μs (and as low as 50 ns). This type is used in applications, where the frequency of operation is greater than 1 kHz, such as choppers and inverters. For both groups, the higher is the voltage rating and the higher is the operating temperature, then the greater is the on-state voltage drop and the longer is the reverse recovery time. For example, for a 600 V, 300 A device with a maximum on-state voltage of 1.5 V, the reverse recovery time $t_{rr} = 1$μs at 25°C at the junction, whereas $t_{rr} = 2$μs at 125°C at the junction. For this same device the data sheet may provide the information that for a softness factor $S = 0.6$ and a $di_D/dt = -25$ A/μs the peak reverse recovery current is $I_{RR} = 30$ A, and in the steady off-state the leakage current $I_R = 35$ mA.

General purpose diodes can be found to have voltage ratings up to 5000 V and current ratings up to 3500 A. For fast recovery diodes, ratings of 3000 V and 1000 A are available. Current densities in both groups of diodes are usually higher than those for thyristors, because the peak operating temperatures can be 5°C to 75°C higher in diodes.

3.8.2. Rectification

Rectification, the uncontrolled conversion from ac to dc power, is one of the principal applications for *pn* diodes. Figure 3.8 depicts different types of ideal rectifier circuits with resistive loads, together with their waveforms and average output voltages. These configurations are similar to the converters described in section 2.3 of Chapter 2 with one exception. Diodes are used instead of controlled switches so that the trigger angle is $\alpha = 0$ always. The choice of the rectifier depends on the source and the load power requirements. For the lowest power a single diode might suffice, whereas for the highest power a three-phase arrangement is necessary.

An analysis of any of these configurations gives an approximate idea of the system performance on both the dc side and the ac side and gives a rough indication of the voltage and current ratings of the diodes. The calculations are simple and quickly done, so it can be quite usual to make a first analysis using ideal elements.

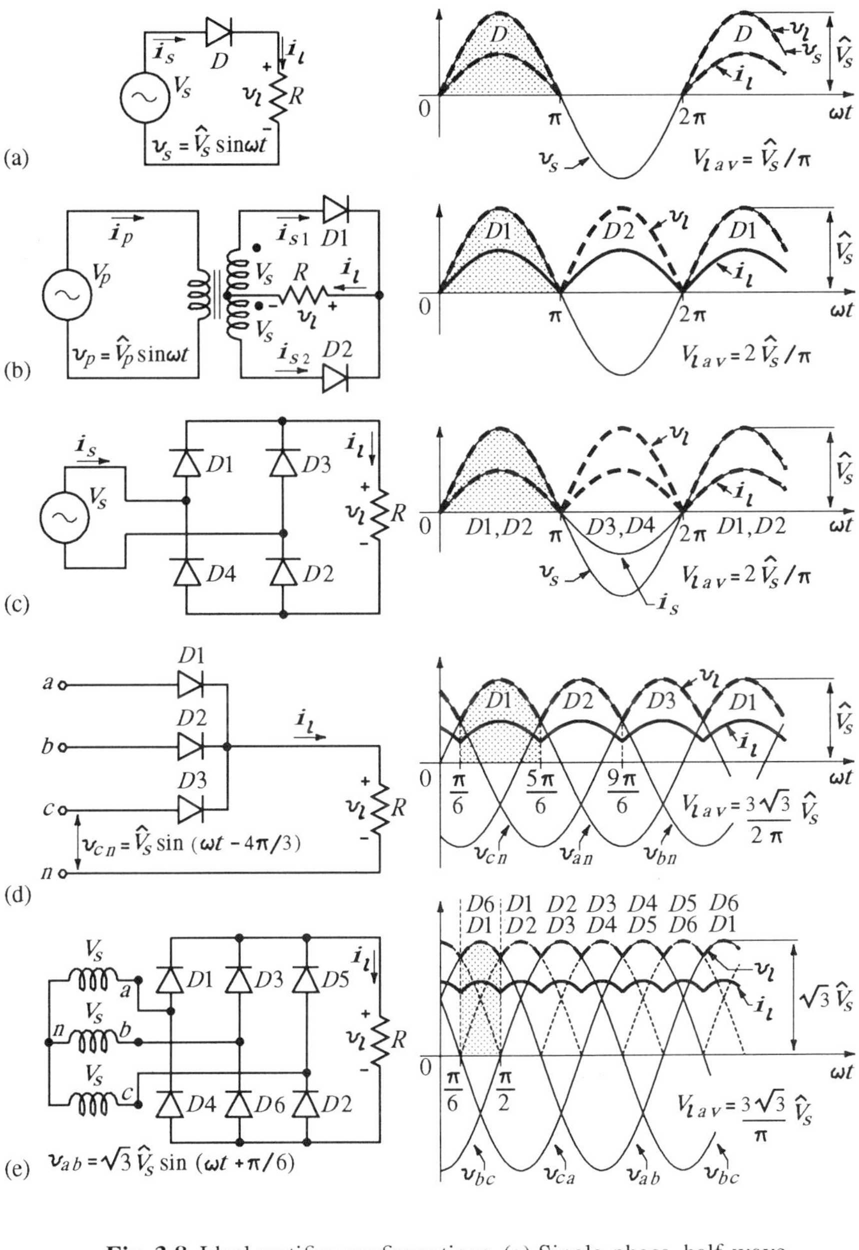

Fig. 3.8 Ideal rectifier configurations. (a) Single-phase, half-wave, (b) single-phase, full-wave, centre tap, (c) single-phase, full-wave bridge, (d) three-phase, half-wave bridge, (e) three-phase, full-wave bridge.

In summary, the output expressions for ac-dc converters using ideal diodes for rectification into a resistive load are as follows.

- For single-phase, half-wave rectification the average load voltage is

$$V_{l\,av} = \frac{\hat{V}_s}{\pi}. \tag{3.8.1}$$

- For single-phase, full-wave, centre-tap rectification the average load voltage is

$$V_{l\,av} = 2\frac{\hat{V}_s}{\pi}. \tag{3.8.2}$$

- For single-phase, full-wave bridge rectification the average load voltage is

$$V_{l\,av} = 2\frac{\hat{V}_s}{\pi}. \tag{3.8.3}$$

- For three-phase, half-wave bridge rectification the average load voltage is

$$V_{l\,av} = \frac{3\sqrt{3}}{2}\frac{\hat{V}_s}{\pi}. \tag{3.8.4}$$

- For three-phase, full-wave bridge rectification the average load voltage is

$$V_{l\,av} = 3\sqrt{3}\frac{\hat{V}_s}{\pi}. \tag{3.8.5}$$

- For all these rectifier configurations, since the load is resistive, the average load current is

$$I_{l\,av} = \frac{V_{l\,av}}{R}. \tag{3.8.6}$$

Output responses for one or two of these ideal configurations will be analyzed.

In reality, modelling of the rectifier and system has to be more detailed. The ac supply and transformer have impedance. The ac side of the rectifier is likely to have an *LC* filter to improve the actual waveshapes for a better quality of power. There are transients associated with diode switching and the diodes take time to change state. DC loads are usually inductive, so that, in some cases, freewheeling diodes are provided for protection. The output of a rectifier has neither constant current nor constant voltage, so filters on the dc side are used to help reduce the magnitude of this problem. If all these details are attended to, the system analysis can become quite complex. We will tackle some of the details in this section in examples. However, the total complex analysis is left to more specialized texts on the subject of rectification. See the bibliography at the end of the section.

Single-phase, half-wave rectifier.

The circuit depicted in Fig. 3.9 allows rectification of an ac supply to an *RL* load by a single diode. There is no freewheeling diode across the load, so that some of the energy stored in the inductance is returned to the source while the diode conducts and the supply voltage is negative, as shown in the waveforms. The current in the load is always discontinuous. All the elements are treated as ideal in order to simplify the determination of the performance of the circuit. Accordingly, the diode D turns on as soon as the supply voltage goes positive, and remains on until the current i_l becomes zero. Because of the delaying action of the inductance, which may, in part, be an added element for current filtering, the current extinguishes at an angle β that is greater than π radians. The voltage v_l is negative for $\pi \leq \omega t \leq \beta$.

For given values of the components, analysis on the dc side should be concerned primarily with the load power, current waveform, and average and rms values of the variables, voltage and current. The current response is

$$i_l = i_{ss} + i_{trans} \tag{3.8.7}$$

where i_{ss} is the ac steady-state component and i_{trans} is the transient component of the load current i_l. If Z is the impedance $(R^2 + \omega^2 L^2)^{1/2}$,

$$i_l = \frac{\hat{V}_s}{Z} \sin(\omega t - \theta) + A e^{-t/\tau}. \tag{3.8.8}$$

The time constant τ is L/R and the phase angle θ is $\tan^{-1} \omega L/R$. The constant A is determined from the initial conditions that $i_l = 0$ at $t = 0$. So, $A = \hat{V}_s \sin\theta / Z$ and

$$i_l = \frac{\hat{V}_s}{Z} \sin(\omega t - \theta) + \frac{\hat{V}_s}{Z} \sin\theta \, e^{-t/\tau}. \tag{3.8.9}$$

From this equation the performance characteristics of the circuit can be studied.

In order to determine the extinction angle β, the current $i_l = 0$ and the angle $\omega t = \beta$ have to be used in eq. (3.8.9). That is,

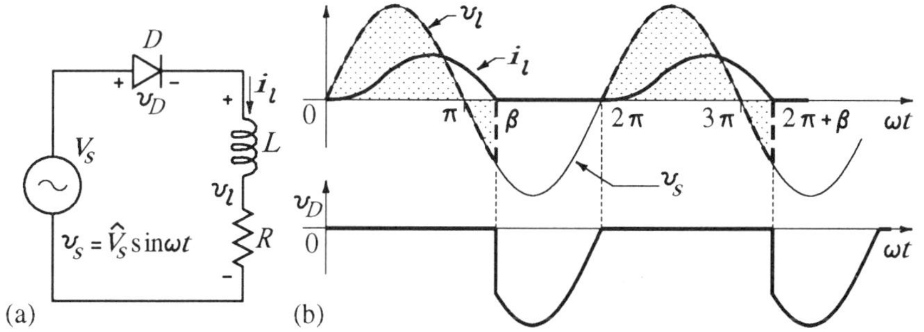

Fig. 3.9 Single-phase, half-wave rectifier.
(a) Circuit diagram, (b) waveforms.

$$0 = \frac{\hat{V}_s}{Z}\sin(\beta - \theta) + \frac{\hat{V}_s}{Z}\sin\theta\, e^{-\beta/\omega\tau}. \tag{3.8.10}$$

This is a transcendental equation in β and must be solved graphically (β versus L/R), or by trial and error calculations, or by numerical methods found in a software math package.

EXAMPLE 3.5

Consider the rectifier circuit shown in Fig. 3.9. If the source voltage is 240V at 50Hz and the load comprises resistance $R = 1\,\Omega$ and inductance $L = 3.18\,\text{mH}$, determine the current rating of the diode for this load. Assume that all elements in the circuit are ideal.

Solution

From eq. (3.8.9) the general current response is
$i_l = (\hat{V}_s/Z)\sin(\omega t - \theta) + (\hat{V}_s/Z)\sin\theta\, e^{-t/\tau}$.
Also, $\hat{V}_s = \sqrt{2} \times 240 = 339.4\,\text{V}$,
$Z = (R^2 + \omega^2 L^2)^{1/2} = (1^2 + 314^2 \times 0.00318^2)^{1/2} = 1.41\,\Omega$,
$\theta = \tan^{-1}\omega L/R = \tan^{-1} 1 = \pi/4\,\text{rad}$, $\tau = L/R = 3.18\,\text{ms}$ and $\omega\tau = 1\,\text{rad}$.
Consequently, $i_l = 240\sin(\omega t - \pi/4) + 169.7e^{-314t}\,\text{A}$.
$i_l = 0$ at $\omega t = \beta$. Upon substitution in the current equation,
$0 = \sin(\beta - \pi/4) + 0.707e^{-\beta}$.
The solution of this transcendental equation is $\beta = 3.94\,\text{rad}$.
The average voltage $V_{l\,av}$ across the load is

$$V_{l\,av} = \frac{1}{2\pi}\int_0^{\beta}\hat{V}_s\sin\omega t\, d\omega t = \frac{\hat{V}_s}{2\pi}[1 - \cos\beta] = \frac{339.4}{2\pi}[1 - \cos 3.94] = 91.7\,\text{V}.$$

Since the average voltage across the inductance L is zero,
$I_{l\,av} = V_{l\,av}/R = 91.7/1 = 91.7\,\text{A}$
The current rating of the diode is greater than 92A.

Voltage filter. Sometimes a capacitor is used as a filter to maintain a minimum change of voltage at the load of a rectifier. Capacitors are cheaper than inductors, especially at low frequencies. Figure 3.10 depicts a half-wave rectifier with a capacitive filter connected on the dc side. The diode conducts only as long as the load voltage v_l (capacitor voltage) is less than the supply voltage v_s, that is, over the conduction interval γ from α to β. During this interval the capacitor is being charged and the power absorbed by the load is delivered by the supply. For the rest of the cycle, the diode blocks and the capacitor delivers power to the load. To have a continuous load current, the load RC time constant must be long compared with the half period of the supply, as shown in Fig. 3.10b. The load current i_l follows the same pattern as the voltage v_l for a resistive load. Over each period of the supply voltage v_s, the charge Q absorbed and the charge delivered by the capacitor C are equal. That is,

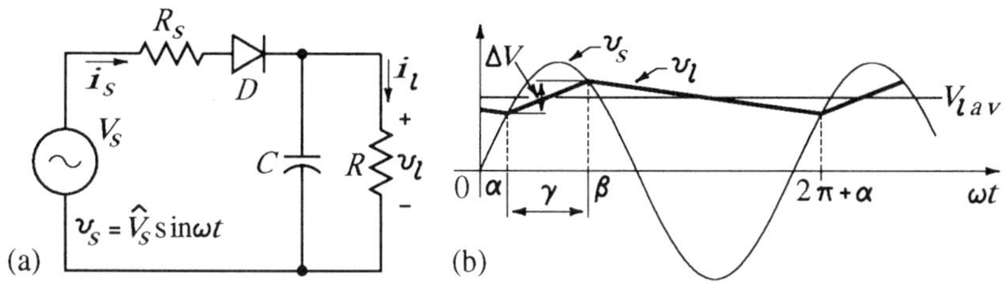

Fig. 3.10 Half-wave rectifier with capacitive filter.
(a) Circuit diagram, (b) waveshapes.

$$Q \times \omega = \int_{\alpha}^{\beta} (i_s - i_l)\, d\omega t = \int_{\beta}^{2\pi+\alpha} i_l\, d\omega t . \tag{3.8.11}$$

For the special cases that the source has a resistance R_s and the filter capacitance has a value $C \to \infty$, the load voltage is a constant V_l,

$$i_s = \frac{v_s - V_l}{R_s} \text{ for } \alpha \le \omega t \le \beta ,\ i_l = \frac{v_l}{R} ,\ V_l = \hat{V}_s \sin\alpha \tag{3.8.12}$$

$$\alpha = \frac{\pi - \gamma}{2} \text{ and } \beta = \frac{\pi + \gamma}{2} . \tag{3.8.13}$$

Upon substitution of these values in eq. (3.8.11)

$$\int_{\alpha}^{\beta} \left[\frac{\hat{V}_s}{R_s} \sin \omega t - \frac{V_l}{R_s} - \frac{V_l}{R} \right] d\omega t = \int_{\beta}^{2\pi+\alpha} \frac{V_l}{R}\, d\omega t . \tag{3.8.14}$$

Evaluation of these integrals provides the relation between the diode conduction angle γ and the circuit parameters

$$\tan\gamma/2 - \gamma/2 = R_s/R . \tag{3.8.15}$$

If the circuit parameters are known, this transcendental equation can be solved for γ, and the rectifier performance can be determined. For the condition that the filter capacitor C is large, eq. (3.8.15) holds true as an approximation. For the special case that the source resistance $R_s \to 0$, while the diode conducts $v_l \to v_s$, and $\beta \to \pi/2$; so, the circuit analysis has to be treated differently.

EXAMPLE 3.6.

A single-phase, half-wave, diode rectifier modulates power from a 240-V, 50-Hz, ac supply, whose source resistance is $R_s = 0.3\,\Omega$, to a resistive load of value $R = 10\,\Omega$. A capacitor, $C = 10{,}000\,\mu\text{F}$, is used to smooth the rectifier output. Estimate **(a)** the output voltage excursion ΔV (the peak-to-peak deviation from the average load voltage $V_{l\,av}$) and **(b)** the current rating of the diode.

Solution

We can refer to Fig. 3.10 for the circuit diagram and the waveforms. Since the value of the capacitor ($C = 10{,}000\,\mu F$) appears to be large and since the load time constant ($\tau = RC = 0.1$ s) is ten times the supply-voltage half period ($T/2 = 1/(2f) = 0.01$ s), in the first instance, we can assume that the load voltage v_l is almost constant. In this case, we can use eq. (3.8.15), that gives the diode conduction angle γ as a function of the resistance ratio R_s/R. That is,
$\tan\gamma/2 - \gamma/2 \approx R_s/R$. For $R_s/R = 0.03$ the solution is $\gamma \approx 0.87$ rad. (50°).
Still assuming that the load voltage is reasonably constant
$v_l \approx V_l \approx V_{l\,av} = \hat{V}_s \sin\alpha$, where $\alpha = (\pi - \gamma)/2 = 1.13$ rad. (65°).
So, $V_l = \sqrt{2} \times 240 \times \sin 1.13 = 307.6$ V.

(a) In the practical case, the capacitor charges over an interval of γ and discharges over the rest of the period $(2\pi - \gamma)$. The increase and decrease of charge are equal.

That is, $C\Delta V \approx \dfrac{V_l\,av}{R} \times \dfrac{2\pi - \gamma}{\omega} = \dfrac{307.6}{10} \times \dfrac{2\pi - 0.87}{314} = 0.53$ C.

Therefore, the voltage excursion $\Delta V \approx 0.53/C = 0.53/10^{-2} = 53$ V.
This is a considerable voltage ripple (± 26.5 V) and stretches the approximation of nearly constant load voltage. More accurate results would be obtained from an iterative procedure, using this result as a starting point to find $v_l(\alpha)$ and $v_l(\beta)$.

(b) over the range $\alpha \le \omega t \le \beta$, the current i_s in the diode is given by
$i_s = (v_s - v_l)/R_s \approx (\hat{V}_s \sin\omega t - V_l)/R_s$, where $\beta = \alpha + \gamma \approx 1.13 + 0.87 = 2.0$ rad.
The average current $I_{s\,av}$ is

$$I_{s\,av} = \frac{1}{2\pi}\int_{\alpha}^{\beta} \frac{\hat{V}_s \sin\omega t - v_l}{R_s}\, d\omega t \approx \frac{1}{2\pi R_s}(\hat{V}_s(\cos\alpha - \cos\beta) - V_l(\beta - \alpha)).$$

That is, $I_{s\,av} = \dfrac{1}{2\pi \times 0.1}(339.4(\cos 1.13 - \cos 2.0) - 307.6(2.0 - 0.87)) = 30.7$ A.

The current rating of the diode must be greater than 30A.

Source impedance and commutation overlap. A rectifier circuit usually has some source inductance and some stray inductance that is lumped together as L_s. The effect of this inductance can be to reduce the average voltage and current on the dc side. Some idea of the reduction can be obtained from a simple analysis of the ideal circuit shown in Fig. 3.11a. Figure 3.11b illustrates the waveforms. The full lines represent the case where $L_s = 0$ and the broken lines indicate the results accounting for a finite value of the inductance L_s. The effect is not present without a freewheeling diode D. In order to emphasize the effect of the inductance L_s, it will be assumed that the current filter $L \to \infty$, so that the load current i_l is a constant I_l.

If there is negligible source and stray inductance L_s, diode DS is on only while it is forward biased by the supply voltage v_s. Diode D is on only while diode DS is reverse biased by the supply voltage. If the load inductance L approaches an infinite value, the load current i_l is a constant I_l, so the diode currents have rectangular waveforms.

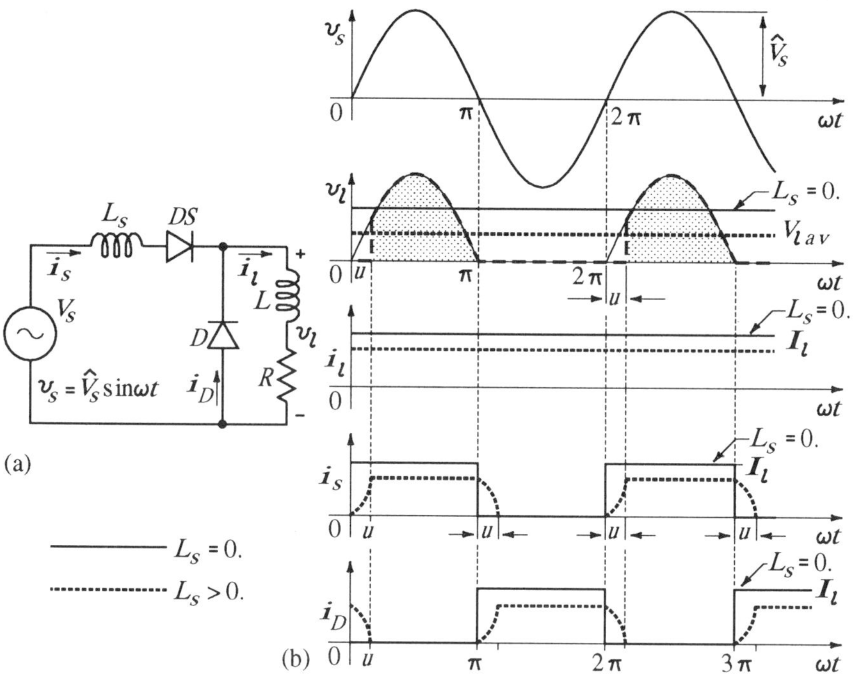

Fig. 3.11 Commutation overlap.
(a) Circuit diagram, (b) waveforms.

Consider the case that L_s is finite and there are steady operating conditions. At $\omega t = \pi$, there is a dynamic transition as the supply voltage changes polarity and diode DS starts to turn off. The source current i_s cannot change from $i_s = I_l$ to zero instantaneously because of the stored energy in L_s. Since the diode D is assumed ideal, it turns on at $\omega t = \pi$, and represents a short circuit. Kirchhoff's voltage law around the loop containing the elements v_s, L_s, DS (short circuit) and D (short circuit) gives $v_s = L_s di_s / dt$. In integral form

$$\int di_s = \int \frac{v_s}{L_s} dt = \int \frac{v_s}{\omega L_s} d\omega t. \qquad (3.8.16)$$

Accounting for initial conditions $i_s = I_l$ at $\omega t = \pi$. The current i_s reduces according to the form

$$i_s(\omega t) = I_l + \frac{1}{\omega L_s} \int_{\pi}^{\omega t} v_s \, d\omega t = I_l - \frac{\hat{V}_s}{\omega L_s}(1 + \cos\omega t). \qquad (3.8.17)$$

This is true over the range that i_s is positive, that is, $\pi \le \omega t \le \pi + u$. At the same time the diode current i_D rises according to Kirchhoff's current law

$$i_D = I_l - i_s = \frac{\hat{V}_s}{\omega L_s}(1 + \cos\omega t). \tag{3.8.18}$$

The current i_s falls to zero and the current i_D rises to I_l at $\omega t = \pi + u$. The angle u is called the *commutation overlap* because this is the interval over which the current commutates, or switches from one branch to the other. Upon substitution in either eq. (3.8.17) or (3.8.18)

$$\cos u = 1 - \omega L_s I_l / \hat{V}_s. \tag{3.8.19}$$

At $\omega t = 2\pi$, there is a dynamic transition as diode DS turns on and diode D turns off. There is an overlap during which both are on while the current i_s rises to the value I_l, constrained by the inductance L_s, and the current i_D falls from the value I_l to zero, constrained by Kirchhoff's law. During this overlap both diodes act as short circuits and the voltage of the supply appears across the inductance L_s, so that the supply current i_s is

$$i_s(\omega t) = \frac{1}{\omega L_s}\int_{2\pi}^{\omega t} v_s \, d\omega t = \frac{\hat{V}_s}{\omega L_s}(1 - \cos\omega t). \tag{3.8.20}$$

The current i_s reaches I_l when $\cos\omega t = 1 - \omega L_s I_l / \hat{V}_s$. This gives the same overlap angle u as in eq. (3.8.19). While the current i_D in diode D is falling from I_l to zero over the interval u, the voltage across the load is zero. This is depicted in the voltage waveform of Fig. 3.11b by the broken line. This means the average load voltage $V_{l\,av}$ reduces from the value $\hat{V}_s/\pi$ with $L_s = 0$ to

$$V_{l\,av} = \frac{1}{2\pi}\int_{u}^{\pi} v_s \, d\omega t = \frac{\hat{V}_s}{2\pi}(1 + \cos u). \tag{3.8.21}$$

Since the load current I_l is given by

$$I_l = V_{l\,av}/R \tag{3.8.22}$$

then, eqs (3.8.19), (3.8.21) and (3.8.22) describe the behaviour of the rectifier. Can the effect of commutation overlap be ignored in the calculations of the circuit performance?

EXAMPLE 3.7

A single-phase, half-wave, diode rectifier modulates power from a 240-V, 60-Hz ac supply, whose source inductance is $L_s = 60\,\mu\text{H}$, to a load, whose resistance is $R = 1.0\,\Omega$ and whose filter inductance is $L \to \infty$. Find **(a)** the average load current, **(b)** the commutation overlap u and **(c)** the average load current if $L_s \approx 0$.

Solution

Refer to Fig. 3.11 for the circuit diagram and waveforms. Since the inductance L approaches an infinite value, the load current i_l is a constant I_l and the voltage across the inductance is virtually zero.

(a) Equation (3.8.21) expresses the average load voltage as

$$V_{l\,av} = \frac{\hat{V}_s}{2\pi}(1+\cos u)$$

but eqs (3.8.22) and (3.8.19) give

$V_{l\,av} = RI_l$ and $\cos u = 1 - \omega L_s I_l/\hat{V}_s$.

Hence, $RI_l = \dfrac{\hat{V}_s}{2\pi}(1 + 1 - \omega L_s I_l/\hat{V}_s)$

$$\text{or,}\quad I_l = \frac{2\hat{V}_s}{2\pi R + \omega L_s} = \frac{2\sqrt{2}\times 240}{2\pi\times 10 + 377\times 60\times 10^{-6}} = 107.7\ \text{A}.$$

(b) The commutation overlap is

$u = \cos^{-1}(1 - \omega L_s I_l/\hat{V}_s)$.

So, $u = \cos^{-1}(1 - 377\times 60\times 10^{-6}\times 107.6/339.4) = 0.12\,\text{rad}.\ (6.9^\circ)$.

(c) If the source inductance is zero, the overlap angle u is zero and the average load voltage is

$v_{l\,av} = \hat{V}_s/\pi = \sqrt{2}\times 240/\pi = 108$ V.

Therefore, the average load current is

$I_{l\,av} = V_{l\,av}/R = 108$ A.

The percentage decrease in current is about 0.4% if the source inductance is taken into account. From an inspection of the equation for average current

$I_l = 2\hat{V}_s/(2\pi R + \omega L_s)$,

if $R > 1.5\omega L_s$, the effect of the commutation overlap u can be neglected.

3.9. DIODES IN PARALLEL AND SERIES

If the current requirement for a load is greater than the current rating of available, individual diodes, the solution is to connect a number of diodes in parallel to share current. If the voltage of reverse bias is greater than the voltage rating of available devices, diodes can be connected in series to share the voltage.

How many diodes are connected in series and parallel is a question that is answered in the same manner as if the devices were thyristors. Section 5.12 describes the precautions that must be taken to protect multi-connected thyristors in the steady state and in the transient state. The same applies to multi-connected diodes. No two diodes have the same I-V characteristics, so sharing will be unequal. For economy it is better to try to match diodes as closely as possible. If a mismatch is apparent then components are added to the circuit to force a reasonably equitable sharing.

EXAMPLE 3.8

Two diodes each with a nominal rating of 2000 V, must withstand a total reverse voltage of 3000 V peak. The reverse characteristics are shown in Fig. EX3.8. (**a**) Find the ratio of voltage sharing at the peak supply voltage. If equal-value resistors R_{SH} are used to force voltage sharing to the extent that diode *D1* has no more than a maximum of 1650 V reverse bias, (**b**) find the value of the resistance R_{SH}.

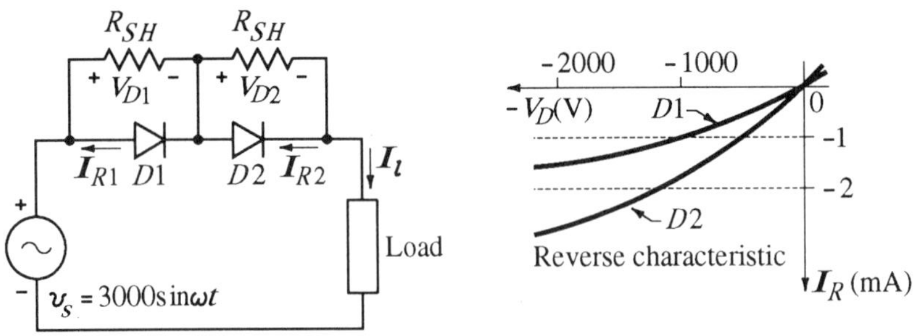

Fig. EX3.8 Diodes in series.

Solution

(**a**) With no forced sharing, the leakage current I_R is the same in both diodes and the peak sharing is such that $V_{D1} + V_{D2} = -3000$ V. From the characteristics $V_{D1} = 1850$ V and $V_{D2} = 1150$ V.
The diode *D1* drops 62% of the voltage at 3000 V.

(**b**) With forced voltage sharing such that $V_{D1} = -1650$ V and $V_{D2} = -1350$ V, then, from the characteristics, $I_{R1} = 1.4$ mA and $I_{R2} = 1.7$ mA.
While the diodes are blocking and the source voltage is –3000 V, the load current I_l is given by
$I_l = V_{D1}/R_{SH} - I_{R1} = V_{D2}/R_{SH} - I_{R2}$.
So $R_{SH} = (V_{D2} - V_{D1})/(I_{R2} - I_{R1}) = 3000/0.3 \times 10^{-3} = 1000\ \text{k}\Omega$.

3.10. SUMMARY

The diode is an important component in power electronic circuits both as a switch and as a component for protection.

The diode is a *pn* device. It has two layers of silicon and has two terminals. Conduction is allowed in one direction and conduction is blocked in the reverse direction. With only two terminals, it is an uncontrolled switch; there are no internal means to turn the diode on or off. The switching action is determined by the voltages and currents in the circuit to which the diode is connected. If the diode is forward biased by a positive voltage at its anode terminal, it will offer almost no impedance to current conduction. So, it will be in the on-state. If the current in the diode is reduced to zero and if the diode is reverse biased by a negative voltage at its anode terminal, it will offer a high impedance to current. So, it will be in the off-state. It takes time to change states. This time is of the order of microseconds. Refer to data sheets in Appendix 1.

These characteristics make the diode ideally suited to the uncontrolled conversion of ac power to dc power. Diode rectification can range from watts to megawatts. In order to accommodate this range, the diode configuration starts from a single device for half-wave rectification and progresses through single-phase, full-wave rectification, three-phase, half-wave rectification, to three-phase, full-wave rectification with multiple bridges.

Diodes find other applications in power-electronic circuits. The main one is to allow freewheeling current in an inductive element after the inductance has been isolated by a controlled switching action.

Schottky diodes are metal-oxide rectifiers and they offer an advantage of low voltage drop in the on-state. They have disadvantages. The low reverse withstand voltage and high leakage current limit their use to circuits operating at less than 100 V.

Protection of diodes is similar to protection of thyristors with respect to overcurrent, reverse overvoltage and thermal considerations.

3.11. PROBLEMS

Section 3.4

3.1 In the circuit diagram depicted in Fig. 3.4 the source voltage V_s has a rectangular waveform ±20 V with a period $T = 1$ ms. The load resistance is $R = 0.5\,\Omega$. From the diode characteristic shown in Fig. P3.1 determine the equivalent circuit parameters. What are the values of the average power absorbed by the load and the diode, calculated for all three models depicted in (a), (b) and (c) of Fig. 3.4?

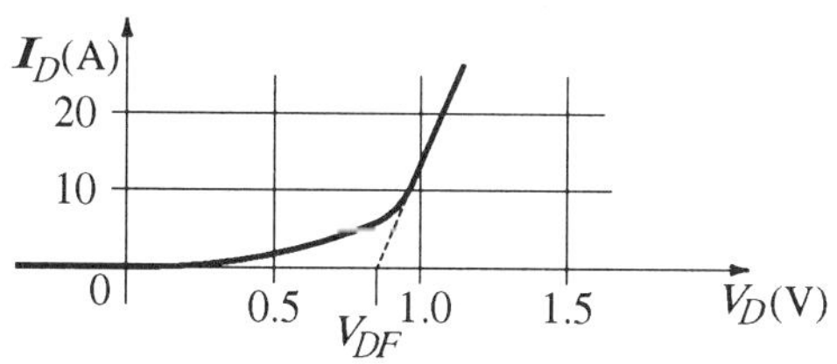

Fig. P3.1

Section 3.7

3.2 During conduction the voltage drop v_D across a power diode is modelled by

$$v_D = V_D + R_D\, i_D$$

where V_D is a fixed component, R_D is the slope resistance of the I-V characteristic and i_D is the diode current. A diode acts as a half-wave rectifier to modulate power from a 240-V, 50-Hz supply to a resistive load of 1.414 Ω. The diode model is represented by $V_D = 1.0\,\text{V}$ and $R_D = 1.0 \times 10^{-3}\,\Omega$. **(a)** Find the average power dissipated in the diode if the transient switching losses can be neglected. **(b)** The junction temperature is not to exceed 150°C and the ambient temperature is 30°C. If the diode thermal resistance, junction to case is specified as $R_{\theta JC}$ 0.1°C/W, calculate the necessary value of the bonding and heatsink thermal resistance. See EXAMPLE 3.3.

3.3 A single diode is used to rectify the current from an ac source of voltage 1000 V at 2 kHz to a resistive load of value 1.0 Ω. For the conditions prevailing and the diode specifications, the reverse recovery charge is $Q_{RR} = 10\,\mu\text{C}$ and the softness factor $S = 0.6$. Find **(a)** the reverse recovery time t_{rr} of the diode and **(b)** the peak reverse current I_{RR} during commutation.

3.4 A diode is used to rectify the current from an ac supply of 240 V to a resistive load that is to absorb an average power of 60 kW. A diode at hand is known to have the specifications 1000 V, maximum junction temperature 180°C, thermal resistance, junction to ambient, $R_{\theta JA} = 0.32\,°\text{C/W}$ and the on-state voltage drop is 1.8 V at 300 A. Determine whether this diode can be used for this application if the ambient temperature is maintained at 30°C.

3.5 A chopper circuit (refer to Fig. 3.7) incorporates a freewheeling diode with a snubber capacitor $C_s = 1.0\,\mu\text{F}$ and a snubber inductor $L_s = 20\,\mu\text{H}$ for a load current $I_l = 300$ A that is ripple free. The dc source has a voltage $V_s = 600$ V. If it can be assumed that the diode reverse recovery time is $t_{rr} = 2\,\mu\text{s}$ and the softness factor is $S = 1.0$, estimate the minimum voltage rating of the diode.

3.6 A chopper circuit, as shown in Fig. 3.7 incorporates a 100-V Schottky diode, that, in reverse bias, can be modelled by a 500 pF capacitor. The dc supply of voltage $V_s = 40$ V with a source inductance $L_s = 5\,\mu\text{H}$ provides a current in the load that is 100A without any significant ripple. Calculate the peak reverse recovery voltage V_{RR}. Does the diode require a capacitive snubber to limit the reverse diode voltage?

Section 3.8

3.7 Figure 3.8 indicated the values of the average voltage across the output for the different rectifier configurations. Prove that these expressions are correct.

3.8 A half-wave, diode rectifier modulates the power from a single-phase, 208-V ac supply to a resistive load, $R = 2\,\Omega$. If all the elements are ideal, find the average power absorbed by the load and estimate the ratings of the diode.

3.9 Consider a half-wave diode rectifier. The source is a single-phase, 120-V, 60-Hz supply and the load is purely resistive with a value $R = 1.2\,\Omega$. If the elements in the circuit are ideal, determine **(a)** the ac source power factor PF and **(b)** the total harmonic distortion THD of the source current.

3.10 A single-phase, half-wave, diode rectifier modulates power from a source voltage of 240 V at 50 Hz to an *RL* load, where $R = 1\,\Omega$ and $L = 2$ mH. If the elements in the circuit are assumed to be ideal, find the current rating of the diode. Refer to Fig. 3.9.

3.11 Consider the rectifier circuit diagram in Fig. 3.9. The source is a single-phase 120-V, 60-Hz supply and the load has resistance $R = 1.2\,\Omega$ in series with inductance $L = 2.4$ mH. If the elements of the circuit are considered to be ideal, determine **(a)** the ac source power factor PF and **(b)** the total harmonic distortion of the source current.

3.12 A single-phase, half-wave, diode rectifier modulates power from a source voltage of 240 V at 50 Hz to an *RL* load, where $R = 1\,\Omega$ and $L = 2$ mH. If the load has a freewheeling diode connected across it, determine the current ratings of the diodes and the power absorbed by the load. Assume all the elements of the circuit are ideal. Is the load current continuous?

3.13 Consider the rectifier circuit diagram in Fig. 3.9. The source is a single-phase, 120-V, 60-Hz supply. The load has resistance $R = 1.2\,\Omega$ in series with inductance $L = 2.4$ mH. A freewheeling diode is connected across the load. If the elements of the circuit are considered to be ideal, determine **(a)** the ac source power factor PF and **(b)** the total harmonic distortion THD of the source current.

3.14 A single-phase, half-wave diode rectifier modulates power from an ac supply of voltage $V_s = 120$ V to a superconducting magnet (resistance $R = 0$) whose inductance is 2 mH. Consider all the circuit elements to be ideal. What is **(a)** the average voltage across the magnetic coil, **(b)** the peak energy stored in the magnetic field and **(c)** the current rating of the diode?

3.15 Consider the rectifier circuit diagram in Fig. 3.9. The source is a single-phase, 120-V, 60-Hz supply. The load has resistance $R = 1.2\,\Omega$ in series with inductance $L \approx \infty$ so that the load current is virtually constant. A freewheeling diode is connected across the load. If the elements of the circuit are considered to be ideal, in steady-state operation determine **(a)** power factor PF of the ac source current and **(b)** the total harmonic distortion THD.

3.16 A single-phase, half-wave diode rectifier modulates power from a 240-V, 50-Hz supply to a resistive load, whose value is $R = 3\,\Omega$. If the circuit components can be considered ideal, find **(a)** the lowest value of capacitance C to maintain the load current continuous and **(b)** for the same condition find the current rating of the diode. See Fig. 3.10.

3.17 A single-phase, half-wave, diode rectifier modulates power from a 240-V, 50-Hz source to a resistive load, $R = 30\,\Omega$, with a capacitive filter, such that the load time constant is 10ms. If all the elements of the circuit are ideal, calculate (**a**) the average load current and (**b**) the current excursion.

3.18 Consider the rectifier circuit diagram in Fig. 3.11. The source is a single-phase, 120-V, 60-Hz supply. This supply has a source impedance whose value is an inductance $L = 5\,\text{mH}$. The load has resistance $R = 1.2\,\Omega$ and inductance $L \approx \infty$, so that the load current is virtually constant. A freewheeling diode is connected across the load. If the circuit is considered to be in steady state, estimate (**a**) the power absorbed by the load, (**b**) the commutation overlap angle u, (**c**) the power factor of the ac supply and (**d**) by how much the power factor of the ac supply increases for zero source impedance.

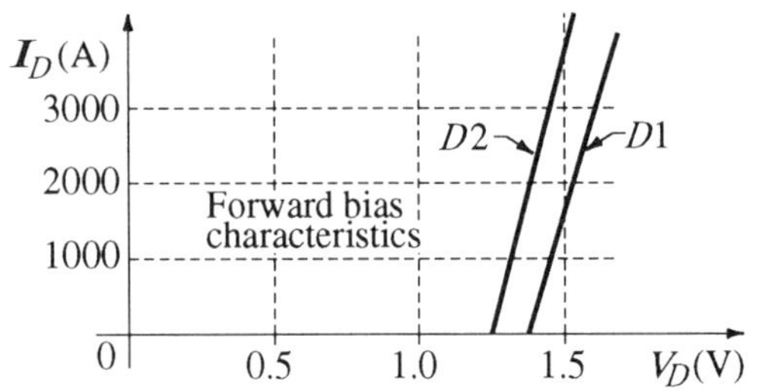

Fig. P3.19

Section 3.9

3.19 Figure P3.19 illustrates the I-V characteristics of two diodes. (**a**) If the two diodes are connected in parallel to share a load current of 5350 A, find the conduction losses. (**b**) If one diode is to conduct no more than 55% of the load current, find the total value of the power dissipation.

3.12. BIBLIOGRAPHY

Baliga, B.J. *Modern Power Devices.* New York: John Wiley & Sons, Inc., 1987.

Davis, R.M. *Power Diode and Thyristor Circuits.* Stevenage: IEE, 1979.

Ghandi, S.K. *Semiconductor Power Devices.* New York: John Wiley & Sons, Inc., 1987.

Neudeck, G.W. *The PN Junction Diode.* Modular Series on Solid State Devices, Volume 2., Massachusetts: Addison-Wesley, 1983.

Roehr, Gill. (ED). *Silicon Rectifier Handbook.* Motorola Inc., 1973.

Schaefer, J. *Rectifier Circuits, Theory and Design.* New York: John Wiley & Sons, Inc., 1975.

Smith, M.W. (ED). *Electronic Data Library, Transistors-Diodes.* General Electric Co., 1982.

CHAPTER 4

THE BJT TRANSISTOR

4.1. INTRODUCTION

The *bipolar junction transistor* is known as a signal amplifier. In the context of power electronics we will discuss this semiconductor device as a power switch.

This transistor, hereafter called the BJT, is a three-layer, three-terminal, silicon device. It is undoubtedly the most common semiconductor device that can have its state controlled completely in an off-state, in an on-state or in any active region between these two extremes.

A description of the name BJT is not as succinct as the diode. The name *bipolar* is to signify that charges of both polarities are involved in the production of current, just like the diode, although the transistor is basically a minority carrier device. *Junction* is used to imply the fact that *pn* junctions are used to construct the *transistor*, whose origin is *trans*-fer res-*istor*. This last name conjures up the idea of an adjustable effective resistance that is changed by a control signal. In data sheets, a common name, that is used in connection with the transistor action, is *transconductance*, rather than transfer resistance.

Figure 4.1a gives a one-dimensional depiction of the BJT structure with three *n-p-n* layers of doped silicon, three terminals, collector C, emitter E and base B, and two *pn* junctions. We know about the uncontrolled action of the *pn* junction from Chapter 3. If two of these *pn*-junction building blocks are constructed together in this special way the result is a controlled switch. The action of the BJT is that under the influence of a base signal the emitter emits charge carriers that are injected into the base. Most of these carriers are collected in the collector region due to the collector bias. The charge flow constitutes the current. There are *pnp* transistors that are used as power switches too, but the *npn* type shown here is more common because it is cheaper.

The circuit symbol for the *npn* BJT is shown in Fig. 4.1b together with the terminal connections. The junctions are not obvious in this symbol, but the arrow indicates the direction of conventional current in a *pn* junction. The currents and voltages associated with the BJT are included in the drawing.

Figure 4.1c illustrates the use of a BJT in a simple circuit that comprises a voltage source V_s, a load R and a BJT whose purpose is to modulate power from the source to the load by acting as a switch. The terminals C and E are connected in series with the main power circuit, whereas the terminals B and E are connected to a low-power *driving circuit* that is used to control the BJT's on and off switching actions. If there is no signal at the base B, the BJT is in the off-state. The transistor blocks the supply voltage V_s which is positive at the collector C, so that the voltage $v_{CE} = V_s$, the voltage $v_l = 0$ and the current $i_C = i_s = i_l = 0$ for ideal elements.

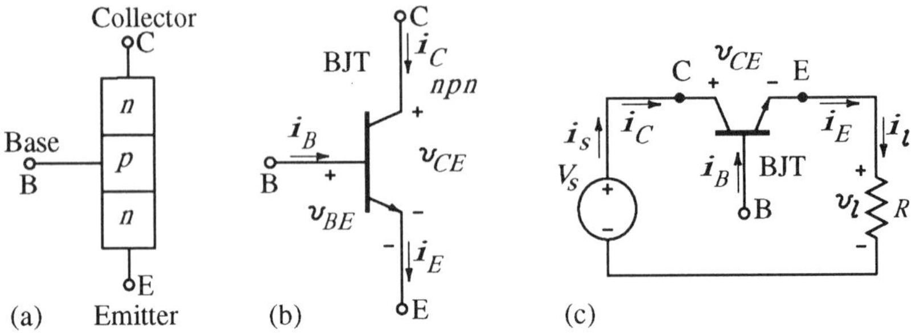

Fig. 4.1 BJT. (a) *npn* structure, (b) circuit symbol, (c) circuit with BJT switch.

Operating in the common-emitter mode[1], if the base terminal is positively biased with respect to the emitter terminal, and if the collector terminal is positively biased with respect to the emitter terminal, then the base current i_B turns on the BJT such that, ideally, $v_{CE} \approx 0$, $v_l \approx V_s$ and $i_C = i_s = i_l \approx V_s/R$. The switch remains on as long as the base is being driven with a current i_B. As soon as the base current is brought to zero the BJT turns off.

With $v_{CE} > 0$ the BJT is turned on and off by means of the base signal. The thyristor cannot be turned off in a similar manner. The facility of turn-off gives an advantage over the thyristor and opens up a broader range of applications.

The BJT is not the perfect switch. Like the diode and the thyristor, the BJT is constrained to have current in one direction. Unlike the diode and the thyristor the BJT cannot block much more than about 20 V in the reverse polarity across the emitter-to-collector terminals. Therefore, BJTs are not used to modulate power from an ac supply unless a diode is connected in series with the switch.

Like the thyristor and the diode, the BJT is charge controlled. Inherently, it takes time to switch the BJT on and off. Even though this power semiconductor device can have switching frequencies of the order of ten times that of the thyristor, it is not the fastest power transistor.

Voltage ratings over 1000 V, current ratings in the range of hundreds of amperes and switching frequencies over 10 kHz mean that BJTs are suited to inverter and chopper applications for powers up to and beyond 200 kW.

In this chapter on BJTs we will describe the steady-state and transient characteristics together with the limitations and protection. The BJT action is more complex than the thyristor and it is more fragile. This becomes evident.

[1] The emitter terminal is indirectly connected to both the base terminal and the collector terminal.

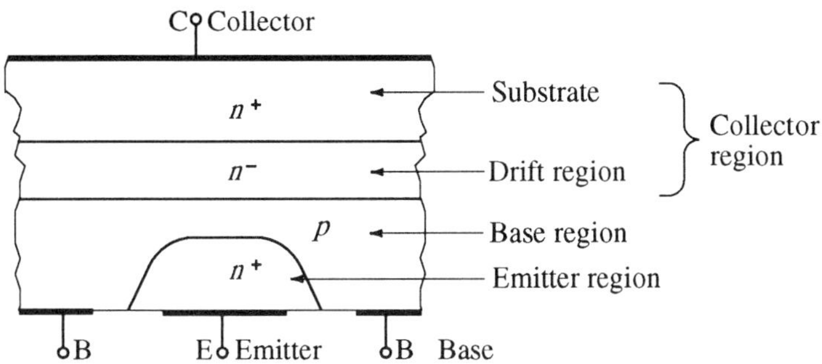

Fig. 4.2 BJT structure.

4.2. BJT STRUCTURE

Figure 4.1a shows a bipolar-transistor's structure in the *npn* form. The BJT power switch is manufactured with a drift region, like the diode, and, in addition, has an interdigitated layering of the base and emitter. Figure 4.2 shows an outline of the BJT with the four-layer structure of one interdigitated section.

The substrate of the silicon wafer is highly doped to produce an n^+ region, that is called the collector. Also part of the collector is an almost intrinsic region. It is a very lightly doped n^- type, thin enough for low on-state resistance, but thick enough to almost double the off-state, collector withstand voltage (its main purpose). Because of the nature of the current in it, this is called the drift region and acts in the same way as in the power diode. The transistor withstand voltage is increased because the widening of the collector-base depletion layer takes place in the drift region rather than in the *p*-type base region. This property allows the added *p* region to be thin, a requirement for a reasonable current gain (collector/base). For an abundance of charge carriers a highly doped n^+-type silicon layer is added by diffusion and forms the emitter region to complete the *npn* power transistor. Interdigitation is formed between the base and emitter regions by interleaving fingers of each. The result is less crowding of the main collector current and high ratings.

This brief description of the structure of the BJT leads the way to develop the steady-state, current-voltage characteristics of the device for different voltage biases at the terminals.

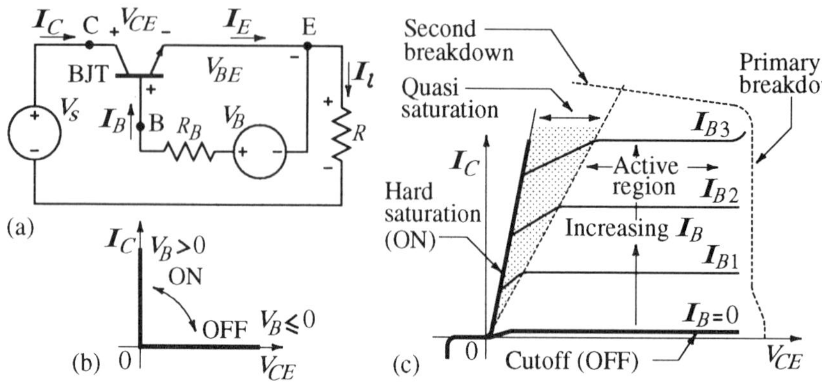

Fig. 4.3 BJT I-V characteristics.
(a) Circuit diagram, (b) ideal, (c) four regions of operation.

4.3. BJT I-V CHARACTERISTICS

The characterization of the BJT, showing current versus voltage, is more complex than the diode. Manufacturers have created many subscripts to append to voltage and current symbols in order to define the limits of these parameters under the many possible operating conditions. For a book that is aimed at introductory general circuit operation and performance, the complexity will be masked to the extent that we analyze switching circuits with ideal elements. The practical realization, that protection against the hazards of semiconductor frailty is a necessity, forces us to take into account some of the parameters.

Figure 4.3a shows a simple circuit diagram that depicts the variables and circuit parameters for the steady-state characterization of the *npn* BJT in the common-emitter mode of connection. In order to cover all the possible regions of BJT operation the two dc sources, V_s and V_B, are adjustable in magnitude and polarity.

4.3.1. Ideal Switch

For those situations whereby a circuit analysis is to be performed under the conditions that the BJT is operating as a switch and the supply voltage V_s is nominally greater than 100 V, all the elements can be assumed to be ideal.

Consider the switch to be off. As long as the base-emitter source voltage $V_B \leq 0$, the switch remains off, no matter what value of positive source voltage V_s is applied to the collector terminal C. In the off-state there is no current I_C and the switch blocks any positive voltage V_s. A change from the off-state to the on-state occurs if the source voltage $V_s > 0$ and if the source voltage $V_B > 0$. The voltage V_B allows current I_B to be injected at the base terminal. This action turns on the switch by creating a zero impedance path between the collector C and emitter E for load current ($I_l = I_C$) driven by V_s. Once on, the switch remains on as long as the base drive is continuous. Interruption of the base current causes the switch to turn off. Table 4.1 illustrates the conditions of the switch and Fig. 4.3b depicts the ideal states graphically.

With a positive voltage at the collector C with respect to the emitter E, application of a positive base drive turns on the BJT and the removal of the base drive turns off the BJT. The BJT cannot withstand a negative voltage that is applied to the collector with respect to the emitter. Accordingly, this condition is avoided.

4.3.2. Nonideal Switch

The ideal I-V characteristic of the BJT has its place in network analysis of switched circuits. If circuit design is of concern, whether it be for a base-driver circuit or whether it be for protection devices, the real characteristics, that are given in the manufacturers' data sheets, must be used. Under the assumption of constant temperature, a simplified set of I-V characteristics is illustrated in Fig. 4.3c.

The I-V characteristics can be divided into four regions of operation. There are the two states ON and OFF that form the boundaries given by hard saturation and cutoff respectively. Operation is either at high current and low voltage, or at low current and high voltage. These are similar to the ideal states shown in Fig. 4.3b. Between these two states there are two more definable regions of operation. One is the quasi-saturation region that is a useful characteristic. The other is the active region that is used for linear amplification, but that is avoided in steady operation for switching applications. We will investigate the four separate regions briefly in order to appreciate the loss mechanisms and switching times of the BJT. There is a fifth region of operation whereby the emitter is biased by a positive voltage with respect to the collector. Since the withstand voltage level is often less than 20 V at this bias, this region is avoided.

Cutoff mode of operation.
Cutoff is synonymous with off-state. The BJT is in the off-state if there is no base drive. A positive voltage at the collector terminal with respect to the emitter terminal is blocked because the collector-base junction is reverse biased. Only a small leakage current I_C exists because of the impact ionization that occurs. As the collector voltage is increased, impact ionization increases until avalanche breakdown (primary breakdown) occurs. The cutoff region is illustrated with $I_B = 0$ in Fig. 4.3c.

In practice, there is a *sustaining voltage* above which the transistor must not be subjected. Otherwise, damage will ensue. The sustaining voltage is less than the avalanche breakdown voltage at the collector-base junction. It is the transistor effect that can cause damage and creates the need for data sheets to have sustaining-voltage values included. Associated with leakage current I_C is the delivery of holes to the base region. The hole flow, which is an internal base-region current I_B, creates an increase of collector current by the transistor amplifying effect ($I_C = \beta I_B$). Below the sustaining-voltage value the collector current is small. Above the sustaining-voltage value the collector current increases to such a level that, with the high value of voltage V_{CE} the BJT overheats and is destroyed.

The sustaining voltage can be increased up to a value close to the avalanche breakdown voltage by applying a negative-bias voltage $-V_B$ to the base terminal with respect to the emitter. A negative current $-I_B$ reduces the net current in the base, that, in turn, reduces the leakage current I_C.

Active region of operation.

As shown in Fig. 4.3a, the source voltage V_B is applied to make the base-emitter junction forward biased. The source voltage V_s makes the collector-base junction reverse biased. These conditions set the BJT in the active region of operation, as shown in Fig. 4.3c.

It is the current I_B in the base that controls the point of operation in the active region, but it is the combined drive voltage V_B and the resistor R_B that control the current I_B. If the load line is fixed, as shown in Fig. 4.4, increasing I_B takes the operating point from X2 to X4 in the active region. The harder the base is driven the lower is the on-state voltage V_{CE} across the BJT. However, if the base current I_B is held constant, the collector current I_C is essentially constant, even though the source voltage V_s or the load resistance R may be slowly adjusted. The I-V characteristics are shown in Fig. 4.3c for increasing base currents I_{B1}, I_{B2} and I_{B3}.

The operation of the BJT at some point in the active region is maintained by the base current I_B. At this point of operation the transistor effect takes place. The emitter region is a source of electrons. The forward bias of the base-emitter junction causes these electrons to be injected into the base region. Since the transistor is bipolar, holes from the base region are injected into the emitter region. These two charge flows constitute the emitter current I_E. Most of the electrons cross the base region, come under the influence of the electric field that is produced by the reverse-biased collector-base junction, and are swept into the collector region to constitute the collector current[2] I_C. Those electrons, that do not cross into the collector region, recombine in the base region or exit via the base terminal to constitute (together with the holes) the base current I_B. Consequently for a given current I_B there is a collector current I_C. Two equations of note from this description of the transistor action are the nodal equation (see Fig. 4.3a)

$$I_E = I_B + I_C \tag{4.3.1}$$

and the equation for the amplification of the control current I_B to the collector current I_C

$$\frac{I_B}{I_C} = \frac{1}{\beta} = \frac{I_E - I_C}{I_C} \tag{4.3.2}$$

where β (also known as h_{FE}) is the transistor current gain. A high gain is

[2] There is a relatively small number of thermally generated holes that are swept by the field into the base region from the space charge layer of the collector-base junction, but these can be ignored.

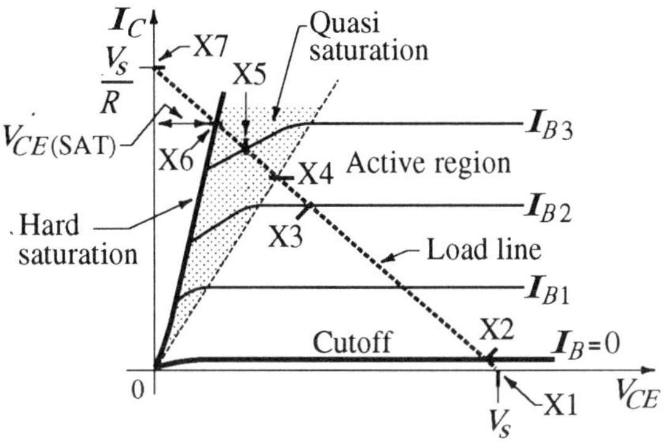

Fig. 4.4 BJT load-line characteristic.

obtained if the difference $I_E - I_C$ is small. This is accomplished in a number of ways. That component of base and emitter current that is produced by the holes in the base region is made small by the emitter being heavily doped, so that the stored hole distribution is small. If the base region is made thin then more of the electrons can pass from the emitter region to the collector region. Further, recombination of minority carriers can be reduced by having a long electron lifetime in the base region because this gives the electron a greater chance to cross into the collector region.

In practice the current gain $\alpha = I_C/I_E \approx 0.95$, so, since $\beta = \alpha/(1-\alpha)$, the gain β of a power BJT is about 20, more or less. This is low compared with the thyristor ratio of gate current to anode current; this can be about 500. Consequently, the base driver of the BJT has to deliver much higher power. The gain β is not a constant. It is a function of collector current I_C, collector voltage V_{CE} and junction temperature. As the voltage V_{CE} increases so does the gain β. However, as the current I_C increases and as the junction temperature increases, the gain increases initially and then decreases. This makes the use of data sheets imperative for an accurate design or analysis of BJT circuits.

The BJT power switch would not be able to tolerate operation in the active region during the normal, steady-state conduction mode, because both the voltage V_{CE} and the current I_C would be high simultaneously. The power dissipation $V_{CE}I_C$ associated with these high values would cause the device to overheat. This region of operation is avoided in the steady state by keeping the BJT in either the cutoff region or the saturation region. The active region cannot be avoided altogether. Switching from one state to the other demands the trajectory of voltage and current through the active region, but the duration in this region can be short.

Saturation mode of operation.

Let the base-emitter junction of the BJT be forward biased by the source voltage V_B, as shown in Fig. 4.3a. In the active region of operation, the applied voltage V_s will drive a current I_C such that the collector voltage V_{CE} reverse biases the base-collector junction. As the base drive increases the base current I_B, the amplification factor β of the transistor means that the collector current I_C increases. Associated with the current increase there is a collector-emitter voltage drop. At some point of base current I_B increase and collector voltage V_{CE} decrease the voltage V_{BE} will become greater than the voltage V_{CE}. This means that the collector-base junction is no longer reverse biased. The onset of forward bias of the base-collector junction is defined as the point of saturation of the BJT. The whole area that is hatched in Fig. 4.3c covers all conditions for which the base current I_B sets the base-collector junction in forward bias. This is called the quasi-saturation (or soft-saturation) mode of operation, because the degree of saturation can vary with the current I_B. While in the quasi-saturated mode the BJT voltage V_{CE} is low but not at its lowest value. An increase in I_B does not produce a significant increase in I_C (β decreases).

The forward bias of the pn^- layer sets up the electric field that causes the injection of holes from the base region into the drift region. The electron flow from the emitter is the main current but the hole flow decreases the net current I_C. Associated with this is the reduction of the current gain β. With the excess carrier build up in the drift region the conductivity is modulated so that the resistance of the n^- layer decreases. Therefore the voltage drop decreases and V_{CE} drops due to both the current and resistance decrease. This characteristic of quasi-saturation is shown in Fig. 4.3c by the changing slope of the I-V characteristics.

Hard saturation.

Hard saturation is the extension of the quasi-saturation mode of operation to its limit. Saturation is reached if the base voltage V_B and its response I_B cause the base-collector junction to be forward biased. As V_B and I_B are increased more, the voltage V_{CE} decreases until it can decrease no further, because the stored charge reaches all the way across the drift region so that it is completely conductivity modulated. The locus of the minimum value of V_{CE}, that is called $V_{CE(\mathrm{SAT})}$, is the region of hard saturation and is shown as the thick line in Fig. 4.3c. At rated current, the voltage $V_{CE(\mathrm{SAT})}$ has a value between 1 V and 2 V.

While in the on-state, it would seem reasonable to assume that it is good practice to operate the BJT in the region of hard saturation, by driving the base hard; that is, inject a high current I_B into the base. In this way, the voltage drop V_{CE} across the transistor is at its minimum value and the power dissipation in the device is kept low during conduction. Hard saturation produces a problem at turn-off. It takes time to remove all those excess carriers in the drift region, so the turn-off time is long. The compromise between conduction losses and turn-off time is to operate the switch in the quasi-saturation region in the on-state.

Region of operation. In what region the BJT operates depends on the external circuitry as well as the internal characteristics of the transistor. Reference can be made to the circuit shown in Fig. 4.3a.

Figure 4.4 shows how to determine the operating point. Superimposed on the I-V characteristics of the BJT is the main power-circuit load line. This line is shown thick and broken. It is a straight line drawn between two points, one on each axis. The points are determined completely by the circuit external to the BJT.

Point X1 is fixed by $V_{CE} = V_s$ and $I_C = 0$, and represents the ideal operating point for the switch being off. The collector current is ideally zero and the voltage across the collector-emitter terminals is ideally equal to the supply voltage V_s.

Point X7 is fixed by $V_{CE} = 0$ and $I_C = V_s/R$, and represents the ideal operating point for the switch being on. The BJT looks like a short circuit in the ideal on-state, so that the voltage across the collector-emitter terminals is zero and the current is determined by the source voltage V_s and the circuit resistance R; that is, $I_C = V_s/R$.

A line can be drawn between the two points X1 and X7. This line is the load line. As long as the voltage V_s and the resistance R remain constant, the operating point of the transistor will be on this line. The actual point of operation on this line depends entirely on the value of the current I_B injected into the base terminal B by the base drive circuit. Some operating points are as follows.

- The point X2 is the actual operating point at cutoff, for which there is no base current, the collector current is the leakage current I_C and the voltage $V_{CE} = V_s - I_C R$.
- The point X3 is the operating point if the base current is I_{B2}. This is within the active region for which both I_C and V_{CE} are relatively high values, creating high power dissipation within the BJT.
- In order to reach the operating point X4, the base current I_B has to be raised so that the base-collector junction is at the onset of being forward biased. This is the edge of the quasi-saturation region. The current I_C is high and the voltage V_{CE} is relatively low.
- If the base current is raised to the value I_{B3}, the operating point X5 is in the mid-saturation region.
- A further increase in base current brings the operating point to X6, the point of minimum voltage called $V_{CE(\text{SAT})}$, below which value the collector voltage cannot fall for the given circuit configuration of V_s and R. This point of operation lies in the region of hard saturation.

As a power switch the BJT steady-state operation is in the cutoff region at X2 or the quasi-saturation region between points X4 and X6 on the load line.

EXAMPLE 4.1

A BJT controls power from a dc source of voltage $V_s = 200$ V to a resistive load of value $R = 4\,\Omega$. For this load condition and with the transistor in hard saturation, the BJT voltage drop is $V_{CE(\text{SAT})} = 1.1$ V. The base resistance R_B is chosen to be $0.5\,\Omega$, such that there is overdrive if the base source voltage drop is $V_{BE(\text{SAT})} = 1.8$ V. For the on-state condition determine **(a)** the forced current gain β_F and **(b)** the power loss in the BJT.

Solution

Refer to Fig. 4.3a for the circuit diagram relating to this problem. Consider the BJT to be on.

(a) For the main circuit
$I_C = (V_s - V_{CE(\mathrm{SAT})})/R = (200 - 1.1)/4 = 49.7$ A.
For the base driver circuit
$I_B = (V_B - V_{BE(\mathrm{SAT})})/R_B = (10 - 1.8)/0.5 = 16.4$ A.
The forced current gain is $\beta_F = I_C/I_B = 49.7/16.4 \approx 3$.

(b) In the steady-state condition the BJT power dissipation P is
$P = V_{CE(\mathrm{SAT})} I_C + V_{BE(\mathrm{SAT})} I_B = 1.1 \times 49.7 + 1.8 \times 16.4 = 54.7 + 29.5 = 84.2$ W.
In overdrive the base power dissipation is not insignificant.
The steady state efficiency η of the BJT is
$\eta = \text{output}/(\text{output} + \text{losses}) = (200 - 1.1)49.7/(198.9 \times 49.7 + 84.2) = 0.99$ (99%).
Power switching by means of a BJT appears to be efficient.

4.4. BJT MODELS

Models of the BJT abound. There are equivalent circuit models and there are computer-oriented models of great detail. The models are mainly for analysis in the active region, so they tend to be too complex for analysis in the switch-mode form of operation.

The simplest model for the BJT power switch is the ideal switch. If there is a base drive such that current I_B is directed into the base terminal and if the collector terminal is positive with respect to the emitter terminal, the BJT is on and appears as a short circuit in the main power circuit. Otherwise, the BJT is off and appears as an open circuit. This kind of model is suitable for system analysis.

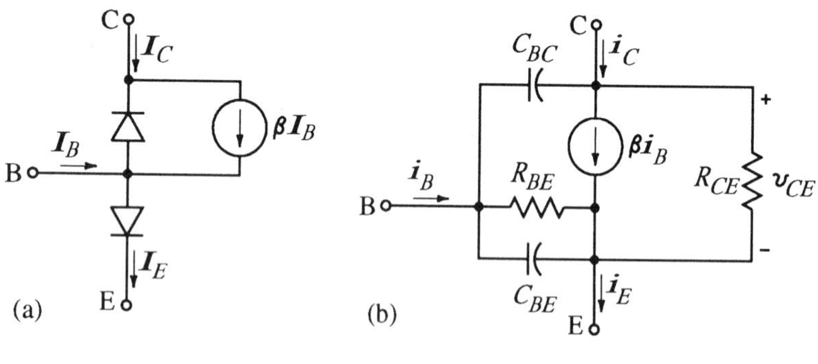

Fig. 4.5 BJT models.
(a) Ideal steady state, (b) transient state.

A simple steady-state model is depicted in Fig. 4.5a. The two ideal diodes represent the two *pn* junctions of the *npn* transistor. The collector current I_C is finite only if there is a base drive current I_B. The current source $I_C = \beta I_B$ accounts for the amplification provided by the transistor effect. Analysis is somewhat limited in this case to the use of Kirchhoff's law

$$I_B + I_C = I_E \tag{4.4.1}$$

and the application of the manufacturer's data relating the short-circuit common emitter current gain β or the short-circuit common base current gain α where

$$\beta = I_C/I_B, \quad \alpha = I_C/I_E \quad \text{and} \quad \beta = \alpha/(1-\alpha). \tag{4.4.2}$$

It takes time to turn the BJT on or off because of the delay to establish a charge distribution in the device and to deplete a region of charge respectively. These effects are modelled by equivalent depletion layer capacitances that are not constant. The fact that there are voltage drops across the base-emitter junction (0.7 V) and across the collector-emitter terminals (1 to 2 V) means that equivalent resistors can be included in the transient model. This is illustrated in Fig. 4.5b. The resistor R_{BE} that accounts for the voltage drop between the base and emitter terminals is straight forward. The resistor R_{CE} is more complicated. The voltage drop between the collector-emitter terminals is far from constant because it accounts for the low resistance values of the two n^+ regions (collector and emitter), the small voltage drop across the drift region and the difference between the two junction voltages ($V_{BE} - V_{BC}$). Much data are required to make use of this model.

The current source βI_B represents the transistor action. Sometimes this effect is described differently. The current source can be described by $g_m V_{BE}$, where V_{BE} is the voltage across the base-emitter junction and g_m is defined as the transconductance (siemens) of the BJT.

4.5. BJT TURN-ON

The BJT is a minority carrier, charge-controlled device, so it takes a finite time for the switch to turn on to the point that the voltage across it is a minimum. If the BJT is off and a positive voltage is applied to its collector with respect to its emitter, the transistor remains off until a source voltage V_B drives a current into the base. Then it goes through the process of turning on

The process of changing states from off to on may take about 1 or 2 μs. We will consider the action of turning on, the losses incurred in the BJT during turn-on and the base driver circuit to turn on the switch.

4.5.1. Turn-on Action

Refer to the BJT transient circuit model shown in Fig. 4.5b. The application of a base drive voltage V_B across the base and emitter terminals of the device will give rise to a base current I_B. There is no change in the condition of the transistor for an interval t_d, during which the capacitor C_{BE} charges up to $V_{BE} \approx 0.7$ V. In fact, the negative charge on the base-emitter space-charge layer must be discharged first, and then charged to become forward biased to $+V_{BE}$ so that carrier injection can commence. After this delay time t_d, if the voltage V_{CE} is positive at the collector terminal, carriers cross from the emitter to the base and then from the base to the collector. The collector current rises in time t_r at a rate that is determined by the magnitude of the base current, and the collector-emitter voltage v_{CE} falls to a value that depends on the magnitude of the base current too. How the voltage falls in relation to the current rise is a function of the external circuit parameters. No matter what their values, the BJT goes from the off-state (cutoff) to the on-state (saturation) by traversing the active region of operation. The turn-on time t_{on} can be defined in terms of the delay time and the current rise time. That is,

$$t_{on} = t_d + t_r\,. \tag{4.5.1}$$

From an oscillogram, the rise time t_r is usually determined as the time it takes for the collector current to rise from 10% to 90% of its final steady value I_C. The delay time t_d, which is measured under test as the interval from $0.1I_B$ to $0.1I_C$, is influenced by the base-drive current, independent of the value of I_C and is a fraction of the rise time t_r. As might be expected the rise time t_r increases with the current I_C. For fast turn-on there is overdrive at the base. Overdrive is the increase of base current above the value that would result in hard saturation. Once the BJT is on, the base current is reduced to a value $I_{B(\text{SAT})}$ that would be just sufficient to maintain saturation with $v_{CE} = V_{CE(\text{SAT})}$. It can be seen from the I-V characteristics in Fig. 4.3c that $I_{B(\text{SAT})}$ is not a constant. It is a function of the collector current I_C. This indicates that the base-drive circuitry has to be relatively complex in order that the base current can follow a profile of collector current. In overdrive, it is usual to designate the current ratio I_C/I_B as the forced current gain β_F. This does not influence the steady value of I_C ($I_C \approx V_s/R$), so the value of β_F is less than the natural gain β.

4.5.2. Turn-on Losses

The calculation of the power dissipated in the BJT during the turn-on process is approximate. The variation of voltage v_{CE} and current i_C is not linear with time, although we make that assumption in order to make some form of calculation. Further, the rise of current i_C is not necessarily in phase with the fall of voltage v_{CE} because of the influence of the external circuit parameters.

One way to make some predictions about the losses is to take two extreme cases to find the limits. One extreme is to consider the power circuit to be purely resistive and the other extreme is to have a highly inductive load.

Resistive load. Figure 4.6a illustrates a circuit with a resistive load R whose power absorption is modulated by a BJT acting as a power switch. The dc source voltage V_s is considered ideal. The base control signal, i_B that turns on the BJT, is assumed to rise much faster than the collector current i_C. Figure 4.6b shows the variation of the base-circuit values of current i_B and base-emitter junction voltage v_{BE} from the moment the base source voltage V_B is applied at $t=0$. The

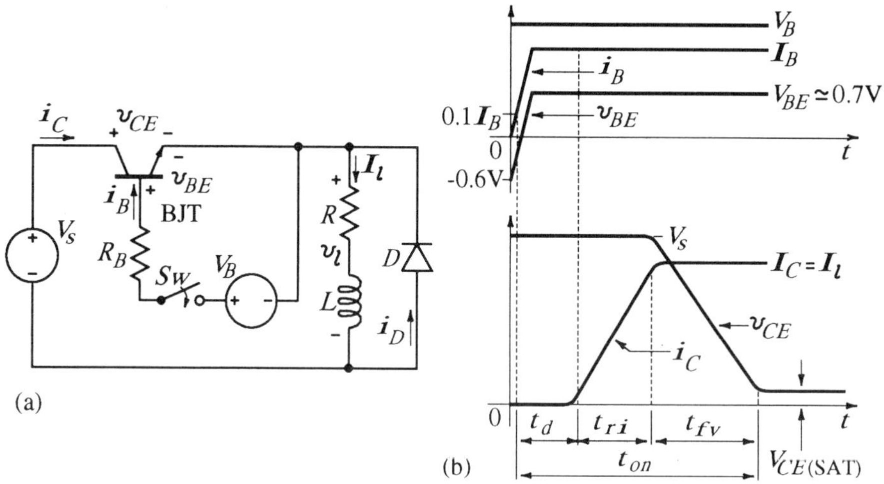

Fig. 4.6 BJT turn-on in a resistive circuit.
(a) Circuit diagram, (b) waveforms.

subsequent rise of collector current i_C is shown to be in synchronism with the collector-emitter voltage v_{CE} fall. This is a reasonable assumption in a linear resistive circuit since Kirchhoff's voltage law yields

$$v_{CE} = -Ri_C + V_s . \tag{4.5.2}$$

With allowances for the official definitions of the delay and rise times, an approximation of the energy dissipation in the BJT during the turn-on time t_{on} is readily made. There is no energy loss during the delay time t_d, because the collector current is $i_C \approx 0$, and because the base current is assumed to contribute a negligible loss. However, over the rise-time interval t_r of the current, Fig. 4.6b demonstrates the existence of a finite voltage v_{CE} and a finite current i_C simultaneously. If $t' = t - t_d$ (see Fig. 4.6b), the time variations of the voltage and current over the interval t_r are

$$v_{CE} = V_s - Ri_C = V_s(1 - \frac{t'}{t_r}) \text{ and } i_C = \frac{I_C}{t_r} t' . \tag{4.5.3}$$

In the steady on-state for $t > t_{on}$ the current I_C is given by

$$V_s = V_{CE(\text{SAT})} + RI_C \approx RI_C . \tag{4.5.4}$$

The instantaneous value of power p absorbed by the BJT during the turn-on interval is expressed by

$$p = v_{CE} i_C = V_s I_C \frac{t'}{t_r}(1 - \frac{t'}{t_r}) . \tag{4.5.5}$$

Consequently, the energy loss W_{on} during the turn-on process is

$$W_{on} = \int_0^{t_r} p\,dt' = \frac{V_s I_C}{6} t_r . \tag{4.5.6}$$

The values V_s and I_C are determined by the the conditions of the external circuit, and t_r is obtained from the manufacturer's data sheet for a given junction temperature. Average power loss is a significant factor, but the switching frequency of the BJT must be known if this power is to be calculated.

EXAMPLE 4.2

A BJT controls power from a dc source of voltage $V_s = 200$ V to a resistive load of value $R = 4\,\Omega$. For this load condition and with the transistor in hard saturation, the BJT voltage drop is $V_{CE(\text{SAT})} = 1.1$ V. If the current rise-time is $t_r = 1.1\,\mu$s for turn-on, estimate **(a)** the energy loss in the BJT during the turn-on process. **(b)** Compare the turn-on loss with the BJT conduction loss over an equivalent interval of time t_r.

Solution

(a) Refer to Fig. 4.6. The steady current in the load with the BJT turned on is

$$i_l = I_l = I_C = \frac{V_s - V_{CE(\text{SAT})}}{R} = \frac{200 - 1.1}{4} = 49.7 \approx 50\text{ A}.$$

During the delay time t_d the loss due to the leakage current is assumed to be small. During the turn-on interval t_r the BJT absorbs electric energy W_{on}.

$$W_{on} = \int_0^{t_r} v_{CE} i_C\,dt', \text{ where } v_{CE} \approx V_s \left(1 - \frac{t'}{t_r}\right) \text{ and } i_C \approx \frac{I_C}{t_r} t'.$$

Thus, $v_{CE} = 200\left(1 - \dfrac{t'}{1.1\times10^{-6}}\right)$, $i_C = 45.2\times10^6 t'$ and

$$W_{on} = \int_0^{1.1\times10^{-6}} 200\times45.2\times10^6 t' \left(1 - \frac{t'}{1.1\times10^{-6}}\right) dt' = 1.82\times10^{-3}\text{ J}.$$

(b) While the BJT is in the on-state it conducts 49.7 A and has a voltage drop of 1.1 V. The on-state energy loss W over an interval of 1.1 μs is

$$W = \int_0^{t_r} V_{CE(\text{SAT})} I_C\,dt = v_{CE(\text{SAT})} I_C t_r = 1.1\times4.97\times1.1\times10^{-6} = 0.06\times10^{-3}\text{ J}.$$

This energy loss is small. The conclusion is that as the frequency of switching becomes high the contribution to losses by the turn-on process can become significant.

Inductive load. The other extreme case for turn-on loss is the one where the load is inductive. Figure 4.7a shows an inductive load that incorporates a freewheeling diode D to provide a path for the load current while the BJT is off. The current

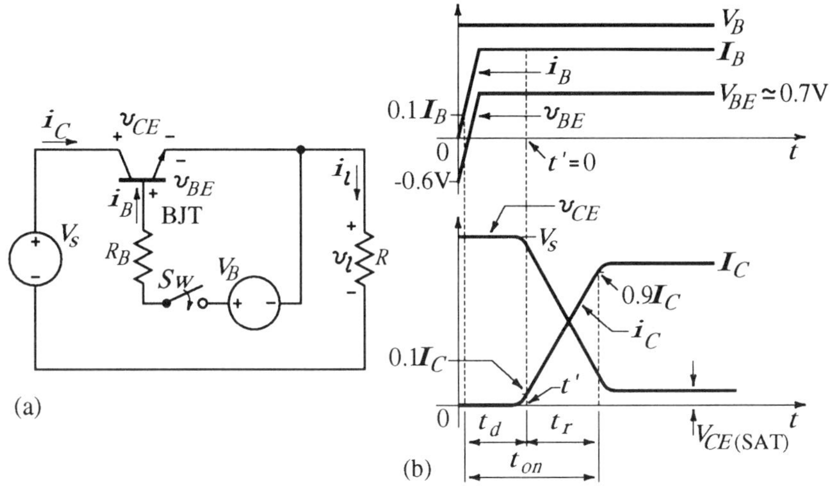

Fig. 4.7 BJT turn-on in an inductive circuit.
(a) Circuit diagram, (b) waveforms.

rise i_C, shown in Fig 4.7b, is due to the transistor turn-on action initiated by driving a current i_B into the base region. However, the fall of voltage v_{CE} across the BJT is initiated by the external circuit.

Let the BJT operate as a chopper and be in steady operation. Consider the BJT to be in the off-state at $t=0^-$, with $i_C \approx 0$, $v_{CE} \approx V_s$, $v_l \approx 0$ and $i_l = I_l$. Chopper action maintains an average voltage $V_{l\,av}$ across the load, depending on the duty cycle m. The instantaneous voltage v_l across the load will be almost V_s if the BJT is on, and will be almost zero if the BJT is off. The transition between zero and V_s occurs when the current in the freewheeling diode falls to zero, at which point the diode begins the process of blocking the reverse voltage across it. During the whole transient process of turn-on the current i_l in the inductive load is assumed to be unchanging. This assumption is reasonably valid if the load time constant L/R is much longer than the period of chopper switching. This gives us the second extreme case.

The initial condition is that, at $t=0$, the switch Sw in the base drive circuit is closed. At $t=0^+$ the base current i_B rises quickly to a relatively high value I_B and charges the base-emitter space-charge layer to a steady forward-bias value $V_{BE} \approx 0.7$ V. During this delay time t_d, the collector current remains $i_C \approx 0$, the collector voltage remains $v_{CE} = V_s$ and the diode current remains $i_D = I_l$.

With the base-emitter junction forward biased the charge carriers (electrons) are injected from the emitter region to the base region. With the collector-base junction reverse biased by the collector-emitter voltage v_{CE}, most of those electrons in the base region are swept into the collector region by the electric field at the junction. This is the collector current i_C, that rises in time t_{ri} to the value

$I_C = I_l$, as shown in Fig. 4.7b. During the whole rise time t_{ri} the voltage across the BJT remains $v_{CE} = V_s$, because, as long as there is any current i_D in the diode ($i_D = I_l - i_C$), the voltage drop across it and the load remains zero. Kirchhoff's voltage law

$$V_s = v_{CE} + v_l \tag{4.5.7}$$

has to be satisfied at all times. Hence $v_{CE} = V_s$ for $t \leq (t_d + t_{ri})$. At the end of the interval t_{ri} the collector current has risen to $i_C = I_l$. According to Kirchhoff's current law

$$i_C + i_D = I_l = \text{constant} \tag{4.5.8}$$

the diode current has fallen to zero, so the diode becomes reverse biased and turns off. It is assumed that the diode has a fast recovery so that the rise of the load voltage v_l and the fall of collector-emitter voltage in time t_{fv} is governed mainly by the transistor action. In Fig. 4.7b the voltage is shown as a linear function. During this interval t_{fv} the BJT is initially in the active region. The voltage v_{CE} and the current i_C are high, so the power dissipation is high. As the voltage v_{CE} falls the collector-base junction becomes forward biased, the gain β falls as the stored charge stretches all across the drift region and the rate of fall of v_{CE} actually decreases (not shown for convenience). If I_B is high enough to produce overdrive at the end of the interval t_{fv} the BJT is in hard saturation and the voltage across the transistor is $V_{CE(SAT)}$, the minimum value. The total turn-on time is best defined for analysis to be the interval from the initiation of the gate signal to the steady conditions that $i_C = I_C$ and $v_{CE} = V_{CE(SAT)} \approx 0$. That is,

$$t_{on} = t_d + t_{ri} + t_{fv}\,. \tag{4.5.9}$$

An inspection of Fig. 4.7b allows us to estimate the energy W_{on} absorbed by the BJT during the turn-on process. We have approximated the curves by linearization, so we may as well approximate the turn-on intervals to be that t_d extends from time $t = 0$ to the time that the current i_C begins to increase, that t_{ri} is the interval from $i_C = 0$ to $i_C = I_l$, and that t_{fv} is the interval from the time $v_{CE} = V_s$ and starts to decrease to the time that $v_{CE} \approx 0$. We will let these intervals be contiguous. The energy lost is

$$W_{on} = \int_0^t p\,dt = \int_0^t v_{CE} i_C\,dt = \int_0^{t_d} V_s \times 0\,dt + \int_0^{t_{ri}} V_s I_l \frac{t'}{t_{ri}}\,dt' + \int_0^{t_{fv}} V_s I_l \left(1 - \frac{t''}{t_{fv}}\right) dt''$$

where $t' = t - t_d$ and $t'' = t - (t_d + t_{ri})$. Thus,

$$W_{on} = 0 + \frac{V_s I_l}{2} t_{ri} + \frac{V_s I_l}{2} t_{fv} = \frac{V_s I_l}{2}(t_{ri} + t_{fv})\,. \tag{4.5.10}$$

If we are given the turn-on times and the circuit steady-state conditions, the determination of the approximate value of the turn-on losses is straight forward. We have not accounted for the dissipation due to the base drive. Is this reasonable?

EXAMPLE 4.3

A BJT controls power from a dc source of voltage $V_s = 200$ V to an RL load with a freewheeling diode connected across it. The value of the load resistance is $R = 4\,\Omega$ and the value of the load inductance is high enough to consider the load current to be virtually constant at 48 A for steady-state operation as a chopper circuit. If the current rise-time is $t_{ri} = 1.1\,\mu$s and the voltage fall-time is $t_{fv} = 1.5\,\mu$s for turn-on, estimate the energy loss in the BJT during the turn-on process. Compare this loss with the loss for the case of a purely resistive load in EXAMPLE 4.2.

Solution

Refer to Fig. 4.7. With the BJT on the current $i_C = I_C = I_l = 48$ A (given). The voltage across the BJT, while it is on, will be neglected. There is negligible loss during the delay time t_d. Hence, the energy loss W_{on} during turn on is

$$W_{on} = \int_0^t p dt = \int_0^{t_{ri}} V_s I_C \frac{t'}{t_{ri}} dt' + \int_0^{t_{fv}} V_s I_C \left(1 - \frac{t''}{t_{fv}}\right) dt''.$$

So, $W_{on} = \dfrac{V_s I_C}{2}(t_{ri} + t_{fv}) = \dfrac{200 \times 48}{2}(1.1 + 1.3) \times 10^{-6}$.

$W_{on} = 11.52 \times 10^{-3}$ J.

This value is roughly six times the value of the turn-on loss of a BJT with a purely resistive load. The highly inductive load is the worst case for switching losses. In any analysis for heatsink design, the worst case should be considered.

4.6. BJT TURN-OFF

If the BJT is on, it means that there is a positive current I_B into the base terminal and there is a positive voltage at the collector terminal with respect to the emitter terminal. The transistor can be turned off while conducting current I_C. Remove the base current ($I_B = 0$) and the transistor effect is lost. Without the positive bias at the base-emitter *pn* junction, there is no electric field to accelerate the carriers from the emitter region to the base region. Accordingly, the BJT turns off. However, it takes a relatively long time to turn off, because blocking is achieved only after the excess charge carriers have recombined naturally. The speed of turn-off is aided by reverse biasing the base-emitter junction, providing a negative base current and sweeping out many of the charge carriers by this means. See Fig. 4.8 for the voltage and current waveforms during turn-off.

During turn-on the base current causes charge to be stored in the drift region. The higher is the base current, the greater is the stored charge, the greater the saturation and the shorter the turn-on time. At the initiation of turn-off with a negative base terminal, there is virtually no change in the values of current i_C and voltage v_{CE} until that stored charge is removed. The harder is the saturation, the more

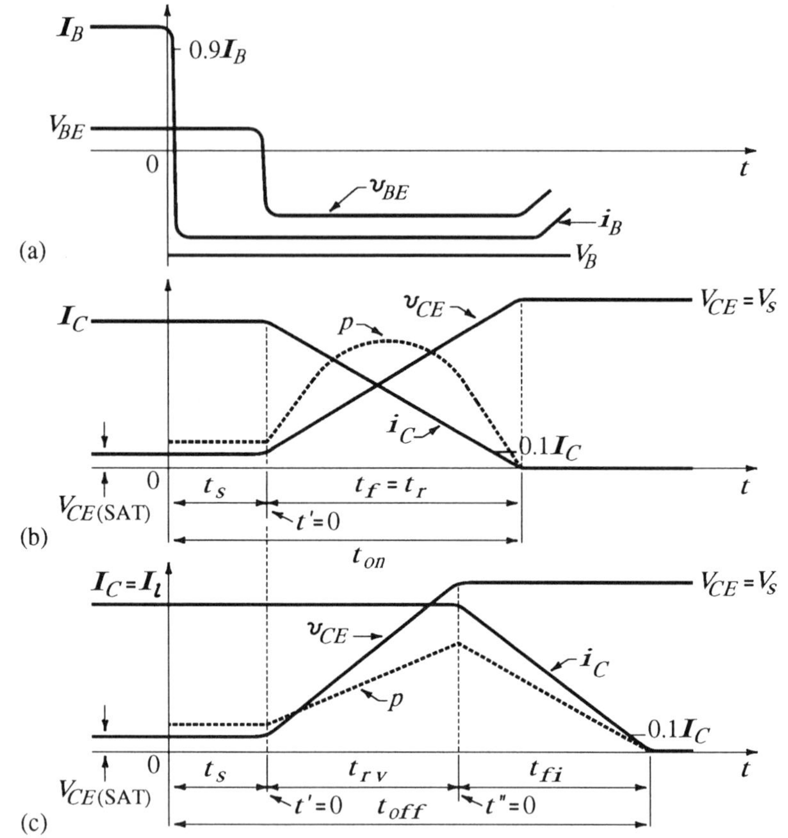

Fig. 4.8 BJT turn-off.
(a) Base signal, (b) resistive circuit response, (c) inductive circuit response.

is the stored charge, the longer it takes to remove, and the longer is the turn-off time. It appears that a short turn-on time is associated with a long turn-off time, so the designer of the chip has a compromise to make.

In the process of turn-off the delay taken to remove the stored charge is called the storage time t_s. After the stored charge has been removed, the condition of the BJT is between the saturation and the active regions. As shown in Fig. 4.8b, the voltage v_{CE} rises quickly in time t_r and the current i_C falls to zero in time t_f. During the fall time the remaining charge carriers are removed. The phase relation of voltage and current depends on the external-circuit parameters. Increasing the negative value of the base current $-I_B$ reduces the storage time t_s and the fall time t_f, but there is a limit to the reverse base-emitter junction voltage $-V_{BE}$. Avalanche breakdown occurs at about 10 V. The ratio of *on* base current to *off* base current ranges from about 0.5 to 2.

4.6.1. Turn-off Losses

The BJT user is given information about the main stages of turn-off. It includes storage time t_s, current fall time and voltage rise time and these values can be associated with base current, collector current and junction temperature. In order to calculate a rough estimate of the losses incurred in the BJT during turn-off, the user has to know the form of the main circuit configuration. We will take the two extreme cases of a purely resistive load and a highly inductive load, in the same way that turn-on losses were studied in the last section.

Resistive load. Consider the circuit shown in Fig. 4.6a. It has a purely resistive load. The BJT acts as a chopper to modulate the power that is absorbed in the load. Let the BJT be on, with $i_B = I_B$, $i_C = I_C \approx V_s/R$, and $v_{CE} = V_{CE(\mathrm{SAT})} \approx 0$. This condition is represented by the point X6, that is, the intersection of the resistive load line and the I-V characteristic in hard saturation (see Fig. 4.4).

At $t=0$ the switch Sw is opened, or, better still, the base current is reversed. Figures 4.8a and 4.8b illustrate the changes during turn-off. Over the storage-time interval t_s the excess charge carriers that are stored in the base and collector regions (in particular the n^- drift region) are swept out by the base current. The base-emitter junction becomes reverse biased, the BJT comes out of the saturation region into the active region, the current i_C and voltage v_{CE} follow the resistive load line down to the cutoff operating point X2, and, ideally, the base current is made to follow the profile of the collector current. After a time interval t_f, the fall time of the current, $i_C = 0$, $v_{CE} = V_{CE} = V_s$ and $V_l = 0$. The total time t_{off} for the BJT to turn off is

$$t_{off} = t_s + t_f. \tag{4.6.1}$$

More precisely, the turn-off time can be defined as the time interval between the base current being $0.9I_B$ and the collector current being $0.1I_C$. This accounts more reasonably with the problems of measurement techniques. For calculation, it is easier to linearize curves and use the limits of on and off values to describe time intervals.

With reference to Figs 4.8a and 4.8b, at time $t=0$, the direction of base current is reversed. After time t_s the stored charge is removed, but up to this point the losses W_s in the BJT are conduction losses that amount to

$$W_s = \int_0^{t_s} p\,dt = \int_0^{t_s} v_{CE} i_C dt = V_{CE(\mathrm{SAT})} I_C t_s \tag{4.6.2}$$

where p is the instantaneous power loss due to the simultaneous existence of v_{CE} and i_C. During the current fall interval t_f the losses W_f amount to

$$W_f = \int_0^{t_f} p\,dt' = \int_0^{t_f} v_{CE} i_C dt' = \int_0^{t_f} \frac{V_s}{t_f} t' I_C \left(1 - \frac{t'}{t_f}\right) dt' = \frac{V_s I_C}{6} t_f \tag{4.6.3}$$

where $t' = 0$ at $t = t_s$. This equation has the same form as the turn-on loss W_{on} expressed in eq. (4.5.6). The total loss W_{off} during the turn-off process is

$$W_{off} = W_s + W_f = V_{CE(\mathrm{SAT})} I_C t_s + \frac{V_s I_C}{6} t_f. \qquad (4.6.4)$$

The storage time t_s can be five times as long as the fall time t_f. From a circuit point of view the average power dissipation P_D in the switch is an important quantity that is used in the heatsink calculations. In terms of switching times and frequency f of switching an estimate of P_D can be obtained from

$$P_D = \frac{1}{T}\int_0^T p\,dt \approx (W_{on} + W_{off}) f + P_C. \qquad (4.6.5)$$

where P_C is the average conduction loss due to dissipation in the steady on-state.

EXAMPLE 4.4

A BJT switch, acting as a chopper with a switching frequency $f = 8$ kHz and duty cycle $m = 0.7$, modulates power from a dc source of voltage $V_s = 200$ V to a resistive load of value $R = 4\,\Omega$. The on-state BJT voltage drop is $V_{CE(\mathrm{SAT})} = 1.1$ V. In either the on-state or the off-state the base drive dissipates 10 W continuously in the BJT. For these conditions, the current rise time during turn-on is $t_r = 0.6\,\mu$s with a delay time $t_d = 0.1\,\mu$s, and the current fall time during turn-off is $t_f = 1.1\,\mu$s with a storage time $t_s = 3\,\mu$s. Determine the efficiency of the BJT converter.

Solution

Figures 4.6 and 4.8b illustrate the turn-on and turn-off characteristics of the BJT for a resistive load. The storage time t_s can be associated with the on-time t_{ON}, and the delay time t_d can be associated with the off-time t_{OFF}. Since the load is resistive, we can consider the voltage v_{CE} to fall linearly from V_s to $V_{CE(\mathrm{SAT})}$ in the same time t_r that the current i_C rises to $I_C = I_l = V_s/R$ at turn-on. Also, the voltage v_{CE} will rise linearly to V_s in the same time t_f that the current i_C falls to zero at turn-off.

No information is given about leakage current during the off-state, so we will assume that the negative base current keeps the leakage current negligibly small.

The aim is to determine the BJT losses.

The period of switching is $T = 1/f = 1/(8 \times 10^3) = 125\,\mu$s.

The duty ratio $m = 0.7$, so the ideal on-time is

$mT = 0.7 \times 125 = 87.5\,\mu$s.

The actual on-time $t_{ON} = mT + t_s - t_d - t_r$;

that is, $t_{ON} = 87.5 + 3 - 0.1 - 0.6 = 89.8\,\mu$s.

During conduction the load current is $I_C = (V_s - V_{CE(\mathrm{SAT})})/R$.

That is, $I_C = (200 - 1.1)/4 = 49.7$ A.

The energy loss W_C during conduction in the steady state is

$W_C = V_{CE(\mathrm{SAT})} I_C\, t_{ON} = 1.1 \times 49.7 \times 89.8 \times 10^{-6} = 4912\,\mu$J.

Equation (4.5.6) expresses the turn-on loss W_{on} as

$$W_{on} = \frac{V_s I_C}{6} t_r = \frac{200 \times 49.7}{6} \times 0.6 \times 10^{-6} = 995\,\mu\text{J}.$$

Equation (4.6.3) expresses the turn-off loss W_f during the fall time as

$$W_f = \frac{V\,I_C}{6} t_f = \frac{200 \times 49.7}{6} \times 1.1 \times 10^{-6} = 1823\,\mu\text{J}.$$

The energy loss W_B created by the base drive is
$W_B = P_B T = 10 \times 125 \times 10^{-6} = 1250\,\mu\text{J}$.
The average power dissipation P_D in the BJT is
$P_D = (W_C + W_{on} + W_f + W_B)/T$.
So, $P_D = (4912 + 995 + 1823 + 1250)/125 = 71.8\,\text{W}$.
The average power P delivered to the load is
$P \approx V_s m\, I_C = 200 \times 0.7 \times 49.7 = 6962\,\text{W}$.
Efficiency is $\eta = \text{output}/(\text{output} + \text{losses}) = 6962/(6962 + 71.8)$. This indicates an efficiency of conversion of almost 99%.

Inductive load. The other circuit to be considered for turn-off loss calculation is shown in Fig. 4.7. The chopper modulates power to an inductive load that has a freewheeling diode. If the time constant *L/R* of the load is long compared with the period of switching the BJT on and off, the load current is considered to be constant throughout the BJT turn-off interval t_{off}. Turn-off action in an inductive load is depicted in Figs 4.8a and 4.8c. Figure 4.8a gives the base signal as the reference at $t = 0$ and Fig. 4.8c illustrates the collector voltage and current responses. These curves are linearized for convenience.

Initially the BJT is on in hard saturation, the base current is reversed and excess charge carriers are swept out of the collector and base regions over an interval t_s. At $t = t_s$ $(t' = 0)$, the base-emitter junction becomes reverse biased, the BJT comes out of hard saturation and, with the transistor effect lost, the collector voltage v_{CE} rises quickly to the source value V_s in a time interval t_{rv}. Although the voltage rise is illustrated as being linear, it is curved in practice. There is a slow rise while the BJT is in the quasi-saturation region and then there is a more rapid rise while the BJT is in the active region of operation.

While the collector voltage v_{CE} rises over the interval t_{rv}, the collector current is maintained at $i_C = I_C = I_l$ by the load inductance. The freewheeling diode *D* cannot conduct because it is reverse biased by the load voltage v_l which is given by $v_l = V_s - v_{CE}$. At the end of the interval t_{rv} the diode *D* is no longer reverse biased. It conducts and its current i_D rises quickly to the value of the load current I_l. Over the same time interval t_{fi}, by Kirchhoff's current law $i_C = I_l - i_D$, the collector current i_C falls to zero, after which the BJT is truly off. There is no current i_C and the transistor blocks.

There is a changing instantaneous power loss p during turn-off. From the simplified geometry of the responses, shown in Fig. 4.8c, the energy loss W_{off} over the turn-off interval t_{off} is expressed by

$$W_{off} = \int_0^{t_{off}} p\,dt \approx \int_0^{t_s} V_{CE(\text{SAT})} I_l\,dt + \int_0^{t_{rv}} \frac{V_s}{t_{rv}} t' I_l\,dt' + \int_0^{t_{fi}} V_s I_l \left(1 - \frac{t''}{t_{fi}}\right) dt''.$$

Therefore, $$W_f = V_{CE(\mathrm{SAT})} I_l t_s + \frac{1}{2} V_s I_l (t_{rv} + t_{fi}) . \tag{4.6.6}$$

In some cases the storage time is defined as the interval from $0.9I_B$ to $0.9I_C$, in which case t_{rv} would be incorporated in the storage time. The fall time t_{fi} is a function of the temperature, the base current and the external circuit parameters. In general, the fall time is a fraction (about one fifth) of the storage time. In other cases the rise time t_{rv} and the fall time t_{fi} are combined and called the crossover time t_c. More specifically from a test measurement point of view, the crossover time can be defined as the interval of turn-off from the time the voltage $v_{CE} = 0.1V_s$ to the time the current $i_C = 0.1I_l$. That is, $t_c \approx t_{rv} + t_{fi}$. Very approximately, $t_s \approx 2t_c$. BJT data are found in Appendix 2.

EXAMPLE 4.5

A BJT chopper modulates power from a dc source of voltage $V_s = 200\,\mathrm{V}$ to an *RL* load, whose resistance is $R = 4\,\Omega$. The inductance L is high enough to consider the load current to be virtually constant if a freewheeling diode is connected across the load. See Fig. 4.7a. The chopper has a switching frequency of 8 kHz and a duty cycle of $m = 0.7$. The on-state BJT voltage drop is $V_{CE(\mathrm{SAT})} = 1.1\,\mathrm{V}$. For these conditions the turn-off storage time $t_s = 3\,\mu\mathrm{s}$, the voltage rise-time $t_{rv} = 1.1\,\mu\mathrm{s}$ and the current fall-time $t_{fi} = 1.3\,\mu\mathrm{s}$. Estimate the average power loss in the BJT due to the turn-off process and compare it with the conduction loss.

Solution

Refer to Fig. 4.8c to view the linearized turn-off waveforms.

Over the storage interval t_s at the beginning of turn-off the instantaneous value of the power p_s dissipated in the BJT is

$p_s = v_{CE} i_C = V_{CE(\mathrm{SAT})} I_C = V_{CE(\mathrm{SAT})} I_l$.

For a constant load current I_l the voltage across the load inductance is zero. Hence the load current is

$I_l = V_{l\,av}/R \approx mV_s/R = 0.7 \times 200/4 = 35\,\mathrm{A}$.

This is reasonable if the on-time t_{ON} of the BJT is long compared with the turn-off time t_{off}. A more accurate prediction of the value I_l can be made by accounting for the effect of the storage time t_s and the delay time t_d as shown in EXAMPLE 4.4.

The power $p_s = V_{CE(\mathrm{SAT})} I_l = 1.1 \times 35 = 38.5\,\mathrm{W}$.

The energy dissipated W_s during the storage time is

$$W_s = \int_0^{t_s} p_s dt = p_s t_s = 38.5 \times 3 \times 10^{-6} = 115.5 \times 10^{-6}\,\mathrm{J}.$$

During the interval of the voltage rise t_{rv}, the value of the instantaneous power p_{rv} dissipated in the BJT is

$$p_{rv} = v_{CE} i_C = \frac{V_s}{t_{rv}} t' \times I_C = \frac{V_s I_l}{t_{rv}} t' = \frac{200 \times 35}{1.1 \times 10^{-6}} t' = 6.36 \times 10^9 t' .$$

This ignores the finite value $v_{CE} = V_{CE(\mathrm{SAT})}$ at $t' = 0$.
The energy dissipated W_{rv} during the interval t_{rv} is

$$W_{rv} = \int_0^{t_{rv}} p_{rv} dt' = \int_0^{1.1\times10^{-6}} 6.36\times10^9 t' dt' = \frac{6.36\times10^{-3}}{2}(1.1)^2 = 3848\times10^{-6}\ \mathrm{J}.$$

During the interval of the current fall t_{fi}, the value of the instantaneous power p_{fi} dissipated in the BJT is

$$p_{fi} = v_{CE} i_C = V_s \times I_C \left(1 - \frac{t''}{t_{fi}}\right) = 200\times35\left(1 - \frac{t''}{t_{fi}}\right) = 7\times10^3\left(1 - \frac{t''}{t_{fi}}\right).$$

The energy dissipated W_{fi} during the interval t_{fi} is

$$W_{fi} = \int_0^{t_{fi}} p_{fi} dt'' = 7\times10^3\left(t_{fi} - \frac{t_{fi}}{2}\right) = 3.5\times10^3\times1.3\times10^{-6} = 4550\times10^{-6}\ \mathrm{J}.$$

The energy dissipated during the storage interval is small in comparison with the other components. The total energy W_{off} dissipated in the BJT during turn-off is
$W_{off} = W_s + W_{rv} + W_{fi} = (115.5 + 3848 + 4550)\times10^{-6} = 8514\times10^{-6}$ J.
The average power P_{off} dissipated due to the turn-off process while the chopper is switching at a frequency f (and period T) is

$$P_{off} = \frac{1}{T}\int_0^{t_{off}} p dt = \frac{1}{T}W_{off} = W_{off}\times f = 8514\times10^{-6}\times8\times10^3 = 68\ \mathrm{W}.$$

This loss is not insignificant and is due to the high frequency of chopper operation.
The instantaneous value of the power loss p_C during BJT conduction is
$p_C = v_{CE} i_C = V_{CE(\mathrm{SAT})} I_l = 1.1\times35 = 38.5$ W.
The average power loss P_C due to conduction is

$$P_C = \frac{1}{T}\int_0^{t_{ON}} p_C dt = \frac{t_{ON}}{T} p_C = m p_C \approx 0.7\times38.5 = 27\ \mathrm{W}.$$

The small effects of delay and storage have been ignored in the calculation of the conduction loss.
The switching loss is more than twice the conduction loss in this particular case.

4.6.2. Current Focusing

During the process of turn-off the current has a tendency to focus or concentrate in the central region of the emitter islands[3]. This phenomenon increases the power dissipation. Figure 4.9a illustrates the focusing action and why it occurs.

At turn-off the base terminal is negatively biased with respect to the emitter terminal by the base-driver source. The base current in the base region is mainly

[3] A single emitter island is shown in Fig. 4.2. but interdigitation provides a great number.

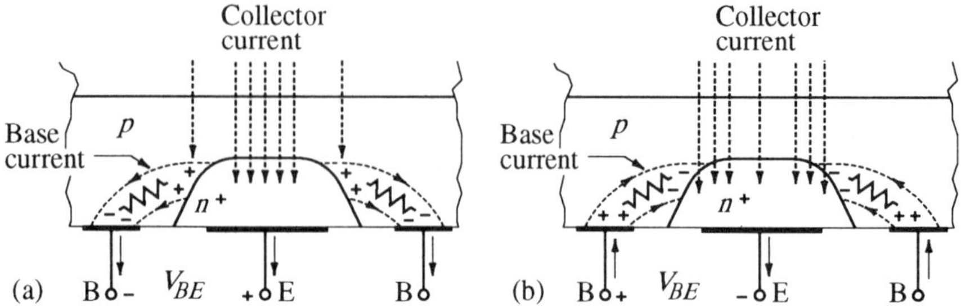

Fig. 4.9 Current focusing.
(a) Turn-off, (b) on-state.

from the emitter lateral periphery to the base metalization as shown by the curved broken lines. Due to the resistivity of the doped silicon the current $-I_B$ causes a potential drop such that the emitter lateral periphery is at a higher potential than the centre, as shown. The main current i_C will take the path towards the low potential at the emitter-base junction. This leads to the focusing and high current densities in small areas.

The same kind of focusing takes place while the BJT is in the steady on-state, as shown in Fig. 4.9b. In this case the base current is reversed and so too is the potential drop across the base region. The lower potential at the emitter lateral periphery makes the collector current crowd towards this area. It is the same phenomenon with the same effect as depicted in Fig. 4.9a except that it is directed towards a different area of concentration. The crowding, or focussing, can lead to overheating and breakdown. This is called *second breakdown.*

A more uniform current distribution is desirable in order to limit local temperature rise. The solution is to reduce the base-region resistivity. A higher level of doping in the base region will lower the resistance, but it also reduces the current gain β. A better way to alleviate the problem is the practice to interdigitate the emitter and base regions. The finger or island concept of the emitter and base reduces the length of the path of the base current and this reduces the resistance and the potential drop. Also the base current density is reduced.

4.7. BJT POWER DISSIPATION

The BJT dissipates power at all times that it is connected to an external circuit with a supply. Whenever the driver supplies base current, the power P_B, that is dissipated due to the base current I_B in the BJT, is given by

$$P_B = V_{BE} I_B \tag{4.7.1}$$

for as long as the base current has a finite value. This dissipation may have different values, $P_{B\,ON}$ for the on-state and $P_{B\,OFF}$ for the off-state. The base power dissipation is small compared with the other components of power dissipation.

As long as the BJT is in the off-state the power dissipation P_{OFF} is associated with the leakage current $I_{C\,\text{leak}}$. That is,

$$P_{OFF} = V_{CE} I_C \approx V_s I_{C\,\text{leak}}. \tag{4.7.2}$$

This is another small component of the total loss.

One of the important components of power dissipation in the BJT is the conduction loss P_{ON} while the switch is in the on-state. Its value is given by

$$P_{ON} = V_{CE(\text{SAT})} I_C. \tag{4.7.3}$$

The value of this power is relatively high.

Switching from the on-state to the off-state creates an energy loss W_{off}, whose value depends on the external circuit parameters. The energy loss is expressed by eq. (4.6.4) and eq. (4.6.6).

Switching from the off-state to the on-state creates an energy loss W_{on}. The relevant expressions for this loss are found in eq. (4.5.6) and eq. (4.5.10).

If the times associated with these loss components are short and the steady switching frequency f of the BJT is high, the important loss calculation is the average power. Such a calculation is important for the heatsink design. The average power dissipated P_D is given by

$$P_D \approx ((P_{B\,ON} + P_{ON})\, t_{ON} + (P_{B\,OFF} + P_{OFF})\, t_{OFF} + W_{off} + W_{on}) f. \tag{4.7.4}$$

Care has to be taken to account for the delay time t_d and the storage time t_s with respect to the off-state time t_{OFF} and the on-state time t_{ON}.

EXAMPLE 4.6

A BJT switch, acting as a chopper with a switching frequency $f = 8$ kHz and duty cycle $m = 0.7$, modulates power from a dc source of voltage $V_s = 200$ V to an RL load with a freewheeling diode. The load resistance is $R = 4\,\Omega$ and the value of the load inductance is high enough to consider the load current to be virtually constant over a period of switching. While in the on-state the BJT voltage drop is $V_{CE(\text{SAT})} = 1.1$ V. While in the off-state the leakage current is $I_{C\,leak} = 0.1$ A. In both the on-state and the off-state the base drive dissipates 10 W continuously in the BJT since there is always a positive or a negative base current. For these conditions data sheets provide the information that during turn-on the delay time is $t_d = 0.1\,\mu$s, the current rise time is $t_{ri} = 0.6\,\mu$s and the voltage fall time is $t_{vf} = 0.6\,\mu$s. During turn-off the storage time $t_s = 3\,\mu$s, the voltage rise time $t_{rv} = 2.6\,\mu$s and the current fall time is $f_{fi} = 1.1\,\mu$s. **(a)** Determine the component values of the BJT energy dissipation. **(b)** If the ambient temperature is 25°C and if the BJT junction-to-case thermal resistance is $R_{\theta JC} = 0.3$°C/W, what is the maximum value of the heatsink thermal resistance $R_{\theta CA}$ (case to ambient) so that the junction temperature does not exceed 150°C.

Solution

Figures 4.7 and 4.8c show the circuit diagram and the waveforms of voltages and currents, and indicate the nomenclature.
By definition the on-time t_{ON} is the interval that $i_C = I_C$ and $v_{CE} = V_{CE(\mathrm{SAT})}$. In like manner the off-time t_{OFF} is the interval that $i_C \approx 0$ and $v_{CE} \approx V_S$.

(a) Period $T = 1/f = (1/8) \times 10^{-3} = 125\,\mu\mathrm{s}$.
$mT = 0.7 \times 125 = 87.5\,\mu\mathrm{s} = t_d + t_{ri} + t_{fv} + t_{ON}$.
$t_{ON} = mT - t_d - t_{ri} - t_{fv} = 87.5 - 0.1 - 0.6 - 0.6 = 86.2\,\mu\mathrm{s}$.
$t_s + t_{rv} + t_{fi} + t_{OFF} = (1-m)T = 0.3 \times 125 = 37.5\,\mu\mathrm{s}$.
$t_{OFF} = 37.5 - 3 - 2.6 - 1.1 = 30.8\,\mu\mathrm{s}$.
On-state $I_C = (V_s - V_{CE(\mathrm{SAT})})/R = (200 - 1.1)/4 \approx 49.7\,\mathrm{V}$.
Off-state $V_{CE} = V_s - I_{C\,leak}R = 200 - 0.1 \times 4 \approx 200\,\mathrm{V}$.
For the delay interval $t_d = 0.1\,\mu\mathrm{s}$:

Energy dissipation $W_d = \int_0^{t_d} V_{CE}\, I_{C\,leak}\, dt \approx V_s I_{C\,leak} t_d$.

So, $W_d = 200 \times 0.1 \times 0.1 \times 10^{-6} = 2\,\mu\mathrm{J}$.
For the current rise interval $t_{ri} = 0.6\,\mu\mathrm{s}$

Energy dissipation $W_{ri} = \int_0^{t_{ri}} V_{CE}\, i_C\, dt = \dfrac{V_s I_C}{2} t_{ri}$.

So, $W_{ri} = 200 \times 49.7 \times 0.6 \times 10^{-6}/2 = 2984\,\mu\mathrm{J}$.
For the voltage fall interval $t_{fv} = 0.6\,\mu\mathrm{s}$:

Energy dissipation $W_{fv} = \int_0^{t_{fv}} v_{CE}\, I_C dt \approx \dfrac{V_s I_C}{2} t_{fv}$.

So, $W_{fv} = 200 \times 49.7 \times 0.6 \times 10^{-6}/2 = 2984\,\mu\mathrm{J}$.
For the on-state interval $t_{ON} = 86.2\,\mu\mathrm{s}$:

Energy dissipation $W_{ON} = \int_0^{t_{ON}} v_{CE}\, i_c dt = V_{CE(\mathrm{SAT})} I_C t_{ON}$.

So, $W_{ON} = 1.1 \times 49.7 \times 86.2 \times 10^{-6} = 4713\,\mu\mathrm{J}$.
For the storage interval $t_s = 3\,\mu\mathrm{s}$:

Energy dissipation $W_s = \int_0^{t_s} v_{CE}\, I_C dt = V_{CE(\mathrm{SAT})} I_C t_s$.

So, $W_s = 1.1 \times 49.7 \times 3 \times 10^{-6} = 164\,\mu\mathrm{J}$.
For the voltage rise interval $t_{rv} = 2.6\,\mu\mathrm{s}$:

Energy dissipation $W_{rv} = \int_0^{t_{rv}} v_{CE}\, I_C dt \approx \dfrac{V_s I_C}{2} t_{rv}$.

So, $W_{rv} = 200 \times 49.7 \times 2.6 \times 10^{-6}/2 = 12{,}930\,\mu\mathrm{J}$.
For the current fall time $t_{fi} = 1.1\,\mu\mathrm{s}$:

Energy dissipation $W_{fi} = \int_0^{t_{fi}} V_{CE}\, i_C dt \approx \dfrac{V_s I_C}{2} t_{fi}$.

So, $W_{fi} = 200 \times 49.7 \times 1.1 \times 10^{-6}/2 = 5470\,\mu\mathrm{J}$.
For the off-state interval $t_{OFF} = 30.8\,\mu\mathrm{s}$:

Energy dissipation $W_{OFF} = \int_0^{t_{OFF}} V_{CE} I_c \, dt \approx V_s I_{C\,leak} t_{OFF}$.

So, $W_{OFF} = 200 \times 0.1 \times 30.8 \times 10^{-6} = 616\,\mu J$.
For the period of the cycle $T = 125\,\mu s$:
The base energy dissipation is $W_B = P_B T = 10 \times 125 \times 10^{-6} = 1250\,\mu J$.
An inspection of the component losses reveals that the turn-off energy dissipation is dominant.

(b) The total energy W_D dissipated over a complete cycle is
$W_D = W_d + W_{ri} + W_{fv} + W_{ON} + W_s + W_{fv} + W_{fi} + W_{OFF} + W_B$.
So, $W_D = 2 + 2984 + 2984 + 4713 + 164 + 12{,}930 + 5470 + 616 + 1250 = 31{,}113\,\mu J$.
The total average power P_D dissipated in the BJT is
$P_D = W_D/T = 31113/125 = 249\,W$.
The source delivers $P_s = V_s I_s \approx V_s \times m I_C = 200 \times 0.7 \times 49.7 = 6962\,W$.
The BJT losses are about 3.6% of the power delivered and more than three times the loss incurred with a resistive load. See EXAMPLE 4.4.
The thermal equation for steady-state operation is
$\Delta T = T_J - T_A = R_{\theta JA} P_D = (R_{\theta JC} + R_{\theta CA}) P_D$. (See section 5.10 for more details.)
Hence the necessary thermal resistance between the BJT case and ambient is

$$R_{\theta CA} = \frac{T_J - T_A}{P_D} - R_{\theta JC} = \frac{150 - 25}{249} - 0.3 = 0.2^\circ C/W.$$

4.8. BJT BASE DRIVE

The base-drive circuit of the transistor has a dual purpose. It is designed to provide current at the base of the BJT to turn on the switch quickly and also to provide current to turn off the switch quickly. In the latter case the base current must be in the reverse direction from that for turn-on.

Figure 4.7a shows a typical BJT circuit diagram. The very simple base driver comprises the dc source voltage V_B, the switch *Sw* and the resistor R_B. Values of these parameters may be of the order of $V_B = 15\,V$ and $R_B = 10\,\Omega$ for a BJT that has a current rating about 20 A. The switch *Sw* might be another BJT that is controlled by a low-level signal source. This circuit is simple, indeed. What is not shown is the circuit for base-current reversal for BJT turn-off. Base-drive circuits can be complex.

In general, there are some ideal requirements related to BJT turn-on and turn-off. For fast turn-on, the base current must rise quickly to a high value. For fast turn-off and lower losses, it is better that the BJT is in a low level of saturation, so that once the switch is in the on-state the base current should be reduced. In fact, while in the on-state, best results are obtained if the value of the base current is controlled to follow the profile of the collector current. As the I-V characteristics show, in Fig.4.3c, the lower is the value of the collector current, the lower is the value of base current to keep the BJT in the quasi-saturation region. Although it is not shown on any of the figures, it is assumed that there is electrical isolation

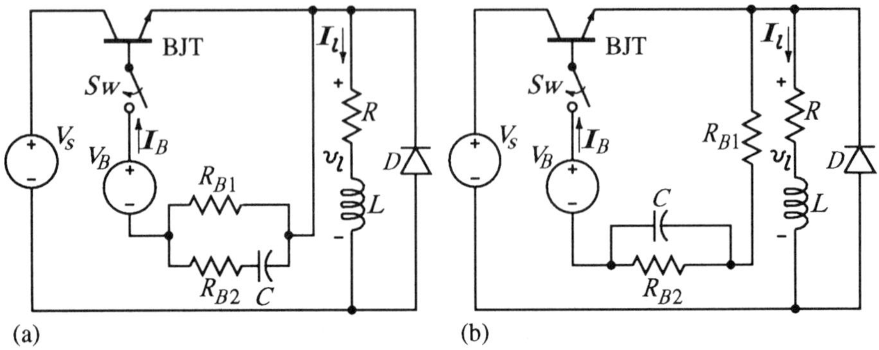

Fig. 4.10 Base drive circuits.
(a) Parallel resistors, (b) series resistors.

between the base circuit and the base-driver source V_B. Further, there is electric decoupling between the signal source and the switch *Sw*, usually by means of an optocoupler.

Two similar base-drive circuits for turn-on are shown in Fig. 4.10. In both cases the results are the same. A control signal will cause the switch *Sw* to close. The source voltage V_B will appear as a step input to the base circuit, so that the capacitor appears as a short circuit initially. The effective resistance for each base circuit is a minimum for this condition, so the base current rises to a maximum, drives the base hard and turns on the BJT quickly. As the capacitor *C* charges up, the current I_B falls. When the capacitor is fully charged it appears as an open circuit. At this point the effective circuit resistance is a maximum, so that the base current I_B is reduced to a level that brings the BJT into the quasi-saturation level. This facilitates turn-off by reducing the storage time t_s. In the design of a base-drive circuit such as this account is taken of turn-on times and *RC* time constants.

EXAMPLE 4.7

A BJT switch, acting as a chopper with a switching frequency $f = 8$ kHz and a duty cycle $m = 0.7$, modulates power from a dc source of voltage $V_s = 286$ V to an *RL* load whose resistance value is $R = 4\,\Omega$. The circuit diagram is shown in Fig. 4.10a. For a fast rise-time t_{ri} of 0.4 μs, the initial forced gain is to be $\beta_F = 5$ and before turn-off the gain is increased to 10. Estimate the values of the base drive parameters R_{B1}, R_{B2} and *C* if the drive supply is $V_B = 15$ V.

Solution

The on-state current is $I_C = I_l \approx V_{l\,av} / R \approx mV_s / R = 200 / 4 = 50$ A.
It is assumed that the load current I_l is constant.
At turn-on *C* is a short circuit, and the base current must be
$I_B = I_C / \beta_F = 50 / 5 = 10$ A.

Therefore, $\dfrac{R_{B1}R_{B2}}{R_{B1}+R_{B2}}=\dfrac{V_B}{I_B}=\dfrac{15}{10}=1.5\,\Omega.$

Just before turn-off the capacitor is fully charged and the base current must be $I_B=I_C/\beta=50/10=5$ A.

Hence, $R_{B1}=V_B/I_B=15/5=3\,\Omega$.

From the equation in R_{B2}, $R_{B2}=3\,\Omega$.

The BJT time in the on-state is approximately given by

$t_{ON}=mT=0.7\times125=87.5\,\mu s$. ($T=1/f, f=8$ kHz.)

For full charge on C in 87.5 µs, $5R_{B2}C<87.5\,\mu s$.

In the limit $C=87.5/15=5.8\,\mu F$.

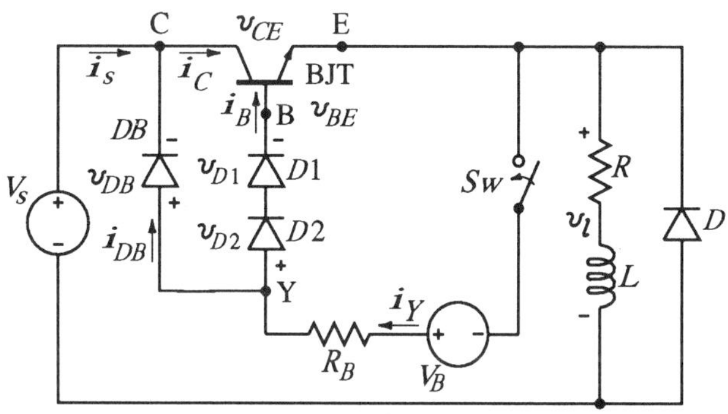

Fig. 4.11 A Baker clamp for antisaturation.

4.8.1. Baker Clamp

A BJT in hard saturation can take a long time to turn off. There is an antisaturation circuit, called a Baker clamp. This combats the problem by allowing some current to bypass the base to keep the BJT out of hard saturation. Figure 4.11 illustrates one of a number of possibilities. The important arrangement is to have $v_{YB}>v_{YC}$. This is indicated by having two diodes in the branch YB and only one diode in branch YC.

At all times we have the voltage between the points Y and E given by the expressions

$$v_{YE}=v_{DB}+v_{CE}=(v_{D1}+v_{D2})+v_{BE} \tag{4.8.1}$$

The significance of this equation is that $(v_{D1}+v_{D2})>v_{DB}$ by design and that v_{CE} undergoes a great change of value at turn-on.

If the BJT is off initially, and then the base-circuit switch *Sw* is closed, the turn-on process begins. At first, the voltage v_{CE} is high so that the Baker diode *DB* is reverse biased and $i_{DB}=0$. All the current i_Y is directed to the base to turn on the BJT quickly and i_C rises to its full-load value I_C. After this the voltage v_{CE} across the BJT falls until the diode DB becomes forward biased and conducts current i_{DB}. Now the voltage V_{CE} across the BJT is clamped to a constant voltage, given by rearranging eq. (4.8.1), so that, in the steady state

$$V_{CE} = V_{BE} + (V_{D1} + V_{D2}) - V_{DB}. \tag{4.8.2}$$

This voltage is a constant because V_{BE} is a constant (approximately 0.7 V) and the voltages across the forward biased diodes are approximately constant. The actual value of V_{CE} can be designed to be any convenient value above $V_{CE(\mathrm{SAT})}$. Since the voltage drops across *pn* junctions are roughly the same, $V_{BE} \approx V_{DB}$. Consequently, attention is given to the value of $(V_{D1} + V_{D2})$ by increasing the number of diodes or using a zener diode. The choice is a compromise. If V_{CE} is made large the storage time t_s is short, the turn-off time is reduced and the turn-off loss W_{off} is reduced. However the on-state loss W_{ON} is increased.

EXAMPLE 4.8

A BJT chopper is to modulate power from a dc source of 200 V to an *RL* load whose resistance is $R = 4\,\Omega$. A Baker clamp is used to keep the BJT out of hard saturation in order to shorten the turn-off time. The circuit diagram is shown in Fig. 4.11. The base drive has a source voltage $V_B = 10$ V and a resistance $R_B = 1.266\,\Omega$. Each diode has a conduction drop of 1 V, and the base-emitter voltage is $V_{BE(\mathrm{SAT})} = 1.7$ V with a current gain of $\beta = 10$. Estimate the increase of the conduction power dissipation if the diode clamp is used.

Solution

Without the diode *DB* there is no clamping.
Base current $I_B = (V_B - 2V_D - V_{BE(\mathrm{SAT})})/R_B$.
So, $I_B = (10 - 1 - 1.7)/1.266 = 4.975$ A.
Therefore, $I_C = \beta I_B = 10 \times 4.975 = 49.75$ A.
$V_{CE(\mathrm{SAT})} = V_s - R\,I_C = 200 - 4 \times 49.75 = 1.0$ V.
Conduction loss is $P_C = V_{CE(\mathrm{SAT})} I_C = 1.0 \times 49.75 = 49.75$ W

With the diode *DB* to clamp the BJT
$V_{CE} = V_{BE(\mathrm{SAT})} + 2V_D - V_{DB} = 1.7 + 2 - 1 = 2.7$ V.
V_{CE} is greater than $V_{BE(\mathrm{SAT})}$ so the BJT is out of saturation.
The main current is $I_s = (V_s - V_{CE})/R = (200 - 2.7)/4 = 49.33$ A.
Base-drive current $I_Y = (V_B - 2V_D - V_{BE(\mathrm{SAT})})/R_B = 4.975$ A.
Collector current $I_C = \beta I_B = \beta(I_Y - I_{DB}) = \beta(I_Y - I_C + I_s)$.
So, $I_C = \dfrac{\beta}{\beta+1}(I_Y + I_s) = \dfrac{10}{11}(4.975 + 49.33) = 49.36$ A.
Conduction loss $P_C = V_{CE}\,I_C = 2.7 \times 49.36 = 133.2$ W.
The conduction loss is increased by 168% with the use of a Baker clamp.

4.9. BJT PROTECTION

Care has to be taken to protect the diode. More care has to be taken to protect the BJT because of its fragility. There is less tolerance to overcurrents in bipolar transistors and greater sensitivity to overvoltages. High frequencies of switching can be used with BJTs so switching losses can be high because of the transition through the active region, so thermal conditions are especially important.

4.9.1. Overcurrent

Overcurrent is associated with the on-state. Diodes can carry peak currents that are many times the average current rating. For a BJT the peak current is the same as the average current rating, and this is the same for all minority carrier devices. The I-V characteristics of the BJT are depicted in Fig. 4.3. They show that, in the on-state, as the collector current I_C increases, the voltage V_{CE} also increases, so the power dissipation increases and the temperature at the junction rises. This appears to be tolerable but care has to be taken.

The BJT has a negative temperature coefficient of effective resistance. As the current I_C rises, the power dissipation increases so the temperature rises. With a temperature rise there is an associated effective resistance decrease. This drop can cause a higher current I_C, more dissipation, a further rise in temperature, and so on. This positive feedback can give rise to thermal runaway that leads to the destruction of the BJT.

Fuses cannot be used directly as a means to limit the current in a BJT because the actions in the transistor are much faster than those in a fuse. Only a few microseconds are available to set up protection. The ideal way to protect the BJT is to monitor the current I_C and the voltage V_{CE}. If they both rise above reference values, the control circuit can inject a negative signal to the base terminal to turn off the BJT. This utilizes the full controllability of the BJT.

There are circumstances such as severe faults when protection must be afforded by a shorting switch that is connected in parallel with the BJT. Detection of a severe fault can provide a signal to turn on the shunt switch, which must have a lower voltage drop and a higher tolerance to transient fault current (the thyristor, for example) than the BJT. The shunt switch diverts the current from the BJT. The BJT is protected. What about the fast acting shunt switch? It must be protected by a fuse blowing to clear the fault. This form of protection is called a crowbar circuit.

4.9.2. Overvoltage

Overvoltage is associated with the off-state. For any *pn* junction there is the likelihood of primary break down, if the reverse-bias voltage, collector to base, exceeds a limit. Primary breakdown is commonly called avalanche breakdown and is associated with impact ionization.

Minority carriers, accelerated by the junction field that is produced by reverse bias, can collide with impurities to break bonds and produce more carriers which can also be accelerated. If the reverse bias becomes too high, the minority carrier flow can avalanche to very high current levels without a decrease in the voltage drop. The resulting high power dissipation usually damages the device. The drift region of the collector allows BJTs to be designed so that the withstand reverse bias at the collector can be over 1000 V. However, the base-emitter junction reverse bias limit is only about 10 V because of the heavily doped emitter region. The limit can be increased by making a thicker base region but this reduces β. Protection against all reverse voltages at the emitter is provided by diodes in reverse parallel across the BJT and is found in such applications as inverters.

Another type of voltage that must be avoided for a BJT in the off-state is the sustaining voltage $V_{CE(SUS)}$ which is less that the avalanche breakover voltage $V_{CE(BO)}$. Any reverse bias voltage will give rise to a leakage current. If the BJT has no base driver, but is blocking a voltage V_{CE}, the leakage current I_C means that there will be charge flow in the base region. The base-region current causes the leakage current to be larger than in a single *pn* junction because of the transistor effect ($I_C = \beta I_B$). If the applied blocking voltage becomes too high, the transistor effect causes the collector current I_C to be increased greatly, without a decrease in V_{CE}. Hence, the large power dissipation $V_{CE}I_C$ will damage the BJT. Data sheets provide information about the values of the sustaining voltage, so that the user can take precautions against exceeding the value. Capacitors across the circuit or a snubber circuit offer transient voltage protection. Steady high values of voltage must be limited by other means such as voltage arresters.[4].

4.9.3. Safe Operating Area (SOA)

A manufacturer includes in the data sheets the values of current I_C and voltage V_{CE}, which, together simultaneously, must not be exceeded, if the BJT is to operate safely within thermal limits. On a graph, as shown in Fig. 4.12, the boundaries of the envelope of current and voltage form a *safe operating area* (SOA). These areas are mainly of concern for power transistors operating in the active region, although pulsed operation is depicted by the broken line.

There are four divisions illustrated for the dc case. Division 1 represents a limit of maximum collector current to prevent melting the bonding connections. A temperature limit is imposed on the BJT junction and division 2 creates a boundary beyond which the product of I_C and V_{CE} would give dissipation that would cause overheating. Division 3 is a limit produced by second breakdown. Second breakdown is a weakness of BJTs. There is a tendency for the collector current to become filamentary (or focused) in some locations due to imperfections or variation of potential (described in Fig. 4.9). The focusing creates high current density. If there is high voltage V_{CE} associated with high current density, then there is overheating and damage, called second breakdown. Second breakdown will not

[4] A voltage arrester is a nonlinear resistor, whose value decreases as the voltage increases.

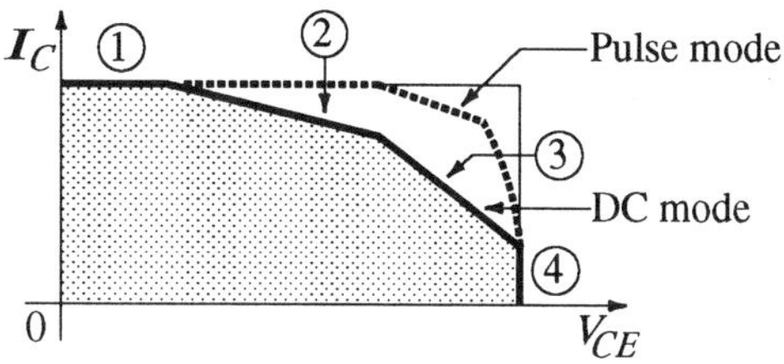

Fig. 4.12 Safe operating area.

occur if operation is kept within the boundary of division 3. Division 4 is the normal breakdown voltage that any junction will experience and offers the maximum voltage boundary.

The SOA is mainly for power transistors that operate in the active region. For BJT switches, the cutoff and saturation are the steady operating regions. Only at turn-on and turn-off is the active region traversed for a short period. Pulsed operation extends the boundary of the SOA to a rectangle in the limit. The curves provided by the manufacturer are much more complex than Fig. 4.12. There are forward bias SOAs and reverse bias SOAs. The latter have smaller areas, because the negative base bias increases the likelihood of second breakdown due to a more uneven distribution of current.

4.9.4. Transients

Transient currents and voltages are suppressed in the same way for transistors as for any semiconductor switch. Inductors limit the rate of change of current and capacitors limit the rate of change of voltage.

The series snubber circuit is an inductor L_s and its use is to limit the rate of rise of transistor current di_C/dt at turn-on. If the collector current i_C rises too quickly as the voltage v_{CE} falls, there is the possibility of second breakdown. The value of the inductance L_s can be calculated from relations obtained in section 4.5 and Fig. 4.7, that describe turn-on. The rate of change of current has two expressions associated with the BJT and one associated with the external circuit. Combined, these are

$$\frac{di_C}{dt} = \frac{I_C}{t_r} = \frac{V_s}{L_s} \tag{4.9.1}$$

where $I_C = I_l$, the load current and L_s is in series with the source. This gives us the requirement for the value of the inductance L_s as

$$L_s = V_s\, t_r / I_l\,. \tag{4.9.2}$$

A similar argument holds for turn-off. The voltage v_{CE} must not rise too quickly as the collector current falls, otherwise second breakdown can occur. A parallel snubber, comprising a capacitor C_s across the BJT, limits the rate of change of the voltage v_{CE}. Referring to section 4.6 and Fig. 4.8 that describe turn-off for an inductive load, there are two expressions for the rate of change of voltage. They are

$$\frac{dv_{CE}}{dt} = \frac{V_{CE}}{t_f} = \frac{i}{C_s}. \tag{4.9.3}$$

Since the final voltage across the BJT is $V_{CE} = V_s$, the source voltage, and since the capacitor current while $v_{CE} < V_s$ is $i \approx I_l$, the constant load current, then the expression for the value of the capacitance is

$$C_s = I_l\, t_f / V_s. \tag{4.9.4}$$

Besides avoiding second breakdown at turn-off, the parallel snubber circuit shifts much of the switching power dissipation from the BJT to the capacitor circuit.

It is also possible to reduce the effects of second breakdown at turn-off by limiting the rate of change of base current $-di_B/dt$, because, as Fig. 4.9a shows, it is the base current that can induce collector current focusing.

The base circuit of a BJT is not as sensitive as the gate circuit of a thyristor. However, there may be a small capacitor across the base and emitter terminals of the BJT to reduce oscillations, and a damper resistance may be connected in parallel. The base terminal is physically connected to the power circuit through the BJT, but the power source for the base drive and the signal source to the base drive will both be electrically isolated to protect the low voltage components.

EXAMPLE 4.9

Figure EX4.9 shows the circuit diagram of a boost converter (see section 2.2.2). The circuit is ideal except for the BJT. That is, the source current I_s is constant and the load voltage V_l is constant. The parallel snubber circuit R_sC_s controls the BJT voltage rise time at turn-off. The current fall time is $t_{fi} = 2\ \mu s$ at turn-off and the crossover time is $t_c = 1\ \mu s$ at turn-on. Assume that the delay time t_d is negligible and let the storage time t_s be incorporated with the on-time t_{ON}. Compare the turn-on and turn-off losses.

Solution

Section 2.2.2 describes the action of a boost converter. The voltage ratio is given to be $V_s/V_l = 1{:}2$, the converter is a dc transformer, so the current ratio is $I_s/I_l \approx 2{:}1$ for power invariance (losses are small compared with the power absorbed by the load).

The load current is $I_l = V_l/R = 600/4 = 150$ A.

Accordingly, the supply current is $I_s = 2 \times 150 = 300$ A.

Over the crossover interval t_c at turn-on, the BJT loss W_c is

$W_c = V_l\, I_s\, t_c/2 = 600 \times 300 \times 10^{-6}/2 = 90{,}000\ \mu J$.

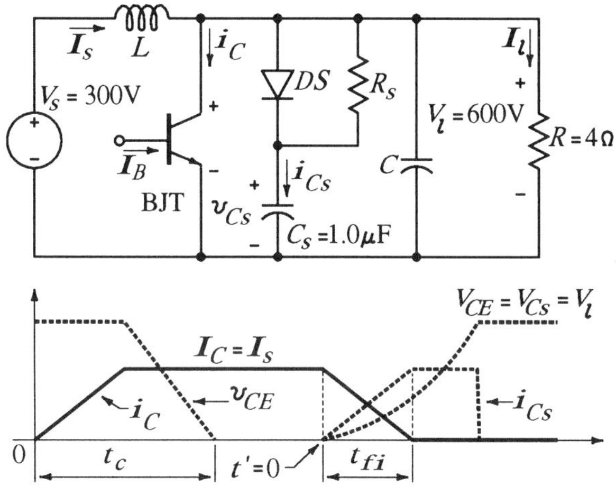

Fig. EX4.9

It is assumed that the capacitor C_s discharges in the resistor R_s.
Over the turn-off interval t_{fi} $(0 \le t' \le 2\,\mu\text{s})$ the current i_C falls linearly from $i_C = I_C = I_s$ to $i_C = 0$ and the capacitor current i_{Cs} rises linearly from $i_{Cs} = 0$ to $i_{Cs} = I_s$. If $t' = 0$ is the initiation of turn-off, the currents are given by

$i_C = I_s(1 - t'/t_{fi}) = 300(1 - 0.5 \times 10^6 t')$,

$i_{Cs} = I_s t'/t_{fi} = 150 \times 10^6 t'$ and $v_{Cs} = v_{CE} \approx \dfrac{1}{C_s}\displaystyle\int_0^{t'} i_{Cs}\,dt' = 7.5 \times 10^{13} t'^2$ V.

Power p dissipated in the BJT during the turn-off interval is
$p = v_{CE}\, i_C = 7.5 \times 10^{13}\, t'^2 \times 300\,(1 - 0.5 \times 10^6\, t')$.
Therefore, the energy loss W_{off} over the same interval is

$$W_{off} = \int_0^{t'} p\,dt' = 22.5 \times 10^{15} \left[\frac{t'^3}{3} - 0.5 \times 10^6 \frac{t'^4}{4} \right]_0^{t'} .$$

For, $t' = t_{fi} = 2\,\mu\text{s}$, $W_{off} = 22.5 \times 10^{15} \times 8 \times 10^{-18} (1/3 - 1/4) = 15{,}000\,\mu\text{J}$.
Without the snubber capacitor C_s, $v_{CE} = V_l$ and the loss W'_{off} over t_{fi} is
$W'_{off} = V_l\, I_s\, t_{fi}/2 = 600 \times 300 \times 10^{-6} = 180{,}000\,\mu\text{J}$.
Therefore, the BJT turn-off loss is reduced ten fold by C_s as well as affording the protection of lower *dv/dt*.

4.10. BJT RATINGS AND APPLICATIONS

The power industry wants to be able to modulate higher and higher power levels more and more rapidly with simpler controls. BJTs can handle less power than thyristors, but switched BJTs can modulate power more rapidly to give better quality. To a certain extent the BJT is simpler to control because it can be turned off with a signal and this is not the case for a thyristor.

The voltage ratings of BJTs have a range that is greater than 1000 V. This rating is the forward blocking voltage. There is no significant reverse blocking voltage as with the conventional thyristor. The unidirectional current ratings can be of the order of hundreds of amperes. Switching times are a few microseconds, so the switching frequency can be of the order of tens of kilohertz. Base currents I_B to control the on and off states are high in comparison with thyristor gate currents I_G ($I_B = 3$ A continuous and $I_G = 0.2$ A pulsed for 40 A devices). Therefore, the power supplies that provide the base current are relatively bulky and costly. A successful attempt to reduce the base driver power requirement has resulted in the cascaded arrangement of BJTs.

Darlington connection. Successful cascading of two or three BJTs, called the Darlington configuration, has led to the manufacture of such a package on a single chip. Figure 4.13 shows a configuration without the normal improvements for optimum operation. The external connections are a collector terminal C, an emitter terminal E and a base terminal B. Internally, the two transistors have a common collector connection and the emitter of *BJT1* is connected directly to the base terminal of *BJT2*. With a positive current signal I_B injected at terminal B and a positive voltage V_s at terminal C, the correct conditions exist for transistor *BJT1* to turn on. With *BJT1* in the on-state, current from its emitter is the base current I_{B2} of transistor *BJT2*. So, transistor *BJT2* turns on and the current I_C from the voltage source V_s exists in the load, that is not shown in the figure. By definition the current gains of the individual transistors and the overall gain β of the Darlington connection are

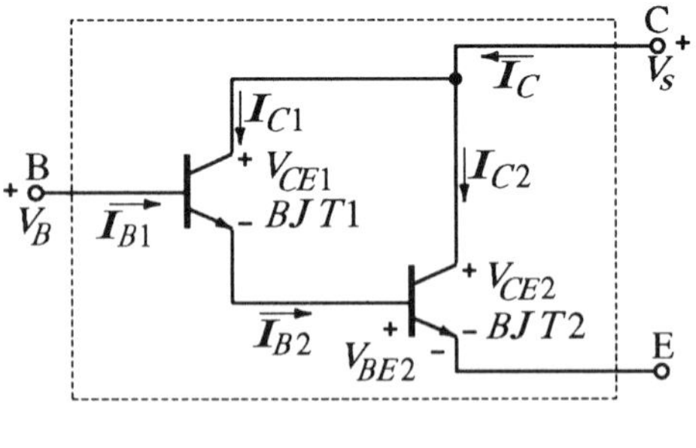

Fig. 4.13 A simplified Darlington connection.

$$\beta_1 = \frac{I_{C1}}{I_{B1}}, \quad \beta_2 = \frac{I_{C2}}{I_{B2}} \quad \text{and} \quad \beta = \frac{I_C}{I_{B1}}. \tag{4.10.1}$$

The nodal equations are

$$I_C = I_{C1} + I_{C2} \quad \text{and} \quad I_{B2} = I_{C1} + I_{B1}. \tag{4.10.2}$$

Hence, the overall gain is

$$\beta = \beta_1 + \beta_2 + \beta_1 \beta_2. \tag{4.10.3}$$

If the main power switch *BJT2* was rated at 400A with a gain $\beta_2 = 10$ and if this switch were used alone, it would require a base driver to supply a current of 40 A to turn it on. In the Darlington configuration with the transistor *BJT1* rated at 40 A and with a gain $\beta_1 = 20$, the overall gain is $\beta = 230$. This reduces the base driver current to less than 2 A, a great improvement over the single BJT.

Penalties have to be paid for overall gain improvement. Darlington configurations have a higher on-state voltage than the equivalent single transistor, so on-state losses are higher. In a single BJT $V_{CE(\text{SAT})}$ is less than $V_{BE(\text{SAT})}$, because the base-collector junction is forward biased. However, in a Darlington connection, the figure shows that

$$V_{CE(\text{SAT})2} = V_{CE(\text{SAT})1} + V_{BE(\text{SAT})2}. \tag{4.10.4}$$

So the transistor *BJT2* cannot be driven into hard saturation and this means an increased on-state voltage.

Another penalty to be paid by using the Darlington configuration is longer switching times. Unless modifications are made, the transistors switch on and off sequentially, and this prolongs the overall change of state. Analysis of the Darlington configuration can be accomplished by treating it as a single transistor with external terminal characteristics.

Parallel and series BJTs. High values of blocking voltages and conduction currents are not found together in single semiconductor devices. For example, a high voltage BJT may block 1400 V but conduct no more than 300 A. A high current BJT may conduct 800 A, but block no more than 100 V. For loads that require higher voltages or higher currents than those available for single devices, thoughts are given to connections of BJTs in series or parallel.

It is very difficult to connect BJTs in series in order to share voltage in the blocking state. The turn-on and turn-off times, and the base drives cannot be matched well enough to reduce the likelihood of prolonged time in the active region and the risk of second breakdown.

It is difficult to connect BJTs in parallel to force equitable current sharing, but with care it is done. The main difficulty arises from the inherent BJT characteristic of a negative temperature coefficient of resistance. If one BJT conducts more current than another, its loss will be greater, its temperature will rise, its effective resistance will drop and its share of the current will increase still further because the BJTs have a common voltage. Despite this problem, with careful matching and with similar techniques to those used for thyristors in parallel (section 5.11.2), multiple BJTs in parallel can operate within a range of 10% of current derating.

4.10.1. Applications

If we consider the BJTs limitation of being unable to block reverse voltage, it is evident that this switch cannot be used alone in ac-dc converters. If we consider the BJTs merits of being able to be switched on and off in a controlled fashion at a high frequency relative to the thyristor, then pulse-width modulation (PWM) of power from a dc source seems to a be a good use. Choppers and inverters up to about 200 kVA fall into this category. The higher frequencies lead to higher efficiencies of converters and smaller electromagnetic components.

Choppers are discussed in section 2.2 under the title dc-dc conversion. Thyristors were symbolised as the switches in these choppers. Within their power limitations, BJTs are better than thyristors in choppers, because the switching frequency is higher and turn-off is easier. Choppers have been used extensively in examples in this chapter. Single-phase inverters for dc-ac conversion can be treated as multiple choppers, so inverters will not be discussed further, even though BJTs are used frequently as the major switching component. A good example of the use of the turn-off capability of the BJT is in ac-dc conversion.

AC-dc Conversion. Figure 4.14a shows an ac-dc converter that uses a BJT to modulate the power delivered to a resistive load. The diode bridge rectifies the alternating voltage of the supply v_s and the BJT controls the dc voltage that appears across the load. In this way the BJT is protected from becoming reverse biased and being damaged.

An important role of the BJT is its ability to control both the firing delay angle α and the extinction advance angle β. This adds another dimension that can be used to improve the quality of waveforms. In the examples of ac-dc converters with the single controlling variable α in section 2.3 it is noticeable how high the harmonic distortion can be and how low the power factor can be. With the added control variable β the power factor can be improved. It must be noted that for a resistive load the power factor is the same for a given power no matter at what value β is set.

Case 1: Resistive load. Figure 4.14 shows the waveforms of the load voltage, load current and ac supply current for control of the delay angle α with the extinction angle $\beta = \pi$, and for control of both the delay angle α and the extinction angle β respectively. As β is decreased in value it can be seen that the supply-current waveform shifts to the left within the frame of a half period. This indicates that the fundamental component of source current is less lagging so that there is an improvement in the displacement power factor DPF.

EXAMPLE 4.10

Consider the circuit illustrated in Fig. 4.14. The BJT modulates power from a 240-V, ac supply to a resistive load whose value is $R = 4\,\Omega$. **(a)** Determine the voltage and current ratings of the diodes and the BJT. **(b)** If the delay angle is set

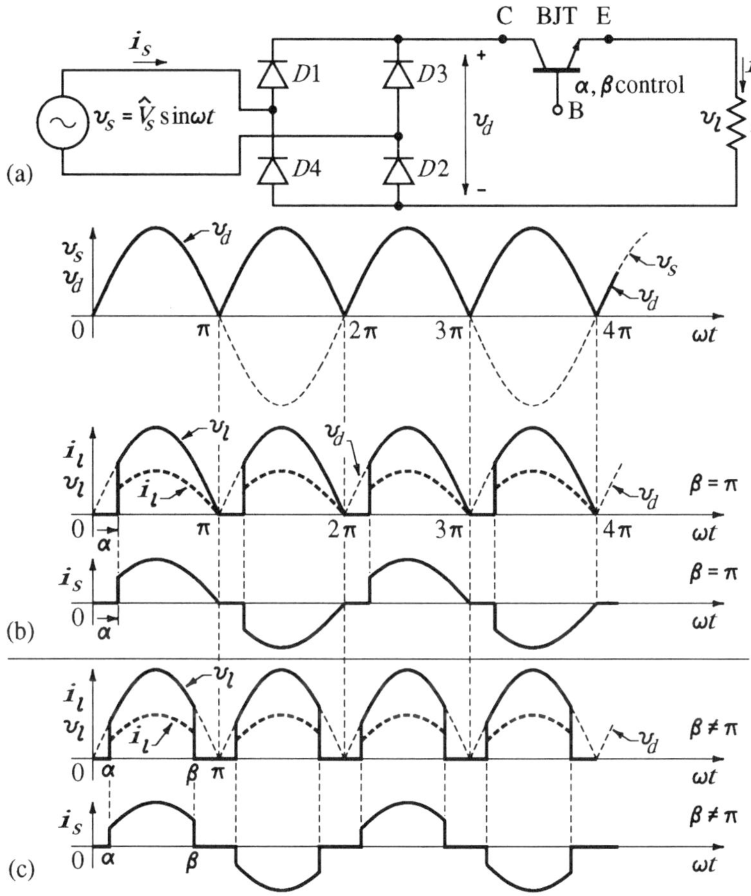

Fig. 4.14 BJT control of ac-dc conversion for R load.
(a) Circuit diagram, (b) α control, (c) α and β control.

at $\alpha = \pi/3$ radians and the extinction angle is set at $\beta = \pi$ radians, determine the power delivered by the ac source and the displacement power factor DPF. **(c)** If the delay angle is set at $\alpha = \pi/6$ radians and the extinction angle β is set so that the same power is delivered as in part (b), determine the value of the angle β and the displacement power factor DPF.

Solution

(a) The diodes must withstand a reverse voltage of $\hat{V}_s$ from the supply and the BJT must withstand a forward voltage of $\hat{V}_s$ from the supply. Therefore, the voltage ratings of both the diodes and the BJT must be greater than $\sqrt{2} \times 240$ V, that is, 340 V.

The current rating of the BJT is associated with the maximum average current that can exist in the load. The maximum load current occurs at $\alpha = 0$ and $\beta = \pi$ radians. That is

$$I_{l\,av} = \frac{1}{\pi R}\int_0^{\pi} v_s\, d\omega t = \frac{2\hat{V}_s}{\pi R} = \frac{2\sqrt{2}\times 240}{\pi\times 4} = 54\text{ A.}$$

Therefore, the current rating of the BJT must be greater than 54 A. The diodes conduct half the average current of the BJT so the current ratings of the diodes must be greater than 27 A.

(b) Refer to Fig. 4.14b. For the conditions $V_s = 240$ V, $R = 4\,\Omega$, $\alpha = \pi/3$ rad and $\beta = \pi$ rad,

$$V_{l\,rms}^2 = \frac{1}{\pi}\int_{\alpha}^{\pi} \hat{V}_s^2 \sin^2\omega t\, d\omega t = \frac{\hat{V}_s^2}{2\pi}\left[\pi - \alpha + \frac{\sin 2\alpha}{2}\right] = 46339\,.$$

The average power P delivered is the power absorbed.
$P = V_{l\,rms}^2/R = 46339/4 = 11.58\times 10^3$ W.

The Fourier coefficient a_1 is $a_1 = \frac{2}{\pi}\int_{\alpha}^{\pi} i_s \cos\omega t\, d\omega t = \frac{\hat{V}_s}{\pi R}\int_{\alpha}^{\pi} \sin 2\omega t\, d\omega t$.

$$a_1 = -\frac{3\hat{V}_s}{4\pi R} = -\frac{3\times\sqrt{2}\times 240}{4\pi\times 4} = -20.25\,.$$

The Fourier coefficient b_1 is $b_1 = \frac{2}{\pi}\int_{\alpha}^{\pi} i_s \sin\omega t\, d\omega t = \frac{\hat{V}_s}{\pi R}\int_{\alpha}^{\pi} \sin^2\omega t\, d\omega t$.

$$b_1 = \frac{\hat{B}_s}{\pi R}\left[\frac{2\pi}{3} + \frac{\sin 2\pi/3}{2}\right] = \frac{\sqrt{2}\times 240}{4\pi}\times 2.527 = 68.25\,.$$

The displacement power factor DPF is
$\text{DPF} = \cos\theta_1 = \cos\tan^{-1} a_1/b_1 = \cos\tan^{-1} -20.25/68.25$. So, $\text{DPF} = 0.95$.

(c) For the condition that $\alpha = \pi/6$ radians and $P = 11.58$ kW, we must find β. For the same power the rms value of the load voltage must be the same. In this case

$$V_{l\,rms}^2 = \int_{\alpha}^{\beta} \hat{V}_s^2 \sin^2\omega t\, d\omega t = \frac{\hat{V}_s^2}{2\pi}\left[\beta - \alpha - \frac{\sin 2\beta}{2} + \frac{\sin 2\alpha}{2}\right].$$

From part (a) $46339 = \frac{240^2}{\pi}\left[\beta - \frac{\pi}{6} - \frac{\sin 2\beta}{2} + \frac{\sin\pi/3}{2}\right]$.

That is, $2\beta - \sin 2\beta = 5.235$.
From a math software package $\beta = 2.1568$ rad.

The Fourier coefficient a_1 is $a_1 = \frac{2}{\pi}\int_{\alpha}^{\beta} i_s \cos\omega t\, d\omega t = \frac{\hat{V}_s}{2\pi R}[-\cos 2\omega t]_{\pi/6}^{2.16}$.

$$\text{So, } a_1 = \frac{\sqrt{2}\times 240}{2\pi\times 4}\left[-\cos(2\times 2.16) + \cos\frac{\pi}{3}\right] = 12\,.$$

The Fourier coefficient b_1 is $b_1 = \dfrac{2}{\pi}\int_{\alpha}^{\beta} i_s \sin\omega t \, d\omega t = \dfrac{\hat{V}_s}{\pi R}\left[\omega t - \dfrac{\sin 2\omega t}{2}\right]_{\pi/6}^{2.16}$.

So, $b_1 = \dfrac{\sqrt{2}\times 240}{\pi \times 4}\left[2.16 - \dfrac{\pi}{6} - \dfrac{\sin(2\times 2.16)}{2} + \dfrac{\sin\pi/3}{2}\right] = 68.33$.

The displacement power factor DPF is
$\mathrm{DPF} = \cos\theta_1 = \cos\tan^{-1} a_1/b_1 = \cos\tan^{-1} 12/68.33 = 0.98$.

By utilizing the turn-off mechanism of the BJT, the displacement power factor is improved. This is more noticeable for lower values of modulated power.

It has been assumed that the elements in the circuit diagram of Fig. 4.14 were ideal. This assumption allows the BJT to be turned off at angle β with no bad effects. However, most supplies have source impedance of an inductive nature, and any conductors contribute to stray inductance. If the BJT interrupts current at any finite value, the stored energy in the inductance gives rise to voltage spikes ($v = L di/dt$). A hundred amperes interrupted in one microsecond will give rise to a voltage spike of about 1000 V if the inductance is 10 μH. What contingencies can be taken to prevent this voltage from doing any damage?

Case 2: RL load. Figure 4.15a illustrates a circuit for ac-dc conversion. The load, in this case has both resistance R and inductance L. If the BJT is to turn-off with a finite current in the load (that is, with an extinction angle $\beta < \pi$), a freewheeling diode must be connected across the load. The best case for the load current quality is to have a high enough inductance so that the ripple is negligible. However, this produces the very worst case of the poorest power factor for any load. Controlling the extinction angle β can provide power factor improvement.

Delay angle α control, $\beta = \pi$. Figure 4.15b shows the voltage and current waveforms for $\beta = \pi$ and an arbitrary firing angle $\alpha = \alpha_1$. The average voltage $V_{l\,av}$ across the load is

$$V_{l\,av} = \int_{\alpha_1}^{\pi} \hat{V}_s \sin\omega t \, d\omega t = \frac{\hat{V}_s}{\pi}(1+\cos\alpha_1). \qquad (4.10.5)$$

The current in the load for $L \to \infty$ is

$$i_l = I_{l\,av} = I_{l\,rms} = I_l = \frac{V_{l\,av}}{R} = \frac{\hat{V}_s}{\pi R}(1+\cos\alpha_1). \qquad (4.10.6)$$

On the ac side the rms value of the source current is

$$I_{s\,rms} = \left[\frac{1}{\pi}\int_{\alpha_1}^{\pi} I_l^2 \, d\omega t\right]^{1/2} = I_l(1-\alpha_1/\pi)^{1/2} = \frac{\hat{V}_s}{\pi R}(1+\cos\alpha_1)(1-\alpha_1/\pi)^{1/2}. \qquad (4.10.7)$$

From the above equations we can find an expression for the power factor PF in

(a)

(b)

(c)

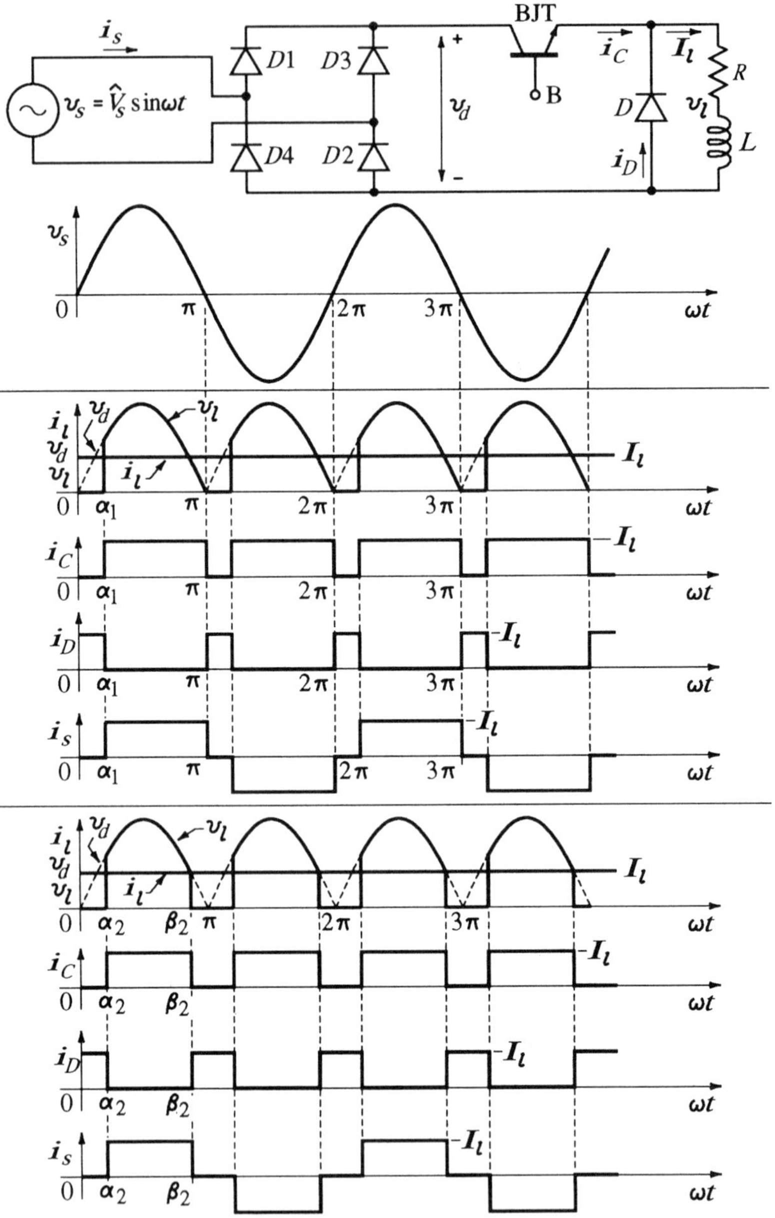

Fig. 4.15 BJT control of *RL* load.
(a) Circuit diagram, (b) α control, (c) α and β control.

terms of the firing angle α_1. The power factor PF is defined as

$$\mathrm{PF} = \frac{P}{V_s I_{s\,rms}} = \frac{I_l^2 R}{V_s I_{s\,rms}} = \frac{V_{l\,av}}{V_s(1-\alpha_1/\pi)^{1/2}} = \frac{\sqrt{2}\,(1+\cos\alpha_1)}{\pi(1-\alpha_1/\pi)^{1/2}}. \tag{4.10.8}$$

For any α the power factor can be calculated. It is independent of the source voltage or load resistance. The function is true only if the load inductance has a high value. At $\alpha_1 = 0$, for the maximum power delivered to the load, the power factor is 0.9.

Delay angle α and extinction angle β control. Figure 4.15c shows the voltage and current waveforms for the converter for a value of β other than π radians. We want to find the relation between α and β for the maximum possible power factor on the ac side. The constraint we will apply is to fix the value of power absorbed by the load. This fixes I_l but allows the source current $I_{s\,rms}$ to be a variable in terms of arbitrary α_2 and β_2, that is,

$$I_{s\,rms} = \left[\frac{1}{\pi}\int_{\alpha_2}^{\beta_2} I_l^2\, d\omega t\right]^{1/2} = \frac{I_l}{\sqrt{\pi}}(\beta_2-\alpha_2)^{1/2}. \tag{4.10.9}$$

Therefore, in terms of the BJT switching angles, the power factor PF is

$$\mathrm{PF} = \frac{P}{V_s} \times \frac{1}{I_{s\,rms}} = \frac{\sqrt{\pi}P}{V_s I_l} \times (\beta_2-\alpha_2)^{-1/2} = k(\beta_2-\alpha_2)^{-1/2} \tag{4.10.10}$$

where k is a constant. For maximum power factor, $d\mathrm{PF}/d\alpha = 0$, so

$$0 = \frac{d\beta}{d\alpha} - 1. \tag{4.10.11}$$

For a fixed power the average load voltage $V_{l\,av}$ is fixed.

$$V_{l\,av} = \frac{1}{\pi}\int_{\alpha_2}^{\beta_2} \hat{V}_s \sin\omega t\, d\omega t = \frac{\hat{V}_s}{\pi}(\cos\alpha_2 - \cos\beta_2). \tag{4.10.12}$$

From this equation

$$\frac{d\beta}{d\alpha} = \frac{\sin\alpha_2}{\sin\beta_2}. \tag{4.10.13}$$

Upon substitution in eq. (4.10.11) we have $\sin\alpha_2 = \sin\beta_2$, that is, for maximum power factor

$$\beta_2 = \pi - \alpha_2. \tag{4.10.14}$$

This symmetry of switching gives the displacement power factor DPF a value of unity, the highest value. Accordingly, we would expect an improved power factor, except at $\alpha_2 = 0$. This result provides a rule that switching for any power must have symmetry for the best power factor on the ac side.

EXAMPLE 4.11

Consider the ac-dc converter illustrated in Fig. 4.15. The BJT modulates power from a 240-V, ac supply to a load whose resistance is $R = 4\,\Omega$. The inductance value is high enough to consider the load current to be virtually constant. **(a)** If the switching angles of the BJT are set at $\alpha = \pi/2$ and $\beta = \pi$, determine the average power and power factor on the ac side. **(b)** If the switching angles of the BJT are set to $\beta = \pi - \alpha$ for the same average power as in part (a), determine the power factor.

Solution

(a) Average power $P = I_l^2 R = \dfrac{V_{l\,av}^2}{R} = \dfrac{\hat{V}_s^2}{\pi^2 R}(1+\cos\alpha_1)^2$

α_1 refers to the case that $\beta = \pi$. $\alpha = \pi/2$ rad.

So, $P = \dfrac{2\times 240^2}{\pi^2 \times 4}(1+\cos\pi/2)^2 = 2.918\times 10^3$ W.

Power factor $\mathrm{PF} = \dfrac{P}{V_s I_{s\,rms}} = \dfrac{V_{l\,av}}{V_s(1-\alpha/\pi)^{1/2}} = \dfrac{\sqrt{2}}{\pi}\dfrac{(1+\cos\alpha)}{(1-\alpha/\pi)^{1/2}} = 0.636$.

(b) For $\beta = \pi$, $\alpha = \alpha_1$ and $P_1 = \dfrac{\hat{V}_s^2}{\pi^2 R}(1+\cos\alpha_1)^2$.

For $\beta = \pi - \alpha$, let $\alpha = \alpha_2$. The average power

$$P_2 = \frac{V_{l\,av}^2}{R} = \frac{1}{R}\left[\frac{1}{\pi}\int_{\alpha_2}^{\pi-\alpha_2} \hat{V}_s \sin\omega t \, d\omega t\right]^2 = \frac{4\hat{V}_s^2}{\pi^2 R}\cos^2\alpha_2 .$$

If the powers P_1 and P_2 are to be the same

$(1+\cos\alpha_1)^2 = 4\cos^2\alpha_2$.

That is, $2\cos\alpha_2 = 1+\cos\alpha_1$.

$\alpha_1 = \pi/2$, so, $\alpha_2 = \pi/3$ radians.

Power factor $\mathrm{PF} = \dfrac{P}{V_s I_{s\,rms}} = \dfrac{V_{l\,av}}{V_s(1-2\alpha_2/\pi)^{1/2}} = \dfrac{2\sqrt{2}}{\pi}\dfrac{\cos\alpha_2}{(1-2\alpha_2/\pi)^{1/2}}$.

$\mathrm{PF} = \dfrac{\sqrt{2}\times\sqrt{3}}{\pi} = 0.778$.

There is power factor improvement using $\beta = \pi - \alpha$. It is 22% better.

4.11. SUMMARY

The BJT switch has advantages over the thyristor in that the BJT can operate at much higher frequencies and has controlled turn-off. Otherwise, it has poorer characteristics, having to traverse the active region whenever it changes states (higher losses), suffers from having a second breakdown phenomenon, cannot block reverse voltage, has lower current and voltage ratings, requires a relatively high current continuous base drive, and does not have a surge current rating much higher than its average current rating. Like the thyristor, the BJT has a low on-state voltage drop.

The voltage ratings of the BJT are up to about 1400 V and the current ratings are up to about 800 A. Choppers and inverters up to about 200 kVA can be built with BJT switches. Darlington transistors are manufactured to provide current gains of about 200, whereas a single BJT has a current gain β between 10 and 20.

4.12. PROBLEMS

Section 4.3

4.1 Consider the circuit diagram in Fig. 4.3a. The BJT acts as a chopper to modulate power from the 240-V dc source to the purely resistive load $R = 1.2\,\Omega$. The frequency of chopper switching is 3 kHz and the duty cycle is $m = 0.8$. If the transistor is driven into hard saturation so that $V_{CE(\mathrm{SAT})} = 2.1$ V, calculate the ratio of the average conduction losses in the BJT to the average power absorbed by the load. Ignore the switching losses.

Section 4.4

4.2 Consider the circuit diagram, shown in Fig P4.2. The bipolar transistor operates as a switch to modulate power from the 600-V, dc supply to the load whose resistance is $R = 4\,\Omega$. **(a)** From a circuit point of view, determine the power in the load while the transistor is on. **(b)** Using the transistor data-sheet characteristics and the fact that the transistor operates with a short-circuit common emitter gain $\beta_F = 10$, determine the power dissipation in the transistor during conduction. The temperature of the transistor junction is assumed constant.

4.3 Consider the circuit diagram in Fig. 4.6a. The BJT acts as a chopper to modulate power from the dc source, whose voltage is $V_s = 240$ V, to a purely resistive load, whose resistance is $R = 1.2\,\Omega$. The frequency of the chopper switching is 3 kHz and the duty cycle is $m = 0.8$. If the current rise-time is $t_r = 1\,\mu$s for turn-on, determine **(a)** the energy loss in the BJT during turn-on and **(b)** the average power dissipated in the BJT due to the turn-on losses. Compare this result with the value of conduction losses found in problem 4.1.

4.4 Consider the circuit diagram in Fig. 4.7a. The BJT acts as a chopper to modulate power from a dc source of voltage $V_s = 240$ V to an *RL* load whose resistance is $R = 1.2\,\Omega$ and whose inductance *L* is high enough to consider the load current to be virtually constant. A freewheeling diode is connected across the load. The switching frequency of the chopper is 3 kHz and the

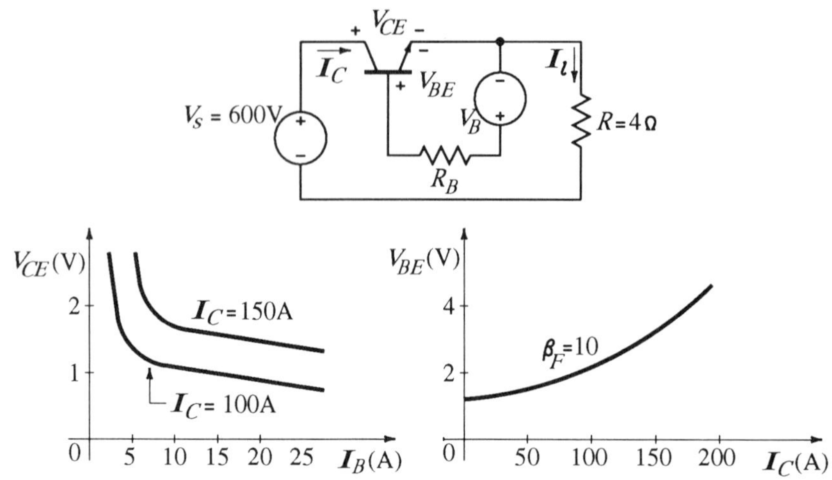

Fig. P4.2

duty cycle is $m = 0.8$. If the current rise-time is $t_{ri} = 1\,\mu$s and the voltage fall-time is $t_{fv} = 1.3\,\mu$s for turn-on during steady-state switching, determine (**a**) the energy loss in the BJT during the turn-on process and (**b**) the average power dissipated in the BJT due to turn-on losses. Compare this result with the turn-on loss for the case described in problem 4.3.

Section 4.6

4.5 Figure P4.2 illustrates a BJT acting as a chopper with a switching frequency f=8 kHz and a duty cycle m=0.7. It modulates power from a 600-V dc source to a resistive load of 4 Ω. While in the on-state the BJT voltage drop is $V_{CE(\mathrm{SAT})}$ =1.8 V and the base drive dissipation in the BJT is 30 W. While in the off-state the BJT leakage current is $I_{C\ leak}$ =0.15 A and the reverse base drive dissipation is 30 W. For this condition data sheets provide the switching times: delay time t_d=0.5 μs, rise-time t_r = 1 μs for turn-on, storage time t_s=4 μs and fall-time t_f=3 μs for turn-off. (**a**) Determine the component values of the BJT energy dissipation. (**b**) If the ambient temperature is 25°C and if the BJT junction-to-case thermal resistance is $R_{\theta JC}$ =0.1°C/W, what is the maximum value of the heatsink thermal resistance $R_{\theta CA}$ (case to ambient) so that the junction temperature does not exceed 150°C?

4.6 A BJT chopper modulates power from a dc source of voltage V_s = 240 V to an RL load, whose resistance is R = 1.2 Ω. The inductance L is high enough to consider the load current to be constant if a freewheeling diode is connected across the load. See Fig. 4.7a. The chopper has a switching frequency of 3 kHz and a duty cycle m = 0.8. The on-state BJT voltage drop is $V_{CE(\mathrm{SAT})}$ = 1.4 V. If the voltage rise-time is t_{rv} = 1.2 μs, if the current fall-time is t_{fi} = 1.5 μs and if the storage time is t_s = 3.2 μs for turn-off during steady-state switching conditions, determine (**a**) the average power dissipated in the BJT due to turn-off losses. (**b**) Compare the result of part (a) with the conduction loss. (**c**) Compare the result of part (a) with the turn-on loss, found in problem 4.4. (**d**) What is the total average power dissipation in the BJT? (**e**) What are the voltage and current ratings for a BJT in this circuit?

4.7 Consider the circuit diagram of a BJT chopper shown in Fig. 4.7 and refer to the waveforms for an inductive load in Fig. 4.8. Let the chopper operate at a frequency $f = 8$ kHz and a duty cycle $m = 0.7$ to modulate power from a 600-V source to an *RL* load. The resistance is $R = 4\,\Omega$ and the inductance L is high enough so that the load current is virtually constant over a period of switching, if the freewheeling diode is connected across the load. While in the on-state the BJT voltage drop is $V_{CE(\mathrm{SAT})} = 1.8$ V and the base drive dissipation in the BJT is 30 W. While in the off-state, the BJT leakage current is $I_{C\,leak} = 0.15$ A and the reverse base drive dissipation is 30 W. For this condition data sheets provide the switching times. At turn-on the delay time is $t_d = 0.5\,\mu$s, the current rise-time is $t_{ri} = 1\,\mu$s and the voltage fall-time is $t_{fv} = 1\,\mu$s. At turn-off the storage time is $t_s = 4\,\mu$s, the voltage rise-time is $t_{rv} = 3\,\mu$s and the current fall-time is $t_{fi} = 1.3\,\mu$s. Determine the efficiency of the BJT converter.

Section 4.7

4.8 Figure P4.8 depicts the basic circuit diagram of a boost converter, whose capacitor C maintains the load voltage virtually ripple free. For a switching duty cycle of $m = 0.7$ and a switching frequency of 8kHz the crossover time for the BJT at turn-on is 1.0 μs and the crossover time at turn-off is 2.0 μs. (Note: crossover time can be defined as the interval between $0.1V_{CE}$ and $0.1I_C$ during a change of state.) The storage time is $t_s = 4\,\mu$s. (**a**) Sketch the BJT voltage v_{CE} and current i_C over the period of one switching cycle and (**b**) estimate the average power dissipation in the BJT if the on-state voltage is $V_{CE(\mathrm{SAT})} = 1.8$ V. Ignore the base drive loss. (**c**) What is the conversion efficiency?

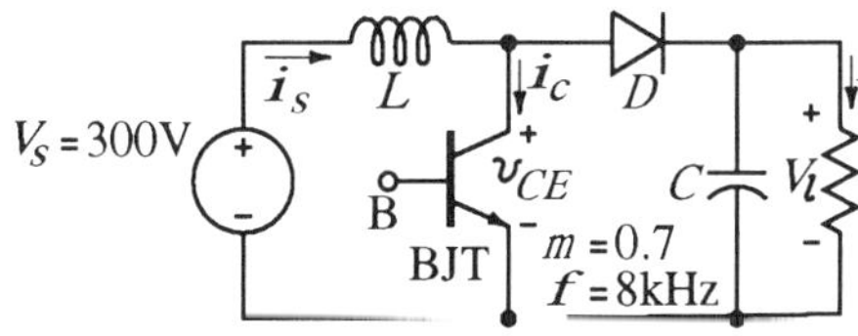

Fig. P4.8

4.9 Consider that the BJT chopper, shown in Fig. P4.2, operates at a duty cycle $m = 0.7$. In the on-state the BJT voltage drop is $V_{CE(\mathrm{SAT})} = 2\,\mathrm{V}$ and in the off-state the BJT leakage current is negligible. For this condition, the turn-on time comprises the delay time $t_d = 0.5\,\mu\mathrm{s}$ and the rise-time $t_r = 1\,\mu\mathrm{s}$, and the turn-off time comprises the storage time $t_s = 4\,\mu\mathrm{s}$ and the fall-time $t_f = 3\,\mu\mathrm{s}$. The heatsink provides a thermal resistance from the junction to ambient of $R_{\theta JA} = 0.1\,^\circ\mathrm{C/W}$. For a junction temperature of 150°C and an ambient temperature of 25°C, estimate the maximum frequency of switching and the power absorbed by the load. Neglect losses due to the base drive.

Section 4.8

4.10 A peaked base current is provided in the BJT illustrated in Fig. 4.10b for fast turn-on without degrading the turn-off performance. The BJT acts as a chopper with a switching frequency of 10 kHz and a duty cycle of $m = 0.5$. The power is from a dc source of 400 V to an RL load whose resistance is 5 Ω. The initial forced gain is to be $\beta_F = 5$ at turn-on and then increased to 10 before the turn-off action. Calculate suitable values of the parameters of the base drive circuit if the base source voltage is 10 V.

4.11 Figure P4.11 shows a BJT chopper circuit with a base drive that gives peaking for fast turn-on without detracting from turn-off performance. The switch frequency is 10 kHz and the duty cycle is $m = 0.8$. For good performance at turn-on the forced gain is $\beta_F = 5$, and for good performance at turn-off the gain is $\beta = 12$ before switching. The ratio of steady positive and negative base current signals is unity. Estimate the parameter values of the base circuit. Assume $V_{BE(\mathrm{SAT})} = 1\,\mathrm{V}$.

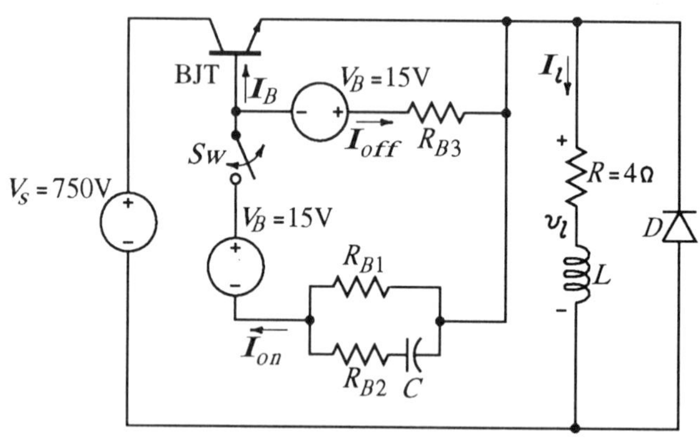

Fig. P4.11

4.12 Consider the circuit shown in Fig. 4.11. Let the source voltages be $V_s = 400$ V and $V_B = 15$ V. Let the resistance values be $R = 4\,\Omega$ and $R_B = 1.015\,\Omega$. The BJT operates with a current gain of $\beta = 10$ and a junction drop $V_{BE(\mathrm{SAT})} = 2.9$ V. The conduction drops across each diode is 1V. Determine the voltage drop V_{CE} across the BJT with and without the Baker-clamp diode *DB*.

Section 4.9

4.13 A BJT chopper modulates power from a dc source of 400 V to an *RL* load with an ideal freewheeling diode. See Fig. 4.7. The load resistance is $R = 4\,\Omega$. The rise-time of the current at turn-on is $t_{ri} = 1\,\mu$s and the fall-time of current at turn-off is $t_{fi} = 2\,\mu$s. A capacitor ($C_s = 0.5\,\mu$F) is connected across the BJT to limit the dv/dt at turn-off. Estimate the changes in the losses during BJT current rise and fall with and without the snubber capacitor.

4.14 Consider the BJT boost converter, shown in Fig. EX4.9. The data sheets for the BJT give a limit of 400 V, 100 A on the curve for the safe operating area at turn-off. Beyond these values second breakdown is likely. If the fall-time for the collector current is $t_{fi} = 2\,\mu$s at turn-off, check that the snubber-circuit capacitor value is conservatively chosen to be 1.0 μF.

Section 4.10

4.15 Consider the circuit diagram of the ac-dc converter that is shown in Fig. 4.14. **(a)** For a resistive load, show that the power factor on the ac side is directly proportional to the rms value of the load current and is independent of the type of control used, α, β, or α and β. **(b)** Determine the power factor as a function of the delay angle α for $\beta = \pi$.

4.16 Consider the circuit diagram of the ac-dc converter shown in Fig. 4.15. Let the inductance *L* be high enough to consider the load current to be virtually constant. If the extinction angle β is fixed at a value $\beta = \pi$ radians determine **(a)** the value of the delay angle α that gives rise to the maximum power factor. **(b)** What is the maximum value of the power factor and **(c)** at what load power does this occur?

4.17 Consider the circuit diagram of the ac-dc converter shown in Fig. 4.15. Let the load inductance have a value such that the load current can be considered virtually constant. If the extinction angle is to be $\beta = \pi - \alpha$ radians, determine **(a)** the value of the delay angle α that gives rise to the maximum possible power factor for this type of switching. **(b)** What is the maximum value of the power factor and **(c)** at what load power does this occur?

4.18 Consider the ac-dc converter illustrated in Fig. 4.15. The BJT modulates power from a 240-V, ac supply to a load whose resistance is $R = 4\,\Omega$. The inductance *L* is high enough to consider the load current to be virtually constant. If the switching angles of the BJT are set at $\alpha = \pi/2$ and $\beta = \pi$, determine the average power and total harmonic distortion THD on the ac side.

4.19 Consider the ac-dc converter depicted in Fig. 4.15. The source voltage is 240 V and the load has resistance $R = 4\,\Omega$. The load current can be assumed to be constant. If the switching angles of the BJT are set so that $\beta = \pi - \alpha$ and if the power absorbed by the load is 2.9 kW calculate the total harmonic distortion THD of the supply current. Compare with problem 4.18 answer.

4.20 Consider the ac-dc converter illustrated in Fig. 4.15 to have a load current that is ripple free. The BJT switching angle is $\alpha = \alpha_1$ for an extinction angle $\beta = \pi$ and a switching angle $\alpha = \alpha_2$ for an extinction angle $\beta = \pi - \alpha$. If the power to the load is to be the same in both cases what is the relation between α_1 and α_2. Check the relation with the results of problems 4.18 and 4.19.

4.13. BIBLIOGRAPHY

Baliga, B.J. *Modern Power Devices.* New York: John Wiley & Sons, Inc., 1987.

Baliga, B.J., and D. Chen. *Power Transistors: Device Design and Applications.* New York: IEEE Press, 1984.

Blicher, A. *Field Effect and Bipolar Power Transistor Physics.* New York: Academic Press, 1981.

Ghandi, S.K. *Semiconductor Power Devices.* New York: John Wiley & Sons, Inc., 1987.

Grove, A.S. *Physics and Technology of Semiconductor Devices.* New York: John Wiley & Sons, Inc., 1967.

Hower, P.L. *Bipolar Transistors, Semiconductor Devices for Power Conditioning.* New York: Plenum Press, 1982.

Neudeck, G.W. *The Bipolar Junction Transistor.* Modular Series on Solid State Devices, Volume 3. Massachusetts: Addison-Wesley, 1983.

Peter, J.M. (ED). *The Power Transistor in its Environment.* Thomson-CSF Sescosem Semiconductor Division, 1978.

____. *Silicon Power Transistor Handbook.* Pittsburgh: Westinghouse Electric Corp., 1967.

Sittig, R., and P. Roggwiller. (EDs). *Semiconductor Devices for Power Conditioning.* New York: Plenum Press, 1982.

Smith, M.W. (ED). *Electronic Data Library, Transistors-Diodes.* General Electric Co., 1982.

Sze, S.M. *Physics of Semiconductor Devices.* Second Edition. New York: John Wiley & Sons, Inc., 1981.

CHAPTER 5

THE THYRISTOR

5.1. INTRODUCTION

The thyristor is a 4-layer, 3-terminal, semiconductor switch. It can handle more power than any other semiconductor switch. The foundations of thyristor characteristics and control are laid out in this chapter. This provides a base for device comparison, so that the right choice of switch can be made for each particular application of power conditioning.

It is important to know how to operate a thyristor in a circuit. Consequently, emphasis will be laid on this aspect, rather than on the physics of the device. We want to use the thyristor not design it. A description of how the switch works leads on to its operating characteristics. This is followed by thyristor turn-on methods and turn-on circuits. Once on, the thyristor has to be turned off. The means of turn-off are described.

Power rating is an important topic for the user. Integral with power ratings are protection, thermal considerations, and multiply-connected switches.

The thyristor is a switch. Its name is compounded from the greek word *thy-* meaning *switch*, and*-istor*. This latter part is derived from *transistor* (*trans*-fer-*res*istor) to indicate that the switch belongs to the transistor family. The action of the thyristor is similar to the electromagnetic relay, the thyratron tube and the mercury arc rectifier. Its purpose is to control power in the wide range of tens of watts to megawatts.

The thyristor is a fast switching device that is a major component of power electronics. It modulates power in ac and dc systems by being switched on and off in a particular sequence. Although the *thyristor* is an internationally known name for a particular switch, it is also considered to be a family of 4-layer devices. The principal ones are the silicon-controlled rectifier (SCR), the GTO and the triac. There is a number of lower power devices that belong to the family. These are used in trigger circuits for turning the SCR on. (See section 5.5.) Some names are the *silicon–controlled switch* (SCS), the *diac*, the *silicon unilateral switch* (SUS) and the *programmable unijunction transistor* (PUT). In this text thyristor and SCR are used synonymously.

Rectification, inversion and regulation of power flow are the duties of the thyristor. It is possible for a single device, that can sit in the palm of the hand, to control 1 MW of power. The thyristor might be small, even after it has been encapsulated. However, it must be kept in mind that by the time that controllers, drivers, protection and cooling have been taken into account, you may have a cubicle on your hands that measures 50 cm × 40 cm × 30 cm and more. In general, what is in the cubicle converts a supply of one kind to a supply of another kind.

In general the good characteristics of the thyristor are: light weight, reliability (long life), fast acting (switching up to 10 kHz), turns on with low signal power, simple turn-on, no moving parts and high power-handling capability.

No switch is perfect. The thyristor has the limitations that turn-off can be complicated, its thermal time constant is short and there is power dissipation (the current density is limited to about 150 A cm^{-2} and the voltage drop across a conducting thyristor is about 1.5 V). These problems will be discussed in greater detail.

It is common to find thyristors that can block thousands of volts of either polarity, or conduct thousands of amperes. This reflects well on the importance of the technology, if we consider that, in the late 1950s, ratings were hundreds of volts and tens of amperes. It is unfortunate that high current and high voltage produce manufacturing difficulties. High voltage requires a thick silicon wafer and high current requires a large diameter wafer. Both requirements give rise to increased impurities and decreased yields. Consequently, one requirement (current or voltage) can be developed, but it is at the expense of the other requirement. One manufacturer offers a thyristor that has a blocking voltage of 3600 V, rated at 1000 A. The same manufacturer offers a thyristor with a current rating of 1250 A but its blocking voltage is 2600 V. Thyristor design involves compromise.

5.2. THYRISTOR STRUCTURE

We started this chapter by saying that the thyristor is a 4-layer, 3-terminal semiconductor switch. Figure 5.1a depicts a symbolic representation of the alternate *p* and *n* layers of silicon that make up the switch with its three junctions J1, J2 and J3. Also shown in Fig. 5.1b is the circuit symbol of the thyristor. It looks like a diode with an added control terminal G (hence the name silicon-controlled rectifier). The terminals A, K and G stand for **A**node, **K**athode (but spelled *Cathode*) and **G**ate. A simple circuit, that employs a thyristor, is illustrated in Fig. 5.1c. It can be used to explain the operation of the thyristor. The thyristor has three terminals. Two terminals (G and K) are used to switch the device on, and two terminals (A and K) allow the conduction of load current.

Ideally the thyristor is off (or blocking) until the anode A and gate G are biased with suitable positive voltages with respect to the cathode K. In this blocking state the thyristor offers ideally infinite impedance to the conduction of current.

With positive voltages at the anode and gate terminals, the thyristor turns on, conducts current, has ideally zero impedance and remains on without any further gate signal.

If there is no gate signal and if the anode current i_A is reduced to zero, the thyristor switches off (or blocks) and resumes the infinite impedance state.

Load current i_A has one direction only, from anode to cathode.

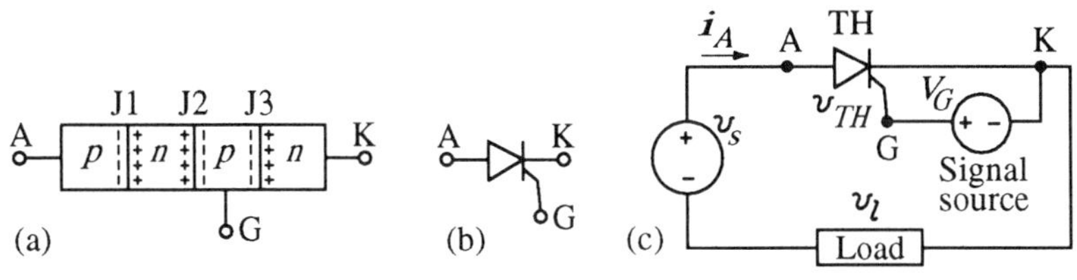

Fig. 5.1 The thyristor.
(a) *pnpn* representation, (b) circuit symbol, (c) simple circuit.

5.3. THYRISTOR MODELS

The idea that the thyristor has two states, on (zero impedance ideally) and off (infinite impedance ideally), serves well in the analysis of many circuits. However, protection of the thyristor necessitates a better knowledge without going into the physics of semiconductor devices. We will use two simple models to describe the operation of thyristor switching, the diode model and the two transistor model.

5.3.1. Diode Model of the Thyristor

An inspection of Fig. 5.1a conveys the idea that a thyristor is equivalent to three diodes in series. That is, there are three *pn* junctions J1, J2 and J3, at which there are the diffusion potential barriers, that were described in Chapters 1 and 2.

Without a gate bias, whatever the polarity of the anode voltage, there is at least one reverse-biased junction. If the anode is positive junction J2 blocks. If the cathode is positive junctions J1 and J2 block. Therefore, under normal conditions, no conduction can take place.

If the anode is positively biased with respect to the cathode, and if the gate is positively biased with respect to the cathode, the *p* layer at the gate is flooded with minority carriers (electrons) and loses its identity as a *p* layer. Therefore, the thyristor becomes equivalent to a single positively biased diode and conducts.

The idea that the gate current causes the inner *p*-layer to be flooded with electrons so that it loses its identity is oversimplification, but it is useful as a short description. The *p* region of the gate circuit is flooded with electrons at the onset of a gate signal. Most of these electrons diffuse towards junction J2, are accelerated across by the electric field of this junction, and are collected in the *n* region between junction J1 and J2. A space charge of electrons is built up in this *n* region, so that holes from the *p* region between the anode and junction J1 diffuse across junction J1 to neutralize the space charge. In turn these holes reach junction J2, are accelerated across the junction and form a space charge in the *p* region between J2 and J3. This space charge causes more electrons to flow across junction J3 from the *n* region to the *p* region. We have come full cycle. This process continues in the fashion of positive regeneration, so that the only limit to the

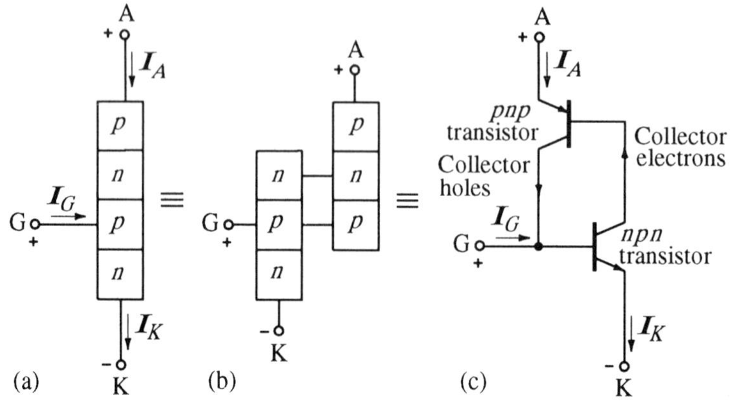

Fig. 5.2 Two-transistor model of the thyristor.

current in the thyristor is that created by the load-circuit resistance.

5.3.2. Two-transistor Model of the Thyristor

The two-transistor model of the thyristor has proved to be successful in both the dc and transient modes of analysis. Consider Fig. 5.2a, which depicts the symbolic form of the 4-layer, 3-terminal device. With very little stretch of the imagination, we can consider splitting the middle *pn* layers into parallel sections and keeping the conducting paths unaltered as shown in Fig. 5.2b. This configuration is the same as a *pnp* transistor connected to an *npn* transistor. The equivalent circuit diagram is shown in Fig. 5.2c. Let us consider the operation of such a circuit. We can use the description of the operation of a transistor found in Chapter 4.

A gate current I_G is supplied to the base of the *npn* transistor. The emitter (cathode K) is negatively biased from on external source. Therefore, the collector of the *npn* transistor feeds electrons to the base of the *pnp* transistor. Its emitter is positively biased from an external source. Consequently the collector current from the *pnp* transistor augments I_G to give increased base current to the *npn* transistor. This regenerative feedback leads to both transistors going into complete saturation with all the junctions forward biased. The removal of the gate signal does not alter this condition. The transistors are on. That is, the thyristor is on and the current is limited only by the load impedance.

Once on, the regenerative feedback maintains the on-state of the switch. The only way to turn off the thyristor is to reduce the anode current I_A virtually to zero.

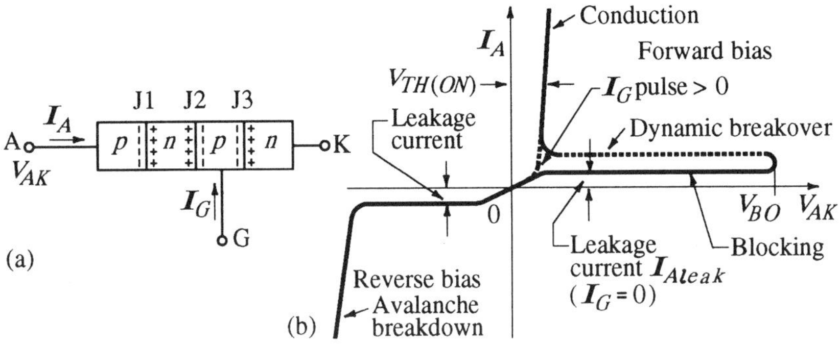

Fig. 5.3 The thyristor.
(a) Structure, (b) I-V characteristics.

5.4. THYRISTOR I-V CHARACTERISTICS

Under steady conditions the thyristor has three states: 1. reverse bias, 2. forward bias and blocking, and 3. forward bias and conducting. Each contributes to the total I-V characteristic. Figure 1.12 of Chapter 1 shows the ideal characteristics, and Appendix 4 gives typical data. Once more let us consider each state in turn.

1. Reverse bias. Figure 5.3a shows the symbolic representation of the thyristor. Reverse bias signifies that the cathode K is positive with respect to the anode A. Under this condition, junctions J1 and J3 can be seen to be reverse biased and junction J2 is forward biased. A reverse biased junction blocks conduction.

There is a reverse leakage current, as shown in Fig. 5.3b. A positive gate bias, while the anode has a negative polarity, increases the reverse leakage current.

A further increase of the reverse bias causes the depletion layers to widen. Junction J1 blocks most of the voltage. If J1 and J2 depletion layers overlap, the result is *punch through* and conduction. Junction J3 cannot block much more than 40 V, so its contribution is small. To guard against this the *n* region between J1 and J2 is wide. In any case, the minority carriers have increased energy to dislodge others and are accelerated across the junction. Eventually a reverse bias will be reached that will cause avalanche breakdown (as in the diode and BJT). This characteristic is shown in the third quadrant of Fig. 5.3b.

2. Forward bias, blocking. If the anode A is positive with respect to the cathode K, it can be seen from Fig. 5.3a that junction J2 is reverse biased. Therefore the current is limited to a forward leakage current. See Fig. 5.3b, quadrant one. A small gate current increases the leakage current $I_{A\,leak}$.

3. Forward bias, conducting. For the thyristor to conduct, the anode A needs to be positive with respect to the cathode K and sufficient charge needs to be injected into the gate region. See that part of Fig. 5.3b with $I_G > 0$. In the conducting mode the voltage drop $V_{TH(ON)}$ across the thyristor is very small (about 1.5 V) and the value of the current is load limited.

There are four ways to inject charge into the gate region to have the thyristor in the conducting mode.

5.5. THYRISTOR TURN-ON

There are four methods for thyristor turn-on: 1. light activation, 2. gate electrical signal, 3. high forward-bias voltage and 4. dv/dt turn-on.

Methods 1. and 2. are conventional means for thyristor turn-on and methods 3. and 4. are to be avoided. High temperature can increase the leakage current to a level that the thyristor turns on, but it is hoped that the cooling arrangement prevents this occurring.

1. Turn-on by light activation.
A beam of light directed at junction J3 of the thyristor can produce sufficient energy to break electron bonds and bring about the necessary minority carriers to switch on the thyristor. The thyristor has to be designed to accept light at the gate region. These devices are available.

The main application for these devices is found in HVDC (high-voltage, direct-current) thyristor converters, in which it is necessary to provide electrical isolation between the gate drivers and the high voltages found at both anodes and cathodes of the series-connected thyristors. Light-activated thyristors are manufactured with ratings as high as 6000 V and 3500 A.

2. Gate turn-on.
If the thyristor is forward biased at the anode by a source voltage V_s, additional minority carriers can be injected into the gate region to switch the device on. Figure 5.4 depicts a simple circuit. If the gate current i_G is large enough, the thyristor TH switches on soon after both the anode A becomes positive with respect to the cathode K and the gate G becomes positive with respect to the cathode K.

An arbitrary choice of a thyristor (type 2N4378) from a catalogue provides the specification, blocking voltage $V_{AK} = 1400\,\text{V}$, rated current $I_A = 110\,\text{A}$, gate current $I_G = 250\,\text{mA}$ and average gate power $P_G = 3\,\text{W}$. From this we find that the current gain I_A/I_G is approximately 500, and the power gain is approximately 50,000. For any device these are good characteristics.

It takes a little time for the thyristor to turn on, because minority carriers are injected into the gate region at the terminal contacts and the conduction region has to expand from the contacts right across the silicon wafer. This is known as *dynamic plasma spread.* An aid to turn-on speed is to have the points of dynamic-plasma initiation distributed over a large area during manufacture. This technique to reduce turn-on time is known as *interdigitation* and involves the physical interleaving of the gate contacts with the cathode *n* region.

Figure 5.5 depicts the transient turn-on characteristic of anode current rise and voltage drop associated with a thyristor that has its anode and gate positively biased. The application of a source voltage V_G across the gate and cathode terminals of the thyristor at $t = 0$ gives rise to a gate current I_G. There is no change in

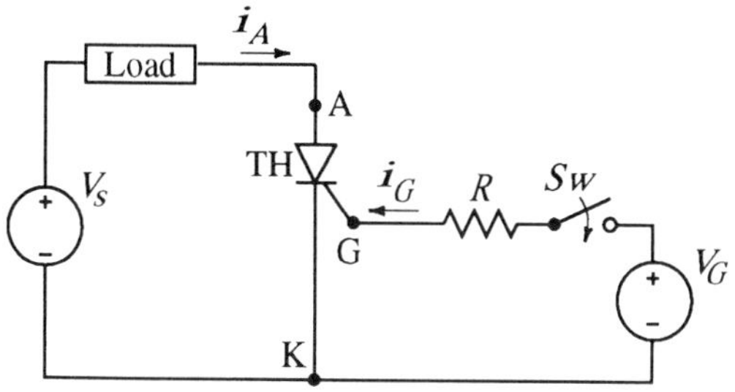

Fig. 5.4 Gate turn-on.

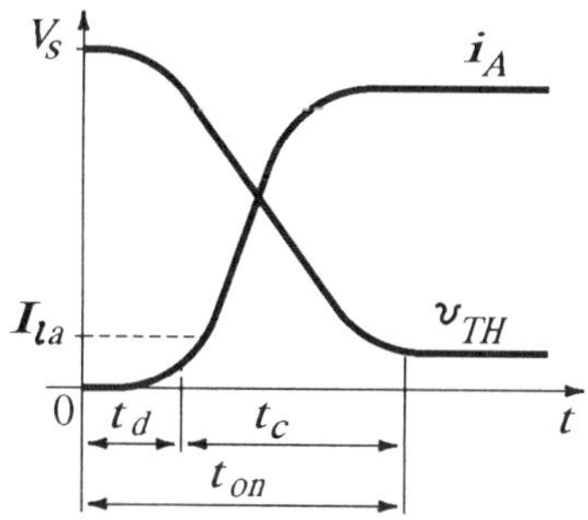

Fig. 5.5 Thyristor turn-on.

the condition of the thyristor over an interval t_d. That is, $v_{TH} \approx V_s$ and $i_A \approx 0$. During this delay time t_d the depletion layer capacitance at junction J1 discharges and recharges with a forward bias of about 0.6 V. At this point carrier injection commences and the transistor effects allow the current i_A to increase. At some stage of anode current rise (the latching current I_{la}) the transistor effect is self-sustaining. The gate signal can be removed, the regenerative effect enables the current to rise in time t_{ri} to its final value. The transistors end up in saturation, or more realistically the thyristor is in the on-state. How the voltage falls with time depends on the type or load.

An all encompassing definition of turn-on time is the interval from gate trigger initiation to the time of both equilibrium charge in the device and steady voltage across the terminals anode to cathode. So the turn-on time is expressed as

$$t_{on} = t_d + t_c \tag{5.5.1}$$

where t_c is the crossover time. Crossover time is the interval from $t = t_d$ to $t = t_{on}$ and covers both the rise time t_{ri} of current i_A and the fall time t_{fv} of voltage v_{TH}. This expression is suitable for the analyst.

In terms of measuring the turn-on time from recordings such as an oscillogram, it is more convenient to change the definition of turn-on time. The delay time t_d can be considered to be the time from when the gate current is ten percent of its final value ($i_G = 0.1 I_G$) to the time when the anode current is ten percent of its final value ($i_A = 0.1 I_A$). The anode current rise time t_{ri} can be considered the interval from the time the anode current is ten percent of its final value to the time the anode current reaches ninety percent of its final value. How the time for the voltage fall is taken into account depends on the load. Two types of load will be taken as examples to account for voltage fall at turn-on.

Turn-on times are not easy to determine, because their values depend on the temperature of the thyristor junction, the geometry of the device, the magnitude and waveform of the gate current and the voltage applied at the anode terminal. It is best to use the information in the manufacturer's data sheets in order to determine turn-on times. Typical times are $t_d = 1\,\mu s$ and $t_{ri} = 3\,\mu s$ for a thyristor rated at 2000 V and 1000 A with a 10 μs gate current pulse of 250 mA at a junction temperature $T_J = 125^\circ C$ and a resistive load.

The thyristor takes longer to turn on than the BJT, so there is a tendency for the energy dissipation during turn-on to be greater. However the gate-drive loss involves a short pulse so the gate energy and gate average power are low compared with the base requirements of the BJT for switching on.

Once on, the thyristor voltage v_{TH} across the anode to cathode terminals has a small value $V_{TH(ON)}$ and is almost constant and relatively independent of the load. A 1500-V thyristor may have a voltage drop of $V_{TH(ON)} = 1.5\,V$ at rated current while a 6000-V thyristor may have a voltage drop of $V_{TH(ON)} = 1.9\,V$ at rated current. Compared with the BJT at high power ratings there is little difference in the conduction voltage drops.

Fast thyristor turn-on requires high current pulses at the gate in order to get charge ($q = \int i\, dt$) in quickly. Figure 5.6 illustrates gate pulses. Figures 5.6a and b show ideal rectangular pulses, whose hatched areas indicate that sufficient charge has been injected for turn-on. The high pulse in Fig. 5.6b is superior to that in Fig. 5.6a. Real pulses are more like those in Fig. 5.6c and d, in which finite rate of rises are depicted. A circuit designer may plan to have gate signals whose rise times are 10 A/μs and whose duration is as short as 2 μs for successful thyristor operation. The duration of the gate pulse is determined by the time it takes for the anode current to reach what is known as the latching current I_{la}. If the gate signal is removed before the anode current rises to the latching value, the thyristor reverts to its off-state. However, once the anode current is greater than the latching-current value, there is sufficient regenerative current (two-transistor model) that the gate pulse can be removed and the thyristor still turns on. The latching current is about $10^{-3} \times I_{A\ rated}$.

The gate contacts have relatively small areas. At the initiation of turn-on only a small cross section of the wafer is conducting. Therefore, the anode current i_A at initiation must be limited. Otherwise, hot spots, created by high currents through small areas of finite resistance, might damage the switch. A typical maximum rate of rise of anode current during turn-on is 30 A/μs. What element in the

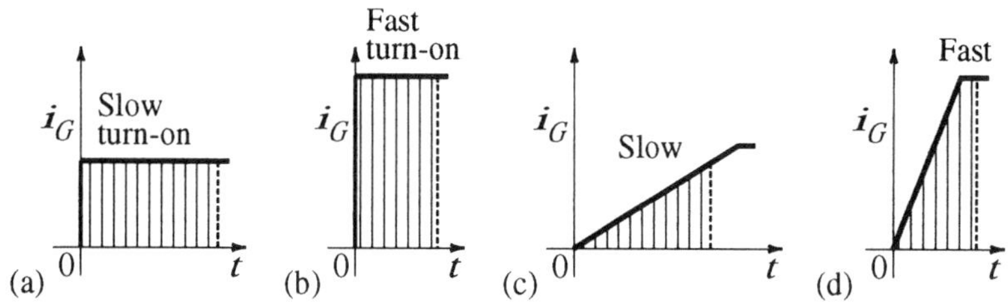

Fig. 5.6 Gate pulses.

circuit would limit the rate of rise of anode current?

EXAMPLE 5.1

A superconductive coil (no resistance) is to be charged from an ideal dc source and a thyristor is to be used as a controller. See Fig. EX5.1. *D* is a freewheeling diode. If the latching current of the thyristor is known to be 4 mA, calculate the duration of the gate pulse to ensure that the thyristor turn-on is successful.

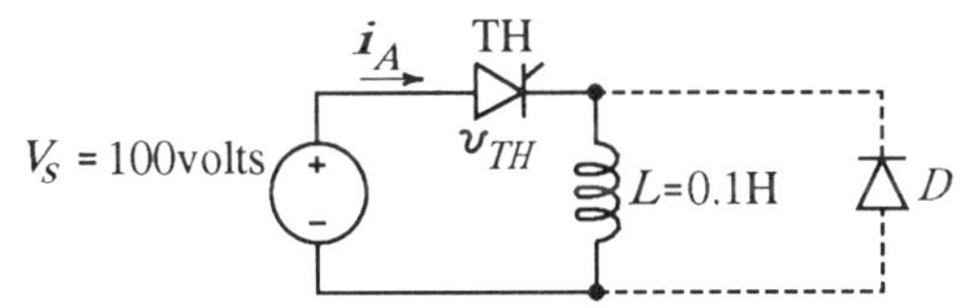

Fig. EX5.1 Thyristor latching.

Solution

With limited data, we must consider the thyristor to be ideal so that, from the onset of the gate pulse, the switch turns on and assumes the zero impedance state. This allows us to find the main current response as a function of time. The gate pulse width must be at least the time for the main current to rise from zero to 4 mA. The gate pulse is initiated at $t = 0$. From Kirchhoff's voltage law
$V_s = v_{TH} + L\,di_A/dt$. But, $v_{TH} = V_{TH(ON)} \approx 0$, so by integration
$i_A = (V_s/L)\,t + \mathrm{C}$, where C is a constant.
At $t = 0$, $i_A = 0$, so $\mathrm{C} = 0$.
Therefore, $i_A = (V_s/L)\,t$.
When $i_A = 4\,\mathrm{mA}$, $t = (4 \times 10^{-3} \times 0.1)/100 = 4\,\mu\mathrm{s}$.
The pulse duration at the gate must be 4 μs for the thyristor to turn on. This value is optimistic since the thyristor does not change states instantly.

---

3. Breakover voltage turn-on.

Let us refer to Fig. 5.3a for the symbolic representation of the thyristor. Consider that there is no gate signal, but that the anode A is positive with respect to the cathode K. The thyristor blocks because junction J2 is reverse biased. However, as the forward bias at the anode is increased, the J2 depletion layer widens. This increases the accelerating voltage for minority carriers. Eventually a forward bias V_{BO} is reached at which avalanche breakdown occurs and turn-on ensues. At this point the current through the switch is limited only by the load.

V_{BO} is the breakover voltage at which conduction occurs. This characteristic is used to turn on 2-terminal, 4-layer, *pnpn* diodes. Otherwise, it is to be avoided. That is, some form of protection against V_{BO} must be provided for thyristors. The simplest protection is to use thyristors whose V_{BO} value is much higher than any expected peak voltage. For example a 1000 V thyristor might well be chosen for a 650 V supply.

4. Turn-on by dv/dt.

Without a gate signal, a thyristor can be turned on by an anode voltage that is changing at a rate greater than or less than 200 V/μs depending on the thyristor.

An inspection of Fig. 5.3a indicates that the depletion layer at a junction is effectively a capacitor C. Its value is of the order of picofarads. The voltage/current characteristic equation for a capacitor is

$$i = C\frac{dv}{dt} . \tag{5.5.2}$$

Therefore, a large dv/dt at the anode will cause a large current i to exist in the gate region. If the current is large enough, the thyristor will turn on.

It is possible to bypass some of this current i by connecting a small resistor between the gate and cathode. In the manufacturing process, a part of the periphery of the gate to cathode boundary can be short-circuited by gold diffusion. This allows the thyristor to withstand higher values of dv/dt without turning on. However, a common method is to use a snubber circuit. This is a capacitor in parallel with the thyristor. The two act as a current divider in the transient state.

EXAMPLE 5.2

The thyristor circuit, illustrated in Fig. EX5.2a, has an R_sC_s snubber circuit to protect against dv/dt turn-on. Consider that the thyristor has just been turned off. Calculate the minimum value of C so that the thyristor will not turn on again due to dv/dt breakdown. The junction capacitance of the thyristor is 20 pF and the minimum value of the charging current required to turn on the thyristor is 4 mA.

Solution

The worst case of dv/dt across TH as it turns off is if V_s is suddenly applied to the anode of the thyristor. The diode makes the R_sC_s snubber circuit unidirectional. The diode D short-circuits the resistor R_s, so that C_s can be most effective to limit dv/dt. The resistor R_s serves its purpose when the thyristor is triggered

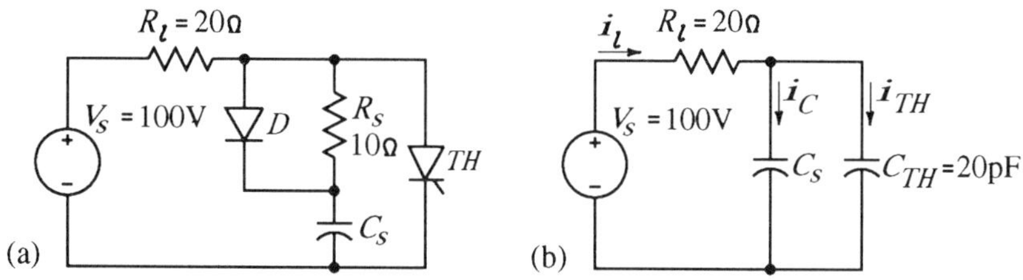

Fig. EX5.2 Snubber circuit to limit dv/dt.

on. The capacitor C_s would discharge through the thyristor at this point and R_s limits the magnitude of discharge current to an acceptable value.

For the sudden application of the source voltage V_s, the equivalent circuit is shown in Fig. EX5.2b. We are given that the charging current through the thyristor must be less than 4 mA, if the switch is to remain off. This indicates that the limit of dv/dt withstand is given by

$$\left.\frac{dv_{TH}}{dt}\right|_{\max} = \frac{i_{TH\,\max}}{C_{TH}} = \frac{4 \times 10^{-3}}{20 \times 10^{-12}} = 200\ \text{V}/\mu\text{s}\,.$$

For a step input, the capacitors appear uncharged and as a short circuit initially. Therefore, the initial current $i_l(0^+)$ from the source to the load is
$i_{l\,(0+)} = V_s/R_l = 100/20 = 5\ \text{A}$.
This means that the current taken initially by the snubber capacitor C_s is given by
$i_C = i_l - i_{TH} = 5 - 4 \times 10^{-3} \approx 5\ \text{A}$.
The rate of change of voltage across C_s is the same as that across C_{TH}. So, for the worst case (at $t = 0^+$),

$$C_s = \frac{i_C}{(dv_{TH}/dt)_{\max}} = \frac{5}{200 \times 10^6} = 0.025\ \mu\text{F}\,.$$

This value would just prevent the thyristor turning on, if the supply were a step input. For safety in practice a value larger than 0.025 μF would be chosen.

5.5.1. Turn-on Losses

During the thyristor turn-on process the values of the voltage v_{TH} and the current i_A can be large at the same time. Consequently the instantaneous power that must be dissipated in the thyristor can have a high value. In order to design a heatsink for the thyristor we must know how much energy is dissipated in the device and how much is contributed during turn-on. The best and worst cases are considered here, the resistive load and the highly inductive load respectively.

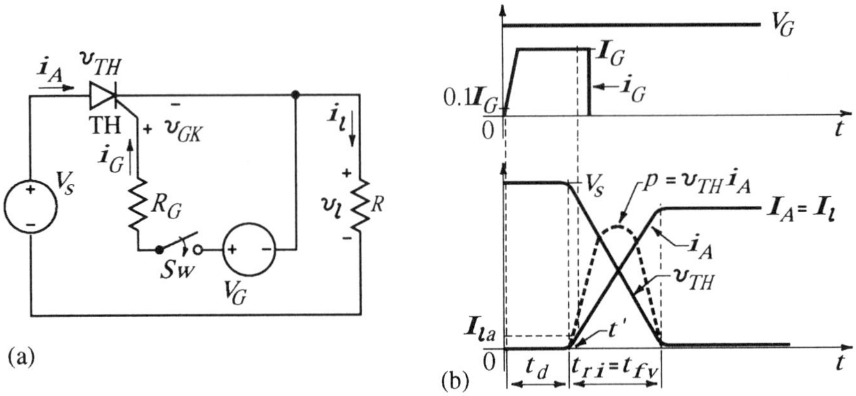

Fig. 5.7 Thyristor turn-on in a resistive circuit.
(a) Circuit diagram, (b) waveforms.

Resistive load. Figure 5.7 depicts a thyristor that modulates power from a dc source, whose voltage is V_s, to a resistive load of value R. The gate drive comprises an equivalent dc source, whose voltage is V_G, a series resistor R_G and an equivalent, unspecified but ideal switch *Sw*.

At $t = 0$, the gate switch *Sw* is closed to initiate the thyristor turn-on process. It is assumed that the gate-current rise-time is negligible compared with the delay time of turn-on. After the delay time t_d, during which $v_{TH} \approx V_s$ and $i_A \approx 0$, the current i_A rises to its final steady value $i_A = I_l \approx V_s / R$. For convenience of loss estimation the rise of anode current is linearized, as shown in Fig. 5.7b, and the time of current change is chosen to be t_{ri}. From Kirchhoff's voltage law for this circuit we have

$$v_{TH} = V_s - i_l R = V_s - i_A R \, . \tag{5.5.3}$$

Thus, the thyristor voltage v_{TH} falls from V_s to virtually zero in a time t_{fv} that is equal to the current rise-time t_{ri}. The voltage falls to an actual on-state value $v_{TH} \triangleq V_{TH(ON)} \approx 1.5\,\text{V}$, but, if the value of the source voltage V_s is high ($V_s > 100\,\text{V}$), then $V_{TH(ON)}$ can be considered negligible in comparison.

With these simplified waveforms an estimate of the power and energy losses during turn-on can be made. Let $t' = 0$ be the time that the anode current begins to rise. This time instant is $t' = t - t_d$. During the delay interval the power loss p_d in the thyristor is small. Its value is

$$p_d = V_{TH} i_A \approx V_s I_{A\,leak} \tag{5.5.4}$$

where $I_{A\,leak}$ is the leakage current for a forward bias voltage V_s at a particular junction temperature. During the rise-time t_{ri} the instantaneous power loss in the thyristor is

$$p = v_{TH} i_A \approx V_s\left(1-\frac{t'}{t_{ri}}\right)\times I_l\frac{t'}{t_{ri}}. \tag{5.5.5}$$

The overall energy loss W_{on} due to the turn-on mechanism is

$$W_{on} = \int_0^{t_{on}} p\,dt = \int_0^{t_{on}} v_{TH} i_A\,dt = \int_0^{t_d} V_s I_{A\,leak}\,dt + \int_0^{t_{ri}} V_s I_l\left(1-\frac{t'}{t_{ri}}\right)\frac{t'}{t_{ri}}\,dt'. \tag{5.5.6}$$

$$W_{on} = V_s I_{A\,leak} t_d + \frac{V_s I_l}{6} t_{ri} = W_d + W_c. \tag{5.5.7}$$

Values of t_d and t_{ri} would be found experimentally, theoretically or from data sheets. V_s and I_l are determined from the specifications of the application.

EXAMPLE 5.3

Consider the circuit diagram in Fig. 5.7a. The source voltage is $V_s = 200$ V and the resistive load is $R = 4\,\Omega$. For this condition the thyristor has a delay time $t_d = 0.5\,\mu s$ and a rise-time $t_{ri} = 3\,\mu s$ at turn-on. The thyristor leakage current is $I_{A\,leak} = 2$ mA and the on-state voltage drop is $V_{TH(ON)} = 1.5$ V. Estimate **(a)** the components of energy loss incurred in the thyristor during turn-on and **(b)** compare the turn-on loss with the thyristor loss during the on-state over an equivalent interval of time t_{ri}.

Solution

Refer to Fig. 5.7b and eq. (5.5.7).

(a) The thyristor loss W_d during the delay interval t_d is
$W_d = V_s I_{A\,leak} t_d = 200\times 2\times 10^{-3}\times 0.5\times 10^{-6} = 0.2\,\mu J.$
The thyristor loss W_c during the rise-time interval t_{ri} is
$$W_c = \frac{V_s I_l}{6} t_{ri} = \frac{V_s}{6}\frac{(V_s - V_{TH(ON)})}{R} t_{ri} = \frac{200}{6}\times\frac{(200-1.5)}{4}\times 3\times 10^{-6} \approx 5000\,\mu J.$$
The total loss $W_{on} = W_d + W_c = (0.2 + 5000)\times 10^{-6} \approx 5000\,\mu J.$
The delay-time loss is so small it can be neglected.

(b) While the thyristor is in the on-state it conducts 49.6 A and has a voltage drop of 1.5 V. The on-state energy loss W over an interval of 3 μs is
$$W = \int_0^{t_{ri}} v_{TH}\, i_A\, dt = \int_0^{t_{ri}} V_{TH(ON)} I_A\, dt = 1.5\times 49.6\times 3\times 10^{-6} = 223\,\mu J.$$
The energy is small but not insignificant. The total thyristor losses depend on the duration of conduction. A comparison with losses in a BJT in EXAMPLE 4.2 indicates that the losses are of the same order.

Inductive load. The longest turn-on time occurs if the load in a thyristor circuit is highly inductive. Figure 5.8a depicts a chopper circuit with an inductive load that incorporates a freewheeling diode. Turning the thyristor on by closing the gate switch *Sw* gives rise to waveforms similar to those shown in Fig. 5.8b. The

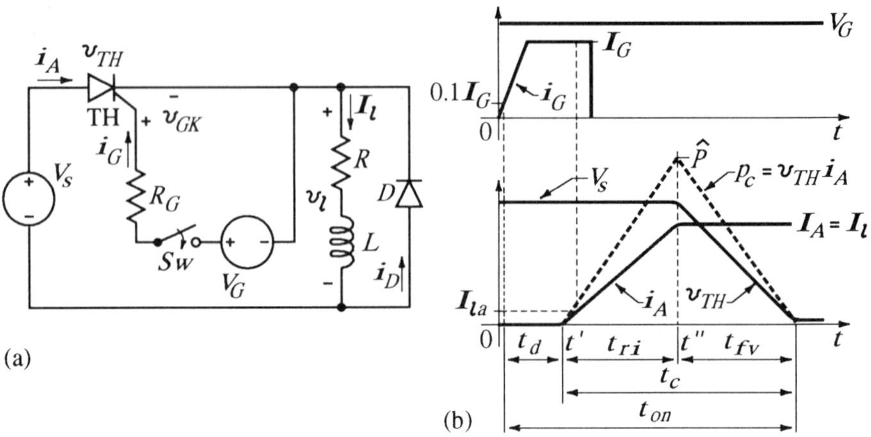

Fig. 5.8 Thyristor turn-on in an inductive circuit.
(a) Circuit diagram, (b) waveforms.

waveforms have been linearized to simplify the analysis.

At the onset of turn-on there is the usual delay time t_d, before the current i_A begins to rise. It is assumed that steady-state switching has been reached and that the inductance is high enough so the the load current can be considered constant. Kirchhoff's current law gives us the expression

$$i_D = I_l - i_A \, . \tag{5.5.8}$$

While i_A is rising, it is less than the load current I_l, so the diode current i_D is finite. As long as i_D is finite the voltage drop across the diode D is virtually zero. From Kirchhoff's voltage law

$$v_{TH} = V_s - v_l \, . \tag{5.5.9}$$

Since $v_l \approx 0$ while the anode current is rising, the voltage across the thyristor remains constant at $v_{TH} = V_s$.

Once the anode current has risen to its final value $i_A = I_A = I_l$ in the rise-time interval t_{ri}, the current i_D in the diode becomes zero and so the diode changes state, turns off and blocks. The voltage v_l across the load begins to rise, so, from eq. (5.5.9), the voltage v_{TH} across the thyristor begins to fall. In the time interval t_{fv}, the thyristor voltage falls from V_s to $V_{TH(ON)}$ (approximately zero) and the turn-on process is complete. The function of the voltage v_{TH} has been linearized.

The linearization of the waveforms has simplified the form of the function of the instantaneous power p_c dissipated in the thyristor during turn-on. The function is a triangular waveform. Adding the component intervals gives the turn-on time t_{on} as

$$t_{on} = t_d + t_{ri} + t_{fv} = t_d + t_c \tag{5.5.10}$$

where t_c is the crossover time.

The energy W_{on} dissipated due to the turn-on process is

$$W_{on} = \int_0^{t_{on}} p_c dt \approx \int_0^{t_{ri}} V_s i_A \, dt' + \int_0^{t_{fv}} v_{TH} I_l \, dt'' = \frac{V_s I_l}{2} t_c . \tag{5.5.11}$$

This relation shows that turn-on loss is a maximum because the crossover time t_c is at a maximum. Compare this with eq. (5.5.7).

EXAMPLE 5.4

A thyristor controls power from a dc source of voltage $V_s = 200$ V to an *RL* load with a freewheeling diode connected across it. See Fig. 5.8. The value of the inductance is high enough to consider the load current to be constant at 48 A for steady-state operation as a chopper circuit. For this condition the thyristor has a delay time $t_d = 0.5\,\mu s$, a current rise time $t_{ri} = 3\,\mu s$ and a voltage fall time $t_{fv} = 2\,\mu s$ at turn on. What is the energy loss in the thyristor during turn-on?

Solution

Ignore the loss during the delay interval. From an inspection of the power waveform in Fig. 5.8b the energy loss in the thyristor is

$$W_{on} = W_c = \int_0^{t_c} p_c dt' = \frac{\hat{P}}{2} t_c = \frac{V_s I_l}{2}(t_{ri} + t_{fv}) = \frac{200}{2} \times 48 \times 5 \times 10^{-6} = 24 \times 10^{-3} \text{ J}.$$

This is nearly five times the loss for a thyristor with a resistive load.

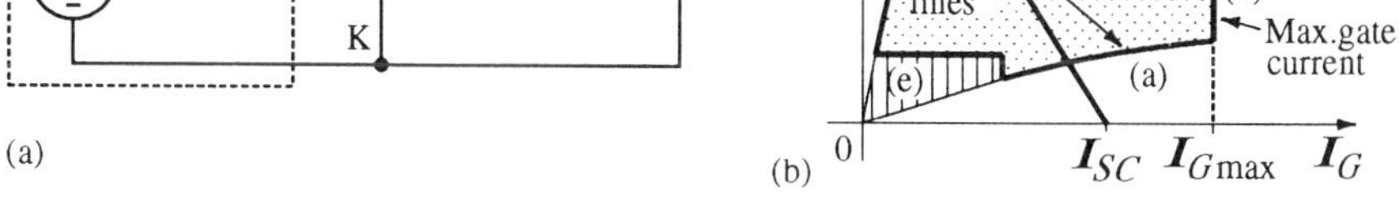

Fig. 5.9 Gate characteristics.

5.5.2. Turn-on Circuits

We will concern ourselves with gate signals for turning on thyristors. So the gate characteristics need to be discussed. The simple gate circuit, shown in Fig. 5.9a, aids this discussion. The question arises, what values of source voltage V_G and external resistance R_G should be used to drive an adequate gate current I_G for thyristor turn-on? The answer is that there is a range of values that gives satisfactory results.

In Fig. 5.9b the gate characteristics of a thyristor are shown. There are five curves to consider. These are associated with the voltage V_{GK} across the gate and cathode terminals, and the steady gate current I_G. Curves (a) are the diode characteristics of the *pn* junction, gate to cathode for the temperature limits. Curve (b) is the maximum voltage line, above which damage could occur. Curve (c) is a hyperbola derived from the fact that the power ($V_{GK}I_G$) into the gate has a maximum permissible value. Curve (d) represents the peak gate current. Curve (e) provides a safety factor; within the boundary, the low values of I_G may not trigger the thyristor, especially at low temperatures (reduced charge mobility). Within the envelope of curves (a), (b), (c), (d) and (e) any combination of V_{GK} and I_G will cause the thyristor to turn on.

The actual current I_G not only depends on V_G and the gate-to-cathode impedance (this is not a constant, but changes during turn-on), but also depends on the value of the resistor R_G. R_G comprises the source internal resistance and any added external value. Consequently, the actual current I_G depends upon the load line, curve (f). The load line is drawn between the gate source, open-circuit voltage V_G on the ordinate axis and the short-circuit current (V_G / R_G) on the abscissa axis. Low impedance sources are required for gate trigger circuits. An example will demonstrate the use of the thyristor specifications to design the gate circuit.

EXAMPLE 5.5

For the given gate-circuit configuration shown in Fig. EX5.5 determine **(a)** the minimum value of the resistance R_G and **(b)** the maximum gate-pulse frequency, if R_G has the value determined in part (a) and if each pulse duration is 5 μs. The specified peak gate power dissipation for the thyristor is 20 W, while its specified average gate power dissipation is not to exceed 2 W.

Solution

(a) Maximum power transfer to the gate occurs if the value of R_G is equal to the resistance of the gate. This is the minimum value of R_G. Maximum power is also peak power, so peak power $= V_G^2 / 2R_G = 2 \times 20$ W. Therefore $R_G = 7.2\,\Omega$.
This result could have been obtained graphically. On the V_{GK} *vs.* I_G graph plot the peak power hyperbola ($V_{GK}I_G = 20$ W). From the open-circuit point on the load line ($V_{GK} = V_G = 24$ V, $I_G = 0$) draw the load line to be a tangent to the hyperbola. This means that the minimum value of R_G is used (since the short-circuit current, $I_{sc} = V_G / R_G$, is a maximum). Symmetry of the hyperbola means that the tangent meets it at half the value of the open-circuit voltage V_G. This is the gate

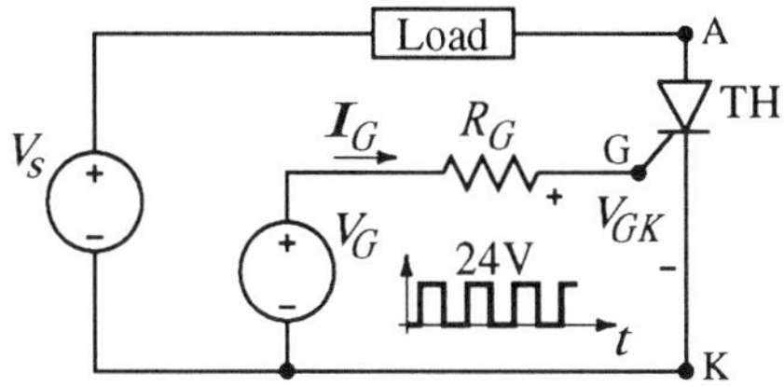

Fig. EX5.5 Gate circuit.

voltage V_{GK} for peak power. Consequently, the gate resistance and R_G have the same value and this also indicates maximum power transfer. From the graph the load line meets the abscissa axis in $I_{sc} = 3.33$ A.
Hence, $R_G = V_G / I_{sc} = 24 / 3.33 = 7.2$ ohms.

(**b**) The voltage pulses at the gate are rectangular, so that the power pulses are rectangular in shape also, if it is assumed that the circuit is purely resistive.
Hence, average gate power is $P_{av} = P_{pk} t_{ON} / T$,
where t_{ON} is the power pulse duration, T is the period of the pulses and P_{pk} is the power amplitude. The maximum pulse frequency f is given by
$f = 1 / T = P_{av} / (P_{pk} t_{ON}) = 2 / (20 \times 5 \times 10^{-6}) = 20 \times 10^3$ pulses per second.

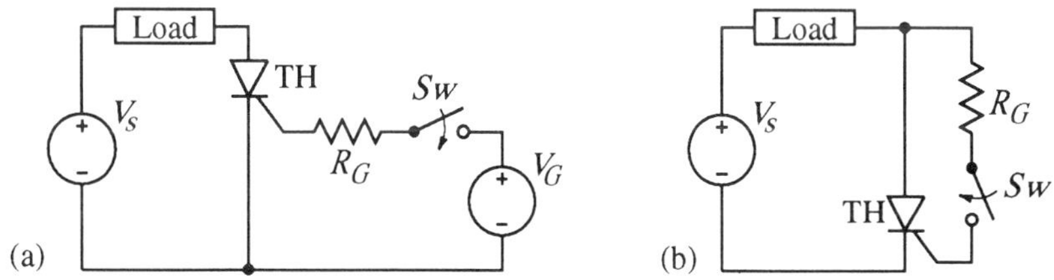

Fig. 5.10 DC gating signals.

There are many kinds of gate circuits. However, it is possible to classify gate signals into three types. These are dc signals, pulse signals and signals for ac applications. Examples of circuits are given for each type.

DC Gate Signals.
DC gate signals are usually simple, and can be obtained from batteries, converters, the power source directly if it is a dc supply, bistable multivibrators (flip-flops), or logic and microprocessor circuits with amplifiers.

Battery source. Figure 5.10a illustrates a gating circuit that has its dc source in the form of a battery. In practice the battery represents any kind of dc supply, since it is uncommon to use a battery except in standby cases. It is assumed that the power circuit is supplied from a dc source of voltage V_s. The thyristor TH blocks until the switch *Sw* is closed to allow the conduction of a dc gate current. The thyristor turns on, since it is forward biased. Once the thyristor is on there is no need for the gate current. Switch *Sw* must be interrupted to remove the gate signal. The switch *Sw* in the driver circuit could be a mechanical relay, a reed switch, a thyristor, or more likely a transistor. The choice is whichever is cheapest and compatible with the controller that provides the signal to the switch *Sw*. If the switch *Sw* is to handle power of the order of watts, the control signal will be of the order of milliwatts. As shown, there is no isolation between the gate signal and its source, and no isolation between the gate circuit and controller if a transistor or thyristor is used for the switch *Sw*. If this is a problem, the use of an opto-isolator integrated circuit (IC) solves it.

Figure 5.10b illustrates a gate circuit that utilizes the main power source. If the control switch *Sw* is closed, a dc gate current exists. The resistor R_G must have a high value to limit the gate current to a safe level for turn-on. Once the thyristor is on, the gate current automatically falls to near zero because the thyristor short-circuits the gate circuit.

Bistable multivibrator. The multivibrator family, consisting of bistable, monostable and astable multivibrators, plays a role in turn-on circuits. It comprises transistor circuits at the fundamental level and logic gates at the intermediary level. As we use the family members only insofar as input/output is concerned, we deal only with the modules in ICs for digital switching circuits.

The bistable multivibrator is better known as a flip-flop. For this application of producing a dc signal to the thyristor gate a *T* flip-flop is used. The *T* indicates that, for every pulse input to the flip-flop, the output *toggles*, that is, it changes state. See Fig. 5.11. Assume the output Q of the flip-flop is low (0). A short pulse at the input causes the output Q to change state and go high (1). The output remains at this state, giving a dc signal to a thyristor gate, until the next input pulse. Every input pulse causes the output to change state. So, to have a gate signal, a pulse is applied to the input of the *T* flip-flop, and, to remove the gate signal, another input pulse is applied. The method is simple and inexpensive. However, the logic circuitry must be isolated from the power circuit.

Gate Pulses.

A single pulse of current, or a train of pulses can be applied to the gate terminal to turn on a thyristor. This enables isolation between the gate and control circuit by means of a transformer or an opto-isolator. An advantage of pulse signals over a dc signal is that, for the same average power, the pulse signals can have a higher magnitude of current. This provides faster turn-on.

A train of pulses is considered to provide more reliable turn-on than a single pulse, if the time of application of the pulse train is longer than the single pulse. This is especially true for inductive loads in the power circuit. The rise of current in an inductive load is slower than in a resistive load, and therefore the current takes longer to reach the latching current value. The gate signal has to be applied for the time that the load current takes to reach the latching value. A train of

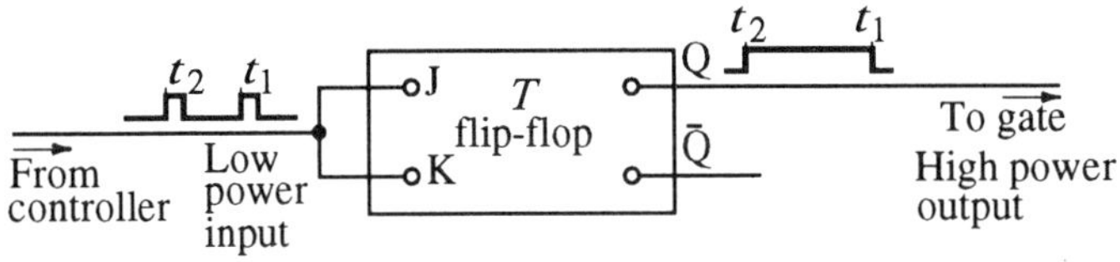

Fig. 5.11 Bistable multivibrator.

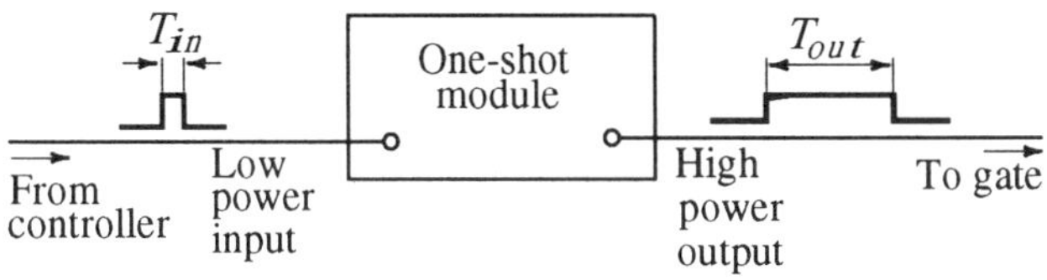

Fig. 5.12 Monostable multivibrator.

pulses would cause less power dissipation than a single pulse of the same magnitude, applied for the same time.

Three types of pulse generator are described. They are the monostable multivibrator, the astable multivibrator and the relaxation oscillator.

Monostable multivibrator. The monostable multivibrator (known as a *one shot*) has the same generic transistor circuit as the bistable multivibrator (flip-flop). The adjustment of resistor and capacitor values in the circuit provides a useful characteristic. It is that, for each low power, short duration T_{in}, input signal, a higher power pulse appears at the output for a longer and adjustable time T_{out}. Figure 5.12 illustrates this in modular form.

Astable multivibrator. By altering a few components in the circuit, the multivibrator can become astable (or free running). Its characteristic is shown in Fig. 5.13. For an input pulse of duration T into the module, the output is a train of pulses of total duration T. The frequency of the pulses is determined by certain RC values connected external to the IC as specified in the data sheets.

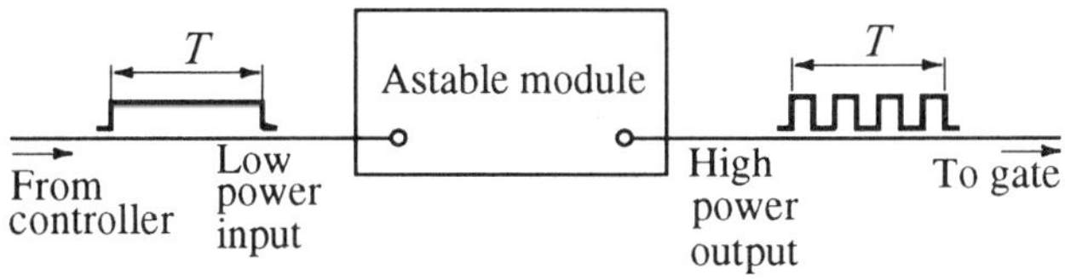

Fig. 5.13 Astable multivibrator.

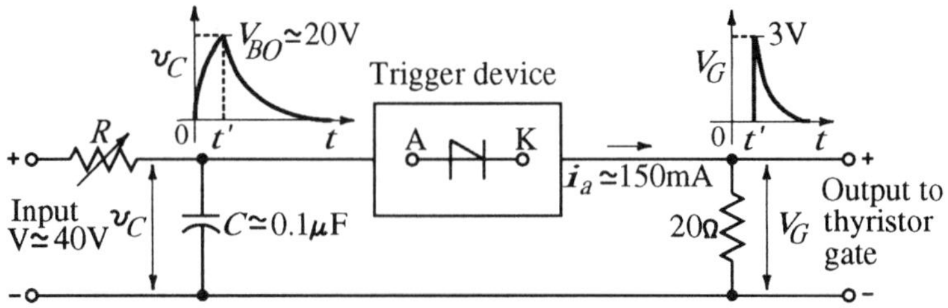

Fig. 5.14 Relaxation oscillator.

Relaxation oscillator. The relaxation oscillator has a dc input and a pulse output. Figure 5.14 illustrates the circuit. The trigger device is assumed to be blocking. The voltage builds up across the capacitor C, until the breakover voltage of the trigger device is reached. At this point the trigger device turns on and the capacitor discharges into the thyristor gate to turn on the thyristor. After the capacitor discharges, the trigger device recovers its blocking state and the cycle repeats. The time between pulses is determined by the charging time constant RC and the breakover voltage of the device.

A 2-terminal, 4-layer diode (a thyristor without a gate terminal) would satisfy the requirements of a trigger device to break over at a particular anode voltage and recover after the current drops below its holding value (about 20 mA).

There is a number of other low power devices that are suitable for use in trigger circuits. These devices include the silicon controlled switch (SCS), the diac, the silicon unilateral switch (SUS), the silicon bilateral switch (SBS), the unijunction transistor (UJT) and the programmable unijunction transistor (PUT).

These devices will be described in the next subsection. However, for particular ratings, the reader must refer to the manufacturers' data sheets.

Silicon controlled switch (SCS).
Similar to the thyristor, the silicon controlled switch, depicted in Fig. 5.15, is a 4-layer, *pnpn,* unidirectional device. Unlike the conventional thyristor, the SCS has an anode gate as well as a cathode gate. Both can be used to turn on or turn off the device.

From the two-transistor model description of the thyristor, positive current at the cathode gate or negative current at the anode gate can turn on a forward-biased device. If the SCS is on, a negative current at the cathode gate (base of the *npn* transistor) turns off the *npn* transistor. Consequently, there is no base current to the *pnp* transistor, which turns off too. A similar action occurs if a positive current is injected at the anode gate.

Typical ratings are as follows: blocking voltage 100 V, rated current 200 mA, holding current 3 mA, anode-to-cathode conduction drop 2 V, gate current 50 μA, turn-on time 2 μs and turn-off time 8 μs.

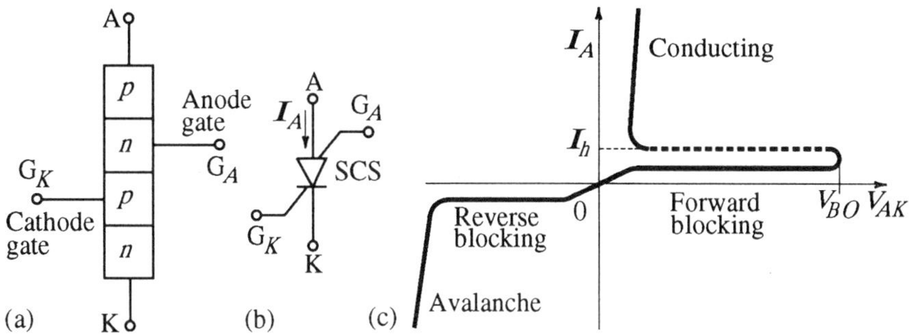

Fig. 5.15 Silicon controlled switch (SCS).
(a) Structure, (b) symbol, (c) characteristics.

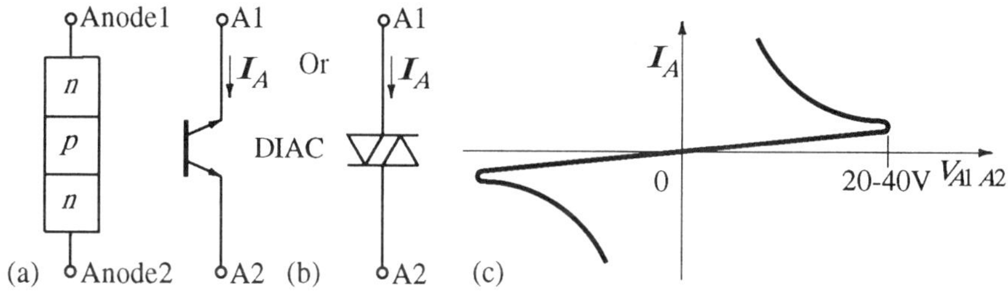

Fig. 5.16 Diac trigger diode.
(a) Structure, (b) symbols, (c) characteristics.

Diac.
Figure 5.16 gives a depiction of a diac switch that can be used in a relaxation oscillator (Fig. 5.14) with an ac supply. The diac is a 3-layer (*npn* or *pnp*), 2-terminal diode that is constructed like a transistor without a base terminal. It is voltage controlled and switches symmetrically with either polarity at about 20 V to 40 V to produce a pulse of about 3 V at the output.

The name diac is compounded from two terminal (*di*) and alternating current (*ac*) control. The device is most suitable as a trigger for a bidirectional switch such as the triac (see Chapter 8).

Silicon unilateral switch (SUS).
The silicon unilateral switch (SUS) is an integrated circuit of transistors, zener diodes and resistors that, together, give it the characteristic of a small thyristor with an anode gate (gate terminal connected to the inner *n* layer).

Figure 5.17 depicts the circuit symbol and the characteristic of the SUS. With no gate signal, the device will withstand a forward-bias voltage up to about 8 V. At this level it turns on ($t_{on} \approx 1\ \mu s$) and produces a peak current ($\hat{I}_A \approx 200$ mA) at a pulse voltage (about 3 V) which is suitable for triggering a power thyristor. The

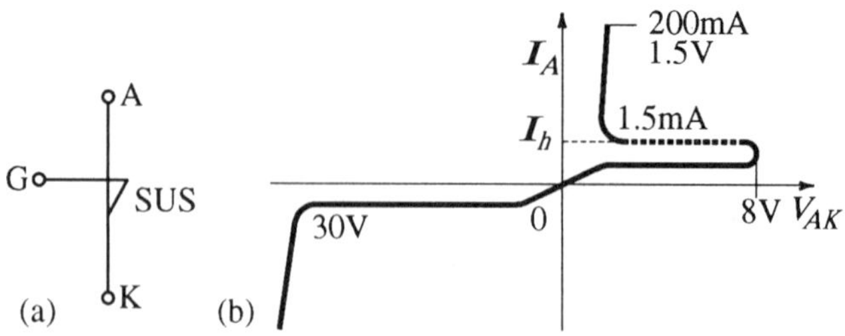

Fig. 5.17 Silicon unilateral switch (SUS).

SUS turns off ($t_{off} \approx 25\,\mu s$) if the current reduces to a level below the holding value ($I_h < 1.5\,mA$). If the switching voltage across the SUS is required to be less than 8 V, a gate signal can be used.

Silicon bilateral switch (SBS).
An integrated circuit comprising two SUSs connected back-to-back with common gates, as shown in Fig. 5.18, allows current in either direction. This device is named the silicon bilateral switch (SBS). It is used to trigger power devices (for example, triacs) used in ac applications.

Unijunction transistor (UJT).
The unijunction transistor (UJT) is a popular device to use in relaxation oscillators for triggering thyristors. Figure 5.19 illustrates its circuit symbol and its characteristic.

The UJT is a 3-terminal, 2-layer, *pn* device. With no emitter bias at the terminal E, the resistance between the two bases ($R_{B2} + R_{B1}$) is of the order of kilo-ohms. R_{B2} is the resistance between terminal B2 and the *p*-region contact, and R_{B1} is the resistance between the *p*-region contact and terminal B1. If the terminals B2 and B1 are positively biased with a voltage V_{BB}, the *p*-region contact with the B2-B1 *n* region creates a voltage divider, such that the voltage between the *p*-region contact and B1 is

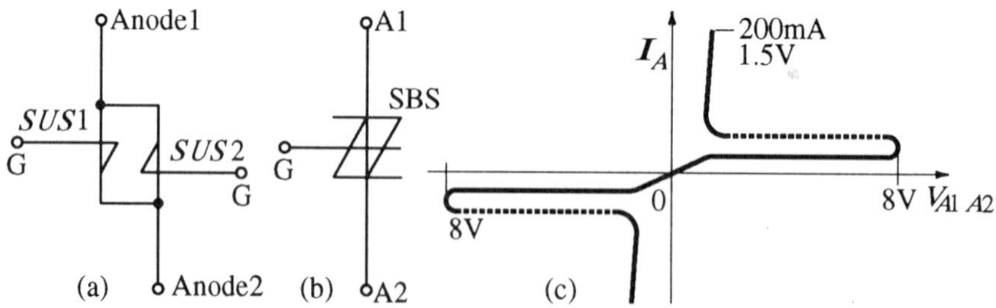

Fig. 5.18 Silicon bilateral switch (SBS).

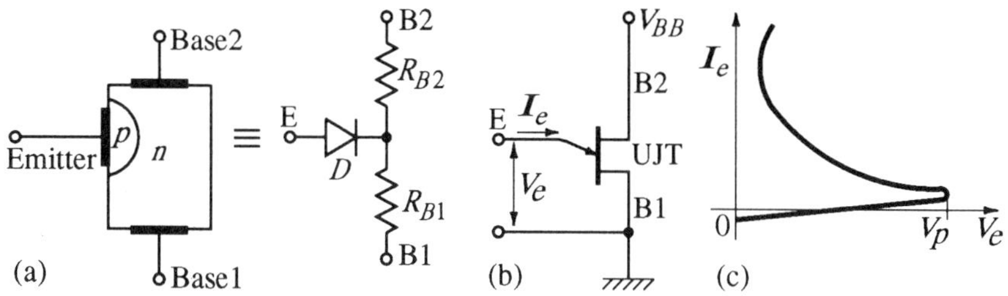

Fig. 5.19 Unijunction transistor (UJT).
(a) Structure, (b) circuit diagram, (c) characteristic.

$$V_{RB1} = \frac{R_{B1}}{R_{B2} + R_{B1}} V_{BB} \triangleq \eta V_{BB} . \tag{5.5.12}$$

The coefficient η is known as the stand-off ratio, because its value determines the reverse bias that the equivalent *pn* diode junction experiences. A typical value of η is 0.56 but depends where the *p* region is located along the *n* bar.

The point of this is that, before current I_e can be injected into the emitter, the emitter voltage V_e must be greater than a value V_p ($V_p = 0.7 + \eta V_{BB}$ volts), where 0.7 V is the *pn* diode barrier voltage. Once the emitter voltage exceeds this value V_p (known as the peak-point voltage), charge (holes) flows into the lightly-doped *n* region near terminal B1 and effectively reduces the resistance R_{B1} to a low value (about 50 Ω). R_{B2} is unaffected by this. The UJT assumes a negative resistance characteristic, V_e reduces to a low value (less than 1V), and, in a relaxation oscillator circuit with the input connected to the emitter E and the output connected to the base terminal B1, a pulse of current is the result.

The resumption to the off-state occurs when V_e falls to a value close to zero volts.

The action of a relaxation oscillator with the UJT acting as the trigger device is as follows, using Fig. 5.20 as reference. Capacitor *C* is charged from the supply at a rate that depends on the time constant $(R_1 + R_2)C$. When the capacitor voltage ($v_C = V_e$) reaches the peak-point voltage V_p, the emitter-base B1 resistance falls rapidly, *C* discharges through *R* and a pulse appears at the gate of the thyristor. As the current pulse falls to zero, the UJT recovers its blocking state and *C* begins to charge again to repeat the cycle. The pulse period depends on the timing circuit (R_1, R_2 and *C*), the stand-off ratio η and the value of *V*. R_B is a resistor to aid temperature stabilization.

Programmable unijunction transistor (PUT).
The programmable unijunction transistor (PUT) is another trigger device for use in thyristor gate circuits. As shown in Fig. 5.21, it is a small thyristor with an anode gate and has the pulse characteristics of a UJT in the relaxation oscillator, with the added advantage of being programmable.

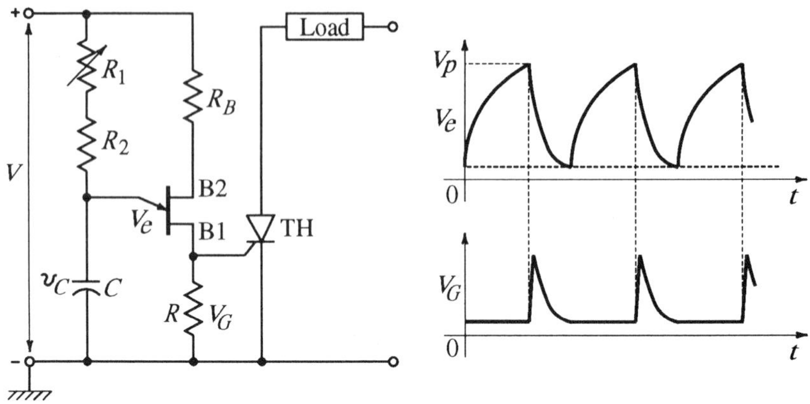

Fig. 5.20 UJT circuit to trigger a thyristor.

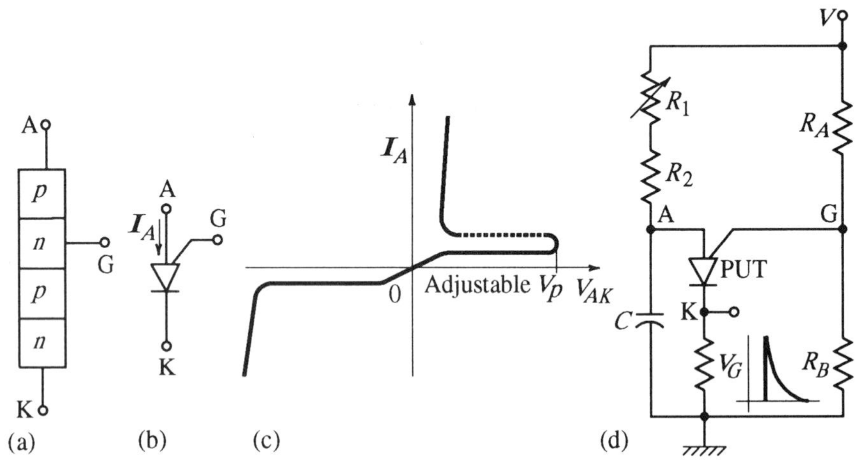

Fig. 5.21 Programmable unijunction transistor (PUT).
(a) Structure, (b) symbol, (c) characteristics, (d) gate circuit.

Operation of the PUT depends on the anode-to-gate voltage. The gate voltage is set by the potential divider and this is the programmable nature of the device. As long as the gate voltage is greater than the anode voltage, the PUT remains in the off-state. Increasing the anode voltage to the point where it is about 0.7 V (the *pn* barrier voltage) above the gate voltage causes the PUT to turn on very quickly (less than 0.1 μs). The anode turn-on voltage V_p is adjusted (or programmed) by changing the gate voltage, that is, by altering the ratio $R_B : (R_A + R_B)$.

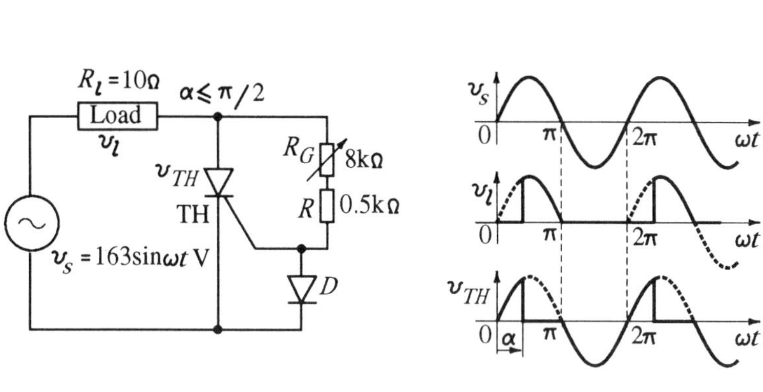

Fig. 5.22 Simple ac phase angle control.

In the relaxation oscillator, illustrated in Fig. 5.21d, R_1, R_2 and C act as an adjustable delay for the peak voltage V_p to be reached. At turn-on, a pulse V_G (up to 6 V) appears at the cathode of the PUT. It is this pulse that can be applied to the gate of a thyristor. As much as 2 A can be applied as a short current pulse. The capacitor C discharges into the cathode resistor, the current drops to zero, the PUT turns off, the anode voltage is less than the gate voltage, the capacitor begins to charge again until V_p is reached and the cycle repeats.

Gate Signals for AC Applications.

Gate signals for ac applications are usually associated with control of the phase angle α, ideally over the range $0 \le \alpha \le \pi$ radians. Fig. 5.22 illustrates a simple control, where the trigger angle α is limited to the range from 0 to $\pi/2$ radians for half-wave rectification.

As the instantaneous value of the ac source voltage v_s increases positively at the anode, the thyristor blocks, but the gate current increases. The value of the gate current depends on v_s and $(R_G + R)$. At some value of v_s the gate current will be high enough to turn on the thyristor. For low values of gate resistance the trigger angle α is small. As the gate resistance is increased, the trigger delay increases up to a maximum of $\alpha = \pi/2$ radians. If a trigger current is not reached by the time the source voltage v_s reaches its maximum value, the thyristor will not turn on, because, thereafter, the gate current reduces.

There is inherent gate protection against overcurrent for low values of α. Once the thyristor is turned on, the gate circuit is virtually short-circuited, so the gate current is never higher than the thyristor turn-on value. The diode D serves to limit the cathode-to-gate voltage while the thyristor is reverse biased.

This principle for controlled turn-on for power modulation in a half-wave rectified circuit can be extended to full-wave rectification.

The range of firing angle control can be increased to π radians by using a time-delay circuit, as shown in Fig. 5.23. The voltage across C lags the supply voltage by an amount equal to a nonlinear function of $1/(\omega C(R_G + R))$. Increasing R_G increases the time taken for the voltage v_C to reach a level so that there is

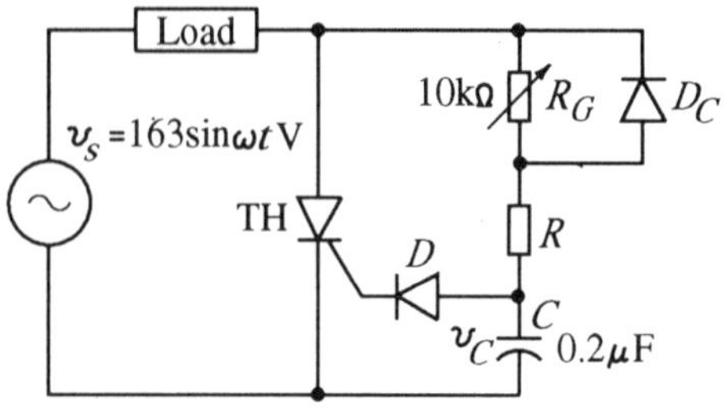

Fig. 5.23 Full-range phase angle control.

sufficient gate current to turn on the thyristor. At turn-on, C discharges through D_C and TH quickly enough so that the phase angle is consistent each cycle.

It is often required to have fast thyristor turn-on. A fast rising gate pulse can satisfy this requirement. A pulse and trigger angle adjustment is provided by a circuit known as a Schmitt trigger.

Schmitt trigger. A Schmitt trigger circuit belongs to the multivibrator family. It provides a precise control of the thyristor firing angle α from 0 to π radians in ac applications. When the input signal to the Schmitt trigger reaches a particular value, there is an output pulse (amplified) that is sustained until the input signal falls below a particular value. These output signals must be produced at an adjustable time with respect to the reference supply voltage. Figure 5.24 shows a Schmitt trigger circuit in block diagram form. The supply voltage provides a synchronizing signal, and a dc signal provides the control for the time taken to reach the trigger voltage V_T. This time is directly associated with the firing angle α. If the dc level is reduced, the input trigger level is delayed.

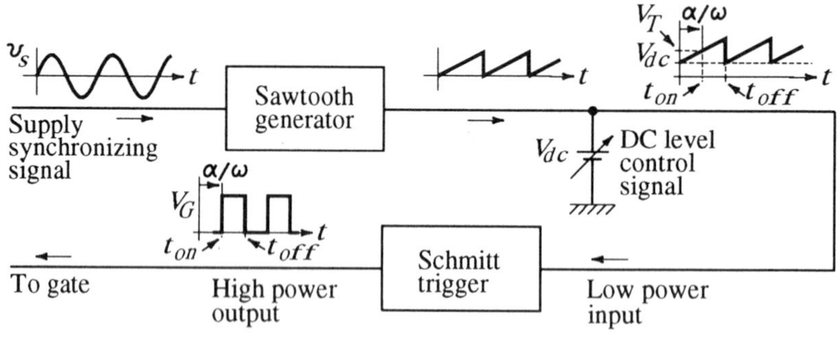

Fig. 5.24 Schmitt trigger circuit.

EXAMPLE 5.6

A heating element is to be controlled by a single thyristor to provide a smooth adjustment of load power from 0 to 800 W. The voltage source is 115 V, 60 Hz. Design a gate circuit using ICs.

Solution

A single thyristor indicates that controlled half-wave rectification is the object.

For 115 V, 800 W, a conservative choice of thyristor is one rated at 400 V peak reverse voltage and 8 A(rms) (type 2N4443). The gate current for this type is 30 mA.

If we are to isolate the trigger circuit from the power circuit, we can use an opto-isolator. An opto-isolator, type 4N33, delivers 40 mA if its gate is driven by 8 mA. This means that the opto-isolator can be driven by other ICs directly.

The input signal to the trigger circuit must be a pulse that is synchronized, shaped and isolated from the power supply. A transformer produces the isolation and a reduced voltage (50 V or less). The shaping is provided by a Schmitt trigger (MC14584).

The pulse to the gate of the thyristor must have an adjustable delay ($0 \leq \alpha \leq 180°$) to enable the load power to be adjusted. Two one-shot multivibrators (one IC, type MC14538) can achieve this. See Fig. EX5.6a for the circuit diagram. Figure EX5.6b depicts the waveforms at particular test points.

The transformer secondary-winding voltage v_s is shown at (i). The zener diode rectifies the alternating voltage waveform, so that the input to the Schmitt trigger is half-wave rectified and clipped, as in (ii) in the figure. A shaped pulse (iii) is the output of the Schmitt trigger and is in synchronism with the positive half wave of the supply. A one one-shot multivibrator can be used as a delay (the firing angle α). It is triggered by the output of the Schmitt trigger. The output pulse (iv) of the first one-shot is on for a time that is determined by the external values of R_1C_1. As this pulse goes positive, it triggers the second one-shot and it is this output pulse (v), amplified and isolated by the opto-isolator that fires the thyristor.

Adjustment of R_1 provides the variation of α from 0 to 180° over the positive half cycle of the supply voltage.

5.6. THYRISTOR TURN-OFF

Once a thyristor has been turned on it will remain on until the anode current I_A has been reduced to near zero. At this point there is no transistor regeneration, recombination[1] occurs and the thyristor reverts to its blocking condition. Commutation is the name given to the turn-off process. *Turn-off* can be defined as the condition that all conduction has ceased and the reapplication of a positive anode voltage will not cause current to be conducted at the anode (unless there is a gate signal applied).

[1] *Recombination* is the process whereby a spatially separated electron and hole collide and mutually annihilate, resulting in zero net charge.

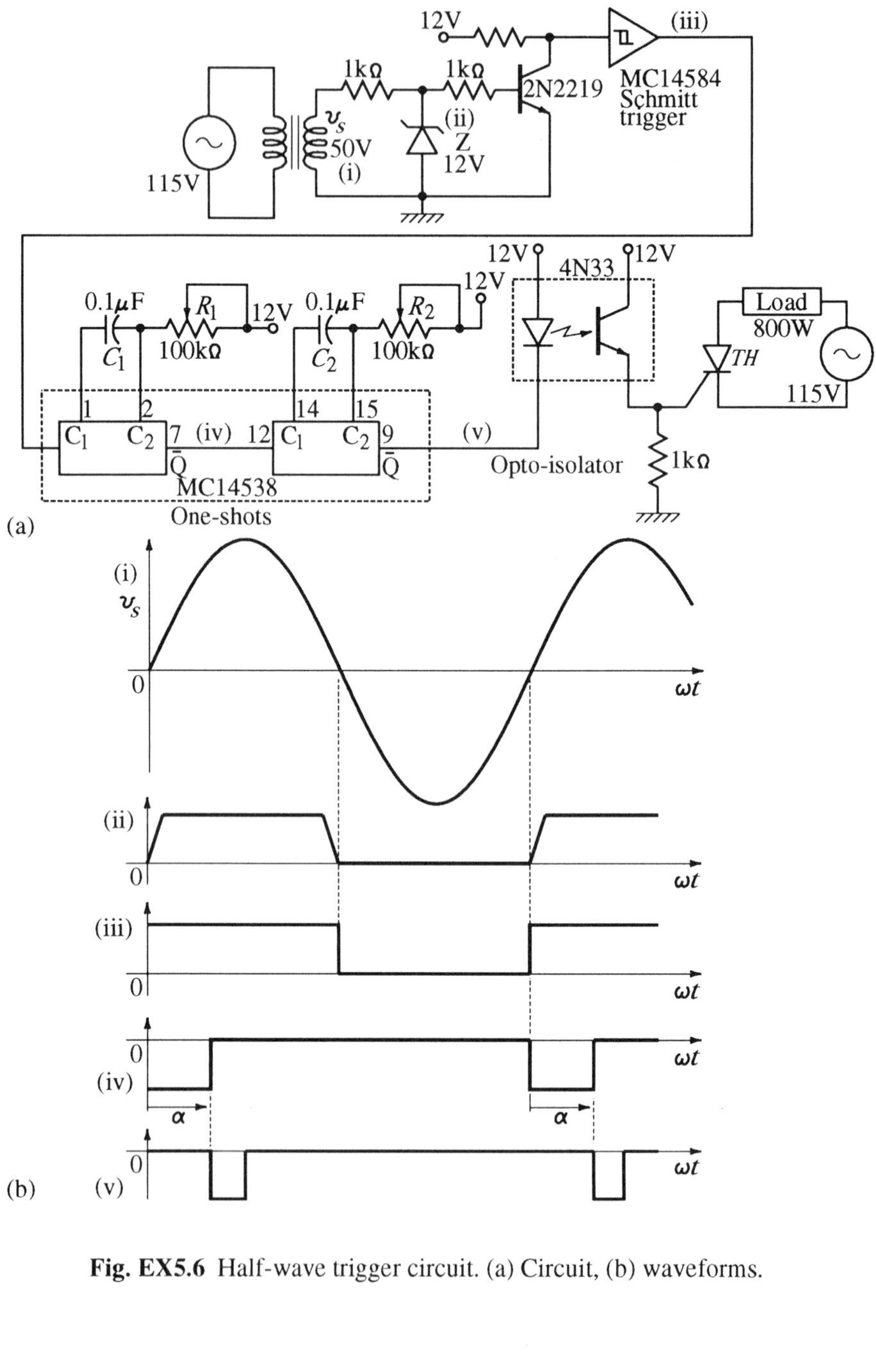

Fig. EX5.6 Half-wave trigger circuit. (a) Circuit, (b) waveforms.

There are three methods of commutation, natural commutation, forced commutation and gate turn-off.

Natural Commutation. Natural commutation describes turn-off by the anode current reducing to a value below what is called the *holding* value I_h ($I_h \leq 0.001 \times I_{A\ rated}$). This occurs in a dc circuit by (a) opening a line switch, (b) increasing the load resistance sufficiently or (c) shunting the thyristor with a sufficiently small resistor. None of these methods is really practical. However, series resonance is a way to bring the current to zero naturally, and this method is referred to again.

Forced Commutation. Forced commutation is the most common method of thyristor turn-off. To reduce the anode current to zero, a reverse bias voltage can be applied across the thyristor. This reverse bias occurs every half cycle if the supply is an ac source. This method is called *ac line commutation* and requires no additional circuitry. At what point the current is reduced to zero in the negative half cycle depends on the circuit impedance.

If the supply is a dc source, forced commutation requires an auxiliary source of energy to bring the thyristor current to zero. The most common auxiliary source is a capacitor, but additional circuitry is needed to charge the capacitor to the correct polarity and to cause it to commutate the thyristor at the right time.

Gate turn-off. The gate turn-off method is like gate turn-on. If we reconsider the two-transistor model of the thyristor, a positive gate signal initiates regenerative feedback to turn on the device. The regenerative feedback is self-sustaining once the thyristor is on, so the gate signal can be removed. A negative gate signal, if it is large enough, can interrupt the regenerative feedback and the thyristor turns off. This has become a well-proven method of commutation. It is described in more detail in Chapter 9 under the heading of GTOs (gate turn-off thyristors).

The turn-off time of a thyristor is important because it determines the maximum frequency of switching. It may not be too important in 60 Hz, ac line-commutation applications, but it becomes extremely important in pulse-width-modulation applications, where the thyristor may be switched on and off to produce 10,000 pulses per second. It is difficult to be exact; however, the definition of turn-off time can be said to be the time from turn-off initiation to the time it takes for the charge to decay to near its equilibrium off-state. The thicker the device wafer, the longer the thyristor takes to turn off. This means that the higher the voltage rating the greater is the turn-off time. The greater the current being conducted, the greater the number of charge carriers and the longer the thyristor takes to turn off. A reverse bias across the device will tend to sweep out charge carriers and shorten the turn-off time.

Typical times for the turn-off of relatively low current (< 50 A) thyristors may be 10 to 100 μs for natural commutation. For the same thyristors, forced-commutation turn-off times can be reduced to 7 to 20 μs. The turn-off time is a matter of manufacturing technique. Thyristors are graded for phase-control applications (normally associated with 50 or 60 Hz supplies), and for inverter applications (fast turn-off). A 1500-A(rms), 2200-V device may have a typical circuit-commutated turn-off time of 250 μs and this would be classified as a phase-control thyristor. On the other hand the same catalogue would describe a 1250-A(rms), 2000-V thyristor with a typical circuit-commutated turn-off time of 50 μs as a fast switching thyristor suitable for inverters and choppers. The turn-off times that are given in a catalogue are typical because turn-off time is temperature dependent. The turn-off time for a particular thyristor at a junction temperature of 125°C can be twice the turn-off time of that at 25°C.

What is the action of turn-off? A reverse bias voltage across the thyristor allows reverse charge flow long enough to remove most of the carriers from junctions J1 and J3, so that these junctions are in the blocking state. This occurs quickly. The thyristor blocks a forward bias voltage after the excess carriers at junction J2 recombine naturally. This latter action takes more time and occurs during both natural and forced commutation.

The turn-off reverse bias voltage is accompanied by a reverse recovery current in the same way as that described for the diode in section 3.6 and Fig. 3.6. There is the reverse recovery charge Q_{RR} to remove, after which junction J1 and J3 block. The loss associated with this reverse recovery is negligible compared with the turn-on losses, so it will be ignored. A noteworthy characteristic of thyristors is that turn-off is not accompanied with losses. While the anode current is finite the voltage v_{TH} is virtual zero and while the voltage is rising the current is zero. This is not the case with transistors.

5.6.1. Turn-off Circuits

A thyristor needs circuit-commutated turn-off (forced commutation) whenever a dc source supplies the power to be modulated.

The capacitor is the main component of the commutation circuit that provides the reverse bias across the thyristor to bring its current to zero. There is a great number of ingenious circuits to accomplish commutation. We will analyze two common circuits, the parallel capacitance turn-off circuit and the *LC* resonant turn-off circuit, to demonstrate the basic principles.

Parallel capacitance turn-off.

Figure 5.25a depicts a circuit diagram for forced commutation. The method is called *parallel capacitance turn-off*. A dc power supply V_s and a load R_l comprise the main circuit. Thyristor $TH1$ modulates the power in the load as a chopper. For turn-off, the components C, R and $TH2$ are used. The circuit operation is as follows.

- Thyristors $TH1$ and $TH2$ are off. Thyristor $TH1$ is turned on so that there is load current i. At the same time, the capacitor C charges to $+V_s$ at plate Y through R, C and $TH1$. Plate X of the capacitor is virtually at ground potential, because thyristor $TH1$ is in the *zero impedance* state. This sets the stage for commutation.
- When it is desired to interrupt the load current, thyristor $TH2$ is turned on. This action puts the capacitor C in parallel with thyristor $TH1$. The voltage across capacitor C reverse biases thyristor $TH1$. If thyristor $TH1$ is reverse biased

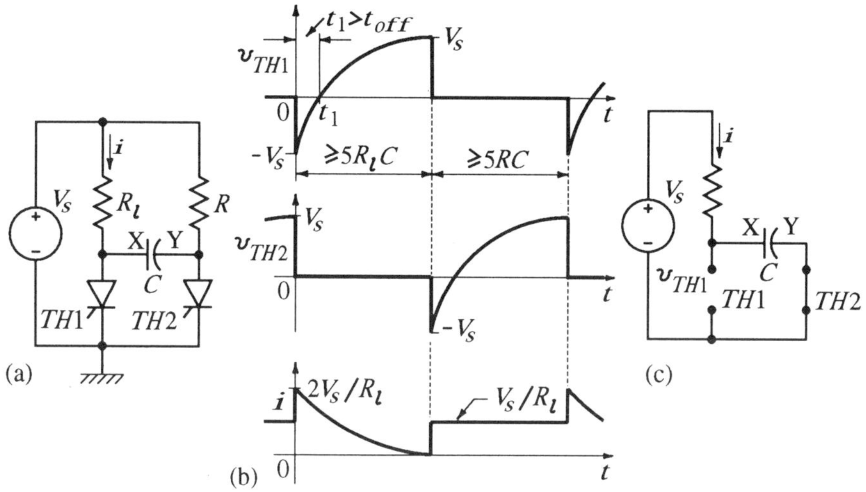

Fig. 5.25 Parallel capacitor turn-off.
(a) Circuit diagram, (b) waveforms, (c) turn-off path.

long enough, it will turn off. Capacitor C discharges and charges, this time becoming positive at X through R_l, C and $TH2$.

- At some particular time thyristor $TH1$ is turned on again. Now, the voltage across capacitor C reverse biases thyristor $TH2$ to turn it off. Then capacitor C becomes positively charged at plate Y so that the cycle can be repeated.

By controlling the times that thyristors $TH1$ and $TH2$ are allowed to conduct, the average power in the load can be adjusted. What we have to do is to determine the value of the capacitance to ensure that commutation is successful.

Analysis. The objective is to find the value of C for given values of the load resistance R_l and the turn-off times[2] t_{off} (or t_q) for the thyristors. The objective is reached by comparing the turn-off time with the time that a thyristor is reverse biased. If the thyristor has not turned off during the period of reverse bias it will conduct again as soon as it becomes forward biased. If both thyristors conduct at the same time, the load current cannot be interrupted.

Let us consider that steady-state switching conditions prevail. The initial conditions are that, at $t=0^-$, thyristor $TH1$ is on, steady current $I=V_s/R_l$ is in the load and the capacitor is at $+V_s$ at plate Y. At $t=0$, thyristor $TH2$ is turned on, thyristor $TH2$ goes into the *zero-impedance* state immediately, the voltage across capacitor C reverse biases thyristor $TH1$, which goes into the *infinite-impedance* state immediately, but does not turn off immediately because charge carriers have to be swept out plus a recombination time is required. These initial conditions are

[2] Typical turn-off times are specified in the manufacturers' data sheets.

depicted in Fig. 5.25b and c.

For these conditions the current i is

$$i(t) = \frac{2V_s}{R_l} e^{-t/R_lC} \tag{5.6.1}$$

which is the response of the R_lC circuit due to a step input of $2V_s$ (V_s from the source and V_s from the capacitor). The voltage across $TH1$ is

$$v_{TH1}(t) = V_s - iR_l = V_s(1 - 2e^{-t/R_lC}) . \tag{5.6.2}$$

Figure 5.25b shows this exponential waveform. Initially $TH1$ is reverse biased. The time t_1 for the voltage v_{TH1} to rise to zero (the time after which $TH1$ becomes forward biased) is given by

$$v_{TH1}(t_1) = 0 = V_s(1 - 2e^{-t_1/R_lC}) . \tag{5.6.3}$$

That is,

$$t_1 = 0.69 R_lC . \tag{5.6.4}$$

However, we must have $t_1 \geq t_{off}$, where t_{off} is the specified turn-off time. So,

$$C \geq \frac{t_{off}}{0.69 R_l} . \tag{5.6.5}$$

In a test circuit, we can start with a value of C that is equal to or greater than the value of this expression and reduce the value until the thyristor does not commutate. The most reliable and economic value of C is a little above this last value. The precise value of C cannot be obtained from this analysis, because of the near ideal characteristics that we have assumed for the thyristors. If the thyristors $TH1$ and $TH2$ have the same turn-off times and if $R \geq R_l$, then the above value of C will allow satisfactory commutation of $TH2$.

It can be noted from the figure and the symmetry of the circuit that the voltages across the thyristors have similar waveshapes. If the capacitor is to charge fully after each switching action, it will take at least five time constants ($5R_lC$) after thyristor $TH2$ has been triggered and at least five time constants ($5RC$) after thyristor $TH1$ has been triggered for this to be accomplished. This means that the minimum period T of switching is

$$T \geq (5R_lC + 5RC) = 5(R_l + R)C . \tag{5.6.6}$$

Therefore the maximum switching frequency of operation for either thyristor is $1/[5(R_l + R)C]$ pulses per second.

A disadvantage produced by the commutation circuit is shown in Fig. 5.25b and eq. (5.6.1). At the initiation of commutation the load current rises to twice its steady value before decaying to zero exponentially. The energy is from C.

An important point is that, if the frequency of switching is increased, the value of C must also be increased to compensate for a lower voltage V_C across C. This increase in the value of C can be calculated by correcting eq. (5.6.1) to read

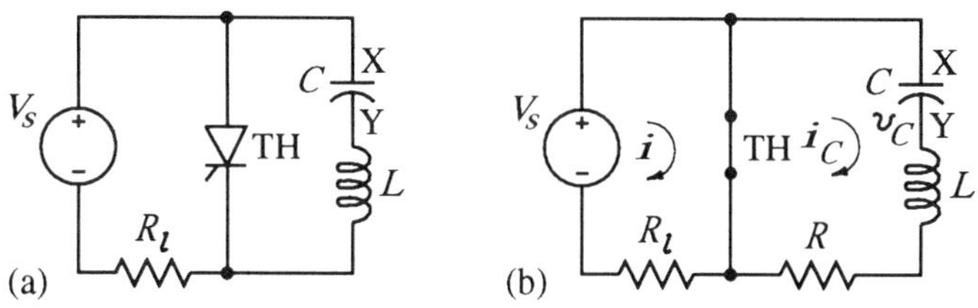

Fig. 5.26 Parallel resonance turn-off.
(a) Circuit diagram, (b) equivalent commutation circuit.

$$i(t) = \frac{(V_s + V_C)}{R_l} e^{-t/R_l C} \tag{5.6.7}$$

and then proceeding with the analysis as before.

EXAMPLE 5.7

Consider a case of parallel capacitance turn-off as illustrated in Fig. 5.25. The load resistance R_l is 5 ohms and the supply voltage is 120 V. Calculate the minimum value of the capacitor C if the manufacturer's specified turn-off time for the thyristor is 15 μs. What is a suitable value of the resistor R, if the thyristor $TH1$ is triggered on every millisecond?

Solution

From eq. (5.6.5) the minimum value of C is calculated from
$C = t_{off} / (0.69 R_l) = 15 / (0.69 \times 5) = 4.3$ μF.
A reasonable value of C would be 5 μF.
The minimum time of one cycle of operation is $(5R_lC + 5RC)$. The value of R provides the only adjustment of this time, since the other parameters are fixed. Let us find the maximum value of R.
In this case $5C(R_l + R) = 10^{-3}$ s. Therefore, $R = 35$ ohms.
The value of R can be less than 35 ohms. This would result in the capacitor charging up faster while $TH1$ is conducting.

Commutation by resonance.
An LC circuit is a resonant circuit. In such a circuit the current tends to alternate as the stored energy in L and C transfer back and forth. This serves the purpose of thyristor commutation in a dc circuit. Figure 5.26a shows a simple thyristor circuit using parallel resonance for commutation.

The operation of the circuit is as follows. Initially the thyristor is off. The capacitor C charges positively to V_s at plate X. To allow a load current i, the thyristor is turned on. This allows the capacitor C to discharge through TH, L and C. Ideally there is no energy loss in this resonant circuit. The plate Y charges to

$2V_s$ (plate X is tied to V_s by the connection to the supply rail, so the capacitor has V_s across it at this moment). Then the capacitor current reverses and the thyristor is reverse biased by the capacitor voltage. Commutation is accomplished if the resonant current i_C reaches a value greater than the load current i for a time greater than t_{off}, the turn-off time of the thyristor. This means that the forward current of the thyristor has been brought to a value lower than the holding current long enough for the thyristor to go into the blocking state. What are the values of L and C so that commutation is successful?

Analysis. To find values of L and C suitable for commutation we determine the circuit response. Assume the initial conditions ($t=0$) that the thyristor is on and the capacitor C has reversed its polarity (plate Y is at $+V_s$ with respect to plate X), so that $v_C(0)=V_s$ and $i_C(0)=0$. These conditions are represented in Fig. 5.26b, in which the resistor R is the equivalent resistance of the inductor L. The voltage equation for the loop C, L, R and TH can be written in the frequency domain as

$$(1/Cs + Ls + R)I_C(s) = V_s/s\ , \tag{5.6.8}$$

or

$$I_C(s) = \frac{V_s/L}{\left(s+\dfrac{R}{2L}\right)^2+\left(\dfrac{1}{LC}-\dfrac{R^2}{4L^2}\right)^2}\ . \tag{5.6.9}$$

The solution in the time domain is

$$i_C(t) = \frac{V_s}{L}\frac{1}{A}e^{-\frac{R}{2L}t}\sin At\ , \tag{5.6.10}$$

where $A^2 = 1/LC - R^2/4L^2$.

This current response is shown in Fig. 5.27a. It is an exponentially damped sine wave. Since the resistance R has little effect on the value of the first current peak and since we are not interested in what happens after that, it is a reasonable approximation to assume that $R \approx 0$. Thus

$$i_C(t) \approx V_s\sqrt{C/L}\,\sin\frac{1}{\sqrt{LC}}t\ . \tag{5.6.11}$$

This approximate response is shown in Fig. 5.27b. The peak value of the current is

$$i_{C\,\max} \approx V_s\sqrt{C/L}\ . \tag{5.6.12}$$

The load current is $i=V_s/R_l$. For commutation to occur the thyristor current must be reduced to zero. That is $i_{C\,\max} \geq i$. In the limit

$$V_s/R_L = i = i_{C\,\max} = V_s\sqrt{C/L}\ . \tag{5.6.13}$$

Therefore

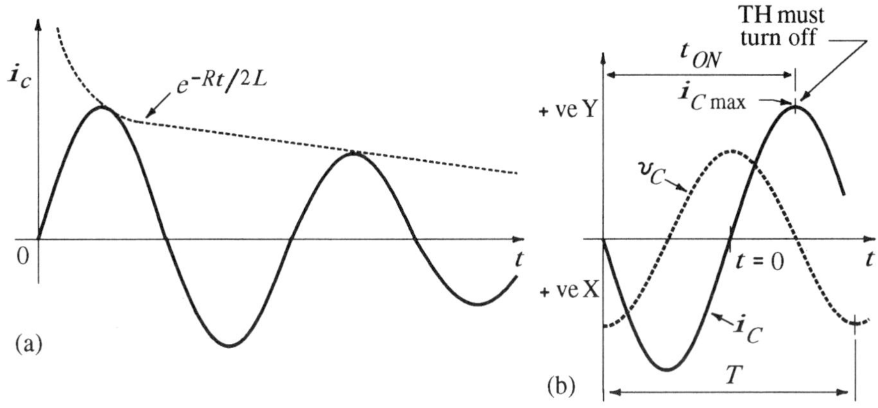

Fig. 5.27 Current response.

$$\sqrt{L/C} = R_l \,. \tag{5.6.14}$$

From eq. (5.6.11) the period of oscillation of the resonant circuit is

$$T = 2\pi\sqrt{LC} \,. \tag{5.6.15}$$

The maximum time t_{ON} that the thyristor conducts is from the time it is turned on to the time that the capacitor current reaches its maximum value in the reverse direction. In terms of the period T this time is $3T/4$ in the limit. Thus

$$t_{ON} = \frac{3\pi}{2}\sqrt{LC} \,. \tag{5.6.16}$$

Equations (5.6.14) and (5.6.16) lead to the solution of the problem. Generally, R_l and t_{ON} are specified. So the values of L and C follow. If the load is inductive, the value of C can be reduced.

This is a simple method for modulating power in the load or controlling the average voltage across the load. The thyristor on-time t_{ON} is fixed and the thyristor off-time t_{OFF} is the variable that is adjusted by the controller. The average load voltage $V_{l\,av}$ is ideally

$$V_{l\,av} = V_s \frac{t_{ON}}{t_{ON} + t_{OFF}} \,. \tag{5.6.17}$$

Some applications require more versatility in the control of the average voltage. Often it is preferable to have the thyristor on-time t_{ON} adjustable. This can be accomplished by using an auxiliary thyristor in the commutation circuit. This is described in the next section.

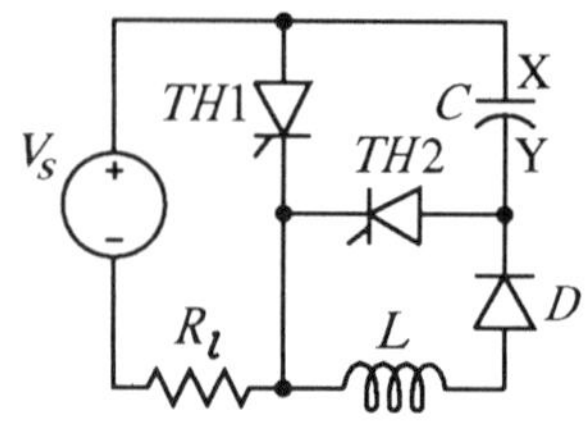

Fig. 5.28 Auxiliary resonance turn-off.

Auxiliary resonant turn-off.
The auxiliary resonant turn-off circuit allows both the on-time t_{ON} and the off-time t_{OFF} of the thyristor to be adjusted by means of the controller. Figure 5.28 depicts the circuit diagram. Thyristor $TH1$ allows modulation of power in the load R_l. The elements L, D and C provide resonance so that C can achieve the correct voltage polarity for commutation. Thyristor $TH2$ is the auxiliary component that is triggered on to initiate the commutation of $TH1$. Since thyristor $TH2$ can be triggered at any reasonable time[3], the time t_{ON} of $TH1$ conduction is adjustable.

The operation of the circuit is as follows. Thyristor $TH1$ is off. Thyristor $TH2$ is triggered on to charge C. At the culmination of the charging process the current is zero and thyristor $TH2$ turns off naturally. Now thyristor $TH1$ can be turned on to allow the conduction of load current. During this conduction period t_{ON} the capacitor discharges through $TH1$, L, D and C. During this part, C reverses polarity. The charge on C is held with the plate Y positive with respect to the plate X, because reverse current is blocked by diode D and thyristor $TH2$. When it comes time for $TH1$ to be commutated, thyristor $TH2$ is fired on. Now the voltage on C reverse biases thyristor $TH1$. $TH1$ turns off. While $TH2$ is on, the capacitor C charges to V_s at plate X with respect to plate Y. Then thyristor $TH2$ turns off. At some time thyristor $TH1$ is turned on and the cycle of operation repeats.

An analysis to determine the values of L and C can be performed. The value of C is found in the same way as in the method of parallel capacitor turn-off. The value of L is determined from resonance analysis. Reversal of the capacitor polarity occurs in a time of $T/2$ seconds, where T is the resonant period, given by $T = 2\pi\sqrt{LC}$. The minimum time that thyristor $TH1$ must be on is $T/2$ to allow C to reverse polarity.

There are two drawbacks to this commutation circuit. The first is that thyristor $TH2$ must always be fired first so that energy can be stored in C. The second is that C charges via the load. There is a way to avoid these. Refer to Fig. 5.29. In this circuit thyristor $TH1$ is turned on initially to allow the conduction of load

[3] A reasonable time is judged by how long the capacitor can hold its charge and is of the order of a few seconds.

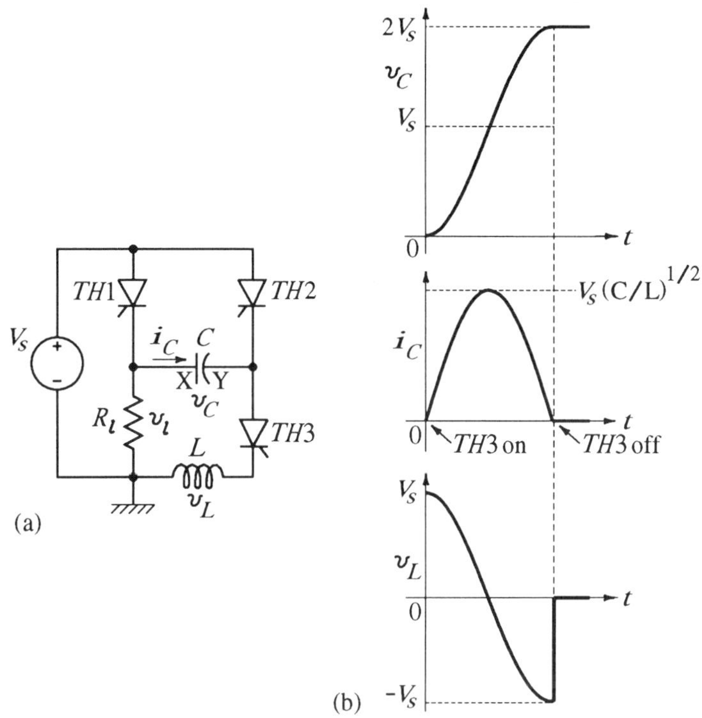

Fig. 5.29 An alternative resonant turn-off circuit.
(a) Circuit diagram, (b) waveforms.

current. Next, thyristor $TH3$ is turned on. The resonant circuit LC, that is directly across the source, has C charged by a current $i = V_s\sqrt{C/L}\,\sin(t/\sqrt{LC})$. So the capacitor charges to $+2V_s$ at plate X with respect to plate Y, but $+V_s$ at plate X with respect to ground as shown in the figure. Note that $V_s = v_C + v_L$ while $TH3$ is on. When the current i_C becomes zero, $TH3$ turns off. To commutate $TH1$, $TH2$ is turned on. At this point the voltage across C causes $TH1$ to be reverse biased. Thyristor $TH1$ turns off, then $TH2$ turns off naturally. The cycle starts to be repeated by triggering $TH1$ on. The explanation has been simplified, because the source is considered ideal. In practice, while $TH3$ is on, source V_s, and R_l and LC are in parallel. After the first cycle, the voltage across C builds up to a voltage that is greater than $2V_s$. During the commutation interval with thyristor $TH2$ on, the capacitor C charges up to V_s volts at plate Y with respect to plate X. Then, with thyristor $TH2$ off and $TH1$ and $TH3$ on, the series resonance causes $v_L = 2V_s$ initially since $v_l = V_s$ and $v_C = +V_s$ at Y. The current i_C is sinusoidal ideally, so, from Fig. 5.29 and the symmetry of the waveforms, after the half-cycle of resonance $i_C = 0$ and $v_L = -2V_s$. Since $v_l = V_s$, then $v_C = 3V_s$ at X with respect to Y. This aids the rapid commutation of thyristor $TH1$.

The use of a single thyristor in a chopper circuit is possible if commutation is accomplished by resonance using *LC* elements in series with the thyristor and the supply. A capacitor and an inductor will cause the current to try to alternate so that at current zero the thyristor will commutate. It will remain off if the reverse bias voltage across the thyristor is maintained long enough for junction J2 to recover its blocking state. In this special case the thyristor on-time t_{ON} is fixed by the resonant frequency. However, there is some adjustment of the off-state time t_{OFF} by means of the switching frequency. If the capacitor is connected across the load resistance R_l, the energy stored in the capacitor during the thyristor off-state can be partly or wholly delivered to the load. Problem 5.16 is a good example of series-resonance commutation. The range of control is not great and peak thyristor currents can be greater than ten times the average value. Unfortunately the analysis is not simple. There are simultaneous differential equations, transcendental equations and non-standard integrals to solve. A math software package greatly facilitates the exercise.

5.7. THYRISTOR POWER DISSIPATION

The thyristor dissipates its losses as heat. Losses are incurred in five ways.

- Leakage current while the thyristor blocks with either polarity produces a power loss $V_{TH}I_{A\,leak}$ that is usually ignored. A 1500-V, 750-A thyristor may have a leakage current of 50 mA. The power loss is 75 W.
- The gate drive produces a power loss in the device. For example for a 1500-V, 750-A thyristor, the peak gate power dissipation may be 150 W with a mean power loss of 3 W.
- There are turn-on losses. The worst case was studied in section 5.5.1. For an inductive load, eq. (5.5.11) was derived to express the energy loss as $W_{on} = V_s I_l t_c / 2$. For the 1500-V, 750-A thyristor that has a turn-on time $t_c = 2\,\mu s$, the maximum loss is $W_{on} = 1.125\,J$. At a switching frequency of 2 kHz this amounts to an average power of 2.25 kW. This makes leakage loss and gate power loss look insignificant.
- On-state losses are important when heat sink design must be considered. For a 1500-V, 750-A thyristor the on-state voltage drop may be $V_{TH(ON)} = 1.7\,V$. In the limit, with continuous conduction of 750 A, the power loss is $P = V_{TH(ON)} I_A = 1.7 \times 750 = 1.275\,kW$. In this particular case the conduction losses are less than the switching losses.
- There are no thyristor turn-off losses like those found in the BJT. This is because the device turns off when the current falls to zero. There are commutation losses due to reverse current in the thyristor sweeping out the charge carriers, but it is not easy to characterize. Commutation usually involves the charging of a capacitor. Once each cycle the capacitor is discharged in the process of reverse biasing the thyristor. A 1500-V dc supply and a 1.0-μF capacitor involves energy, $W_C = CV_s^2/2 = 1.0 \times 10^{-6} \times 1500^2/2 = 1.125\,J$. If this energy is switched at a frequency of 2 kHz, the commutation circuit has to handle 2.25 kW of power. Some of this is dissipated in the thyristor, some is

absorbed by the load, and in a good design most of it could be fed back to recharge the capacitor.

The main thyristor losses are associated with switching and conduction losses. For the 1500-V, 750-A thyristor chosen as an example, the total loss can be about 3.5 kW. Consideration has to be given to this by any switch user. However, in terms of the total power controlled, up to 1125 kW, the losses amount only to 0.3%. The thyristor is a very efficient power modulator.

5.8. THYRISTOR RATINGS

Associated with the characteristics of thyristors there are recommended ratings. These ratings are given in the manufacturers' data sheets and cover voltages and currents for both the gate and anode circuits and typical turn-off times. Table 5.1 depicts the ratings of a fast-switching thyristor.

Table 5.1 Thyristor ratings.

$I_{A\,rms}$ A	$I_{A\,av}$ A	V_{DRM} V	V_{RRM} V	$t_{q\,\min}$ µs	I_G mA	V_{GK} V	di/dt A/µs	I_{ASM} A
1400	900	2000	2000	60	200	1.5	1000	13000

Voltage ratings. Two important voltage ratings found in a data sheet are the repetitive peak off-state voltage V_{DRM} and the repetitive peak reverse voltage V_{RRM}. These are the maximum instantaneous voltages that may be applied in the forward or reverse direction respectively between the anode and cathode connections. They indicate the blocking voltages that must not be exceeded. In conventional thyristors these two voltages have the same value usually. Two other voltages that have significance are the dc value of the breakover voltage V_{BO} and the dc value of the reverse breakdown voltage $V_{(BR)R}$.

Other voltage ratings of note are the non-repetitive peak off-state voltage V_{RSM}, the critical rate of rise of anode voltage that may turn the thyristor on, and the reapplied rate of rise of anode voltage following commutation. The latter is used as a test condition to measure turn-off time for commutation. The on-state voltage drop $V_{TH(ON)}$ is important in terms of losses.

High voltage ratings mean thick silicon wafers, slow turn-off and high on-state voltage drop.

Current ratings. The current ratings are determined by the temperature at which the thyristor is allowed to operate. The maximum junction temperature is typically 125°C. We can define the cooling medium temperature and find the value of the current to produce this maximum temperature. For example, from Table 5.1, 1400 A(rms), or 900 A(av) is the repetitive forward current rating ($I_{A\,rms}$ and $I_{A\,av}$). However, a non-repetitive on-state surge current (I_{ASM}) of 13,000 A peak value is allowed.

Also, Table 5.1 gives the maximum rate of rise of anode current at turn-on that will not damage the device, the typical gate voltage and current (V_{GK} and I_G) that will turn the switch on, and the minimum turn-off time ($t_{q\,\min}$).

Power rating. There is no standard associated with power rating of a thyristor. In the state of no conduction there is a rated voltage associated with blocking. In the state of conduction there is a rated current associated with the maximum temperature rise that can be permitted. However, the temperature rise is associated with the power dissipated in the thyristor and there are the five components of power dissipation, that were discussed in the last section.

EXAMPLE 5.8

Consider a thyristor whose ratings are given in Table 5.1. Estimate the components of power loss in this thyristor and the efficiency of operation.

Solution

At its limit the thyristor is capable of controlling power from the source to the load to a value associated with the peak blocking voltage V_{DRM} and the rated current $I_{A\,rms}$. If the source were an ac supply, if the load were resistive and if the thyristor trigger angle α were $0°$ for half wave, maximum load power, the average power to the load would be

$P_{\max} = V_{l\,rms} I_{l\,rms} = (\hat{V}/\sqrt{2})\, I_{l\,rms} \approx (2000/\sqrt{2}) \times 1400 = 1.98 \times 10^6$ W.

where the subscript l denotes load parameters. The actual maximum power would be less than this by at least 30% for the safe operation of the thyristor.

If the source were a dc supply of 2000 V, capable of supplying 900 A (average value), the maximum average load power would be $2000 \times 900 = 1.8 \times 10^6$ W.

While the thyristor is in the on-state, the forward conduction current i_A causes a power loss in the device. The instantaneous power loss p_{ON} is $p_{ON} = V_{TH(ON)} i_A$, where $V_{TH(ON)}$ is the voltage drop between the anode and cathode of the thyristor. The maximum average power loss P_{ON} due to forward conduction is

$$P_{ON} = \frac{1}{2\pi} \int_0^{2\pi} V_{TH(ON)} i_A \, d\omega t .$$

The voltage $V_{TH(ON)}$ can be assumed constant at a value of 1.5 V during conduction (from 0 to π radians), so that

$$P_{ON} = \frac{1.5}{2\pi} \int_0^{\pi} i_A \, d\omega t = 1.5 \times I_{A\,av}$$

where $I_{A\,av}$ is the average value of the current. The maximum average current $I_{A\,av}$ is given in Table 5.1.

Accordingly, $P_{ON} = 1.5 \times I_{A\,av} = 1.5 \times 900 = 1350$ W.

The thyristor has to dissipate this power without its junction temperature rising above 125 °C.

To estimate the maximum power loss P_{on} in the thyristor during the time of turn-on, we must make some assumptions regarding the turn-on time, the rise of current i_A and the fall of thyristor voltage v_{TH}. Let the turn-on time t_{on} be 5 μs, let the current rise linearly from zero to its final value during the turn-on process and let the voltage across the thyristor fall linearly from its blocking value to zero in the same time. That is, the load is considered to be resistive. Also, let the source voltage be a dc one, whose value is the rated blocking value of the thyristor. In this example, the value from the table is 2000 V. Further, let the switching frequency be 1000 pulses per second (a period T of 1 ms) and let the load be such that the current is the thyristor rated value no matter what the duty cycle of switching. That is, the current always reaches a value of 900 A at the end of turn-on. This is quite artificial, but it does give some idea of the turn-on loss in comparison with the conduction loss. The average power loss P_{on} for this case of switching is

$$P_{on} = \frac{1}{T}\int_0^{t_{on}} v_{TH} i_A \, dt = \frac{1}{T}\int_0^{5\times10^{-6}} (-400\times10^6 t + 2000)(180\times10^6 t)\, dt = 1500\,\text{W}.$$

In this worst case situation, the conduction loss and the turn-on loss are comparable. In general the average turn-on loss is much less. For example, if the switching frequency were 60 pulses per second, the loss would be reduced to $1500\times(60/1000) = 90$ W.

The forward leakage loss of a thyristor is small and comparable to the reverse leakage loss. Consider the thyristor to be capable of blocking a 2000-V dc source voltage and capable of handling a direct current of 900 A. A general rule of thumb is that the thyristor latching current is about one thousandth of the full-load current and the leakage current is about one thousandth of the latching current. This would indicate that the leakage current of this thyristor would be of the order of 1 mA. Therefore, the leakage loss is about 2 W. This is insignificant compared with the conduction loss.

Table 5.1 shows that a gate voltage and current of 1.5 V and 200 mA would be adequate to turn on the thyristor. Thus, the gate power loss is 0.3 W. This is small compared with the forward conduction loss.

To limit the temperature rise of a thyristor it is necessary to minimize all losses, but it can be seen from this example that conduction losses play a significant role. If the switching frequency is high, turn-on losses must also be taken into account.

The efficiency η of operation of the thyristor can be defined as

$$\eta = \frac{\text{output}}{\text{input}}\times 100\% = \left(1 - \frac{\text{losses}}{\text{input}}\right)\times 100\%$$

where the losses would be the thyristor conduction and turn-on losses and the input would be the power from the source. For a 2000-V dc source delivering an average current of 900 A to a load by means of thyristor modulation at 1000 pulses per second, the conduction losses P_{ON} would be about $1.5\times900 = 1350$ W, the turn-on losses P_{on} would be about 1500 W, so that the efficiency of thyristor operation is given by

$$\eta = \left(1 - \frac{2850}{1.8\times10^6}\right)\times 100 = 99.84\%.$$

For medium sized thyristors the efficiency of operation is high.

5.9. THYRISTOR PROTECTION

Thyristors and other semiconductor switches do not have the capability of withstanding overloads to any degree. This is because there is very little mass associated with the silicon wafers and hence there is very little thermal capacity to absorb the energy of voltages and currents beyond rated values. In other words the thermal time constant of a thyristor is short.

Any electric energy in the device will be converted to heat and the temperature will rise rapidly. The temperature of the silicon junction has to be limited to about 125°C, beyond which the thyristor is likely to be damaged. In this section we are going to consider that the ratings of the thyristor are fixed by the device and its cooling arrangements, so that we have to ensure that the ratings are not exceeded. Consequently, we will describe the elements that are added to the circuit to protect the thyristor.

Protection can be divided into a number of areas as is shown in Fig. 5.30. There are two main areas. The first concerns the main power circuit that involves the anode and cathode connections, while the second concerns the gate circuit of the thyristor. The delineation is because of power levels. Gate circuits may have ratings of the order of watts, and, in terms of the power gain of the device, the main power circuit may handle up to one million times this value. Each of the areas of concern have to be given consideration with respect to both steady-state overload conditions and transient surges.

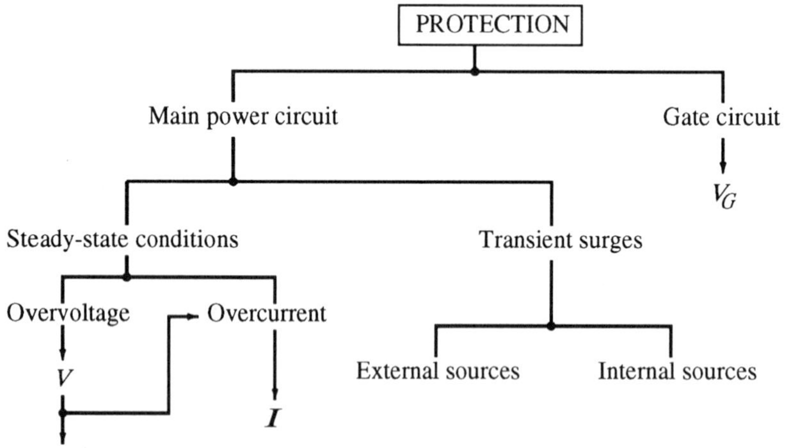

Fig. 5.30 Protection.

5.9.1. Main Power Circuit

Ratings.
Under steady-state conditions the thyristor must withstand both forward and reverse voltages while in the blocking state. Also, the current must not exceed the rated value while the thyristor is in the on-state.

Overvoltage. If the thyristor experiences a reverse bias voltage greater than the device's rated value, the device will break down. A simple and effective way to protect the thyristor against high values of reverse bias voltage is to connect a diode in series with the thyristor. In this way the reverse voltage is shared between the blocking diode and thyristor. The inclusion of a diode in the power circuit decreases the efficiency of the circuit in the conduction state because of the voltage drop across the diode and the power dissipated in it. Some circuit designers will avoid using diodes by choosing thyristors with higher voltage ratings, as much as 30 to 50% higher than that expected for the application.

The same philosophy of over rating may be applied to the forward voltage withstand value. If by some chance there is a forward voltage greater than the withstand value, breakover occurs. The change of state from blocking to conduction changes the problem from one of overvoltage to one of overcurrent and has to be handled this way. In order to limit the likelihood of turn-on by overvoltage a thyristor, whose rating is 30 to 50% higher than that specified, can be chosen.

Overcurrent. The current rating of a device is that which raises the temperature of the junction to its maximum limit. Therefore, an overcurrent will cause overtemperature and a malfunction. There is a number of ways to prevent overcurrent.

One way to protect the thyristor against overcurrent is to use fuses. Special fuses are designed for semiconductor circuits. These fuses must be fast acting, faster than the thyristors which have short thermal time constants, without giving rise to high arc voltages (less than 1.5 times the peak circuit voltage) when they blow. Such fuses are expensive and would not be used in low power circuits, because it would be cheaper to replace a *blown* thyristor than a blown fuse. A term that is used by fuse manufacturers is I^2t, which is proportional to the energy dissipated in a fuse. A fuse will break the circuit at a given I^2t and this value has to be less than the I^2t that would damage the thyristor.

There is a nondestructive means to protect the thyristor against overcurrent, and that way is to monitor the anode current. If the current tends to rise above a specified value, there is action taken by the controller to reduce that current. In the case of a dc supply that is modulated by a chopper (see Fig. 5.31) the thyristor can be switched so that its duty cycle is reduced to bring the value of the current to a safe level. Alternatively the thyristor can be switched off at some maximum value of current and switched on again at some minimum value of current so that the average value is within a safety margin. This is appropriate for a dc motor whose uncontrolled starting current could be as much as 20 times the full-load value, whereas an economic value might be only 2 times the full-load value.

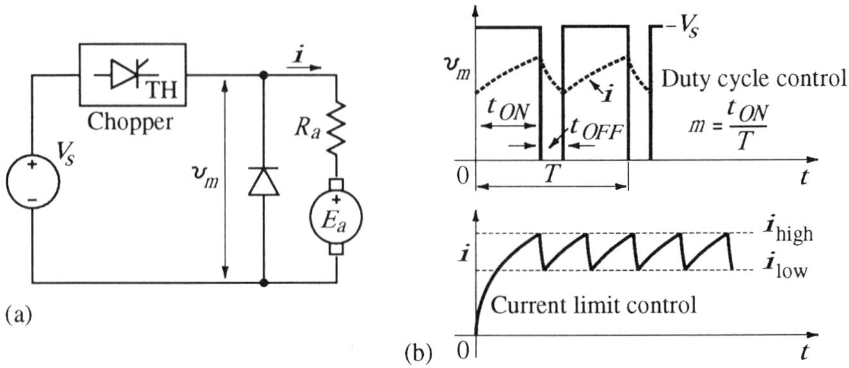

Fig. 5.31 Chopper circuit.
(a) Circuit diagrams, (b) waveforms.

In the case of overcurrent protection in a thyristor circuit involving phase-angle control of an ac supply, a current monitor would allow the controller to increase the delay angle α in order to reduce the current.

Transients.
Power surges appear whenever stored energy is forced to discharge. The occurrences are usually brought about by switching actions either external to the thyristor circuit or internally. The worst case is a lightning discharge, but even EMI (electromagnetic interference) from a thermostat can be damaging indirectly. A general approach to protect thyristors against transient surges is to store the surge energy quickly (usually in a capacitor) and then to dissipate that energy slowly (usually in a resistor). External and internal sources of surges will be treated separately. Fig. 5.32 indicates the possible components that can be added to the main power circuit to protect the thyristor.

External surges. Transient surges of energy occur from lightning strikes on transmission lines, from the action of switching off the main contactor of the distribution system and from changes of load on the system. These surges can contain high values of energy so that the protective components have large values.

If lightning strikes, it is hoped that most of the energy is absorbed externally. However, some of the energy may find its way to the thyristor circuit. Protection in the form of a surge arrester may be used. This is a nonlinear semiconductor resistor, whose normal operating resistance has a high value so that its presence in the circuit is unobtrusive. If the line voltage should rise due to a surge, then the property of the arrester is to have a low value of resistance; the higher is the voltage the lower the value of resistance. In this way, neither the voltage nor the current associated with the surge is experienced by the thyristor. The arrester suppresses the surge by absorbing and dissipating the energy.

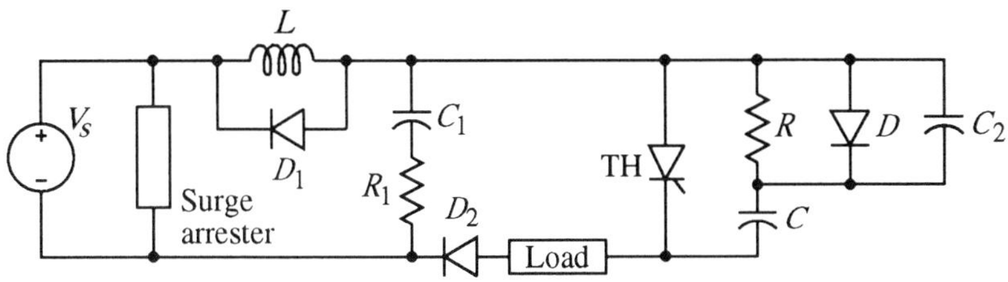

Fig. 5.32 Power circuit components for protection.

In the event that the main contactor is opened, energy that is stored in the transformer finds its way to the thyristor and load unless a capacitor C_1 is connected across the line to absorb the energy quickly.

The same capacitor C_1 can be used to absorb the energy that would be transferred to the thyristor circuit from the leakage inductance of the main transformer and any smoothing inductance L at any time that the load is suddenly reduced. This energy arises from that stored in the inductances, that is, $(1/2)\times(I_{before}^2 - I_{after}^2)L_{total}$. There would be energy oscillating between L_{total} and C_1. But this can be damped by the resistor R_1.

Internal surges. Any switching in the thyristor circuit usually generates small amounts of energy transfer that might manifest themselves in the form of high voltages or high currents. Fig. 5.32 shows a number of components that gives protection against internal surges. The capacitor C provides a path during thyristor turn-off. Nominal values of C and R are $1\,\mu\text{F}$ and $10\,\Omega$, respectively. The resistor R prevents high values of di/dt when the thyristor is turned on and the capacitor C discharges. The diode D and capacitor C form the snubber circuit (see section 5.5) against a high dv/dt being applied to a blocking thyristor. High rates of change of current di/dt in the thyristor at turn-on are limited by the inductor L. The free-wheeling diode D_1 across the inductor provides a path for the current in the inductor when the thyristor is commutated. Thus, the energy in the inductor is dissipated slowly. Without this diode the voltage induced $L\,di/dt$ across the inductor could be damagingly high, because the current would be reduced to zero in a very short time. Although the inductance L has a value of the order of 0.1 mH, if a current was brought to zero from 1000 A in $10\,\mu\text{s}$, the inductor voltage would be about 10 kV and this is to be avoided. The capacitor C_2 has a small value and its purpose is to absorb the energy associated with any high frequency oscillations arising from the *ringing* of *LC* components during switching.

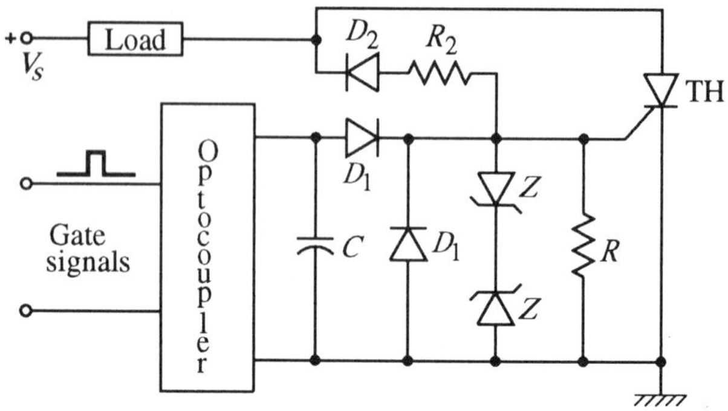

Fig. 5.33 Thyristor gate-circuit protection.

5.9.2. Gate Protection

The gate circuit operates at low power levels (from milliwatts to watts), so that filtering and screening is required to prevent electromagnetic interference (EMI) causing spurious triggering of the thyristor. For those thyristors that have separate gate and cathode leads for trigger pulses, it is a simple safeguard to twist these leads together. Any EMI induces the same voltage in each lead, so the net effect is zero, since the two induced voltages in a twisted pair are equal and cancel each other.

Figure 5.33 shows added components to protect the gate circuit of the thyristor. The components R_2 and D_2 allow current bypass if the anode electrode is negative with respect to the gate. An increased dv / dt limit at the anode is provided by the component R, because it diverts current from the gate region and prevents turn-on. In large thyristors a partial short circuit is created by the manufacturer by gold diffusion at the edge of the silicon wafer between the inner p layer and the outer n layer, and this accomplishes the same purpose as R.

The zener diodes Z clip the voltages to a safe level at the gate in both forward and reverse biasing. Diodes D_1 protect the gate circuit against reverse voltages, and, if they are used, they make one of the zener diodes redundant. Line transients can be diverted from the gate region by the capacitor C, and R would allow the absorbed energy to be dissipated in D_1 and R.

The optocoupler gives the required isolation between the controller, which is the source of the gate signals, and the power circuit, so that the high voltage of the power source does not damage the low voltage components in the controller.

It appears that many components have been used to protect one thyristor and the obvious conclusion is that safety of operation is costly. At low power levels this is so and the use of higher rated thyristors would eliminate a number of added components. However at high power levels the protection circuits become a much smaller proportion the total cost.

5.10. THERMAL CONSIDERATIONS

There is finite resistance between any two of the terminals of the thyristor, and any current will cause electric power to be converted to thermal power within it. The tendency is for the temperature of the thyristor to rise. How much the temperature rises depends on how efficiently the heat is taken away from the junction of the thyristor through the casing, the heatsink and out to the ambient. Since the junction temperature is limited to about 125 ℃ (above which the thyristor does not operate as a switch), there is a link between the size of the heatsink and the maximum average current that can be handled.

One way to determine the thyristor current rating is to choose a heatsink and then to find the value of the current that raises the junction temperature to its limiting value. However, the manufacturers tend to give the current ratings of thyristors without specifying a heatsink. Consequently, from a knowledge of the power that must be dissipated for given device and ambient temperatures, it is usual to use heatsink data sheets to determine the type and size of an appropriate heatsink. Economics comes into play also. A small, inexpensive thyristor would be matched by a small, inexpensive heatsink, and for larger thyristors the heatsink could become more complex, such that forced air cooling and possibly liquid cooling would be worthwhile.

We will not discuss heatsink design. Rather, we will describe simple models that allow an analytical study of the relations among the power, the temperature and the thermal properties of the thyristor unit.

There are two aspects of this study. One is the consideration of steady-state conditions in which the thyristor has been conducting for a long time with the current either as a train of pulses or continuous. The other is the consideration of seldomly recurring pulses or groups of pulses of current for which the heat capacity of the device comes into play. The two considerations are treated differently.

5.10.1. Thermal Resistance

In the steady state a particular current will raise the junction temperature of the thyristor to its maximum permitted value. That particular current is the rated value for the thyristor. It is given as an average value and an rms value. The average value is associated with the power dissipated in the thyristor ($V_{TH(ON)}I_{A\,av}$ watts) and the rms value is pertinent to the load power ($I^2_{A\,rms}R_{load}$ watts). The power dissipated in the thyristor finds its way, thermally, from the junction, through the casing and the heatsink to ambient. A given system has a thermal resistance and it is this that determines the junction temperature for a given power under steady conditions. If we give the definition of thermal resistance, we can quantify the relation between temperature and power.

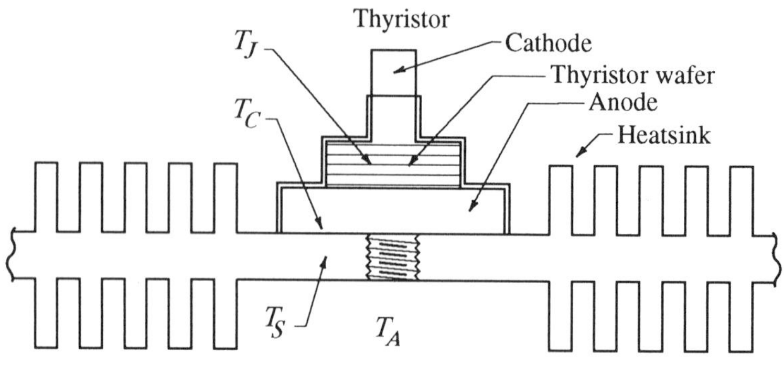

Fig. 5.34 Section of thyristor and heatsink.

- *Thermal resistance* is defined as the temperature difference between two points divided by the power dissipation.

This definition holds under the conditions of thermal equilibrium for which the storage parameter (heat capacity), does not come into play. To derive a simple model of the thermal interactions we can inspect Fig. 5.34 which depicts an impression of a thyristor and heatsink.

The symbols used in this model are

ΔT = temperature difference between two points (°C)
T_A = ambient temperature (°C)
T_C = case temperature (°C)
T_J = junction temperature (°C)
T_S = heatsink temperature (°C)
R_θ = thermal resistance (°C/W)
$R_{\theta JC}$ = thermal resistance, junction to case (°C/W)
$R_{\theta CS}$ = thermal resistance, case to sink (°C/W)
$R_{\theta SA}$ = thermal resistance, sink to ambient (°C/W)
$R_{\theta JA}$ = thermal resistance, junction to ambient (°C/W)
P = power (W)
$V_{TH(ON)}$ = voltage drop across the thyristor, anode to cathode (V)
I_A = constant direct current in the thyristor (A).

The thermal model of the thyristor will be considered to be linear. That is, there is considered to be a point source of power generation at the junction. The power source is from the voltage drop $V_{TH(ON)}$ across the thyristor and the current I_A. This power, in the form of heat, flows from the junction, through the case and heatsink to ambient and creates a temperature gradient from the junction to ambient. If we consider that each of the junction, case, sink and ambient is at a constant temperature, then we can say that the temperature drops are due to the power flow and equivalent thermal resistances. Figure 5.35 shows an equivalent circuit that satisfies this approximate model.

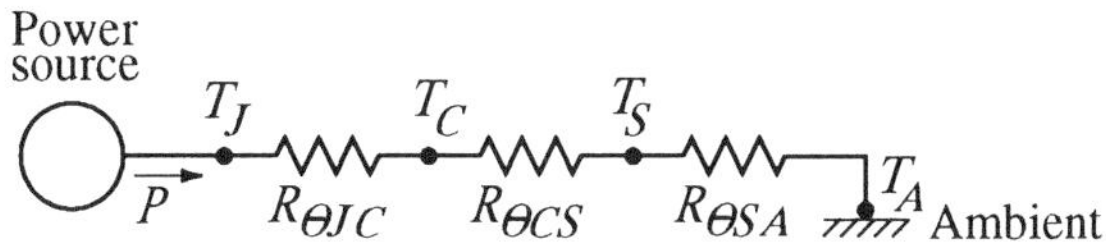

Fig. 5.35 Steady-state thyristor thermal model.

From the definition of thermal resistance

$$R_\theta \triangleq \frac{\Delta T}{P} \tag{5.10.1}$$

and an inspection of Fig. 5.35, we can write

$$P = \frac{T_J - T_C}{R_{\theta JC}} = \frac{T_C - T_S}{R_{\theta CS}} = \frac{T_S - T_A}{R_{\theta SA}} = \frac{T_J - T_A}{R_{\theta JA}} \tag{5.10.2}$$

and
$$P = V_{TH(ON)} I_A \tag{5.10.3}$$

where $R_{\theta JA} = R_{\theta JC} + R_{\theta CS} + R_{\theta SA}$. It is from eq. (5.10.2) and data sheets for the thyristor and heatsinks that the type and size of heatsink is derived.

EXAMPLE 5.9

A thyristor unit has the following data. The maximum forward voltage drop $V_{TH(ON)}$ is 1.5 V while the maximum continuous current I_A is 50 A. The maximum values of thermal resistance, junction to case $R_{\theta JC}$ is 0.55°C/W, and case to sink $R_{\theta CS}$ is 0.11°C/W. If the junction temperature is not to exceed 120°C and if the maximum ambient temperature is 60°C, determine the maximum value of the thermal resistance, sink to ambient $R_{\theta SA}$ that is permitted for long thyristor conduction periods.

Solution

Since the thyristor is subject to long conduction periods, it can be assumed that, for the maximum current of 50 A, the maximum, steady-state temperature of 120°C has been reached at the junction. The power generated at the junction is $P = V_{TH(ON)} I_A = 1.5 \times 50 = 75$ W.

From the definition $R_\theta = \Delta T / P$,

$R_{\theta JA} = (T_J - T_A)/P = (120 - 60)/75 = 0.8$ °C/W.

However, $R_{\theta JA} = R_{\theta JC} + R_{\theta CS} + R_{\theta SA}$.

Therefore, $R_{\theta SA} = R_{\theta JA} - R_{\theta JC} - R_{\theta CS} = 0.8 - 0.55 - 0.11 = 0.14$ °C/W.

For the given power dissipation and thermal resistance, a type and size of heatsink could be obtained from manufacturers' data sheets.

5.10.2. Transient Thermal Impedance

The term transient thermal impedance implies dissipation (thermal resistance) and energy storage (heat capacity) associated with heat flow from the junction of the thyristor to ambient. For transient thyristor currents that comprise short-time overloads, it is the thermal capacity (made up from the mass of the device) that tends to prevent the rapid rise of the junction temperature. Thus, it is possible for current pulse ratings to be greater than continuous current ratings for the same temperature rise. We would like to know how large these pulses may be, so we need a model. If we define transient thermal impedance, we can quantify the relation between temperature and power.

- *Transient thermal impedance* is defined as the change of temperature difference between two points at the end of a time interval divided by the step change in power dissipation occurring at the beginning of the time interval.

This is more complex than the definition of thermal resistance. In this definition the power pulse is specified as being rectangular. This is limiting since pulses have other shapes such as sinusoidal or triangular. Further, it is a temperature difference change over a time interval that must be noted. Thus, if the temperature difference ($T_1 = T_{J1} - T_{S1}$) between the junction and sink at the beginning of the time interval were known, and the temperature difference ($T_2 = T_{J2} - T_{S2}$) between the junction and sink at the end of this time interval were known, then the change of temperature difference is $\Delta T = T_2 - T_1$. This is not a simple matter to determine. The situation is simplified if it is possible to keep the heatsink at a constant temperature during the transient interval of time. In this case ΔT becomes the change of junction temperature $T_{J2} - T_{J1}$, or if the junction and sink temperatures were the same at the beginning of the time interval, then ΔT is also the difference of temperature between the junction and the sink at the end of the time interval. This is what we tend to assume when we calculate the values of power pulse magnitudes over short intervals.

The added symbols used in these transient thermal calculations are

$Z_{\theta(t)}$ = transient thermal impedance (°C/W)

$Z_{\theta JS(t)}$ = transient thermal impedance, junction to sink (°C/W).

From the definition for transient thermal impedance

$$Z_{\theta(t)} \triangleq \frac{\Delta T}{P}. \tag{5.10.4}$$

In reality the equivalent circuit that would represent a linear model of transient thermal impedance between any two points (from among the junction, case, sink and ambient) would have distributed components of thermal resistance and heat capacity to represent the thyristor unit. This is shown in Fig. 5.36. In the steady state the equivalent capacitors would be charged, the temperature gradient would be fixed and the series sum of the distributed thermal resistances would be $R_{\theta JS}$. We want to look for a simpler solution and we have one[4] if we choose an

[4] See the GE SCR Manual, 6th edition, 1979, page 36.

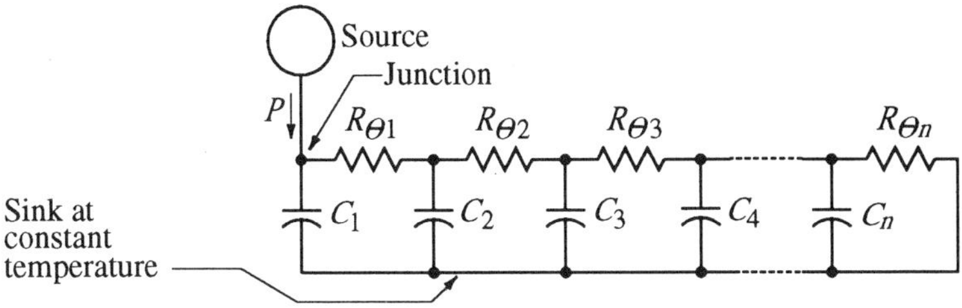

Fig. 5.36 Distributed element model for transient thermal conditions.

empirical solution based on the definition of transient thermal impedance.

The definition of transient thermal impedance calls for a step change of power P over an interval of time Δt. Figure 5.37 depicts this together with the temperatures T_J and T_S at the junction and sink respectively. Data regarding the transient thermal impedance $Z_{\theta JS(t)}$ and the time interval are provided by the manufacturer. It is usual for the time interval to be small such that the heatsink temperature remains constant, although this would require the heatsink to be large. An arbitrary relation between transient thermal impedance and time is given in Table 5.2.

Table 5.2 Transient thermal impedance.

$Z_{\theta JS(t)}$ (°C/W)	0.02	0.06	0.1	0.15	0.2	0.4
Time (s)	0.001	0.005	0.018	0.05	0.12	0.55

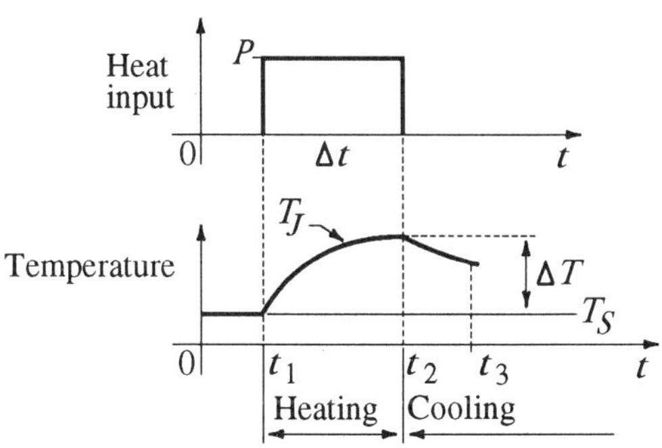

Fig. 5.37 Transient thermal considerations.

Figure 5.37 shows the temperature rise ΔT of the junction for the power pulse P over a time $\Delta t = t_2 - t_1$. At the end of the power pulse the junction of the thyristor will cool. Thus, at the end of the power pulse, that is, at time t_2, we can calculate the temperature rise of the junction from eq. (5.10.4), in which P is known, and $Z_{\theta JS(t)}$ is obtained from the table for the known time $\Delta t = t_2 - t_1$. It is assumed that the temperature of the sink is transiently maintained constant over the time interval Δt.

EXAMPLE 5.10

A thyristor unit has the following data. The maximum forward voltage drop during conduction is 1.5 V and the maximum junction temperature T_J is 120 °C. Table 5.2 holds the data for $Z_{\theta JS(t)}$. The heat sink is transiently maintained at 60 °C. (**a**) Determine the duration of a non-recurrent rectangular pulse of current I_A of magnitude 500 A and (**b**) calculate the value of the maximum continuous current that can be applied to the load.

Solution

(**a**) Since the heat sink is maintained at 60°C, the change in temperature difference ΔT between the junction and the sink is the change in junction temperature ΔT_J. Initially the junction is assumed to be at the same temperature as the sink and finally the junction is at 120°C. Thus,
$\Delta T_J = 120 - 60 = 60 = PZ_{\theta JS(t)}$.
The pulse of power is $P = V_{TH(ON)} I_A = 1.5 \times 500 = 750$ W.
Therefore, $Z_{\theta JS(t)} = 60/750 = 0.08$ °C/W.
From Table 5.2 for $Z_{\theta JS(t)} = 0.08$ °C/W the time interval is 0.011 s.

(**b**) From Table 5.2 we find that at 0.55 s the transient thermal impedance levels off to a constant value 0.4 °C/W. This is the steady-state value for continuous current and is the thermal resistance $R_{\theta JS}$. From the definition for continuous power $P = V_{TH(ON)} I_A = 1.5 \times I_A = \Delta T / R_{\theta JS} = 60/0.4$.
Hence, the continuous current is $I_A = 60/(1.5 \times 0.4) = 100$ A.

We have described how to determine the thermal effects of single pulses of power. This same empirical technique can be used to estimate temperature changes due to multiple pulses of power, or to find the junction temperature some time after a pulse has been extinguished. Consider the temperature profile of the junction as shown in Fig. 5.37. Let us find the temperature at the junction of a thyristor at some time t_3. This time t_3 is within the cooling state since no electric power is being converted to the thermal form in the thyristor. We can still use the pulse technique with the transient thermal impedance data, for, although negative power has no realization with respect to cooling, it can be applied as a useful mathematical facility.

Consider Fig. 5.38. An actual pulse of power P is applied over the time interval from t_1 to t_2. The temperature response at the junction is shown in Fig. 5.38b. The time of interest is t_3, so, to use transient thermal impedance within the

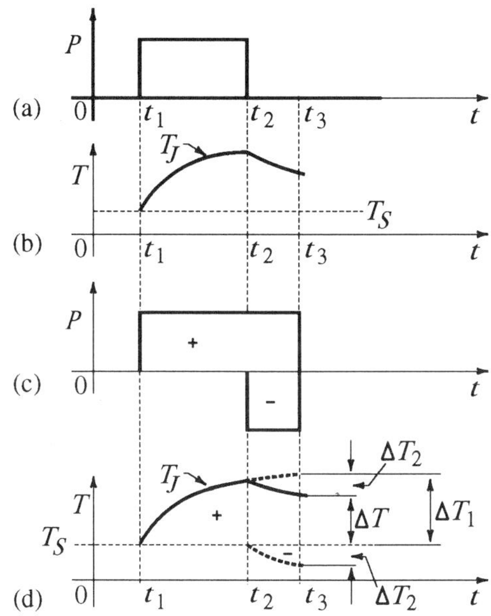

Fig. 5.38 Cooling considerations.

concepts of the definition and the given data (as in Table 5.2), the pulse of power must be extended to time t_3 as shown by the positive pulse drawn in Fig. 5.38c. The junction temperature would rise as shown in Fig. 5.38d. A change of temperature difference ΔT_1 over the time interval $\Delta t_1 = t_3 - t_1$ is given by

$$\Delta T_1 = PZ_{\theta JS(\Delta t_1)}. \tag{5.10.5}$$

Since the heat sink temperature is maintained constant and since the initial value of the junction temperature is the same as the heat sink (the initial temperature difference is zero), ΔT_1 is the same as the final temperature difference between the junction and sink, as shown in Fig. 5.38d.

There is no power applied over the period from t_2 to t_3, so we have to negate the effect of the first pulse over this interval. Let a negative pulse of power *P* be applied over the interval t_2 to t_3, as shown in Fig. 5.38c. We can determine the change of temperature difference ΔT_2 from the equation

$$\Delta T_2 = PZ_{\theta JS(t_3 - t_2)}. \tag{5.10.6}$$

This negative pulse of power *P* is applied as though the temperature of the junction at time t_2 is the same as the heat sink, so, hypothetically, the junction temperature would fall as shown by the decaying curve in Fig. 5.38d and ΔT_2 is the temperature difference between the junction and sink at time t_3.

Both the positive and negative pulses have intervals that end at the same time t_3, so superposition can be used to give the actual pulse of power and give the actual temperature difference ΔT between the junction and heat sink at t_3. That is

$$\Delta T = PZ_{\theta JS(t_3-t_1)} - PZ_{\theta JS(t_3-t_2)} = \Delta T_1 - \Delta T_2 = (T_J - T_S)_{t_3}. \quad (5.10.7)$$

The time intervals are known, so $Z_{\theta JS(t)}$ can be found from the data of $Z_{\theta JS(t)}$ *vs t*. The values of P and T_S are known. Therefore, the value of the junction temperature T_J at time t_3 follows from eq. (5.10.7).

Calculations associated with multiple pulses can be made using this same technique as long as the end of all intervals occurs at the same point in time at which the temperature is required.

EXAMPLE 5.11

A thyristor unit has the following data. The maximum forward voltage drop during conduction is 1.5 V and the maximum junction temperature T_J is 120 ℃. Table 5.2 holds the data for $Z_{\theta JS(t)}$. The heat sink is transiently maintained at 60 ℃. The maximum magnitude of a non-recurrent rectangular pulse of current has a duration of 11 ms. (**a**) Determine the value of the junction temperature 50 ms after the initiation of the pulse. (**b**) Make an approximate estimate of the time that must elapse before another current pulse of the same magnitude and duration can be applied to the load.

Solution

(**a**) From Example 5.10 the non-recurrent pulse of duration 0.011 s has a magnitude of 500 A. To find the temperature at a time of 0.050 s after the initiation of the pulse of current (or power), consider a positive pulse of 500 A over an interval of time 0.050 s and a hypothetical negative pulse of the same magnitude over an interval of time 0.039 s (that is, 0.050 – 0.011 s). The final temperature difference ΔT between the junction and sink at the time 0.050 s is given by
$\Delta T = T_J - T_S = T_J - 60 = P(Z_{\theta JS(0.05)} - Z_{\theta JS(0.039)})$.
That is, $T_J - 60 = 1.5 \times 500(0.15 - 0.138)$.
Thus, $T_J = 69$ ℃.

(**b**) The idea of a non-recurrent pulse is theoretical. The temperature of the junction must drop to the temperature of the sink before another pulse is generated. Theoretically, this takes an infinite time. Practically, it is possible to apply a pulse after a finite time without exceeding the maximum junction temperature. We can assume that the temperature decay can fall to 1°C above the value of the heat-sink temperature before a pulse is initiated a second time.

Using the technique of negative pulses, let a pulse of current of magnitude 500 A be applied from time $t = 0$ to time $t = t'$, where, at time t', the temperature of the junction has fallen to 61°C. The first pulse is on for 0.011 s from time $t = 0$. Therefore, a negative pulse is applied over the interval from $t = 0.011$ s to t'.
From $\Delta T = P(Z_{\theta JS(t')} - Z_{\theta JS(t'-0.011)})$,
$\Delta T = 1$ ℃ and $P = 1.5 \times 500 = 750$ W.

So, $Z_{\theta JS(t')} - Z_{\theta JS(t'-0.011)} = 1/750 = 1.333 \times 10^{-3}$ °C/W.
The difference in time for $Z_{\theta JS(t')}$ and $Z_{\theta JS(t'-0.011)}$ is 0.011 s. Therefore, from a curve generated from Table 5.2, this condition gives $t' \approx 0.55$ s. A practical estimate is that the pulse period must be greater than one second.

EXAMPLE 5.12

A thyristor unit has the following data. The maximum forward voltage drop during conduction is 1.5 V and the maximum junction temperature T_J is 120 ℃. Table 5.2 holds the data for $Z_{\theta JS(t)}$. The heat sink is transiently maintained at 60 ℃. An anode current waveform comprises a rectangular pulse of magnitude 200 A for 18 ms, a dwell of 27 ms and then a pulse of magnitude 500 A for 5 ms. This is repeated seldomly. Determine the maximum junction temperature.

Solution

The maximum junction temperature can occur either at the end of the first current pulse (t = 18 ms), or at the end of the second current pulse (t = 50 ms).
At $t = 18$ ms, $\Delta T_{18} = T_{J18} - T_S = T_{J18} - 60 = P_1 Z_{\theta JS(18)} = 1.5 \times 200 \times 0.1 = 30$ ℃.
Therefore, $T_{J18} = 90$ ℃.
To determine the junction temperature at $t = 50$ ms we must apply a positive 200 A pulse for an interval from 0 to 50 ms, a negative 200 A pulse for an interval from 18 ms to 50 ms, and a positive 500 A pulse for an interval from 45 ms to 50 ms. The net pulses are as specified. Thus,
$\Delta T_{50} = T_{J50} - T_S = T_{J50} - 60 = P_1 Z_{\theta JS(50)} - P_1 Z_{\theta JS(32)} + P_2 Z_{\theta JS(5)}$.
That is, using Table 5.2,
$T_{J50} - 60 = 1.5 \times 200 \times 0.15 - 1.5 \times 200 \times 0.123 + 1.5 \times 500 \times 0.06 = 53.1$.
So, $T_{J50} = 113.1$ ℃.
The maximum temperature occurs at the end of the second pulse.

5.11. THYRISTORS IN SERIES AND PARALLEL.

If the load power rating of a circuit is greater than the handling capability of an individual thyristor, then a number of thyristors in series and parallel can be used. No two thyristors have exactly the same characteristics. So multiply-connected thyristors have problems associated with switching on at the same time, switching off at the same time and sharing voltage or current equally, both in transient and steady-state conditions.

If thyristors are to be connected in series and parallel, it is usual to add components to the circuit to ensure reasonable voltage and current sharing among the thyristors. We will look at typical values of these components for the separate conditions of a series arrangement and a parallel arrangement of thyristors.

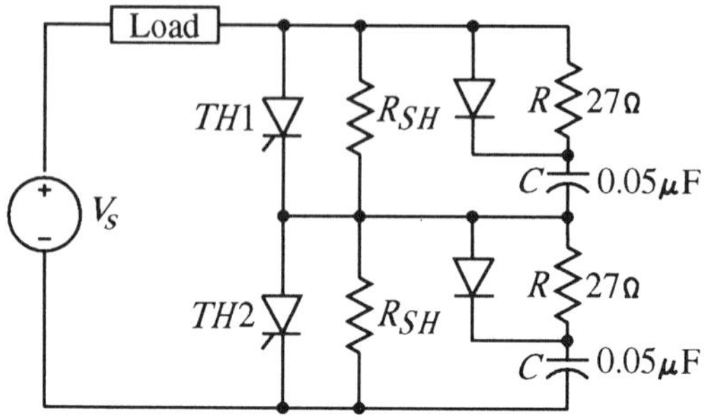

Fig. 5.39 Thyristors in series.

5.11.1. Thyristors in Series

Thyristors are connected in series to share the voltage while they are blocking. The condition for the need to have a series arrangement is if the supply voltage V_s is greater than the breakover voltage V_{BO} of any one individual thyristor.

Protection circuitry is needed for the voltage sharing among the thyristors to be divided reasonably. Figure 5.39 shows the protection circuitry for two thyristors in series. For steady-state voltage sharing, a resistive voltage divider is used. The values of resistance could be matched with the thyristors but economics dictates that the values of the resistors R_{SH} should be the same. This does not give equal voltage division, but if the effects of R_{SH} swamp the effects of the thyristors then reasonable division can be expected. A capacitor voltage divider is used for transient voltage sharing since capacitors tend to oppose a change of voltage. Resistors are added in series with the capacitors to limit the discharge currents when the thyristors are turned on. Typical values of R and C are shown. The values of R_{SH} have to be calculated.

Steady-state voltage sharing.
Consider two thyristors $TH1$ and $TH2$ that are connected in series as illustrated in Fig. 5.40a. Let the breakover voltage of each thyristor be V_{BO}. When the two thyristors are blocking the supply voltage V_s, there will be voltage drops V_1 and V_2 across the thyristors such that

$$V_s = V_1 + V_2 . \tag{5.11.1}$$

However,

$$V_1 + V_2 < 2V_{BO} \tag{5.11.2}$$

since no two thyristors have the same characteristics. This is demonstrated by inspecting the I-V characteristics in Fig. 5.40b. The differences in the characteristics are exaggerated to emphasize the need for a resistive voltage divider for improved voltage sharing. The leakage current $I_{A\,leak}$ through the thyristors must

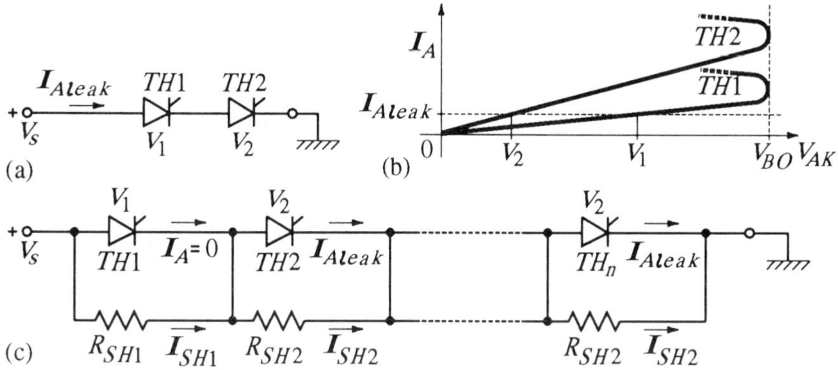

Fig. 5.40 Voltage sharing.

be the same. Where this value intersects the characteristics determines the voltage drops across the thyristors. One voltage drop (V_2 in this case) is low. It is less than half the supply voltage V_s, so eq. (5.11.2) holds.

To improve upon this situation of disproportionate voltage sharing, resistors R_{SH} can be added in parallel with the thyristors, as shown in Fig. 5.40c. The calculation of the value of R_{SH} is as follows.

The situation of taking the worst case makes the values of R_{SH} pessimistic (greater safety), and makes the calculation easier. There are five assumptions.

1. There are n thyristors in series.
2. The value of R_{SH} is a maximum to limit the current.
3. Consider the case that thyristor $TH\,1$ has negligible leakage current and the others have the maximum leakage current $I_{A\,leak}$. This is the worst case. It also indicates that the voltages V_2 across all thyristors except $TH\,1$ are the same.
4. The source voltage V_s is the maximum that can be applied when R_{SH} is used.
5. The value of V_1 across thyristor $TH\,1$ is the peak voltage rating of the thyristors. This is the maximum value that is possible.

We want the value of R_{SH} as a function of the known parameters V_s, V_1, $I_{A\,leak}$ and n. Kirchhoff's voltage law gives us the equation

$$V_s - V_1 = (n-1)V_2 . \tag{5.11.3}$$

Kirchhoff's current law provides

$$I_{SH1} = I_{A\,leak} + I_{SH2} \tag{5.11.4}$$

and Ohm's law gives

$$I_{SH1} = V_1/R_{SH} \text{ and } I_{SH2} = V_2/R_{SH} . \tag{5.11.5}$$

Eliminating V_2, I_{SH1} and I_{SH2} from these equations yields

$$R_{SH} = \frac{nV_1 - V_s}{(n-1)I_{A\ leak}}. \tag{5.11.6}$$

The choice of n is made after the rated voltage V_1 is known.

EXAMPLE 5.13

A 10-kV, dc source supplies a load whose power is modulated by thyristors. The available thyristors are rated at 1 kV, 1000 A. What is the value of parallel-connected resistors R_{SH} that form the voltage divider to force voltage sharing?

Solution

We know that ten 1-kV thyristors in series would not block a voltage of 10 kV because of the slight differences in I-V characteristics, even if a resistive voltage divider were used. We do not want too many thyristors in series because that increases the cost. Let us choose $n = 14$ (it could well have been $n = 13$ or $n = 15$). Information about the leakage current has not been given. However the latching current is about one thousandth of the rated current and the leakage current is about one thousandth of the latching current. Therefore, an estimate of the leakage current $I_{A\ leak}$ is about 1 mA. If we use these values in eq. (5.11.6) we get

$$R_{SH} = \frac{nV_1 - V_s}{(n-1)I_{A\ leak}} = \frac{14 \times 10^3 - 10^4}{13 \times 10^{-3}} = 307\ \Omega.$$

Now, $V_s - V_1 = (n-1)V_2$, so $V_2 = 9 \times 10^3/13 = 692$ V.
Hence, $I_{SH2} = V_2/R_{SH} = 692/(307 \times 10^3) = 2.26$ mA.
It appears that the currents in the resistors must be more than twice the value of the thyristor leakage currents to swamp the leakage effects.

Transient voltage sharing.
Capacitors tend to oppose any change of voltage. Therefore, a capacitive voltage divider will tend to maintain reasonable voltage sharing during a transient voltage surge while the thyristors in a string are blocking. Fig. 5.39 shows this arrangement. The values are nominal. The *RC* circuit is essentially a snubber circuit.

Gate firing series thyristors.
For the sake of speed it is good to turn on all the thyristors in a series string at the same time. Otherwise it is not so critical because, if all thyristors are not on, the remaining switches can turn on either by a delayed gate signal or a breakover voltage V_{BO}, whichever comes first.

The firing circuit must have isolated outputs since the blocking thyristors in series have all gates and all cathodes at different potentials. Optocouplers or transformers accomplish this isolation. Figure 5.41a illustrates simultaneous firing of two thyristors in series using a transformer with two secondary windings. A pulse at the primary winding will enable pulses to appear at the secondary windings. Just as the blocking characteristics of thyristors are slightly different, so are the gate characteristics. The addition of resistance R (50 to 100 Ω) to each gate circuit prevents any thyristor with a low gate impedance from shunting

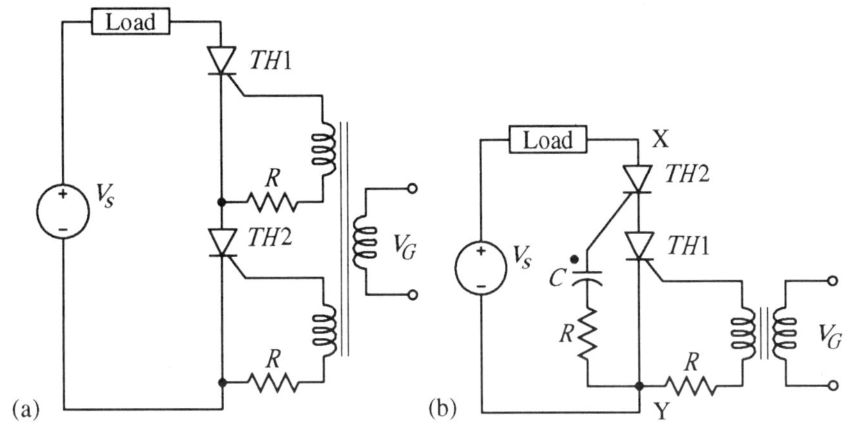

Fig. 5.41 Firing series thyristors.
(a) Isolated gate signals, (b) slave firing.

current from other gates in the string.

Figure 5.41b shows slave firing. In the blocking state of the thyristor there is potential difference between X and Y while the two thyristors block. Therefore, the capacitor C charges up positively at the dot and remains charged while thyristor $TH1$ blocks. A gate signal via the transformer to thyristor $TH1$ turns on this switch. As the voltage across $TH1$ falls, the capacitor voltage causes the gate-to-cathode voltage of thyristor $TH2$ to rise until the current is high enough to turn on thyristor $TH2$. A problem with slave firing is that thyristor $TH2$ has to block a higher voltage while thyristor $TH1$ is turning on.

5.11.2. Thyristors in Parallel

Thyristors are connected in parallel to share the load current while they are conducting. The need for parallel connection is if the load current is greater than the rated current of available single thyristors. How many thyristors are connected in parallel depends on the economics and should be as small a number as possible. The actual number is always greater than the value of the load current divided by the value of the rated current of one thyristor, because the characteristics of the thyristors vary somewhat. This entails the forced sharing of current in both steady and transient states.

Multiply-connected switches in parallel do have their own problems. If the thyristors are not switched on at the same time, those that turn on first must be able to carry the complete load current momentarily. If the thyristors do not switch off at the same time, those that remain on must carry the load current momentarily. The thyristors should share currents as equally as possible in both the steady-state and the transient conditions. Use has to be made of resistors and inductors to this end.

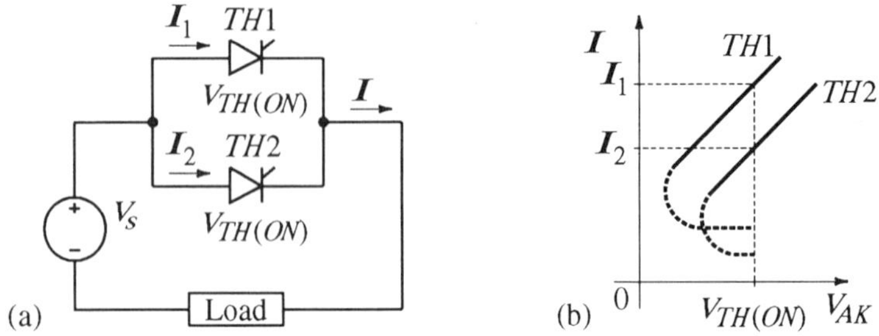

Fig. 5.42 Thyristors in parallel.
(a) Two thyristors sharing the load current, (b) on-state characteristics.

Steady-state current sharing.
Consider Fig. 5.42a in which two thyristors in parallel are illustrated. The ratings of the thyristors are the same. However, the currents in the thyristors will not be the same, because the characteristics of the thyristors in the on-state will differ somewhat. This difference is shown in an exaggerated form in Fig. 5.42b. Because the same voltage is across each thyristor in parallel the current in $TH\,1$ is I_1 and the current in $TH\,2$ is I_2. The latter current could be less than $I/2$. A simple solution to the problem, where one thyristor ($TH\,1$ in this case) could carry a current greater than its rated value because $I = I_1 + I_2$, is the derating of the thyristors. That is, if the load current I were 1000A, then two thyristors each rated at 900 A would ensure, in most cases, that neither would carry a current greater than its rated value. The inequality of the two characteristics are swamped in this way but it is an expensive means to share current.

A better way to share current and to swamp the inequality of the thyristors' characteristics is to use series resistors for steady-state operation and inductors for transient situations.

For steady-state current sharing the most economic way to force current sharing is to use the same values of resistors. Figure 5.43 shows a circuit diagram for current sharing. Consider thyristors $TH\,1$ and $TH\,2$ to be in the on state. The requirement is to have the value of the current I_2 close to the value of current I_1. A generally accepted practice is to aim for the difference in values to be no greater than 20%. If V_1 and V_2 are the minimum and maximum allowable potential differences across $TH\,1$ and $TH\,2$ respectively, we wish to find the value of R as a function of V_2, V_1, I_2 and I_1. Kirchhoff's voltage law provides us with the equation

$$RI_1 + V_1 = RI_2 + V_2 . \tag{5.11.7}$$

Therefore,

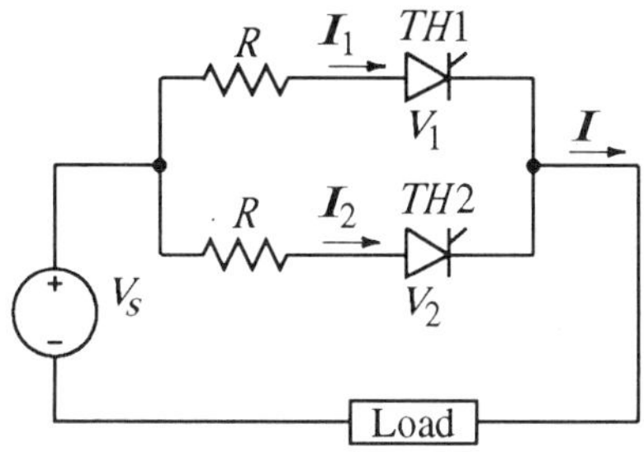

Fig. 5.43 Current sharing with resistors.

$$R = \frac{V_2 - V_1}{I_1 - I_2}. \tag{5.11.8}$$

There is some latitude in arriving at a value of R.

EXAMPLE 5.14

Consider a dc source of 1000 V and a load whose resistance is 1 Ω. The only high current thyristors available to modulate the load power have ratings of 700 A. What is the value of the resistance to be connected in series with each thyristor to obtain reasonable current sharing?

Solution

We can choose arbitrarily the minimum and maximum voltages that we will allow across the two thyristors to share the maximum load current of 1000 A. Let these values be $V_1 = 1.45$ V and $V_2 = 1.55$ V. For a maximum current difference of 20% to be shared by the thyristors, $I_1 = 600$ A and $I_2 = 400$ A.
Hence, $R = (V_2 - V_1)/(I_1 - I_2) = (1.55 - 1.45)/(600 - 400) = 0.1/200 = 0.5\text{m}\Omega$.
The voltage drops across the two resistors are 0.3 V and 0.2 V. There is extra power loss, but it is less than a fifth of the power dissipated in the thyristors.

Transient current sharing.
Resistors force a more equal current sharing through thyristors that are connected in parallel, but they do not compensate for unequal latching currents, unequal holding currents and unequal turn-on times. If one thyristor turns on first, it will be forced to carry the full load current momentarily. This is alright if the di/dt or the surge rating is not exceeded. Otherwise, the current rise in the thyristor must be delayed until other thyristors turn on. This can be accomplished by the use of reactors.

Consider Fig. 5.44. Two thyristors are connected in parallel to share the load current i_l. Ideally the two instantaneous currents i_{A1} and i_{A2} are the same, but the characteristics of the thyristors differ sufficiently to make it impossible for

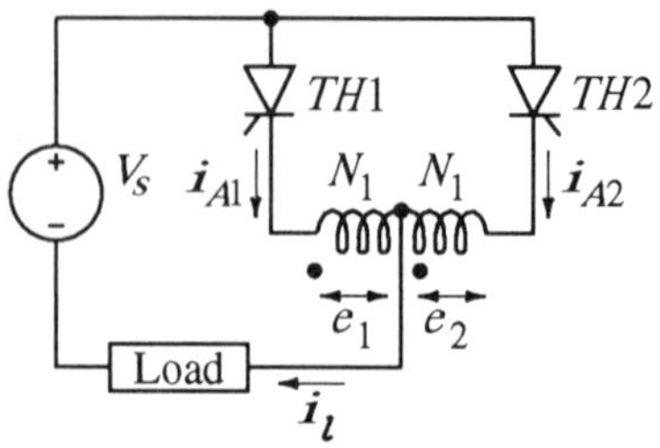

Fig. 5.44 Transient current sharing.

$i_{A1} = i_{A2}$. The reactor, shown in the figure, tends to equalize the currents while there are changes. The reactor is centre-tapped and is bifilar[5] wound, so that any flux linking one half of the winding will link the other half.

Let the current i_{A1} increase faster than current i_{A2} due to one or other of the characteristic differences of the thyristors, so that i_{A1} dominates. The changing current i_{A1} in the left half of the winding produces a changing flux ϕ_{a1} in the core of the reactor. In turn, the changing flux ϕ_{a1} provides an induced emf $(e_1 = N_1\, d\phi_{a1})$ in the winding. At the same time, the flux ϕ_{a1} induces an emf e_2 in the right half of the winding of equal magnitude to e_1, since the turns N_1 of each half of the winding are the same.

Lenz's law states that the polarity of the emf e_1 is such that it opposes that which produced it (current i_{A1}). That is, e_1 is positive at the dot. Accordingly the rise of i_{A1} tends to reduce. The induced emf e_2 is also positive at the dot, and this tends to increase the rate of rise of i_{A2}. So, in general, a reactor of this sort tends to equalize any current changes and a more equal transient current sharing is achieved.

Gate firing of paralleled thyristors.

Simultaneous firing of thyristors in parallel is desirable in order to attempt to prevent unequal current sharing. A common signal for firing can be used in the primary of a pulse transformer as shown in Fig. 5.45a. Here the resistors R_G (50 to $100\,\Omega$) are required to prevent current being shunted by a low gate impedance. Figure 5.45b shows a circuit diagram for slave firing that can be used for two or three thyristors in parallel, where the rated surge current will not be exceeded. Let thyristor $TH1$ be triggered by a gate pulse from the transformer. As $TH1$ turns on its cathode voltage will begin to rise. Since the cathode of $TH1$ is tied to the gate electrode of $TH2$ that gate potential begins to rise also until thyristor $TH2$ turns on. When both thyristors are on there is very little potential difference between the cathode of $TH1$ and the gate of $TH2$, so the power dissipation in the gate circuit of $TH2$ is small.

[5] A bifilar winding comprises two coils wound together on the same core so that any influence (such as a change of flux) affects each coil in the same way.

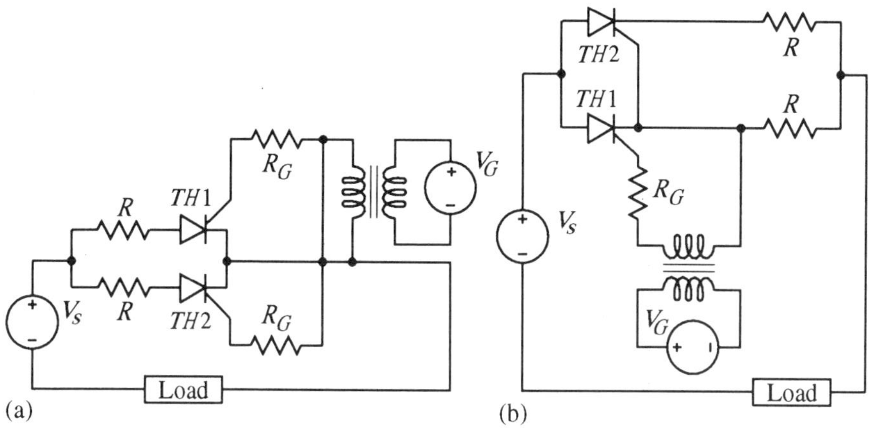

Fig. 5.45 Triggering thyristors in parallel.
(a) Simultaneous firing, (b) slave firing.

5.12. SUMMARY

The thyristor is the most important of all the semiconductor switches for modulating the highest powers at thousands of volts and thousands of amperes. Its structure is relatively robust and simple in its *pnpn* form, so that its use is relatively straightforward.

Like most power-electronics switches the thyristor conducts current in only one direction, but unlike most switches it has the attribute of being able to block voltage of either polarity. This is important if ac power is to be controlled.

The turn-on mechanism is simple. A short current pulse at the gate of the thyristor will allow a change of state from off to on in a few microseconds if the anode is positively biased. The switch will remain in the on-state without a gate signal. The current and power gains associated with turn-on are very high.

The turn-off mechanism can be considered a disadvantage compared with transistors. If the anode current is brought to zero the thyristor will turn off. In ac applications the current falls to zero each cycle naturally. However, a thyristor in a dc circuit has to have its current forced to zero by external circuitry. This is extra complexity. Forced commutation allows turn-off to be faster than natural commutation but, because the thyristor is a minority carrier device, the time of turn-off is relative long. The turn-off process can be over 200 μs for large switches. This means that the thyristor is the slowest semiconductor switch on the market. The slow switching speed limits the frequency of operation to about 1 kHz, so the thyristor is admirably suited to the 50 or 60 Hz power frequencies. At a price, thyristors are manufactured to operate up to 10 kHz, so that the thyristor can be used in high power choppers and inverters. In the quest for better quality power (low power factor and good waveforms) even 10 kHz is a low frequency.

The thyristors have the highest voltage and current ratings of all the semiconductor switches, but there are times when they are not enough. It is fortunate that these devices lend themselves readily to be connected in series and parallel configurations to meet the requirements of today's demands. The high voltage levels put stress on gate circuits and their controllers. This has been overcome by the use of fibre optics and light-activated thyristors.

Protection is an important subject for all semiconductors. Thyristor protection must be given its share of attention. Current is limited by junction temperature, but, as there is no second-breakdown phenomenon, surge currents can be high (as much as ten times rated valued). The rate of change of current di/dt has a limit at turn-on and the rate of change of voltage dv/dt has a limit at turn-off. Snubber circuits are used to help accommodate these limitations. Avalanche breakdown defines the voltage boundary and series connections and voltage arresters are used to expand or protect against this limit respectively.

For the highest power modulation the thyristor is best. Controlled turn-on offers no problem. It is fast controlled turn-off at high power levels that is receiving the attention of the power electronics industry.

5.13. PROBLEMS

Section 5.5

5.1 A thyristor modulates power from a 600-V, dc source to an *RL* load whose values are shown in Fig. P5.1. If the thyristor latching current is $I_{la} = 400$ mA, find the maximum value of the resistance *R* for the thyristor to turn-on with a gate pulse of 100 μs duration.

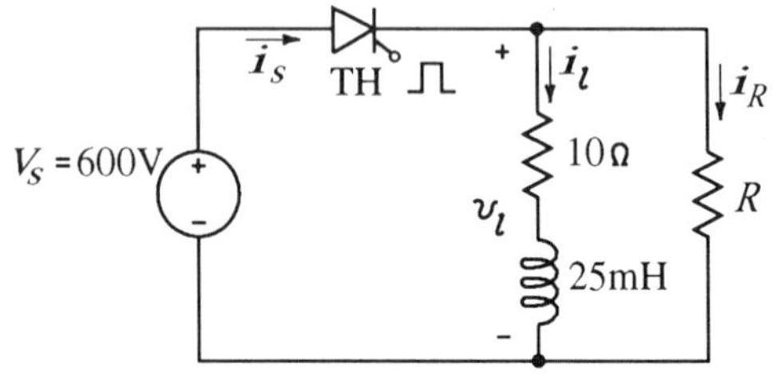

Fig. P5.1

5.2 A thyristor is used to modulate power from a dc source, whose voltage is $V_s = 600$ V, to a resistive load, whose value is $R = 3\,\Omega$. If the thyristor has a latching current that is $I_{la} = 100$ mA and an equivalent junction capacitance of 200 pF, calculate the minimum value of a snubber capacitor across the thyristor to prevent turn-on if the supply circuit breaker is closed. What is the dv/dt withstand of the thyristor?

5.3 A thyristor acts as a chopper to modulate power from a dc source, whose voltage is $V_s = 1000$ V, to a resistive load, whose value is $R = 2\,\Omega$. Refer to Fig. 5.7. From the thyristor data sheet the delay time is $t_d = 0.3\,\mu s$ and the rise-time is $t_r = 2\,\mu s$ for turn-on. The on-state voltage drop is $V_{TH(ON)} = 1.7$ V and the latching current is $I_{la} = 100$ mA. If the chopper operates at a switching frequency of 5 kHz and a duty cycle $m = 0.8$ determine **(a)** the minimum pulse width of the gate signal for successful operation, **(b)** the average power loss in the thyristor due to the turn-on process and **(c)** the average power loss in the thyristor due to the on-state conduction. Assume that the thyristor can be switched off appropriately.

5.4 Consider the circuit diagram in Fig. 5.8a. The dc source voltage is 1000 V, and the *RL* load, whose resistance is $R = 2\,\Omega$, has an inductance high enough for the load current to be considered constant. If the thyristor has a current rise-time $t_{ri} = 2\,\mu s$ and a voltage fall-time of $t_{fv} = 3\,\mu s$, find the average value of the turn-on power loss in the thyristor for the condition of a duty cycle $m = 0.8$ and a switching frequency of 5 kHz.

5.5 A heating element is to be controlled by two thyristors, back-to-back, to provide a smooth adjustment of power from 0 to 1000 W. The voltage source is 115 V, 60 Hz. Design the gate circuits using Schmitt trigger and one-shot ICs like that shown in Fig. EX5.6a.

5.6 Consider the gate circuit illustrated in Fig. 5.9. The source for triggering the thyristor provides a train of rectangular pulses of magnitude $V_G = 20$ V. The specified peak gate power is 150 W and the specified average gate power dissipation is 10 W. Determine **(a)** the minimum value of the added gate-circuit resistance R_G and **(b)** the maximum possible gate pulse frequency if each gate pulse width is 100 μs and the resistance R_G is that value found in part (a).

5.7 The gate characteristic of a 1500-V, 700-A thyristor has the form $V_{GK} \approx 10 I_G$. Calculate **(a)** the value of a gate-driver source resistance R_G if the allowable peak gate-power dissipation is 150 W and the gate-source voltage is $V_G = 40$ V and **(b)** the value of the gate current.

Section 5.6

5.8 Refer to Fig. 5.25. A thyristor $TH1$ modulates power from a dc supply of voltage $V_s = 400$ V to a resistive load, whose value is $R_l = 4\,\Omega$. The commutation circuit comprises the thyristor $TH2$, the capacitor $C = 10\,\mu F$ and the resistor $R = 4\,\Omega$. Determine **(a)** the required turn-off time t_{off} of the thyristor *TH1*, **(b)** the maximum frequency of the chopper switching, **(c)** the power absorbed by the load for the condition of part (b) and **(d)** the voltage and current ratings of the thyristors for maximum frequency switching.

5.9 A thyristor chopper circuit, such as that depicted in Fig. 5.25 has parallel capacitance commutation. The ratings of the inverter type thyristor are 1500 V and 750 A and the specified forced turn-off time of each thyristor is 20 μs. Calculate **(a)** the values of *C* and *R* if the load resistance is $R_l = 2\,\Omega$, the dc source voltage is $V_s = 1000$ V and the maximum switching frequency is 4000 pulses per second and (**b**) the maximum value of the source current.

5.10 Consider the parallel capacitance commutation circuit illustrated in Fig. 5.25a. The dc source voltage is $V_s = 1000$ V, the resistor values are $R_l = R = 2\,\Omega$ and the turn-off time of each thyristor is 20 μs. If each thyristor is switched on at a rate that the capacitor C charges no more than 800 V determine (**a**) the minimum value of the capacitance C for successful commutation, (**b**) the frequency of switching and (**c**) the on-time t_{ON} of each thyristor.

5.11 The switch-mode thyristor circuit depicted in Fig. 5.26 employs parallel resonance turn-off. The dc source voltage is $V_s = 1000$ V and the load resistance is $R_l = 2\,\Omega$. If the values of the commutation-circuit elements are $C = 30\,\mu$F and $L = 2$ mH, determine (**a**) the ideal pulse width of the load current and (**b**) the maximum frequency of the train of pulses.

5.12 A thyristor chopper circuit modulates power from a dc source, whose voltage is 1000 V, to an *RL* load whose resistance is $R = 2\,\Omega$. The inductance is high enough to consider the load current constant if a freewheeling diode is connected across the load. The chopper operates at a switching frequency of 1 kHz with a duty cycle $m = 0.8$. Determine the values of the *LC* elements that are used in a parallel-resonance commutating circuit.

5.13 Consider the auxiliary resonance turn-off circuit depicted in Fig. 5.28. The main thyristor *TH1* acts as a chopper to modulate the power from a dc source of voltage $V_s = 1000$ V to a load, whose resistance is $2\,\Omega$. The chopper frequency is to be 300 Hz and the duty cycle is to be $m = 0.7$. Both thyristors have a forced commutation turn-off time $t_{off} = 150\,\mu$s. Determine (**a**) the values of the commutation elements C and L and (**b**) the average currents in the thyristors and the diode.

5.14 A thyristor chopper modulates power from a dc supply of voltage V_s to an *RL* load with a freewheeling diode connected across it. The inductance L is high enough to consider the load current I_l to be constant. The thyristor utilizes auxiliary resonance turn-off as shown in Fig. 5.28. Determine (**a**) the turn-off time t_{off} available to the main thyristor *TH1* in terms of V_s, I_l and C and (**b**) the maximum frequency of chopping.

5.15 A circuit comprises a 1000-V dc supply, inductance of value $L = 10$ mH, a thyristor and capacitance of value $10\,\mu$F, all in series. If the thyristor is turned on with a short gate pulse, determine (**a**) the current response in this series resonant circuit, (**b**) the on-time t_{ON} of the thyristor before series-resonance commutation occurs and (**c**) the final capacitor voltage. (**d**) Can the thyristor be switched on again? Assume zero initial conditions for the inductor and the capacitor.

5.16 A thyristor chopper modulates power from a 1000-V dc supply, whose source inductance is 100 μH, to a 20-Ω resistive load. Series-resonance commutation is employed by connecting a 10-μF capacitor across the load. If the chopper switching frequency is 800 Hz, (**a**) what turn-off time must be specified for the thyristor, (**b**) what is the average power absorbed by the load and (**c**) what is the average thyristor current? A math software package is useful to solve the transcendental equations and the integrals of this problem.

5.17 In the circuit diagram in Fig. P5.17 the 1000-V dc source, the 10-Ω load resistor and thyristor *TH1* comprise the power circuit and chopper. The other components are used to commutate thyristor *TH1*. (**a**) Explain the action of the commutation circuit. (**b**) What is the time available for thyristor *TH1* to commutate? (**c**) What is the maximum possible frequency of operation of the chopper? For this particular operation of maximum frequency, determine (**d**) the chopper duty cycle m and (**e**) the average power delivered to the load.

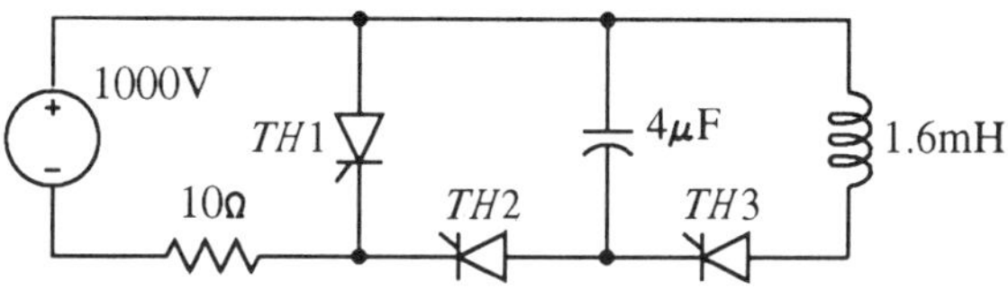

Fig. P5.17

Section 5.7

5.18 A thyristor chopper modulates power from a 1000-V dc supply to an *RL* load that has a freewheeling diode connected across it. The resistance is $R = 2\,\Omega$ and the inductance is high enough to consider the load current to be constant. The thyristor has a delay time $t_d = 0.5\,\mu s$, and a crossover time $t_c = 3.5\,\mu s$ at turn-on. In the on-state the voltage drop is $V_{TH(ON)} = 1.6$ V. If the chopper operates at a duty cycle $m = 0.8$ and a frequency of 400 Hz, determine the thyristor losses as a percentage of the average load power.

5.19 A thyristor provides half-wave controlled rectification from a 60-V, 60-Hz supply to a load whose effective resistance is $R = 1\,\Omega$. The I-V characteristic of the thyristor in the on-state is given by the expression

$$V_{TH(ON)} = 0.001 I_A + 1.6 \text{ volts.}$$

For a delay angle $\alpha = \pi/3$ radians, determine the average power dissipation in the thyristor due to on-state conduction.

Section 5.8

5.20 A single-phase thyristor bridge inverter operates at 400 Hz and converts power from a 1000-V dc supply to a 0.1-mH load. Determine the voltage and current ratings of the thyristors and diodes in the circuit.

Section 5.9

5.21 A 1500-V, 750-A thyristor chopper operates at a switching frequency of 5 kHz to modulate power from a 1000-V dc source to a 2-Ω resistive load. The thyristor has a di/dt limit of 1000A/μs at turn-on and a withstand dv/dt limit of 800V/μs at turn-off. Determine suitable values of the series and parallel snubber elements that will protect the thyristor against current and voltage transients.

5.22 A 1500-V, 750-A thyristor chopper operates at a maximum switching frequency of 400 Hz to modulate power from a 1200-V dc source to a 2-Ω resistive load. The thyristor has a di/dt limit at turn-on of 100A/μs and a withstand dv/dt limit of 300V/μs at turn-off. Determine suitable values of the series and parallel snubber elements that will protect the thyristor against current and voltage transients.

Section 5.10

5.23 A 1500-V thyristor has a maximum on-state voltage drop $V_{TH(ON)} = 1.5$ V. Its thermal resistance, junction to heatsink is $R_{\theta JS} = 0.4$°C/W. The device is connected to a heatsink whose thermal resistance, heatsink to ambient, is $R_{\theta SA} = 0.02$°C/W. If the ambient temperature is 30°C and if the junction temperature must not exceed 120°C, determine **(a)** the maximum average power that can be dissipated in the thyristor and **(b)** the average continuous current rating of the thyristor.

5.24 A thyristor chopper modulates power from a 1200-V dc supply to an *RL* load that has a freewheeling diode connected across it. The load resistance is $R = 4\,\Omega$ and the inductance L is high enough to consider the load current to be constant. The thyristor has a delay time $t_d = 0.6\,\mu$s, a current rise-time $t_{ri} = 1.4\,\mu$s and a voltage fall-time $t_{fv} = 2.1\,\mu$s at turn-on. In the on-state, the voltage drop across the thyristor is $V_{TH(ON)} = 1.7$ V. The chopper operates at a frequency of 4 kHz and has a maximum duty cycle $m = 0.7$. If the thyristor has a thermal resistance, junction to heatsink, $R_{\theta JS} = 0.04$°C/W and if the ambient temperature is 30°C, calculate the value of the heatsink to ambient thermal resistance $R_{\theta SA}$ in order that the junction temperature does not rise above 125°C.

5.25 Consider the circuit diagram in Fig. P5.25. The two thyristors are identical and have a transient thermal impedance characteristic as shown in Table P5.1. Each thyristor has a maximum permissible junction temperature of $T_J = 120$°C and the on-state voltage drop across each thyristor is 1.5 V. **(a)** If the heatsinks can be maintained below a temperature of 65°C, what is the maximum continuous current that each thyristor can conduct? **(b)** For how long can one of the thyristors conduct a short-circuit current of magnitude $I_A = 1500$ A, before the fault is cleared, if the heatsink is transiently maintained at 65°C?

Table P5.1

$Z_{\theta JS(t)}$ (°C/W)	0.002	0.005	0.0095	0.0225	0.035	0.0415	0.045
Time (s)	0.001	0.006	0.02	0.1	0.4	1.0	4.0

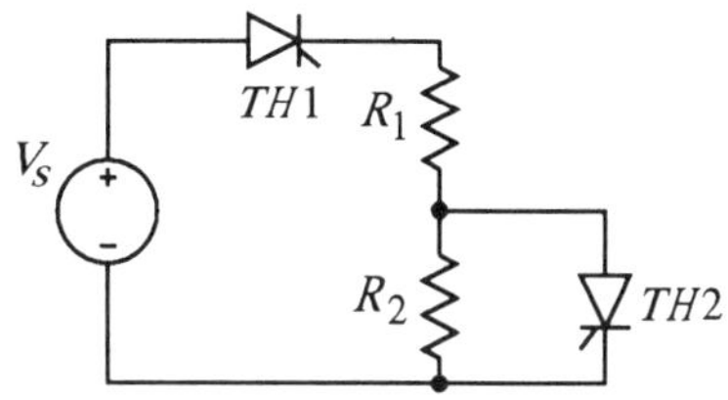

Fig. P5.25

5.26 The transient thermal impedance of a particular thyristor is given in Table P5.1. The on-state voltage drop across the thyristor is $V_{TH(ON)} = 1.7$ V and its maximum permissible junction temperature is 120°C. Consider the case that the thyristor has been conducting a constant current of 400 A for 10s. There is then a step change of load so that the current becomes 1000 A. (For example, the thyristor *TH*2 in Fig. P5.25 is switched on.) How long can this current be maintained before the rated junction temperature is exceeded, if the heatsink is transiently maintained at 60°C?

5.27 A thyristor, whose transient thermal characteristic is shown in Table P5.1, and whose on-state voltage drop is $V_{TH(ON)} = 1.6$ V, has its heatsink temperature maintained below 60°C. The thyristor conducts a current pulse of 800 A for 100ms, followed by a dwell of 100ms and then conducts a current pulse of 100 A for 200ms. If the thyristor's junction temperature is the same as the heatsink initially, calculate the junction temperature at the end of each pulse.

5.28 Consider the circuit diagram in Fig. P5.25. The thyristors, which are identical, have a transient thermal characteristic as shown in Table P5.1, have their heatsink temperatures maintained below 50°C, and have an on-state drop of $V_{TH(ON)} = 1.7$ V. The dc supply voltage is $V_s = 1200$ V and the load resistances are $R_1 = R_2 = 1\,\Omega$. Let thyristor *TH*1 be on for an interval of 6 seconds before thyristor *TH*2 is switched on. Determine the on-state interval of thyristor *TH*2 before it must be switched off, if the maximum permissible junction temperature is 125°C.

5.29 The transient thermal impedance of a particular thyristor is given in Table P5.1. The on-state voltage drop across the thyristor is $V_{TH(ON)} = 1.5$ V. As a chopper, the thyristor switches at a frequency of 20 Hz, modulating power from a 1000-V dc source to a resistive load of $0.8\,\Omega$. If the duty cycle is $m = 0.5$ and if the heatsink is transiently maintained at 50°C, estimate **(a)** the average junction temperature $T_{J\,av}$ and **(b)** the maximum junction temperature $T_{J\max}$.

Section 5.11

5.30 A string of three thyristors in series, each with a peak voltage rating of 1600 V, uses resistors of value 8kΩ to force a more equitable voltage sharing. Consider the worst case that one thyristor has negligible leakage current and the other two thyristors have a maximum leakage current of 50mA. What dc supply voltage can the combination block?

5.31 Consider the circuit diagram in Fig. 5.43. The dc source voltage is $V_s = 1200$ V and the effective resistance of the load is 1 Ω. If the on-state voltage drops across the thyristors are 1.6 V and 1.7 V, determine **(a)** the value of resistance R so that current sharing is within a difference of 15%. **(b)** In such a case as above what is the conduction power loss due to the resistors?

5.32 Two thyristors in parallel share the current from a dc source of 1500 V to a resistive load of 1.0 Ω. The on-state I-V characteristics of the thyristors are

$$V_{TH(ON)}1 = 0.4\times10^{-3} I_{A1} + 1.4 \text{ volts}$$
$$V_{TH(ON)}2 = 0.5\times10^{-3} I_{A2} + 1.5 \text{ volts.}$$

Determine **(a)** the current sharing, **(b)** the value of resistance that must be connected in series with each thyristor so that the current values of the switches do not differ by more than 10% and **(b)** what is the increase in the power losses?

5.14. BIBLIOGRAPHY

Baliga, B.J. *Modern Power Devices.* New York: John Wiley & Sons, Inc., 1987.

Bird, B.M., K.G.King. *An Introduction to Power Electronics.* New York: John Wiley & Sons, Inc., 1983.

Bose, B.K. (ED). *Modern Power Electronics-Evolution, Technology and Applications.* New York: IEEE Press, 1992.

Csaki, F., I.Hermann, I.Ipsits, A.Karpati, P.Magyar. *Power Electonics.* Budapest: Akademiai Kiadl, 1979.

Davis, R.M. *Power Diode and Thyristor Circuits.* Stevenage: IEE, 1979.

Dewan, S.B., A.Straughen. *Power Semiconductor Circuits.* New York: Wiley Interscience, 1975.

Dubey, G.K. *et al. Thyristorised Power Controllers.* Wiley Eastern Ltd., 1986.

Fisher, Marvin J. *Power Electronics.* Boston: PWS-Kent Publishing Co., 1991.

Ghandi, S.K. *Semiconductor Power Devices.* New York: John Wiley & Sons, Inc., 1987.

Gottlieb, Irving M. *Power Control With Solid State Devices.* 1985.

Grafham, D.R., F.B.Golden (ED). *SCR Manual.* 6th ed. New York: General Electric, 1979.

Heumann, K. *Basic Principles of Power Electronics.* Springer Verlag, 1986.

Hoft, R.G. *Semiconductor Power Electronics.* New York: Van Nostrand, 1986.

Kassakian, J.G., M.F. Schlect, and G.C. Verghese. *Principles of Power Electronics*. Reading, Mass.: Addison-Wesley Publishing Co.Inc., 1991.

Kloss, Albert A. *A Basic Guide to Power Electronics*. Chichester: John Wiley & Sons, Inc., 1984.

Lander, Cyril W. *Power Electronics*. 2nd ed. McGraw-Hill Book Co., 1987.

Laster, Clay. *Thyristor Theory and Application*. Tab Books Inc., 1986.

Lee, R.W. *Power Converter Handbook*. Peterborough, Canada: Canadian General Electric, 1979.

Locher, Ralph E., M.W. Smith (ED). *Electronic Data Library, Thyristors-Rectifiers*. General Electric Co.

Mohan, N., T.M. Underland, and P. Robbins, *Power Electronics*. New York: John Wiley & Sons,Inc., 1989.

Ohno, E. *Introduction to Power Electronics*. Oxford: Clarendon Press, 1988.

Ramamoorty, M. *An Introduction to Thyristors and Their Applications*. London: Macmillan, 1978.

Ramshaw, Raymond. *Power Electronics: Thyristor Controlled Power For Electric Motors*. London, Chapman & Hall, 1975.

Rashid, M.H. *Power Electronics*. Englewood Cliffs, N.J.: Prentice Hall,Inc., 1988.

Shepherd, W. *Thyristor Control of AC Circuits*. London: Bradford University Press, 1976.

Shepherd, W., and P. Zand. *Energy Flow and Power Factor in Non-sinusoidal Circuits*. Cambridge: Cambridge University Press, 1979.

Sequier, G. *Power Electronic Converters*. McGraw-Hill, 1986.

Sittig, R., and P. Roggwiller, (ED). *Semiconductor Devices for Power Conditioning*. New York: Plenum, 1982.

Tarter, R.E. *Principles of Solid State Power Conversion*. Howard W. Sams, 1985.

Taylor, Paul D. *Thyristor Design and Realization*. Chichester: John Wiley & Sons, Inc., 1987.

Thorborg, Kjeld. *Power Electronics*. Hemel Hempstead: Prentice-Hall International (UK) Ltd., 1988.

Williams, B.W. *Power Electronics, Devices, Drivers, and Applications*. New York: John Wiley & Sons, Inc., 1987.

Wood, P. *Switching Power Converters*. New York: Van Nostrand, 1982.

CHAPTER 6

THE MOSFET

6.1. INTRODUCTION

The Metal-Oxide-Semiconductor-Field-Effect-Transistor (MOSFET) was originally a signal amplifier, but its characteristics were suitably developed in the early 1980s so that it could be added to the range of power switches. Whereas thyristors are suited to very high-power, low-frequency switching applications, the power MOSFET best fits low-power, high-frequency uses. It is the fastest power switch.

The MOSFET has three terminals, namely, the drain D, the source S and the gate G. A small positive voltage at G together with a positive voltage at D with respect to S turn the switch on. No voltage at the gate means the switch is off.

Examples of the uses of MOSFETs in power electronic circuits are in switched-mode power supplies, in which the high frequency of switching means small components and low size and cost, and in small motor drives, in which pulse width-modulation can be effective for speed control.

The BJT, like the thyristor, is a current controlled switch that requires minority carriers to be established and to be swept out during the respective turn-on and turn-off actions. In contrast, the MOSFET is voltage controlled for switching on and off, so there is not the inertia to state changes that is found in BJTs. This is the reason for the increased speed of switching capability of the MOSFET. The switching speed of the MOSFET can be one order of magnitude greater than the BJT. BJTs are switched at frequencies greater than 10kHz, but power MOSFETS can be switched at frequencies greater than 100kHz.

Ratings of power MOSFET switches are tens of amperes and hundreds of volts, whereas ratings of thyristors can be thousands of amperes and thousands of volts. The power to be controlled by MOSFETs is only a few kilowatts but there is still a large market for applications at this level.

The electric-circuit symbol for an *npn* MOSFET is shown in Fig. 6.1a. The only difference between the symbols for the *npn* and the *pnp* power MOSFETs is the direction of the arrow. We will concern ourselves with the *npn* device, keeping in mind that *pnp* switches have reverse voltage polarities and currents in the opposite direction from the *npn* switch.

As shown in the circuit diagram of Fig. 6.1b, the main terminals D and S connect or disconnect the power supply (voltage V_s) from the load (resistance R). So what is between D and S performs the function of being ideally a short circuit or an open circuit. The switching action is controlled by the gate driver, a supply of low voltage V_G, that can be connected across the MOSFET terminals G and S by means of the generic switch *Sw*.

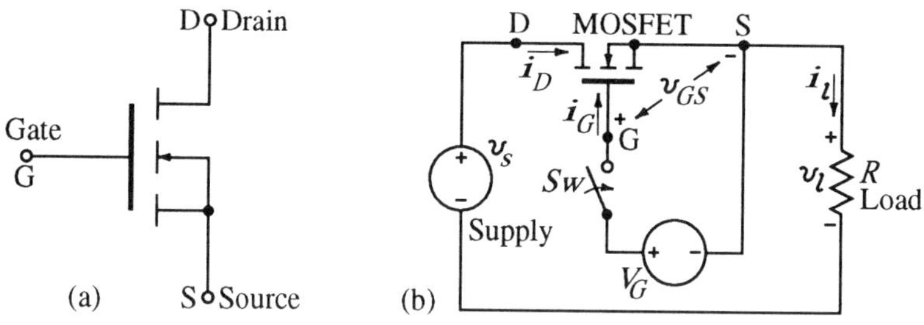

Fig. 6.1 Power MOSFET (*npn*).
(a) Circuit symbol, (b) circuit with MOSFET switch.

Operation of the MOSFET as a switch depends on the polarity of voltage at the terminals. Consider the device to be off initially. In order to turn the device on for the conduction of current in the load, the drain terminal D is made positive with respect to the source terminal S, and the switch *Sw* is closed so that the gate terminal G is made positive with respect to the source terminal S. The MOSFET remains on as long as the drive voltage V_G is applied to the gate terminal. A drain current i_D exists and its value is the same as the supply current and the load current i_l, as shown in Fig. 6.1b, as long as the gate remains positively biased. The gate current i_G has a transient nature. It exists during the dynamic switching action, but in the steady on and off conditions the gate current is virtually zero.

As soon as the gate-drive voltage V_G is disconnected from the gate terminal G the MOSFET turns off. The current i_D drops to zero and, theoretically the device can block a voltage of any bias applied to the drain and source terminals. Because of the design constraints, the MOSFET blocks a forward bias voltage applied to the drain as long as no gate voltage is applied. However, a reverse bias voltage (positive at the sink with respect to the drain) cannot be blocked under any circumstances. This is because there is a parasitic diode within the device structure. This latter characteristic prevents the MOSFET being used alone for controlled rectification applications. However, the main advantage of the device is that both the on and off actions are controlled by the voltage V_G at the gate and the process of switching can be accomplished in nanoseconds.

The name MOSFET is derived directly from the structure of this particular type of *npn* switch. The drain terminal has a metallized contact to one *n*-type region of the transistor and the source terminal has a metallized contact to the other *n*-type region. It is the gate terminal contact that makes the MOSFET different from the other switches that have been described so far. The gate terminal is brought out from a metal contact, but the contact is electrically isolated from the middle *p* region of the transistor by means of a **m**etal-**o**xide-**s**emiconductor (SiO_2). This is the **MOS** part of the name.

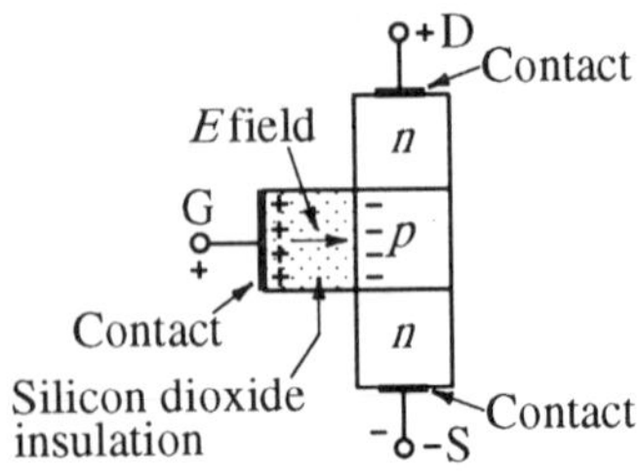

Fig. 6.2 Metal-oxide-semiconductor-field-effect transistor.

Figure 6.2 gives a symbolic representation of the transistor, showing the doped regions of silicon, the contacts and the layer of insulation at the gate. In this elementary form, if there is no gate bias, there is always one *pn* junction to block voltage, no matter what the polarity of the drain-to-source voltage. This is the off-state condition of the transistor switch.

If a gate bias is applied such that the voltage at the gate is positive with respect to the source, there will be an electric field across the metal oxide that acts as the dielectric of an almost perfect capacitor. There will be separation of charge, just as shown. This is the initiation of a **f**ield **e**ffect on the **t**ransistor (**FET** being the rest of the name MOSFET) that allows conduction in the on-state of the switch. The electron charge in the *p* region grows to such an extent, that a channel between the drain and the source appears to be *n* type. With three equivalent *n* regions and no *pn* junctions, the ensuing low resistance between D and S gives a path for current. Removal of the gate bias will be followed by a discharge of the gate capacitor, the removal of the negative charge in the *p* region and the reversion of a blocking state.

It is to be noted that there are MOSFETs that are normally on with no gate signal and that are turned off with the application of a gate signal. This type is called a depletion mode transistor and will not concern us here.

6.2. MOSFET STRUCTURE

There are many types of MOSFET. There are the planar, lateral and vertical types, there are the depletion and enhancement-mode types and there are the *p*-channel and *n*-channel types. A depletion-mode MOSFET is on until a gate voltage is applied. An enhancement-mode MOSFET is off until a gate voltage is applied. We will choose to use only those types of MOSFET that describe the device operation in a simple way or that are types commonly used as power switches.

Figure 6.3a shows the basic structure of a planar lateral MOSFET. Its relative simplicity facilitates a description of the device operation, although this type is used only as a low power signal device. Fabrication of the device is as follows.

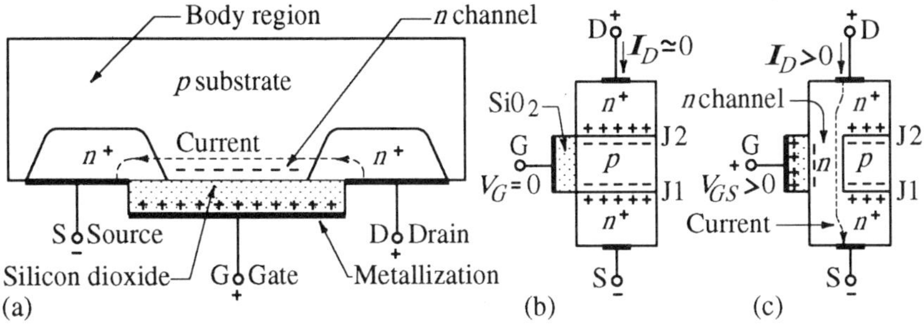

Fig. 6.3 Structure of a planar lateral MOSFET.
(a) Basic structure, (b) device off, (c) device on.

Starting with a *p*-type substrate, that is called the body region, the drain D and source S are created by the diffusion of two highly-doped *n*-type regions. The result is an *npn* transistor. An insulating, metal-oxide layer of silicon dioxide is grown, as shown. Metallization of appropriate surfaces forms the conduction contacts for the drain D, source S and gate G electrodes. Finally the device is packaged.

In this configuration the MOSFET can be connected in a circuit between a supply and a load, as shown in Fig. 6.1b. If the switch *Sw* is open, there is no voltage bias at the gate G. It does not matter what the polarity of the drain D is with respect to the source S, there can be no current between the drain D and source S, other than a small leakage current. Figure 6.3b illustrates this condition. There are two opposing *pn* junctions J1, J2 between D and S so that there is always one of them blocking. This is the switch's off-state.

The MOSFET turns on if the drain D terminal is positive with respect to the source S terminal and if the gate G terminal is made positive with respect to the source S terminal. Refer to Figs. 6.3a and 6.3c. The positive voltage at the gate means that there is an electric field across the gate oxide layer. This insulation at the gate prevents the flow of minority carriers, so there is no possibility of BJT operation in this *npn*-type device. However, the oxide layer, the gate metallization and the *p* substrate form a capacitor, so the electric field forces charge displacement. Positive charge accumulates on the gate metallization surface and negative charge is distributed at the substrate surface. This is shown in the figure. The electric field repels the holes in the *p* region just above the oxide layer and leaves a space charge of negatively charged immobile acceptor ions. As the gate voltage V_G increases, the space charge layer widens, to create what is called an *n-type channel* or an *inversion layer*[1]. This channel links the drain and source and leaves a path between D and S without a *pn* junction. Electrons can flow in this continuous *n*-like layer with very little resistance if V_G is large enough. In this

[1] The channel created between the drain and source is also called an inversion layer.

condition the MOSFET is in the on-state and the drain current I_D is limited virtually by the value of the supply voltage and the load impedance.

If the MOSFET is on, it will remain on as long as the gate voltage is applied. As soon as the gate voltage is removed, there is no electric field to support the n channel. The inversion layer collapses, the pn junctions are re-established and the device reverts to blocking the conduction of current. The MOSFET is off.

The MOSFET just described is a majority-carrier device. There is no delay like that associated with the removal or recombination of minority carriers. This allows switching speeds up to a frequency of 1MHz. However, this planar type has a long, narrow n channel, so its resistance is relatively high and its on-state voltage drop is high. It is not suitable for power switching.

A MOSFET design that is good for power switching is the **v**ertical-**d**iffused MOSFET (VDMOS). Figure 6.4 depicts one full cell of a vertically oriented n-channel MOSFET. There are many cells in parallel both to give a high current capability and to provide short current paths for low resistance and a small voltage drop while the device is switched on.

Fabrication of this device begins with a heavily doped n-type substrate. This is the drain region. It has a large cross-sectional area that allows the conduction of high current. By epitaxial growth the drain region is extended with a lightly doped n layer, that is called the *drift* region[2]. If the MOSFET is in the off-state, the blocking pn junction can have its space-charge layer expand considerably by means of the n^- drift region. This allows a high positive voltage at the drain terminal to be blocked.

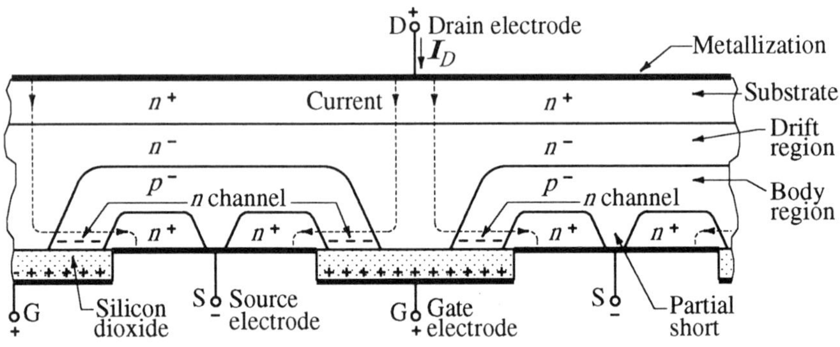

Fig. 6.4 On-state of vertical 4-layer MOSFET.

[2] The drift region is named because charge flow in this region is mainly drift current.

The body regions are formed by diffusing lightly doped p-type material on the surface of the drift region. There can be many thousands of these regions,[3] each one of which is the beginning of the formation of a cell. Following this is the creation of the heavily doped n-type material of the source regions by diffusion. There are as many of these regions as there are body regions, although, in Fig. 6.4, it looks as though there are twice as many. In each cell it appears that there are two n^+ regions. In fact there is one region but an attempt is made to represent in two dimensions that the body region of a cell has a partial resistive contact to the source terminal. This partial short is to enhance the performance of the MOSFET, but it does affect the operation of the device if the bias is such that the source is positive with respect to the drain[4].

An insulating layer of silicon dioxide covers the surfaces of adjacent body regions and a contact is made to form the gate electrode. There is an overlap of the insulation over the n^+ part of the cell for improved performance.

For simplicity the drawing in Fig. 6.4 shows each cell body region (gate) and source region to have contacts brought out individually. In reality the gate contact is covered with insulation to allow all the sources to be put in parallel by metallization over the whole of the surface. This cell structure gives a high degree of interdigitation. In this state the device can be packaged.

Figure 6.4 indicates the conditions for the MOSFET to be in the on-state. The drain D is positively biased with respect to the source S and the positive bias at the gate results in a low resistance n channel at the surface between the p region and the gate oxide layer. This low resistance path is created for the drain current I_D that now has thousands of short parallel paths of large cross-sectional area provided by the cell structure[5].

The creation of a vertically oriented MOSFET in order to have good characteristics of high blocking voltage, low on-state resistance and high current, together with high withstand dv/dt at the drain terminal, gives rise to the structure schematically depicted in Fig. 6.4. Within this structure two other devices can be recognized, namely, an *npn* junction transistor and a *pn* diode. Both of these devices are parasitic and have external terminals D and S. The base of the BJT does not have a connection brought out to the external circuit. Figure 6.5a depicts the two parasitic devices within the structure of the MOSFET.

The *npn*-type structure between the drain D and the source S has the form of a BJT, whose collector is connected to the drain and whose emitter is connected to the source. Because the base B of this parasitic transistor is unconnected in the simple form of the MOSFET, it would appear that there is no supply to drive the base and turn on the BJT. This is not so. During turn-off, the MOSFET can have a high dv/dt applied across its terminals D and S. This dv/dt is shared across DB and BS. If the voltage across BS rises to a sufficiently high value, the resulting minority carriers injected into the base region would turn on the BJT. Once on,

[3] There can be over 1000 cells per square millimetre.

[4] For this condition there is no reverse-biased *pn* junction to block the voltage.

[5] The shape of the cell gives rise to one trade name, the HexFET.

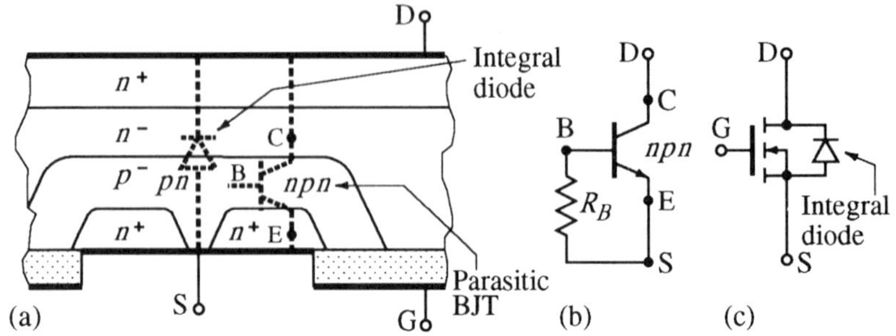

Fig. 6.5 Parasitic devices in the MOSFET structure. (a) Parasitic BJT and diode, (b) turn-on protection, (c) MOSFET with integral diode.

the transistor would stay on because the base terminal is not accessible to initiate turn-off. This uncontrollable action can be and is prevented from interfering with the required operation of the MOSFET. A small region of the *p*-type body region is designed to have a partial short-circuit to the source contact. This path has a relatively low resistance, shown as R_B in Fig. 6.5b. With a low resistance between B and S, the voltage across it cannot rise to a high enough value to turn on the BJT. With this protection, the MOSFET can withstand high dv/dt (in some cases over 5000 V/μs) and the effect of the parasitic BJT can be ignored.

There is an important side effect that is caused by creating a partial short between the *p*-type body region and the source contact. Between the source S and the drain D there is now a parasitic *pin* diode, for which there is no ready means for elimination. Because of this, the MOSFET can never block a voltage that is positive at the source with respect to the drain. Designers have optimized the performance of this *integral diode* so that it has a short recovery time[6] during the transition between on-state and off-state. In this way the disadvantage of no reverse blocking is turned to advantage in those applications such as converters in which the switches must have reverse-connected diodes to conduct reverse current while the switch is off. As shown in Fig. 6.5c, the circuit symbol of the MOSFET is often depicted with this built-in integral diode.

6.3. MOSFET I-V CHARACTERISTICS

A MOSFET power switch has an ideal steady-state characteristic illustrated in Fig. 6.6b. A test circuit diagram is shown in Fig. 6.6a. If the switch *Sw* is open, there is no signal applied to the MOSFET gate G. The MOSFET, as a switch, is off for this condition. As a result, the value of the resistance R_{DS} between the drain D and the source S is infinitely high, so the drain current I_D is zero and the voltage V_{DS} assumes the value of the supply voltage V_s.

[6] The recovery time can be of the order of 100ns.

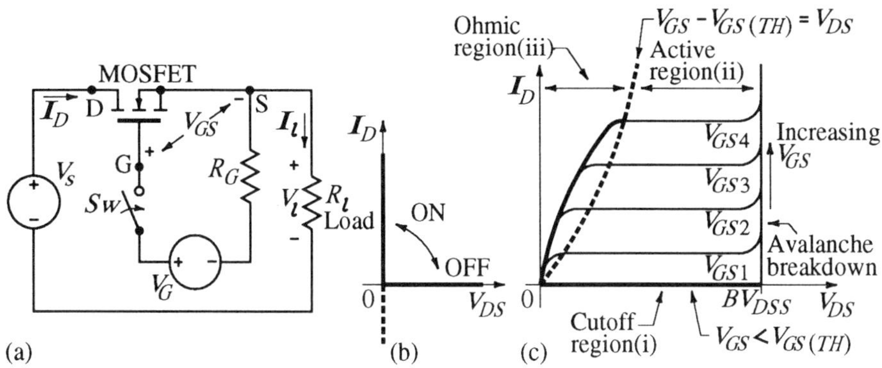

Fig. 6.6 MOSFET I-V characteristics.
(a) Test circuit, (b) ideal switch, (c) real switch.

Once the switch *Sw* of the gate circuit is closed, the voltage V_{GS} at the gate creates an *n* channel in the body region and the MOSFET changes state. That is, the device turns on so that the resistance R_{DS} becomes zero and the drain current I_D assumes a value $I_D = I_l = V_l/R_l = V_s/R_l$, since the voltage V_{DS} across the MOSFET is zero. If the supply-voltage polarity is reversed the MOSFET looks like a forward biased *pin* junction because of the integral diode and it will have a characteristic depicted by the broken line in Fig. 6.6b. The current is in the reverse direction and its value is $I_D = V_s/R_l$.

These ideal characteristics are the ones to use if system calculations are to be carried out. However, if attention is to be directed towards the performance of the MOSFET, reference must be made to the manufacturer's data sheets. The steady-state characteristics from the data sheets are less than ideal and are similar to those illustrated in Fig. 6.6c. There are three separate regions shown, but only two of them apply to the action of the MOSFET being off or on.

Region (i). Referring to Fig. 6.6a, if the gate-circuit switch *Sw* is open, the MOSFET is off and a small leakage current is the response to a forward bias voltage at the drain terminal. The off-state is depicted by the cutoff region in Fig. 6.6c. If the switch *Sw* in the gate circuit is closed there is no significant change from the off-state until the gate voltage V_{GS} is raised to a value that is usually greater than 2 V. The value of V_{GS} at which a change of state occurs in the MOSFET is called the threshold value $V_{GS(TH)}$. Below the threshold value the MOSFET does not conduct. Above the threshold value the MOSFET does conduct. Accordingly, the cutoff region (or off-state) exists as long as $V_{GS} < V_{GS(TH)}$. This condition holds for any value of drain voltage V_{DS} until the avalanche breakover voltage BV_{DSS}[7] is reached. At this value of V_{DS}, current is conducted without a reduction in voltage across the device, so damage could be done. It is to be assumed that the

[7] The symbol could be derived from **B**reakover voltage between **d**rain and source in the steady state.

MOSFET will be operated such that the drain-to-source voltage V_{DS} is maintained below the value BV_{DSS}.

Region (ii). The usual operation of the MOSFET as a transistor amplifier is in the active region. For a particular value of gate voltage V_{GS}, that is greater than the threshold value $V_{GS(TH)}$, a channel is created in the body region so that current can be conducted between the drain and source. Initially, as the drain voltage is increased the channel resistance remains roughly constant, so the drain current I_D increases also. As the drain voltage V_{DS} is increased further, a situation is reached where the voltage drop in the channel causes a reduction in the E-field at the gate. This reduction decreases the channel width[8] and the resistance R_{DS} increases. The characteristic under these conditions is that, as the drain voltage is increased, the pinch-off effect on the n channel alters the resistance R_{DS} such that the drain current I_D remains almost constant. Figure 6.6c shows the constant current curves for increasing values of V_{GS}. In the active region of the power switch the voltage V_{GS} controls the magnitude of the current I_D, so the voltage V_{DS} and the current I_D can both be high at the same time. The associated power loss $V_{DS}I_D$ can be large, so in steady-state switching operation this region is avoided except as a transient mode of operation.

Operation of the MOSFET in the active region gives rise to a useful characteristic. For a particular value of gate voltage V_{GS}, Fig. 6.6c indicates that there is a particular drain current I_D. A plot of I_D against the independent control variable V_{GS} is shown in Fig. 6.7. It is called the *transfer characteristic.* The actual curve is shown as a full line and a linearized characteristic is shown as a broken line. There is no current until the gate voltage reaches the threshold value $V_{GS(TH)}$, then the current rises almost linearly with voltage. The slope of this section of the curve is defined as the transconductance G (with units siemens) and is described by

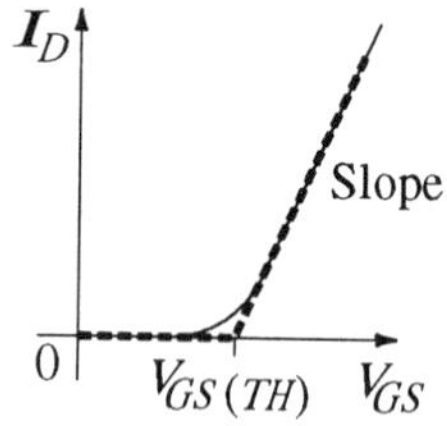

Fig. 6.7. MOSFET transfer characteristic for the active region.

[8] This reduction in the channel width is called the pinch-off effect.

$$G = \frac{I_D}{V_{GS} - V_{GS(TH)}}. \tag{6.3.1}$$

The value of the transconductance G is given in the manufacturers' data sheets and the value generally lies between 1 S and 10 S.

Region (iii). The third region is the region for operation of the MOSFET switch in the on-state. If a plot is made of the drain current I_D for particular values of drain voltage V_{DS}, such that $V_{DS} = V_{GS} - V_{GS(TH)}$, the result is the broken line shown in Fig. 6.6c. This broken line forms a natural boundary between the active region and the on-state region, that is also called the *ohmic region.* A definition of the ohmic region is that part of the characteristic that satisfies the condition that $V_{GS} - V_{GS(TH)} \geq V_{DS}$. The broken line encompasses that part of the characteristic that exhibits almost constant, low drain-to-source resistance that is termed $R_{DS(ON)}$. It represents the transistor resistance in the on-state. A typical value of $R_{DS(ON)}$ for a 500-V, 10-A MOSFET might be 0.5 ohm. There is no pinch-off effect for the particular condition $V_{DS} = V_{GS} - V_{GS(TH)}$. In order to ensure the MOSFET is in the on-state (ohmic region) for all required current values I_D, it is usual to drive the transistor at a higher value of V_{GS} than that required in the active region. That is, at some value between 10V and 15V, V_{GS} puts the transistor into the necessary ohmic region. For protection V_{GS} is never allowed to exceed 20 V.

The two regions of the MOSFET characteristics, cutoff and the ohmic region that define the turn-off and turn-on states of the switch are not perfect, but they are quite close to the ideal requirements. See Appendix 3 for data.

6.4. MOSFET MODELS

There are three separate regions concerning the operation of a MOSFET. These regions have been identified as cutoff, active and ohmic regions. Each region has to be considered separately for model development.

Region (i). The off-state of the MOSFET is defined by $V_{GS} < V_{GS(TH)}$ and $V_{DS} \geq 0$. This is the cutoff region (i) depicted in Fig. 6.6c. There is no drain current I_D other than leakage because, of the two *pn* junctions, one is blocking. This gives rise to a MOSFET equivalent circuit model of two opposing diodes as shown in Fig. 6.8a. The diodes can be treated as being ideal or real.

Region (ii). The conditions for the MOSFET to operate in the active region, as shown in Fig. 6.6c, are $V_{GS} > V_{GS(TH)}$ and $V_{GS} - V_{GS(TH)} < V_{DS}$. A dependent current source can model the MOSFET, as shown in Fig. 6.8b. The value of the current I_D depends entirely upon the value of the gate voltage V_{GS} and the transfer characteristic of the device (see Fig. 6.7). It is given by

$$I_D = G(V_{GS} - V_{GS(TH)}) \tag{6.4.1}$$

where the value of the transconductance constant G is obtained from the data sheets. From this, a system analysis can be carried out to determine the circuit responses to the supply voltage V_s and the control voltage V_{GS}. If the MOSFET

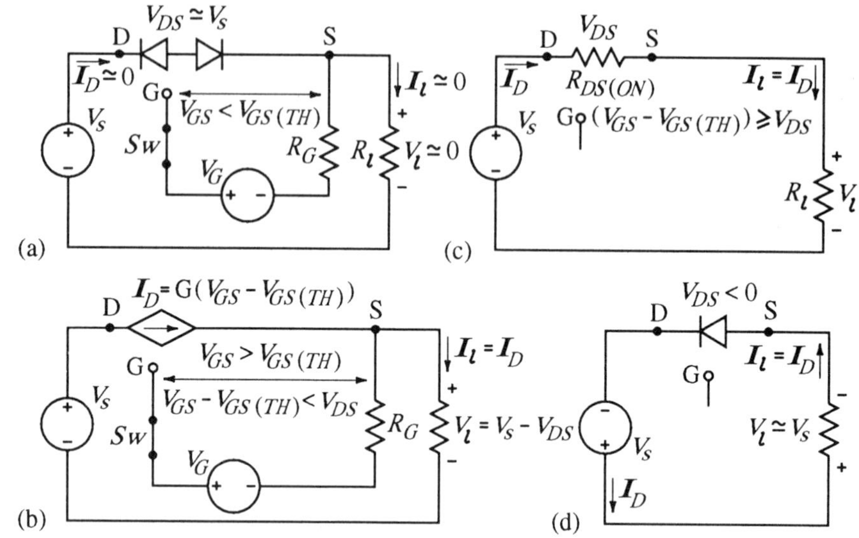

Fig. 6.8 MOSFET models. (a) Cutoff region (i), (b) active region (ii) (c) ohmic region (iii) (d) reverse bias.

is to be employed as a switch this steady-state equivalent circuit is not used.

Region (iii). The MOSFET on-state is defined by the conditions that $V_{DS} > 0$ and $V_{DS} \leq (V_{GS} - V_{GS(TH)})$. This is the ohmic region that is illustrated in Fig. 6.6c. For the best results the value of V_{GS} should be high enough so that the device is operating on the heavy, unbroken curve of the characteristic, or to the left of it. In this way the resistance $R_{DS(ON)}$ of the device between the drain and source is at its lowest value. The model of the MOSFET is an equivalent resistance as shown in Fig. 6.8c. If $R_{DS(ON)} \ll R_l$ then $V_l \approx V_s$ and V_{DS} is of the order of a few volts.

Region of reverse bias. The MOSFET has an integral diode as a result of designing the device for good blocking conditions in the forward direction. However, this integral diode allows conduction in the reverse direction (from source S to drain D). Therefore, the model of the MOSFET is a diode, as shown in Fig. 6.8d, for the condition that $V_{DS} < 0$ (reverse bias). This model holds regardless of the value of the gate voltage V_{GS}.

Models *a, c* and *d* of Fig.6.8 are the ones that are used for the power MOSFET switch. In the ideal case the impedance between the drain and source terminals is zero or infinite for the on-state and off-state respectively.

EXAMPLE 6.1

Figure EX6.1 illustrates the transfer characteristic pertaining to the active region, and the steady-state I-V characteristic pertaining to the ohmic region of a 200-V, 20-A power MOSFET switch. The MOSFET modulates power from a dc supply of 150 V to a resistive load of 7.3 Ω. Determine **(a)** the transistor on-state resistance $R_{DS(ON)}$, **(b)** a suitable value of the gate voltage V_{GS} to maintain the MOSFET in the on state and **(c)** the on-state conduction losses of the MOSFET.

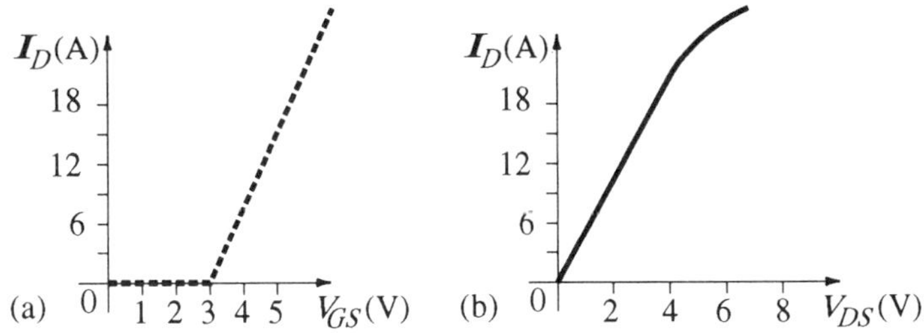

Fig. EX6.1 MOSFET steady-state characteristics. (a) Transfer characteristic, (b) ohmic region.

Solution

A representative circuit diagram for this problem is shown in Fig. 6.8c.

(a) $R_{DS(ON)}$ is the inverse of the slope of the $I_D - V_{DS}$ characteristic in Fig. EX6.1b.
That is, $R_{DS(ON)} \approx 4/20 = 0.2\,\Omega$.

(b) The load current I_l is given by

$$I_l = \frac{V_s}{R_{DS(ON)} + R_l} = \frac{150}{0.2 + 7.3} = 20\text{ A}.$$

The value of power P_D dissipated by the MOSFET while it is on is
$P_D = I_D^2 R_{DS(ON)} = 20^2 \times 0.2 = 80$ W.
Compare this with 3000 W delivered - over 97% efficiency.

(c) The condition for the MOSFET to operate in the ohmic region is $V_{DS} \leq (V_{GS} - V_{GS(TH)})$. From the transfer characteristic in Fig. EX6.1a, the value of the threshold gate voltage is $V_{GS(TH)} = 3$V. From the I-V characteristic at $I_D = 20$A, the voltage drop V_{DS} across the transistor is $V_{DS} = 4$V. Accordingly, the value of the gate voltage is obtained from the condition, $V_{GS} \geq V_{DS} + V_{GS(TH)} \geq 4 + 3 = 7$V.
A suitable value of V_{GS} would be 10 V.

6.5. MOSFET TURN-ON

With the drain terminal D of the MOSFET positively biased, a positive gate signal will turn the device on. The transition between the blocking state and the conducting state will be described in this section. From the user's point of view it is important to determine the energy loss during turn-on. Consequently, from descriptions of the turn-on action and the dynamic model of the MOSFET, turn-on losses will be estimated for particular cases.

6.5.1. Turn-on action.

Figure 6.9 depicts a partial cell of a power MOSFET. It shows the device in the on-state with a low resistance path between the drain and source terminals. The low resistance is formed by a continuous path of effective n -type material, if the drain and gate terminals are both positively biased with respect to the source terminal. The turn-on action is dynamic.

Initially, there is no voltage applied to the gate ($v_{GS}=0$). This is the off-state. A positive voltage applied to the drain D is blocked by a reverse biased *pn* junction between the drift and body regions. Consequently, a very high resistance is offered to the conduction of current.

If a low value of voltage is applied to the gate ($v_{GS} < 2$ V), there are two results. The electric field of the gate voltage across the oxide insulation attracts free electrons to the interface between the drift region and the oxide layer. This free electron layer is called the *accumulation layer.* Further, the same electric field repels holes from the body-region interface to leave a net negative charge of immobile acceptors (encircled in the figure). The *pn* junction remains reverse biased and blocks current.

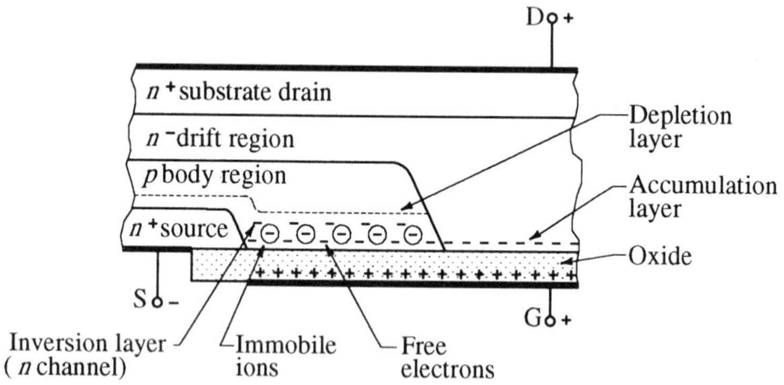

Fig. 6.9 MOSFET at turn-on.

If the value of the voltage at the gate is increased further ($v_{GS} > 2\,\text{V}$), the depletion layer widens. The electron-hole pairs, that are generated by thermal ionization, come under the influence of the increased electric field. The holes are repelled and combine with electrons in the body of the p region. The electrons are attracted to the body-oxide interface. There, the electrons form what is aptly called the inversion layer. The field effect is to create this *inversion* layer (or n channel) that provides a continuous low resistance n region between the drain D and the source S. The value of the gate voltage, at which the n channel comes into being and allows conduction of drain current, is the threshold value $V_{GS(TH)}$.

A further increase of gate voltage ($v_{GS} > V_{GS(TH)}$) causes the inversion layer to widen and this enhances the conductivity of the device. During the time of gate voltage increase the MOSFET is operating in the active region. The characteristic of drain current versus gate voltage is shown in Fig. 6.7. The drain current i_D rises as the gate-voltage increase widens the inversion layer and reduces the resistance R_{DS}. The relationship is almost linear. It is this characteristic that demonstrates that the MOSFET is voltage controlled.

The limit of conductivity enhancement by increasing the gate voltage is reached when there is no further increase in drain current. This is termed operation in the ohmic region. The MOSFET is fully on and its resistance $R_{DS(ON)}$, between drain and source, is at its minimum. The transition from the active region to the ohmic region occurs, by definition, when the drain-to-source voltage v_{DS} falls to a value given by

$$v_{DS} = v_{GS} - V_{GS(TH)}. \tag{6.5.1}$$

Operation in the ohmic region means that turn-on is complete and steady on-state conditions prevail. Any further increase of gate voltage v_{GS} does not alter any of the MOSFET or the circuit conditions significantly. The voltage v_{DS} and resistance $R_{DS(ON)}$ are very close to their minimum values and the drain current i_D is essentially determined by the supply voltage and the load impedance.

It is usual to overdrive the gate voltage v_{GS} in order to ensure operation well into the ohmic region. This minimizes on-state conduction losses. A value of gate voltage of between 10 V and 15 V is common to sustain the on-state. Removal of the gate voltage turns off the MOSFET. The time to turn on the MOSFET is of the order of tens of nanoseconds, and this is why the device can be switched on and off at a very high frequency.

Dynamic circuit model.

Oxide and depletion layers of the MOSFET create capacitors, that, together with resistance in the circuit, give rise to the transient nature of the voltages and currents during the turn-on process. This process is not the same as that found in thyristors and BJTs in which minority carriers must be established. It is a matter of charging effective capacitors that gives rise to the low resistance channel called the inversion layer. In constant-capacitor circuits, charging-time calculations offer no difficulty. However, in a MOSFET the depletion layer capacitances can vary with voltage more than tenfold during turn-on. This makes analysis of charging time difficult unless approximations are made.

The oxide dielectric insulating the gate terminal from the transistor, and the depletion layers at the junctions give rise to effective capacitors C_{GS} and C_{GD} between the gate and the two main terminals, S and D respectively. The reverse-biased junction between the drain and the source creates a capacitance C_{DS}.

The change of state from off to on requires the MOSFET to traverse the active region of operation. Therefore, the dynamic model for turn-on would incorporate the transfer characteristic that involves a current source $i_D = G\,(v_{GS} - V_{GS(TH)})$ and a resistance R_{DS}. These can be incorporated into an equivalent circuit as shown in Fig. 6.10. If the values of the components can be deduced from data sheets, the time for the drain current to rise to its steady value can be estimated. This time is not necessarily the full turn-on time. The crossover time can be associated with the turn-on time, and the crossover time involves not only the rise-time of the drain current but also the fall-time of the transistor voltage v_{DS}. Calculation of the fall-time requires a knowledge of the system load characteristics.

Capacitors tend to oppose changes of voltage. Initially the MOSFET is off and there is no voltage across the gate-to-source terminals so that there is no voltage across the gate-to-source equivalent capacitors C_{GS} and C_{GD}. Closing the switch *Sw* of the gate-drive circuit, as shown in Fig. 6.10, initiates capacitor charging and the establishment of a rising voltage v_{GS} across the gate-to-source terminals. In order to turn on the MOSFET quickly and so minimize the switching energy losses, the capacitors must charge quickly. This is accomplished by using a low value of the drive-circuit resistance R_G. In this way a peak transient current of the order of one ampere may result. Although current and charge are important for speed of switching, the device remains voltage controlled. While the gate voltage v_{GS} is rising and the device is operating in the active region the drain current i_D rises according to the transfer characteristic (see Fig. 6.7) with a virtually linear relationship once the gate threshold voltage has been exceeded. Therefore, the time it takes for the drain current to rise to the steady load value is the time it takes for the gate voltage to rise to its appropriate value on the transfer-characteristic curve. The values of the equivalent capacitors change with voltage, so that a calculation of the rise-time of the voltage v_{GS} gives a very approximate estimate. Typical values of turn-on time are best taken from data sheets, although we will make an attempt at calculations. Once the MOSFET turns on and the effective capacitors are fully charged, the equivalent circuit reverts in form to the steady-state model, shown in Fig. 6.8c. With a constant gate voltage $V_{GS} = V_G$ the capacitors are no longer relevant, the dependent current source can be eliminated because operation is in the ohmic region, and what is left of the equivalent circuit is the on-state resistance $R_{DS(ON)}$, that has a low value. In the on-state, the MOSFET appears to be a simple resistance.

Switching times are directly associated with the time it takes to change the gate voltage v_{GS}. The value of the gate voltage v_{GS} is a time function of the gate resistance and MOSFET capacitance. Since the capacitance of the device is not a function of temperature, then switching times are independent of the temperature.

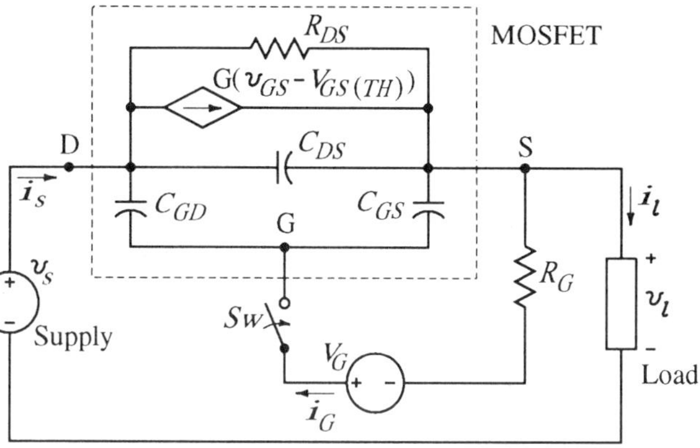

Fig. 6.10 MOSFET dynamic circuit model.

Data sheets for MOSFETs provide typical values of input, output and transfer capacitances for zero gate voltage and moderate drain-to-source voltage. From this information it is possible to calculate, very approximately, the values of the capacitors shown in Fig. 6.10. The input capacitance C_{iss} is seen to be two series capacitors, C_{GD} and C_{DS}, in parallel with gate-to-source capacitor C_{GS}. The capacitance C_{DS} can be ignored since its value is generally significantly larger than the value of the capacitance C_{GD}. This leads to the input capacitance being approximately equal to the two capacitors, C_{GD} and C_{GS}, in parallel. That is

$$C_{iss} \approx C_{GD} + C_{GS} . \tag{6.5.2}$$

The output capacitance C_{oss} can be seen to be the capacitance C_{DS} in parallel with the two capacitors, C_{GS} and C_{GD}, in series. Since capacitance C_{GD} is significantly less than C_{GS}, the output capacitance can be expressed in the form

$$C_{oss} \approx C_{DS} + C_{GD} . \tag{6.5.3}$$

The transfer capacitance C_{rss} is defined as

$$C_{rss} = C_{GD} , \tag{6.5.4}$$

the capacitance between the gate and the drain. From a knowledge of C_{iss}, C_{oss} and C_{rss}, values of the equivalent circuit capacitors can be calculated. As the drain-to-source voltage is reduced, the input capacitance may change from 2500 pF to 3500 pF, the output capacitance may change from 1000 pF to 2500 pF and the transfer capacitance may change from 200 pF to 2500 pF. For turn-on the input capacitance influences the speed of change of states, so C_{GD} and C_{GS} are important. The gate-to-drain capacitance ($C_{GD} = C_{rss}$) can change more than ten-fold; the greatest change occurs near the ohmic region of operation. The gate-to-source capacitance varies little and can be considered constant. These changing values indicate the problems associated with an analysis of turn-on time.

Sometimes, in analysis, a *Miller capacitance* is used. This fictitious capacitance accounts for the charge required during the changing drain voltage v_{DS} during turn-on. It is placed in parallel with C_{GS} and models the effect of C_{GD}. There are some who use this model for analysis. We wont.

6.5.2. Turn-on Losses

The MOSFET goes through a transition, from the off-state to the on-state, during which the voltage and current of the device can be high simultaneously. The associated losses in the switch must be examined.

MOSFET circuit with a resistive load. If the load shown in the circuit diagram of Fig. 6.10 is purely resistive, the turn-on time of switching is shortest insofar as the main power circuit has influence. The simple gate circuit, shown in the figure, influences turn-on time by the magnitude of the resistance R_G.

The initial conditions are that the MOSFET is in the off-state and the dc supply applies a positive voltage to the drain terminal. At time $t=0$, the gate-drive switch *Sw* is closed and the effective capacitors of the device begin to charge because of the gate current response i_G. A low value of resistance R_G means a high initial current i_G and a rapid charge. The time taken for the gate voltage v_{GS} to reach the threshold value is called the delay time $t_{d(on)}$. There is no drain current during this interval. This is illustrated in Fig. 6.11. As the MOSFET charges further, the gate voltage v_{GS} continues to increase above the threshold value and the drain current will rise according to the relation i_D versus v_{GS} of the transfer characteristic (eq. (6.4.1)). The drain current is the same as the load current, so the load voltage rises in proportion to the current. Thus, the drain voltage v_{DS} falls by the same amount, since the supply voltage V_s is constant. The time taken for the drain current to reach its final steady value I_D from the time the gate voltage was the threshold value $V_{GS(TH)}$ is defined as the rise-time t_{ri}. Consequently, the fall-time t_{fv} of the voltage v_{DS} is also time t_{ri}. The total turn-on time t_{on} of the MOSFET is

$$t_{on} = t_{d(on)} + t_{ri} = t_{d(on)} + t_c \tag{6.5.5}$$

where t_c is the crossover time from current rise to voltage fall. If we linearize the rise of gate voltage v_{GS}[9], then the changes of drain current and drain voltage are linear. This simplifies the calculation of turn-on loss in the MOSFET. Figure 6.11 shows a response that is linear. Once the drain current has risen to its steady value, the operation of the device is in the ohmic region ($v_{DS} \leq v_{GS} - V_{GS(TH)}$). Any further increase in gate voltage is termed overdrive because there is no significant change of drain current I_D or drain voltage V_{DS}.

If the gate-drive power is neglected, the instantaneous value of the MOSFET power loss p is given by the product $p = v_{DS} i_D$. With the approximation that $v_{DS} \approx 0$ in the on-state, the energy loss W_{on} over the interval t_{ri} is

[9] A constant current in a constant-capacitance circuit gives a voltage response that is linear.

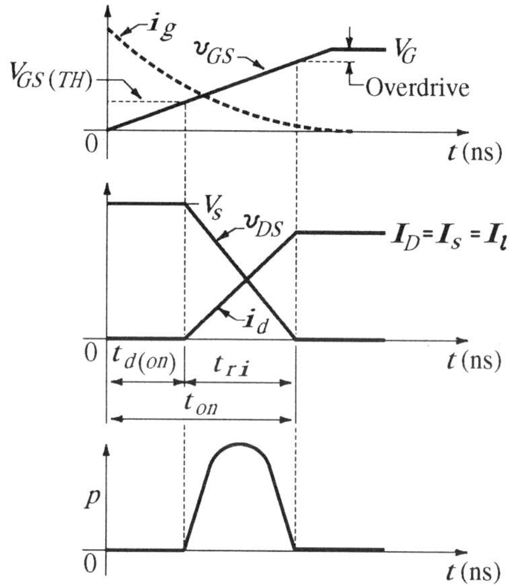

Fig. 6.11 MOSFET turn-on for a resistive load.

$$W_{on} = \int_0^{t_{ri}} p\, dt = \int_0^{t_{ri}} v_{DS} i_D dt = \int_0^{t_{ri}} \left[-\frac{V_s}{t_{ri}} t + V_s \right] \left[\frac{I_l}{t_{ri}} t \right] dt. \tag{6.5.6}$$

This gives the expression for the energy loss to be

$$W_{on} = \frac{V_s I_l}{6} t_{ri}. \tag{6.5.7}$$

As well as being proportional to the system parameters, supply voltage and steady load current, the turn-on loss is proportional to the rise-time t_{ri} of the drain current i_D and in turn this is dependent on how fast the transistor capacitances can be charged by the gate-drive supply V_G. A low value of drive circuit resistance R_G helps to keep the time t_{ri} short and the losses low.

EXAMPLE 6.2

A chopper circuit uses a power MOSFET to modulate power from a dc supply of 150 V to a resistive load of 7 ohms. The data sheet for the 250-V, 30-A MOSFET gives the information that for a gate drive voltage $V_G = 10$ V, with no overdrive, the turn-on time t_{on} comprises a typical delay time $t_{d(on)} = 30$ ns and a current rise-time $t_{ri} = 150$ ns. The total allowable average power dissipation is $P_D = 100$ W, and the on-resistance is $R_{DS(ON)} = 0.12$ ohm. Determine **(a)** the value of the gate threshold voltage $V_{GS(TH)}$, **(b)** the energy loss W_{on} during the turn-on interval, **(c)** the energy loss W_{ON} during the steady on-state interval and **(d)** the maximum power that is possible to be dissipated due to turn-off, if the chopper is operated at a switching frequency of 30 kHz and a duty cycle $m = 0.8$.

Solution

The appropriate chopper circuit for this example is shown in Fig. 6.6a and the dynamic response curves are illustrated in Fig. 6.11. From the data, the steady load current is

$$I_l = I_D = \frac{V_s}{R_{DS(ON)} + R_l} = \frac{150}{0.12+7.0} = 21.07\text{ A}.$$

(a) Since there is no gate overdrive, the gate voltage v_{GS} reaches its final value $v_{GS} = V_G = 10\text{ V}$ at the same instant that the drain current reaches its final steady value $i_D = I_D = I_l = 21.07$ A. In the absence of further information we can assume that changes occur linearly during turn-on. From Fig. 6.11, the threshold value of gate voltage is given by

$V_{GS(TH)}/V_G = t_{d(on)}/t_{on} = t_{d(on)}/(t_{d(on)} + t_{ri})$.

Therefore, $V_{GS(TH)} = V_G t_{d(on)}/(t_{d(on)} + t_{ri}) = 10 \times 30/(30+150) = 1.67$ V.

(b) The energy loss W_{on} during turn-on is given by eq. (6.5.7) to be

$$W_{on} = \frac{V_s I_l}{6} t_{ri} = \frac{150 \times 21.07}{6} \times 150 \times 10^{-9} = 79.01 \times 10^{-6}\text{ J}.$$

(c) At a switching frequency of 30 kHz the period of switching is $T = 1/(30 \times 10^3) \approx 33.3 \times 10^{-6}$ s, which is nearly two thousand times longer than the turn-on time. Accordingly, the steady on-time t_{ON} of the transistor is given by

$t_{ON} \approx mT = 0.8 \times 33.3 \times 10^{-6} = 26.67 \times 10^{-6}$ s.

The energy loss W_{ON} in the MOSFET during the steady on-state is

$W_{ON} = I_D^2 R_{DS(ON)} t_{ON} = 21.07^2 \times 0.12 \times 26.67 \times 10^{-6} = 1421 \times 10^{-6}$ J.

The loss incurred during the on-state is 18 times the loss during turn-on.

(d) The average power P that is dissipated during the turn-on and on-state operation is

$P = (W_{on} + W_{ON})f = (79.01 + 1421) \times 10^{-6} \times 30 \times 10^3 = 45$ W.

The average turn-off power P_{off} that can be dissipated is

$P_{off} = P_D - P = 100 - 45 = 55$ W.

This value appears to be within reasonable bounds.

MOSFET circuit with an inductive load. As far as switching losses are concerned the worst case occurs if the load has a virtually constant current. That is, the inductive-load time constant is long compared with the switching period of the power MOSFET and the load must incorporate a freewheeling diode. The freewheeling diode maintains load current during the intervals that the transistor interrupts the supply current. The main components of such a circuit are shown in Fig. 6.12 and the power MOSFET model is illustrated in Fig. 6.10.

Figure 6.13 depicts the turn-on characteristics, but the waveshapes have been linearized for simplicity of calculation.

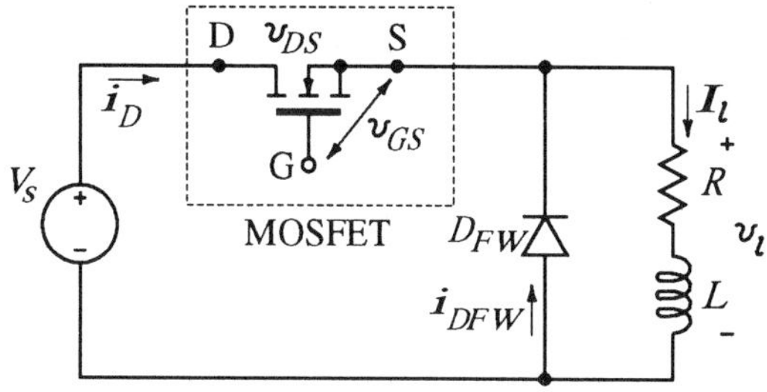

Fig. 6.12 MOSFET circuit with a constant current load.

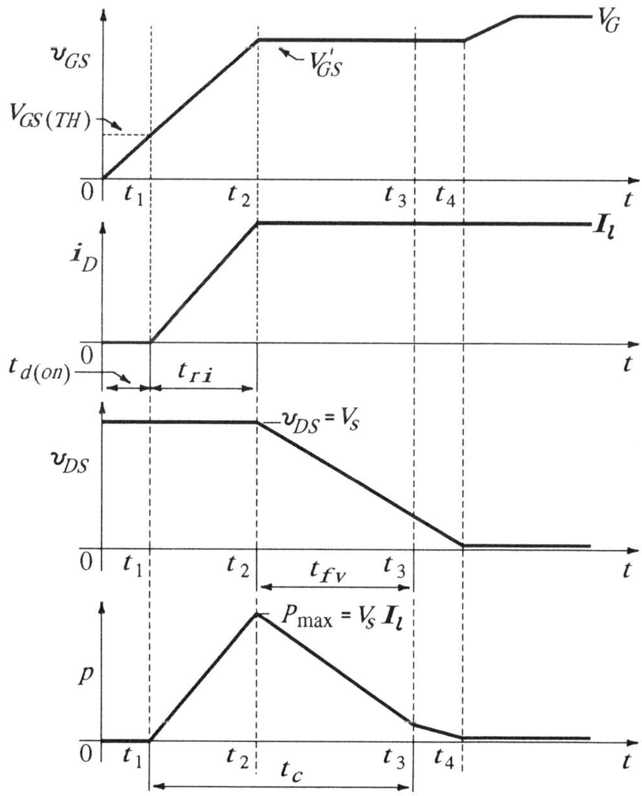

Fig. 6.13 MOSFET turn-on for a constant current load.

Steady-state operation is assumed to prevail. The power MOSFET has been switching on and off enough times to allow the load current I_l to be constant. Prior to time $t=0$, the device capacitors are discharged, the voltage across the device is $v_{DS}=V_s$, the drain current is $i_D \approx 0$, the diode current is $i_{DFW}=I_l$ and the load voltage $v_l=0$ (because the diode D_{FW} is conducting).

At time $t=0$ the gate-drive switch Sw is closed to initiate turn-on. The response is a gate current that starts to charge the device capacitors. As a result, the gate voltage v_{GS} begins to rise. The gate voltage rises to the threshold value $v_{GS}=V_{GS(TH)}$, at which point the MOSFET changes state from operation in the cutoff region to operation in the active region. This change occurs at a time t_1 and since this is the interval during which there is no drain current it is called the delay time $t_{d(on)}$.

For a time interval $t_1 \le t \le t_2$, as the gate voltage v_{GS} rises, the drain current i_D increases. The current rises according to the transfer characteristic $i_D=G\,(v_{GS}-V_{GS(TH)})$, where G is the device transconductance. The time constant of the gate circuit is $\tau_1=R_G C_{iss} \approx R_G(C_{GD}+C_{GS})$ and this influences the change of drain current. During this time interval, the freewheeling diode is conducting, so that the voltage across the load remains zero. Thus, the transistor sustains the full supply voltage across it. This means that the power dissipated by the MOSFET is $p=v_{DS}i_D=V_s i_D$, and this is shown in Fig. 6.13. The power dissipated p rises linearly.

The drain current rises until it reaches its steady value $i_D=I_D=I_l$ at a time $t=t_2$. The rise-time t_{ri} of the drain current is the interval

$$t_{ri}=t_2-t_1. \tag{6.5.8}$$

Time $t=t_2$ is another transition time. Up to this time the diode current i_{DFW} has been falling according to Kirchhoff's current law

$$i_{DFW}=I_l-i_D. \tag{6.5.9}$$

At time $t=t_2$, $i_D=I_l$ so the diode current $i_{DFW}=0$ and the diode goes into the blocking state. That is, it turns off and no longer constrains the load voltage v_l to be zero. The load voltage rises as the MOSFET voltage v_{DS} falls. According to Kirchhoff's voltage law

$$v_l=V_s-v_{DS}. \tag{6.5.10}$$

How the voltage v_{DS} falls depends on the device parameters. The rate of fall is given by

$$\frac{dv_{DS}}{dt}=\frac{d}{dt}(v_{DG}+v_{GS})=\frac{dv_{DG}}{dt}+\frac{dv_{GS}}{dt}. \tag{6.5.11}$$

For some time after $t=t_2$ (up to time $t=t_4$) the gate voltage is clamped at a constant value $v_{GS}=V'_{GS}$, given by the equation of the transfer characteristic.

$$I_D = I_l = G\,(V'_{GS} - V_{GS(TH)}). \tag{6.5.12}$$

Since the drain current is constant, so too must be the gate voltage while operation remains in the active region. If the gate voltage is constant then the rate of change of gate voltage dv_{GS}/dt must be zero. Therefore, the drain voltage, expressed by eq. (6.5.11), becomes

$$\frac{dv_{DS}}{dt} = \frac{dv_{DG}}{dt} = \frac{i_G}{C_{GD}}. \tag{6.5.13}$$

The gate current i_G is a constant while the gate voltage is constant because by Kirchhoff's voltage law

$$V_G = i_G R_G + v_{GS} = i_G R_G + V'_{GS}. \tag{6.5.14}$$

Using this result in eq. (6.5.13)

$$\frac{dv_{DS}}{dt} = \frac{V_G - V'_{GS}}{R_G C_{GD}} \approx \text{constant}. \tag{6.5.15}$$

The fall of drain voltage is nearly linear in the active region because the value of capacitance C_{GD} does not change too much. Up to the time $t = t_3$ the MOSFET operates in the active region. After this, operation starts to be in the ohmic region, v_{DS} is small and the value of the capacitance C_{GD} increases suddenly and radically. The interval in the active region is defined as the voltage fall time t_{fv}, where

$$t_{fv} = t_3 - t_2. \tag{6.5.16}$$

For a linear fall of drain voltage the power p dissipated in the transistor changes linearly according to the expression

$$p = v_{DS} I_D \approx -\frac{V_s I_D}{t_{fv}} t' + V_s I_D. \tag{6.5.17}$$

where $t' = t - t_2$ up to the limit $t = t_3$. The falling drain voltage v_{DS} with the gate voltage v_{GS} clamped at V'_{GS} is known as the *Miller effect*.

During the interval $t_3 \le t \le t_4$, the gate voltage v_{GS} is still clamped at a value V'_{GS} initially and eq. (6.5.15) can still be used, but the value of C_{GD} has changed to a degree that dv_{DS}/dt is small. Since the values of the voltage v_{DS} and the power p are small at this stage it is usual to ignore this as a transient interval by lumping it with the steady on-state.

At some time $t > t_3$, operation of the power semiconductor switch is in the ohmic region, the drain voltage is $v_{DS} = I_D R_{DS(ON)}$ and $v_{DG} \le 0$. The gate voltage v_{GS} becomes unclamped and rises to V_G according to the values of the drive-circuit resistance R_G and the input capacitance C_{iss}. In this time interval the value of $C_{iss} = (C_{GS} + C_{GD})$ increases considerably due to the change of C_{GD} compared with the value in the interval t_{ri} that covers the rising current.

From an inspection of the linearized waveforms in Fig. 6.13 a simple definition of turn-on time t_{on} can be approximated to equal the time t_3 by which time the MOSFET approaches a steady condition for current I_D and voltage v_{DS} in the ohmic region. It is common to define a crossover time t_c that covers the interval from the initial rise of current i_D to the fall of voltage v_{DS} so that $t_c = t_3 - t_1$. In this way turn-on time can be defined for any type of load. The turn-on time for the special case just described can be expressed by

$$t_{on} \approx t_{d(on)} + t_c = t_{d(on)} + t_{ri} + t_{fv} . \tag{6.5.18}$$

This is the worst case for energy loss during the switching action. There is full voltage V_s across the MOSFET while the current rises and then there is full current I_D in the device while the voltage v_{DS} falls. For the worst case the power-dissipation waveform is nearly triangular, so that the energy loss W_{on} during turn-on can be expressed by the area of the triangle as

$$W_{on} = P_{\max} \times t_c /2 = V_s I_l (t_{ri} + t_{vf})/2 . \tag{6.5.19}$$

The delay-time loss is ignored. For any other kind of load connected to this chopper circuit, the energy loss during turn-on will have a value less than that calculated from eq. (6.5.19).

In the dynamic circuit of the MOSFET shown in Fig. 6.10 the internal inductance was neglected. The value may be about 5 nH and this tends to decrease the turn-on loss, but at the expense of turn-off loss. More rigorous ways of analyzing the turn-on process are to be found in the bibliography.

EXAMPLE 6.3

A 250-V, 30-A power MOSFET modulates power from a dc supply to a 7 Ω load that absorbs an average power of 2.8 kW at virtually constant current. Figure EX6.3 gives details of the chopper circuit. The switching frequency of the chopper is 30 kHz. Typical values of the transistor capacitances are given by the data sheets as input capacitance $C_{iss} = 2500$ pF, output capacitance $C_{oss} = 600$ pF and transfer capacitance $C_{rss} = 200$ pF. The MOSFET transconductance is $G = 9$ siemens, the gate threshold voltage is $V_{GS(TH)} = 3$ V, and the on-state resistance is $R_{DS(ON)} = 0.12$ ohm. Calculate **(a)** the turn-on time t_{on}, **(b)** the overdrive of the gate signal and **(c)** the average power dissipated in the MOSFET due to the turn-on transient.

Solution

Figure EX6.3 depicts a constant current load if the time constant L/R is roughly more than 10 times the chopper switching period $1/f$.

The figure also shows a gate drive with a constant current source to charge the MOSFET input capacitance C_{iss} when the switch *Sw* is opened. The gate voltage v_{GS} will rise linearly to a maximum value of $V_{GS} = 10$ V, that is set by the break-over voltage of the zener diode Z. The linear rise of gate voltage gives rise to the MOSFET current, voltage and power responses as shown in Fig. 6.13. The values of the capacitances have to be assumed to be constant during turn-on in the active

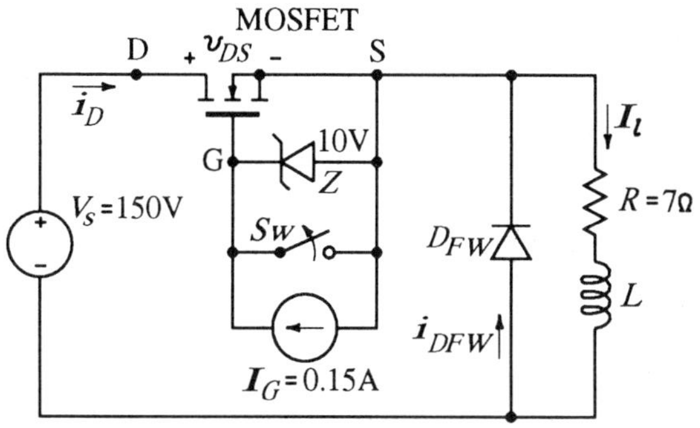

Fig. EX6.3 MOSFET chopper with a constant current gate drive.

region, so the results are approximate. Conversion to the capacitances illustrated in Fig. 6.10 requires the use of eqs (6.5.2) to (6.5.4). Thus,
$C_{GS} \approx 2300$ pF, $C_{GD} \approx 200$ pF and $C_{DS} \approx 400$ pF.
The steady drain current $I_D = (P/R)^{1/2} = (2800/7)^{1/2} = 20$ A.

(a) Over the interval $t_2 = t_{d(on)} + t_{ri}$, the constant gate drive current causes the gate voltage v_{GS} to rise linearly to V'_{GS}, where
$V'_{GS} = V_{GS(TH)} + I_D/G = 3 + 20/9 = 5.33$ V (according to the transfer characteristic).
From $I_G = (C_{GS} + C_{GD})\, dv_{GS}/dt = (C_{GS} + C_{GD})\, V'_{GS}/t_2$,
$t_2 = t_{d(on)} + t_{ri} = (C_{GS} + C_{GD})\, V'_{GS}/I_G = 2500 \times 10^{-12} \times 5.33/0.15 = 88.8 \times 10^{-9}$ s.
During that part of the turn-on interval t_{fv}, $i_D = I_D = 20$ A, v_{GS} is clamped to V'_{GS} (so $dv_{GS}/dt = 0$ and the current in C_{GS} is zero) and the current I_G is charging C_{GD}.
$dv_{DS}/dt = d(v_{DG} + v_{GS})/dt = dv_{DG}/dt = I_G/C_{GD}$.
If C_{GD} is considered constant, the drain voltage v_{DS} will fall linearly from $v_{DS} = V_s = 150$ V to $v_{DS} = I_D R_{DS(ON)} = 200 \times 0.12 = 2.4$ V (the on-state voltage drop). Hence, $(V_s - I_D R_{DS(ON)})/t_{fv} = I_G/C_{GD}$ or
$t_{fv} = (V_s - I_D R_{DS(ON)}) C_{GD}/I_G = (150 - 2.4) \times 200 \times 10^{-12}/0.15 = 196.8 \times 10^{-9}$ s.
The turn-on time is $t_{on} = (t_{d(on)} + t_{ri}) + t_{fv} = (88.8 + 196.8) \times 10^{-9} \approx 286 \times 10^{-9}$ s.
The data sheet gives a typical turn-on time of 200 ns.

(b) The gate signal v_{GS} rises to a value equal to the zener breakover value, that is $V_{GS} = 10$ V. The gate voltage to turn-on the MOSFET is $V'_{GS} = 5.33$ V. Therefore the overdrive gate voltage is $V_{GS} - V'_{GS} = 10 - 5.33 = 4.67$ V.

Percentage gate overdrive is $\dfrac{V_{GS} - V'_{GS}}{V'_{GS}} \times 100 = \dfrac{4.67}{5.33} \times 100 = 87.6\,\%$.

(c) The profile of the power dissipated during turn-on is shown in Fig. 6.13. Making the approximation that the drain voltage v_{DS} is approximately zero at the end of the turn-on interval, the energy loss W_{on} during turn on is
$W_{on} = P_{\max} t_c/2 = V_s I_l (t_{ri} + t_{fv})/2$.
The delay time $t_{d(on)} = (V_{GS(TH)}/V'_{GS}) t_2 = (3/5.33) \times 88.8 \times 10^{-9} = 50 \times 10^{-9}$ s.

Therefore the crossover time is
$t_c = t_{ri} + t_{fv} = t_{on} - t_{d(on)} = (286 - 50) \times 10^{-9} = 236 \times 10^{-9}$ s.
The switching frequency $f = 30$ kHz, so the average power dissipated P_{on} due to turn-on is
$P_{on} = W_{on} f = V_s I_l t_c f/2 = 150 \times 20 \times 236 \times 10^{-9} \times 30 \times 10^3 / 2 = 10.6$ W.
A typical value of the total average power dissipation is 150 W. Therefore the turn-on power dissipation is relatively small.

6.6. MOSFET TURN-OFF

In general terms, the turn-off action of a power MOSFET is the reverse of the turn-on action. If the device is on, it means that there is an *n*-channel in the *p*-body region and this is supported by a gate voltage V_{GS} of sufficient magnitude. Discharge of the input capacitance C_{iss} (C_{GS} and C_{GD}) to a level that the gate voltage $v_{GS} \leq V_{GS(TH)}$, the threshold level, results in the removal of the conducting channel and reversion to the off-state. The time it takes to discharge the capacitance to $V_{GS(TH)}$ is the turn-off time t_{off}, because there are no minority carriers to remove from the device. Since the gate threshold voltage $V_{GS(TH)}$ is about 2 V there is no need for a gate reverse bias unless extra fast turn-off is required.

There are three intervals relating to turn-off. There is the delay time $t_{d(off)}$ during which time the gate voltage v_{GS} falls, but there is no change in drain voltage v_{DS} or the drain current i_D. Also, there are the rise of the drain voltage v_{DS} in time t_{rv} and the fall of the drain current i_D in time t_{fi}. The overall turn-off time t_{off} depends on the type of load as well as the input capacitance of the MOSFET.

The dynamic circuit model, that describes the turn-off behaviour, is the same as that depicted in Fig. 6.10. However, for turn-off, the gate-drive supply V_G would be reduced to zero and switch *Sw* would be closed to provide a path to discharge the equivalent input capacitance energy in the resistance R_G. During the delay interval $t_{d(off)}$ the resistance $R_{DS} = R_{DS(ON)}$ would be in play, because operation is in the ohmic region. Following this the dependent current-source model is used.

6.6.1. Turn-off Losses.

The MOSFET goes through a transition, from the on-state to the off-state, during which both the drain voltage and current can be high enough to create substantial power dissipation in the device. We will look at two extreme cases.

MOSFET circuit with a resistive load. Figure 6.10 illustrates a MOSFET circuit with a resistive load. The initial conditions are that the gate voltage the device is on with a steady drain current $i_D = I_D = I_l$ and the drain voltage $V_{DS} = I_D R_{DS(ON)}$. At time $t = 0$ turn-off is initiated by reducing the gate supply voltage V_G to zero to leave the gate circuit comprising an equivalent capacitance $C_{iss} \approx C_{GS} + C_{GD}$ and a resistance R_G. From this time onwards the gate voltage v_{GS} will decay from a value V_G to zero as the capacitance discharges. The decay is not simply exponential in nature because the capacitance is not constant. For

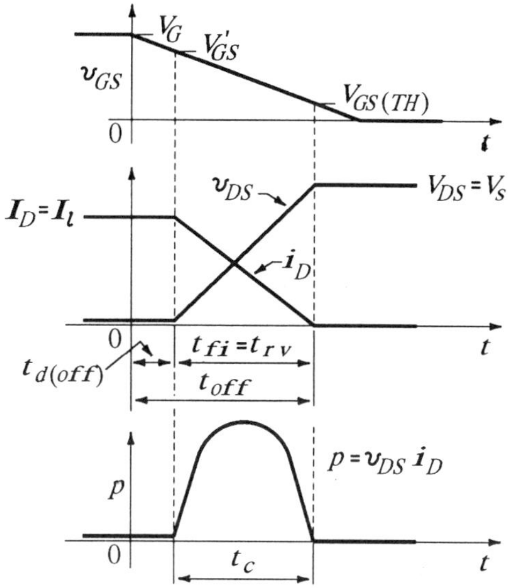

Fig. 6.14 MOSFET turn-off for a resistive load.

simplicity the decay is assumed to be linear.

Figure 6.14 shows the turn-off waveforms. While the gate voltage is in overdrive ($v_{GS} > V'_{GS}$), operation of the MOSFET is in the ohmic region and the on-state current $i_D = I_D = I_l$ is sustained. This interval is called the delay time $t_{d(off)}$. At V'_{GS} the operation of the device is at the transition between the ohmic region and the active region, so that this value of gate voltage just maintains the drain current I_D.

As the gate voltage falls further ($v_{GS} < V'_{GS}$), the transfer characteristic for the active region comes into play. That is, $i_D = G(v_{GS} - V_{GS(TH)})$. If the gate voltage v_{GS} decreases linearly with time, so does the drain current i_D. When the gate voltage v_{GS} reaches the threshold value $V_{GS(TH)}$, the drain current has reached zero and turn-off is complete. The time interval for the gate voltage v_{GS} to fall from V'_{GS} to $V_{GS(TH)}$ is the fall time t_{fi} of the drain current.

Since the load is resistive the load current i_l and the load voltage v_l are in phase, so, from Fig. 6.10, the drain voltage v_{DS} is given by

$$v_{DS} = V_s - v_l = V_s - i_l R_l = V_s - i_D R_l. \tag{6.6.1}$$

The drop in drain current is accompanied by a linear rise of drain voltage as shown in Fig. 6.14. The rise of voltage v_{DS} from the on-state value $v_{DS} = I_D R_{DS(ON)}$ to the off-state value $v_{DS} = V_s$ occurs over an interval t_{rv}, which is the same as the fall of current interval t_{fi} for a resistive circuit.

The turn-off time t_{off} of the MOSFET is given by

$$t_{off} = t_{d(off)} + t_{fi} = t_{d(off)} + t_{rv}\,. \tag{6.6.2}$$

The time interval t_c from the initial rise of the drain voltage to zero drain current is the crossover time, that is given by

$$t_c = t_{fi} = t_{rv}\,. \tag{6.6.3}$$

The instantaneous value of the power dissipated during the crossover time[10] is $p = v_{DS} i_D$. With time t measured from the beginning of the crossover interval t_c, the drain current i_D is described by

$$i_D = -\frac{I_D}{t_c} t + I_D = -\frac{I_l}{t_c} t + I_l\,. \tag{6.6.4}$$

Over the time interval $t_c = t_{rv} = t_{fi}$, the drain voltage v_{DS} is expressed by

$$v_{DS} \approx \frac{V_s}{t_c} t\,. \tag{6.6.5}$$

These two equations can be used to determine an estimate of the energy loss W_{off} during the crossover time of turn-off. That is,

$$W_{off} = \int_0^{t_c} p\, dt = \int_0^{t_c} v_{DS} i_D dt \approx \frac{V_s I_l}{6} t_c\,. \tag{6.6.6}$$

The values of supply voltage V_s and steady load current I_l are known normally, and the value of t_c can be found in the data sheets. A rough estimate of t_c can be determined from a gate circuit analysis, if the input capacitance C_{iss}, the gate voltage V'_{GS} and the threshold voltage $V_{GS(TH)}$ are known. Very approximately

$$v_{GS}(t) \approx V'_{GS} e^{-t/\tau} \tag{6.6.7}$$

where the time constant $\tau = R_G C_{iss}$. From this, the crossover time t_c is given by

$$t_c \approx R_G C_{iss} \ln(V'_{GS} / V_{GS(TH)})\,. \tag{6.6.8}$$

The crossover time t_c and the turn-off time t_{off} are shortest for a resistive load.

MOSFET circuit with an inductive load. The turn-off time of a MOSFET is longest, if the power circuit has an inductive load with a freewheeling diode, such that the load current is virtually constant. Figure 6.12 illustrates such a circuit in which the MOSFET modulates power to the load by chopping action. Initially, the device is in the steady on-state and at time $t = 0$ turn-off is initiated by removing the gate-drive supply voltage V_G and providing a path of resistance R_G for the discharge of the MOSFET capacitance.

---

[10] It is usual to lump the power loss during the turn-off delay time with the power loss during the steady on-state.

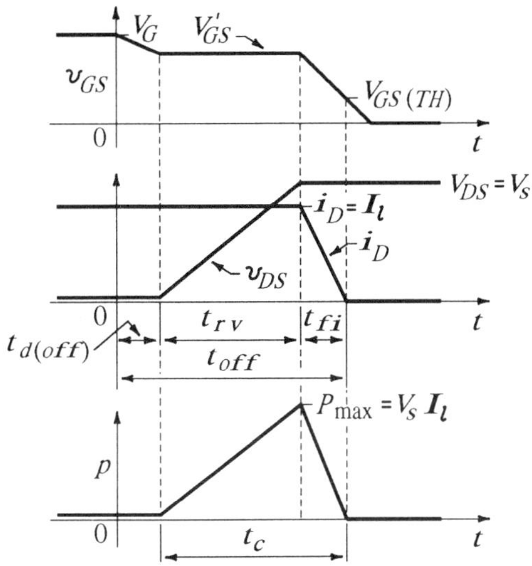

Fig. 6.15 MOSFET turn-off for a constant current load.

Figure 6.15 shows the waveforms of the turn-off action which is the reverse of the turn-on action for such a circuit (see Fig. 6.13). The gate voltage v_{GS} is shown again to be idealized for simplicity of explanation and ease of analysis. During the first interval of turn-off the gate voltage is in overdrive ($v_{GS} > V'_{GS}$), operation is still in the ohmic region ($v_{DS} = R_{DS(ON)} I_l$) and the on-state current is $i_D = I_D = I_l$. This interval is the delay time $t_{d(off)}$.

When the gate voltage falls to the value $v_{GS} = V'_{GS}$ the MOSFET operation starts to be in the active region. Over the next interval the gate voltage is constant at V'_{GS} to satisfy the transfer characteristic $i_D = G(v_{GS} - V_{GS(TH)})$. The load current is constant at I_l and, while the drain voltage v_{DS} is less than the supply voltage V_s, the freewheeling diode across the load is reverse biased, so the MOSFET must conduct the full value $i_D = I_D = I_l$. The value of V'_{GS} for $i_D = I_l$ is given by the equation

$$i_D = I_l = G(V'_{GS} - V_{GS(TH)}). \tag{6.6.9}$$

While the gate voltage $v_{GS} = V'_{GS}$ and the drain current i_D are unvarying, the drain voltage v_{DS} rises from the on-state value (assumed to be approximately zero) to the full blocking value $v_{DS} = V_s$. This time is denoted by t_{rv} in Fig. 6.15 and can be determined as follows. From the dynamic model shown in Fig. 6.10 and the conditions that $V_G = 0$ and the switch *Sw* closed, the rate of change of drain voltage is given by

$$\frac{dv_{DS}}{dt}=\frac{d}{dt}(v_{DG}+v_{GS})=\frac{dv_{DG}}{dt}=-\frac{i_G}{C_{GD}}. \tag{6.6.10}$$

Kirchhoff's voltage law for the loop encompassing the gate circuit gives

$$0=i_G R_G+V'_{GS}. \tag{6.6.11}$$

Consequently, the rate of change of drain voltage is expressed by

$$\frac{dv_{DS}}{dt}=\frac{V'_{GS}}{R_G C_{GD}}. \tag{6.6.12}$$

If C_{GD} is assumed constant while operation is in the active region, the solution of this differential equation is

$$v_{DS}=\frac{V'_{GS}}{R_G C_{GD}}t \tag{6.6.13}$$

and the interval t_{rv} follows from this for $v_{DS}=V_s$. That is,

$$t_{rv}=V_s R_G C_{GD}/V'_{GS}. \tag{6.6.14}$$

Once the drain voltage v_{DS} reaches the value V_s, the gate voltage v_{GS} is no longer clamped, because the freewheeling diode is not reverse biased and can conduct. Consequently, the drain current i_D falls to zero in time t_{fi} that is the same time as the gate voltage v_{GS} falls from the value V'_{GS} to the threshold value $V_{GS(TH)}$. At this point the MOSFET is off. The equations governing the fall of drain current i_D are

$$i_D=G(v_{GS}-V_{GS(TH)}) \tag{6.6.15}$$

from the transfer characteristic and

$$v_{GS}(t)\approx V'_{GS}e^{-t/R_G(C_{GS}+C_{GD})}. \tag{6.6.16}$$

In this last equation at time $t=t_{fi}$, the gate voltage is $v_{GS}=V_{GS(TH)}$, so the current fall interval is

$$t_{fi}\approx R_G(C_{GS}+C_{GD})\ln(V'_{GS}/V_{GS(TH)}). \tag{6.6.17}$$

Combining the above analyses, the crossover time t_c is given by

$$t_c=t_{rv}+t_{fi} \tag{6.6.18}$$

and the turn-off time t_{off} is

$$t_{off}=t_{d(off)}+t_{rv}+t_{fi}=t_{d(off)}+t_c. \tag{6.6.19}$$

The energy loss in the MOSFET during the delay time $t_{d(off)}$ is lumped with the steady on-state loss. The energy loss W_{off} during the crossover interval is, from an inspection of Fig. 6.15,

$$W_{off} = \int_0^{t_c} p\, dt \approx \frac{P_{\max} t_c}{2} = \frac{V_s I_l}{2} t_c . \qquad (6.6.20)$$

The fall of drain current is approximated to be linear in this calculation.

If the MOSFET and circuit parameters are known, rough estimates of the turn-off time and losses can be made from eqs (6.6.9) through (6.6.20).

More rigorous analyses of the turn-off process can be found in the bibliography.

EXAMPLE 6.4

A 250-V, 30-A power MOSFET modulates power from a dc supply to a load that absorbs an average power of 2.8 kW. The chopper frequency of switching is 30 kHz and the duty cycle is $m = 0.8$. Typical values of the transistor capacitances are given to be input capacitance $C_{iss} = 2500$ pF, output capacitance $C_{oss} = 600$ pF and transfer capacitance $C_{rss} = 200$ pF. The MOSFET transconductance is $G = 9$ S, the gate threshold voltage is $V_{GS(TH)} = 3$ V and the on-state resistance is $R_{DS(ON)} = 0.12$ ohm. A dc supply of voltage $V_G = 10$ V, a switch *Sw* and a resistance $R_G = 20$ ohms comprise the equivalent drive circuit. Calculate the turn-off time of the MOSFET and the average power dissipation during this turn-off action for **(a)** a purely resistive load $R = 7$ ohms and **(b)** a virtually constant current load and the same R.

Solution

Figure 6.10 represents the general configuration of the circuit, except that the switch *Sw* in the gate drive circuit is closed and the voltage V_G is zero for the turn-off action. This leaves a resistance $R_G = 20$ ohms to dissipate the energy stored in the transistor capacitances. The initial voltage v_{GS} is 10 V at the initiation of turn-off.

(a) Figure 6.14 shows the general form of the MOSFET waveforms during turn-off for a resistive load. There is not enough data to account for changes in the value of the input capacitance C_{iss} so it has to be assumed constant. The discharge of the capacitance ($C_{iss} \approx C_{GD} + C_{GS}$) in the resistor R_G is described by the time variation of the gate voltage v_{GS} as

$v_{GS} = V_G e^{-t/R_G C_{iss}}$ or $t = R_G C_{iss} \ln (V_G / v_{GS})$

where time t is measured from the initiation of turn-off.

Thus the delay time $t_{d(off)}$ is given by $t_{d(off)} = R_G C_{iss} \ln (V_G / V'_{GS})$,

where V'_{GS} is the lowest voltage to sustain the steady current $I_D = I_l$ in the transistor while it is still on. Since it can be assumed that the on-time of the chopper is long compared with the turn-on and turn-off times, we can approximate the average power absorbed by the load to be

$P = I_{l\,rms}^2 R = m I_l^2 R$, so $I_l^2 = P/(mR) = 2800/(0.8 \times 7)$, or $I_l = 22.36$ A.

Since $I_D = I_l$ in the on-state, the supply voltage V_s is

$V_s = I_l R + I_D R_{DS(ON)} = 22.36 \times (7 + 0.12) = 159.2$ V.

The value of V'_{GS} is given by the transfer characteristic

$I_D = G(V'_{GS} - V_{GS(TH)})$, or $V'_{GS} = I_D / G + V_{GS(TH)} = 22.36/9 + 3 = 5.48$ V.

Hence, $t_{d(off)} = R_G C_{iss} \ln(V_G / V'_{GS}) = 20 \times 2500 \times 10^{-12} \ln(10/5.48) = 30 \times 10^{-9}$ s.
The turn-off time t_{off} is given by
$t_{off} = R_G C_{iss} \ln(V_G / V_{GS(TH)}) = 20 \times 2500 \times 10^{-12} \ln(10/3) = 60.2 \times 10^{-9}$ s.
The crossover time t_c is given by
$t_c = t_{off} - t_{d(off)} = (60.2 - 30) \times 10^{-9} = 30.2 \times 10^{-9}$ s.
Energy dissipation $W_{d(off)}$ during the turn-off delay time is
$W_{d(off)} = I_D^2 R_{DS(ON)} t_{d(off)} = 22.36^2 \times 0.12 \times 30 \times 10^{-9} = 1.8 \times 10^{-6}$ J.
During the crossover interval t_c the energy dissipated W_{off} is
$W_{off} \approx V_s I_l t_c / 6 = 159.2 \times 22.36 \times 30.2 \times 10^{-9} / 6 = 17.8 \times 10^{-6}$ J.
Therefore the average power P_{off} dissipated due to the turn-off action is
$P_{off} = (W_{d(off)} + W_{off}) f = (1.8 + 17.8) \times 10^{-6} \times 30 \times 10^3 = 0.588$ W.

(b) A constant current load can be achieved if a large inductance is connected in series with the load resistance and a freewheeling diode is connected in parallel with both. This is exemplified in Fig. 6.12. The turn-off waveforms for such a case are illustrated in Fig. 6.15. We have to calculate the turn-off delay time $t_{d(off)}$ the drain-voltage rise-time t_{rv} and the drain-current fall-time t_{fi} in order to calculate the turn-off time t_{off}.
The steady load current I_l is expressed in the power absorption equation
$P = I_l^2 R$, so $I_l^2 = P/R = 2800/7$, or $I_l = 20$ A.
This means that the steady on-state current I_D of the MOSFET is $I_D = I_l = 20$ A.
The supply voltage V_s is given by
$V_s = I_D R_{DS(ON)} + I_l R = 20(0.12 + 7) = 142.4$ V.

Delay time. At the end of the delay interval $t_{d(off)}$ the gate voltage $v_{GS} = V'_{GS}$ is given by the transfer characteristic $I_D = G(V'_{GS} - V_{GS(TH)})$.
So, $V'_{GS} = I_D / G + V_{GS(TH)} = 20/9 + 3 = 5.22$ V.
The gate voltage decays according to the equation
$v_{GS} \approx V_G e^{-t/R_G C_{iss}}$ where $C_{iss} = C_{GD} + C_{GS} = 1500$ pF.
Hence,
$t_{d(off)} = R_G C_{iss} \ln(V_G / V'_{GS}) = 20 \times 2500 \times 10^{-12} \ln(10/5.22) = 32.5 \times 10^{-9}$ s.

Voltage rise. During the rise-time t_{rv} of the voltage v_{DS}, the gate voltage $v_{GS} = V'_{GS}$ because the drain current i_D is constant.

$$\text{So, } \frac{dv_{DS}}{dt} = \frac{d}{dt}(v_{DG} + v_{GS}) = \frac{dv_{DG}}{dt} = -\frac{i_G}{C_{GD}} = -\frac{i_G}{C_{rss}} = \frac{V'_{GS}}{R_G C_{rss}}.$$

If the transfer capacitance is assumed to be constant, the rate of rise of the drain voltage v_{DS} is constant. The drain voltage rises from the value $v_{DS} = I_D R_{DS(ON)}$ to the value $v_{DS} = V_s$ in time t_{rv}, so
$t_{rv} = R_G C_{rss}(V_s - I_D R_{DS(ON)}) / V'_{GS} = 20 \times 200 \times 10^{-12} \times 140 / 5.22 = 107.2 \times 10^{-9}$ s.

Current fall. During the drain current fall from $i_D = I_D = I_l$ to zero, the gate voltage falls from $v_{GS} = V'_{GS}$ to $v_{GS} = V_{GS(TH)}$. The gate voltage falls according to the equation
$v_{GS} = V'_{GS} e^{-t/R_G C_{iss}}$, where $t = 0$ at the beginning of the interval t_{fi}.
The fall-time t_{fi} of the current is given by
$t_{fi} = R_G C_{iss} \ln(V'_{GS} / V_{GS(TH)}) = 20 \times 2500 \times 10^{-12} \ln(5.22/3) = 27.7 \times 10^{-9}$ s.

Turn-off time. The turn-off time t_{off} comprises the sum of the above intervals.
$t_{off} = t_{d(off)} + t_{rv} + t_{fi} = (32.5 + 107.2 + 27.7) \times 10^{-9} = 167.4 \times 10^{-9}$ s.
This time is almost three times the turn-off time for the same MOSFET with a purely resistive load of the same value.
Energy dissipation $W_{d(off)}$ during the turn-off delay time is
$W_{d(off)} = I_D^2 R_{DS(ON)} t_{d(off)} = 20^2 \times 0.12 \times 32.5 \times 10^{-9} = 1.56 \times 10^{-6}$ J.
During crossover $t_c = t_{rv} + t_{fi} = 134.9 \times 10^{-9}$ s, the energy dissipated W_{off} is
$W_{off} \approx V_s I_l t_c / 2 = 142.4 \times 20 \times 134.9 \times 10^{-9} / 2 = 192.1 \times 10^{-6}$ J.
Therefore, the average power P_{off} dissipated due to the turn-off action is
$P_{off} = (W_{d(off)} + W_{off}) f = (1.56 + 192.1) \times 10^{-6} \times 30 \times 10^3 = 5.81$ W.
This value is about ten times the average power loss for turn-off for the same MOSFET with a resistive load.

6.7. MOSFET GATE CIRCUITS

The gate drive requirements of a power MOSFET are probably the simplest of all the power semiconductor switches. A gate-to-source voltage of about 10 V has to be established to turn on the MOSFET so that it is in the ohmic region of operation. Then, the gate-to-source voltage has to be reduced to below about 3 V for the device to turn off and be in the cutoff region of operation. Accordingly, gate voltage controls the on and off conditions. However, it is gate-circuit current that controls the speed of the switching action, because it is charge on the device capacitance that creates the *n* channel for conduction.

Figure 6.16a shows a circuit diagram of a gate drive that utilizes a constant supply voltage V_G, a switch *Sw* and two resistors. Upon closing the switch *Sw* the gate current i_G exists until the MOSFET input capacitance C_{iss} charges up and the gate voltage rises to a steady value $v_{GS} = V_{GS}$, at which point $i_G = 0$. The time of turn-on is a function of the values of R_G, R_1, R_2 and C_{iss}. The lower is the resistance $(R_G + R_1)$, the shorter is the charge-up time of the *RC* circuit. Since the internal resistance R_G of the gate supply is not adjustable normally, the resistor R_1 is made to have a low value for fast switching and a reasonable overdrive voltage V_{GS}. Turn-on can be speeded up by connecting a capacitor C_1 across the resistor R_1. This has the advantages of a high initial current.

The MOSFET in this particular circuit turns off if the switch *Sw* is opened. The gate supply is disconnected and the MOSFET capacitance is discharged into the resistor R_2. When the gate voltage decays to the threshold value $v_{GS} = V_{GS(TH)}$, the transistor turns-off. The speed of turn-off increases, if the value of the resistance R_2 is reduced.

Since current in the gate circuit governs switching speed, a current source would appear to be useful. Figure 6.16b depicts such a case. Upon opening the switch *Sw*, a constant gate current $i_G = I_G$ is injected at the transistor gate terminal until the zener breakdown voltage is reached. This zener voltage would be chosen so that the MOSFET would be operating in the ohmic region at the breakdown value. A typical value of zener voltage is 12 V. The gate current i_G would then be

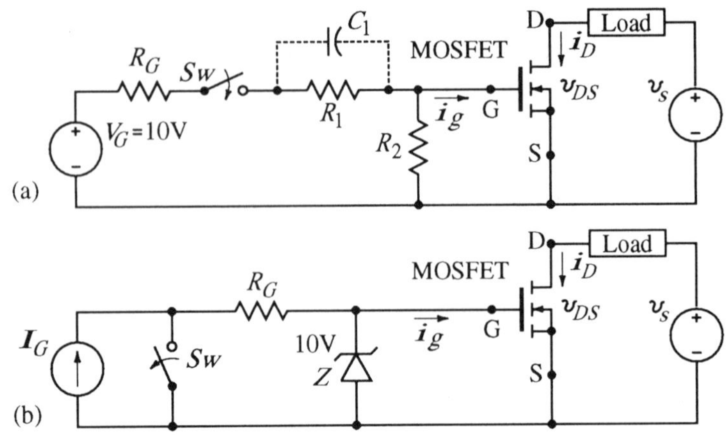

Fig. 6.16 Simple gate drives for MOSFETS.
(a) Voltage drive, (b) current drive.

zero and the zener diode would conduct the source current I_G until the switch *Sw* is closed.

Upon closure of the switch *Sw*, the switch conducts both the source current and the MOSFET gate discharge current. There must be some resistance R_G in the circuit to dissipate the transistor gate energy. When the gate voltage falls to the threshold value the MOSFET reverts to its off-state.

In the discussions on turn-on and turn-off times we have shown that $dv_{DS}/dt = f(i_G)$ and $di_D/dt = f(i_G)$. Two conclusions are drawn from these equations. First, high speed switching is achieved by designing the drive to deliver a high gate current (as much as 1 A). Second, the switching times are a function of both the supply voltage V_s and the steady load current ($I_D = I_l$).

The circuits in Fig. 6.16 are basic. In reality, the switch *Sw* in the gate circuit would be a fast-acting transistor and most probably it would be in a specially designed chip that included interfacing between it and a control signal from TTL logic or a microprocessor-based controller.

EXAMPLE 6.5

A 250-V, 30-A power MOSFET is used as a chopper to modulate power from a 140-V, dc supply to a purely resistive load whose value is $R = 7\,\Omega$. The MOSFET data are the input capacitance $C_{iss} = 2500\,\text{pF}$, the transconductance $G = 9\,\text{S}$, the gate threshold voltage $V_{GS(TH)} = 3\,\text{V}$ and the on-state resistance $R_{DS(ON)} = 0.12\,\Omega$. The equivalent gate-drive supply has a voltage $V_G = 10\,\text{V}$ and a source resistance $R_G = 5\,\Omega$. If the turn-off time of the MOSFET is to be limited to 50 ns, estimate **(a)** the values of the gate drive resistors, **(b)** the turn-on time and **(c)** the maximum current rating of the gate driver.

Solution

Consider the gate-drive circuit without the capacitor C_1 as illustrated in Fig. 6.16a. It is assumed that the current change occurs in the same time interval as the voltage change for both state changes, on and off.

(a) Let the MOSFET be in the on-state and at $t=0$ let the switch Sw be opened. The on-state current I_D is $I_D = \dfrac{V_s}{R + R_{DS(ON)}} = \dfrac{140}{7.12} \approx 20$ A.
The minimum value of gate voltage V'_{GS} to sustain I_D is given by
$I_D = G(V'_{GS} - V_{GS(TH)})$, so $V'_{GS} = (20/9) + 3 = 5.22$ V.
During the turn-off process, the gate voltage v_{GS} is given by
$v_{GS} = V_{GS}e^{-t/\tau_{off}}$, where $\tau_{off} = R_2 C_{iss}$. Let the overdrive voltage be $V_{GS} = 8$ V.
At turn-off, $t_{off} = \tau_{off} \ln(V_{GS} / V_{GS(TH)})$, so that
$R_2 = t_{off} / (C_{iss} \ln V_{GS} / V_{GS(TH)}) = 50 \times 10^{-9} / (2500 \times 10^{-12} \ln 8/3) = 20.4\ \Omega$.
While in the steady on-state, the MOSFET gate voltage is given by
$v_{GS} = V_{GS} = R_2 V_G / (R_G + R_1 + R_2)$.
Therefore, $R_1 = R_2 (V_G / V_{GS}) - R_G - R_2 = 20.4 \times 10/8 - 5 - 20.4 = 0.1\ \Omega$.

(b) If the turn-on process is initiated at $t=0$, the turn-on time is contained in
$V'_{GS} = V_{GS}(1 - e^{-t_{on}/\tau_{on}})$, where $\tau_{on} = R'_G C_{iss}$.
R'_G is the Thevenin equivalent resistance of the gate circuit.
That is, $R'_G = (R_G + R_1) R_2 / (R_G + R_1 + R_2) = 5.1 \times 20.4 / 25.5 = 4.08\ \Omega$.
Hence,
$t_{on} = \tau_{on} \ln V_{GS} / (V_{GS} - V'_{GS}) = 4.08 \times 2500 \times 10^{-12} \ln 8/2.78 = 10.7 \times 10^{-9}$ s.

(c) The maximum gate current occurs at the initiation of turn-on. At this point the input capacitance C_{iss} acts as a short circuit and the gate current i_G is
$i_G = V_G / (R_G + R_1) = 10/5.1 \approx 2$ A.
This is a high value of current, although the average power associated with this value is very low.

Manufacturers' data sheets on MOSFETS often have information on gate charge associated with the input capacitance C_{iss} versus the gate voltage v_{GS}. This information enables some aspects of the gate-drive design to be determined. In particular, the drain-voltage fall-time t_{fv}, C_{iss} and the power rating of the gate drive can all be calculated from the charge curve.

Figure 6.17 shows an arbitrary charge curve that applies to the case that the power-circuit load has virtually constant current. That is, the load is highly inductive and the load has a freewheeling diode connected across it. Typical switching characteristics are like those illustrated in Fig. 6.13 and Fig. 6.15. In the charge curve an idealized drain voltage v_{DS} has been superimposed to reflect the switching characteristics.

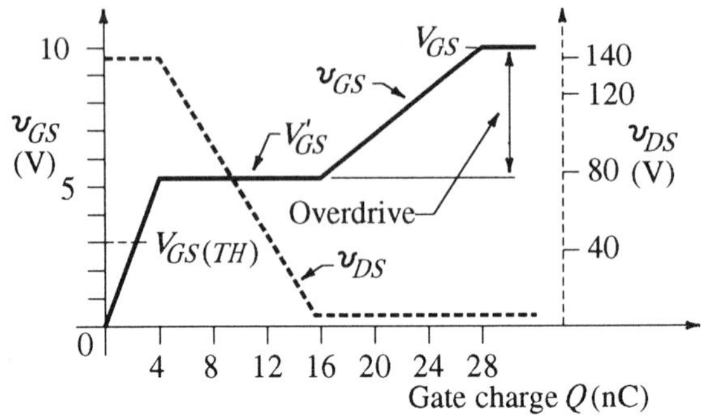

Fig. 6.17 MOSFET charge versus gate voltage curve.

EXAMPLE 6.6

A power MOSFET is used as a 50-kHz chopper to modulate power from a dc supply to an inductive load whose current is virtually constant. The equivalent gate-drive supply is a current source of 0.3 A. If the charge versus gate-voltage curve is as shown in Fig. 6.17, find **(a)** the idealized value of the input capacitance C_{iss} for the on-state, **(b)** the minimum average power that must be supplied by the gate-drive supply and **(c)** the fall time t_{fv} of the drain voltage v_{DS} during turn-on.

Solution

(a) Refer to Fig. 6.17. The gate current i_G can be expressed by $i_G = C_{iss}\, dv_{GS}/dt$. In the overdrive region ($V'_{GS} \le v_{GS} \le V_{GS}$), because of the linearity,

$I_G = C_{iss} \Delta v_{GS}/\Delta t$, or,

$C_{iss} = I_G \Delta t/\Delta v_{GS} = \Delta Q/\Delta v_{GS} = (28-16)\times 10^{-9}/(10-5.22) = 2500\times 10^{-12}$ F.

(b) The energy W_G into the gate terminal to charge the input capacitance during turn on is

$$W_G = \int_0^t v_{GS} i_G dt = \int_0^t v_{GS} dq = \text{area of charge curve.}$$

In terms of numbers,

$W_G = (5.22\times 4/2 + 5.22\times(28-4) + (10-5.22)\times(28-16)/2)\times 10^{-9}$

That is, $W_G = 164.4\times 10^{-9}$ J.

If the chopper switches at $f = 50$ kHz, the average power P_G into the gate is

$P_G = W_G \times f = 164.4\times 10^{-9}\times 50\times 10^3 = 8.2\times 10^{-3}$ W.

If the MOSFET is rated at 250 V, 30 A, then the power gain is

$P/P_G = 250\times 30/(8.2\times 10^{-3}) \approx 9\times 10^5$. Compare this with the thyristor.

The energy stored in the input capacitance is

$$W = \frac{1}{2} C_{iss} V_{GS}^2 = 2.5\times 10^{-9}\times 100/2 = 125\times 10^{-9} \text{ J.}$$

Where is the excess energy, $W_G - W$, dissipated?

(c) The drain voltage v_{DS} falls from V_s to near zero during the interval that $i_D = I_D = I_l$ and $v_{GS} = V'_{GS}$. During this interval t_{fv}, the charge accumulated by the transfer capacitance $C_{rss} = C_{GD}$ is $\Delta Q = I_G t_{fv}$. Hence, from Fig. 6.17, while $v_{GS} = V'_{GS}$.
$t_{fv} = \Delta Q / I_G = (16-4)\times 10^{-9}/0.3 = 40\times 10^{-9}$ s. Compare this with the thyristor.

6.8. MOSFET PROTECTION

All semiconductor switches are sensitive and fragile compared with their electromechanical counterparts. Therefore, protection against overvoltages, overcurrents and transients is a part of any design associated with systems that incorporate MOSFETs.

Voltage control at the gate of the MOSFET simplifies protection compared with that needed for current controlled devices like thyristors and BJTs. If overcurrents and damaging transients can be detected quickly enough, the MOSFET can respond by turning off, all that is needed is to remove the gate-drive voltage. Manufacturers have begun to make MOSFETs with built in current and temperature sensors and gate-drive chips have the ability to remove the drive voltage in response to these protection signals.

6.8.1. Overvoltages

If a MOSFET is in the on-state, system overvoltages do not affect the device directly from a voltage point of view, because the transistor offers itself as a virtual short circuit. A system overvoltage may produce overcurrents, but that is another problem.

Overvoltages that affect the MOSFET directly are those that appear across the drain-to-source terminals while the device is in the off-state, and those that appear across the gate-to-source terminals at any time.

If there is no gate signal applied to the MOSFET, the transistor normally remains off. However, if the voltage at the drain terminal exceeds the breakover value BV_{DSS} (see Fig. 6.6), avalanche breakdown occurs at the n^-p^- junction (see Fig. 6.4) and the device operates in the active region. In this region, high values of voltage v_{DS} can exist with high values of current i_D. The resulting high energy dissipation normally damages the MOSFET. This situation is avoided locally if the supply voltage V_s of the MOSFET power circuit is not allowed to be greater than 80% of the semiconductor's breakover voltage. The user has less control over the generation of overvoltages in other parts of the system. Protection against external overvoltages is usually afforded by connecting a voltage arrester across the power switch. The voltage arrester is a nonlinear resistor whose value decreases with voltage. The aim is to have the arrester turn on before the MOSFET experiences an overvoltage in the vicinity of BV_{DSS}.

Overvoltages V_{SD} at the source terminal do not exist because of the integral reverse diode of the MOSFET. The limiting reverse bias voltage V_{SD} is about 1 V, so the protection problem is one of overcurrent. The current in the integral diode is designed to be the same as the rated forward current.

If it is so desired that current in the MOSFET should not reverse if a reverse bias voltage should exist, a diode in series with the transistor would do the blocking. This is a simple form of protection whose only drawback is an added, small voltage drop in the forward direction during conduction.

The MOSFET is sensitive to static charge and can absorb energy in the device capacitances. To prevent high voltages building up in the handling stages it is common practice to employ grounding techniques. This provides a path for static charge to flow.

The gate-circuit voltage v_{GS} must not exceed 20 V of either polarity. Otherwise, the oxide layer of the gate region will break down and the device will be destroyed. A simple way to limit the gate voltage is to connect across the gate-to-source terminals a zener diode whose rating is less than 20 V.

If a semiconductor switch has a voltage withstand rating that is lower than the supply voltage, it is usual to connect switches in series to handle the higher voltage. MOSFET current ratings are usually less than 50 A. For this low current level, it is unusual for the application voltage levels to be in excess of 1200 V. Single MOSFETs can control voltages up to 1200 V, so that it is not usually necessary to have these devices connected in series. For those cases where MOSFETs are connected in series, the same protection as is afforded thyristors is applied to MOSFETs. Parallel resistors force steady voltage sharing and parallel-connected capacitors force transient voltage sharing more equitably.

6.8.2. Overcurrents

The rated current of a power MOSFET switch is usually the continuous value that maintains the junction temperature at 150°C, if the case temperature is kept at 25°C. Therefore, an overcurrent is a current that causes the junction temperature to exceed 150°C. If the junction temperature does rise above 150°C, the temperature coefficient of resistance associated with $R_{DS(ON)}$ goes from positive to negative in a nonuniform manner. The result is a nonuniform current distribution, so there will be overheating and damage.

Most users do not operate the switch at rated value. If a supply voltage V_s is increased, or if the load impedance is decreased inadvertently, the steady current will increase. The natural response is damage by overheating, or a protection circuit will cause the MOSFET to be turned off. A natural form of protection is to operate the MOSFET below 70% of rated current for design conditions. Then, there is a margin for a change of circuit conditions that might increase the steady current value. In this way one would expect the MOSFET to have a long life.

Gating protection. Thyristor circuits have fast-acting fuses to protect against overcurrents. That is because the thyristor is relatively slow to turn-off, and requires complex commutation circuitry. However, the MOSFET turns off

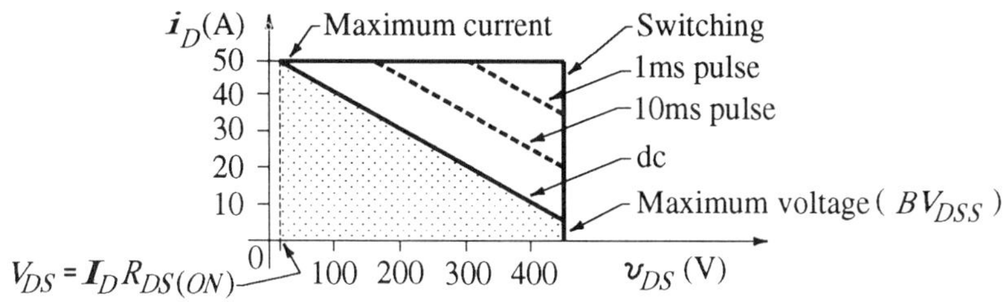

Fig. 6.18 MOSFET safe operating area.

simply and quickly by the removal of the gate voltage V_G. Overcurrent can be prevented by detecting current and using a control circuit to remove the gate voltage and turn off the MOSFET. It is to be noted that any current that is interrupted quickly can give rise to Ldi/dt voltages, so further protection may be necessary.

Safe operating area. In the design of a circuit using MOSFETs, account is taken of the manufacturer's data to match the MOSFET with the application. Data sheets provide information on how much current the MOSFET can tolerate for different operating conditions (that is, for different voltages and time of on-state). The MOSFET does not suffer from the second-breakdown phenomenon, from which BJTs must be protected. Consequently, the current handling capabilities of MOSFETs are directly related to the power dissipation that raises the device's junction temperature to its maximum value (150°C). The information is given in the form of a safe operating area (SOA). An arbitrary example of an SOA is illustrated in Fig. 6.18.

The SOA shows the limits of the values of voltage v_{DS} and current i_D that can occur simultaneously in the MOSFET without the temperature of the junction exceeding 150°C, if the case temperature is maintained at some fixed value, such as 25°C. There is a number of curves that relate to different conditions, but they are all concerned with the same maximum average power that can be dissipated. The rectangular curve is the boundary of switching transition between the cutoff and ohmic regions of operation for switching times limited to a defined maximum value, such as 1μs. As long as the voltage and current are within this rectangle there is no malfunction of the device. At the other extreme, there is the case that the MOSFET is in the active region or ohmic region of operation for a long time. This is the dc case. Values of voltage and current must be within the hatched area for safe operation. Between these extremes there are the cases of pulsed operation. The figure shows boundaries of two arbitrary cases of the conduction of pulses for 10 ms and 1 ms duration.

In order to take advantage of the high currents indicated by the curves in the SOA, the case temperature of the MOSFET package must be maintained below a maximum value that is specified in the data sheet. Heat sinks of good design perform the function of dissipating the heat generated by the losses in the MOSFET. The analysis associated with heat sinks and MOSFETs is similar to that in section 5.10 on the thermal considerations applied to thyristors. The difference between the two is the power level of dissipation. For a thyristor the heatsink may have to dissipate 1000 W or more, whereas for a MOSFET the power level may be only 100 W.

EXAMPLE 6.7

A 250-V, 30-A MOSFET has the following data sheet specifications. The on-state resistance is $R_{DS(ON)} = 0.12\,\Omega$ for the maximum junction temperature of 150°C, the thermal resistance junction-to-case is $R_{\theta JC} = 1$°C/W and the thermal resistance case-to-ambient is $R_{\theta CA} = 29$°C/W without a heatsink. **(a)** Calculate the maximum steady current that can be carried by the MOSFET, if no heatsink is used and the ambient temperature is 25°C. **(b)** Calculate the thermal resistance of a heatsink that allows the MOSFET to operate at rated current value.

Solution

Switching losses are ignored in this example and the on-state current I_D of the MOSFET is assumed to be constant.

(a) The thermal equation for steady state is $\Delta T = P_D R_{\theta JA}$.
That is, $T_J - T_A = I_D^2 R_{DS(ON)} \times (R_{\theta JC} + R_{\theta CA})$.

Thus, $I_D^2 = \dfrac{T_J - T_A}{R_{DS(ON)}(R_{\theta JC} + R_{\theta CA})} = \dfrac{125}{0.12 \times 30} = 34.7$. So, $I_D = 5.9$ A.

Any value of current I_D above this would cause damage if no heatsink were used.

(b) With an adequate heatsink to allow a current $I_D = 30$ A,

$$R_{\theta CA} = \frac{T_J - T_A}{I_D^2 R_{DS(ON)}} - R_{\theta JC} = \frac{150 - 25}{30^2 \times 0.12} - 1 = 0.16\ \text{°C/W}.$$

Overcurrent protection by the use of a heatsink is very important.

Current sensing. Sensing overcurrent by means of a voltage across a resistor in the power circuit can be expensive if it is to be accurate. Some manufacturers have built sensors into the MOSFET. The positive temperature coefficient of resistance of a MOSFET gives a uniform current distribution in the device. If a small fraction of the total number of cells in the MOSFET is isolated for sensing, the current in them is in direct proportion to the bulk current in the rest of the cells. Any change in the conditions of the bulk of the cells, such as temperature rise, is reflected in the cells used for sensing. The sensed current is transformed to a voltage signal, that, at a certain level, can disable the gate drive to turn off the MOSFET.

MOSFETs in parallel. If a MOSFET power circuit demands a load current that is greater than the rating of single available MOSFETs, the devices can be connected in parallel to share the load current. Load sharing is accomplished in such an equitable fashion with MOSFETs that there is no need to add resistors in series with the devices to force current sharing. The reason for this good characteristic is that the MOSFET has a positive temperature coefficient of on-state resistance $R_{DS(ON)}$, and this is so because the device is a majority carrier device.

As the device temperature rises the amplitude of the lattice vibration increases. This causes the majority carriers to have more collisions with the lattice. Since the resistance R_{DS} is proportional to the number of collisions per unit time, the resistance rises with temperature. That is, the temperature coefficient of resistance is positive for the MOSFET.

If one MOSFET of a group in parallel has a lower resistance $R_{DS(ON)}$ it will tend to have a larger current I_D than the others. A larger current means a greater dissipation $I_D^2 R_{DS(ON)}$ that raises the temperature and, therefore, increases the resistance to limit the current. The characteristic is self stabilizing to give an even distribution of current over all the transistors in parallel. Another good characteristic of MOSFETs, that have the same rating, is that the transconductance G of individual devices varies very little. This means that MOSFETs in parallel can share the same gate drive. Transient sharing of current will be good, since the drain currents will rise in proportion to the same gate voltage.

EXAMPLE 6.8

Two 250-V, 20-A MOSFETs have the following data.

$$R_{DS(ON)1} = 0.15 + 0.866 \times 10^{-3} T_1 \, \Omega, \quad R_{DS(ON)2} = 0.21 + 10^{-3} T_2 \, \Omega,$$

where the subscripts 1 and 2 refer to the individual transistors and T is the junction temperature. The two devices are connected in parallel and have a common heatsink that maintains the case temperature at 40°C. Both MOSFETs have a thermal resistance $R_{\theta JC} = 1$°C/W. Find the current sharing by the two MOSFETs for the highest possible load current.

Solution

Consider the case that the MOSFETs are on a long time. Switching losses are not a concern. The data indicate that the on-state resistances have positive temperature coefficients of resistance.

If I is the load current and P_D is the on-state power dissipation of the MOSFETs we can write down the describing equations. The load current is

$$I = I_1 + I_2 . \tag{i}$$

The voltage drop across each MOSFET is the same.

$$I_1 R_{DS(ON)1} = I_2 R_{DS(ON)2} . \tag{ii}$$

The transistors' power dissipation is

$$P_{D1} = I_1^2 R_{DS(ON)1} = (T_1 - T_c) / R_{\theta JC} . \tag{iii}$$

$$P_{D2} = I_2^2 R_{DS(ON)2} = (T_2 - T_c) / R_{\theta JC} . \tag{iv}$$

The temperature dependence of the resistances is repeated

$$R_{DS(ON)1} = 0.15 + 0.866 \times 10^{-3} T_1 \quad \text{(v)}$$

$$R_{DS(ON)2} = 0.21 + 10^{-3} T_2 . \quad \text{(vi)}$$

There are six equations and six unknowns, $I_1, I_2, R_{DS(ON)1}$, $R_{DS(ON)2}$, T_1 and T_2 associated with the MOSFETs. For any load current *I*, solution of these simultaneous equations is best done by a mathematics program that has an equation solver. Manipulation of these equations to a single equation $I_1 = f(I)$ produces a polynomial of the third order and that is laborious to solve by hand.

This problem specifies the value of one variable implicitly. The highest possible load current signifies the maximum temperature of one of the MOSFETs. Current sharing is inversely proportional to resistance, so the lower resistance $R_{DS(ON)1}$ carries the higher current. The power dissipated is I^2R. Since the resistances do not differ greatly, the MOSFET with the higher current will dissipate the most power and will operate at the higher temperature. That is, $T_1 = 150°\text{C}$.
With this information eq. (v) and eq. (iii) give
$R_{DS(ON)1} = 0.28\Omega$, $I_1 = 19.82$ A and $I_1 R_{DS(ON)1} = I_2 R_{DS(ON)2} = 5.55$ V.
Thus the manipulation of eqs (iv) and (vi) gives
$T_2^2 + 170T_2 - 39 \times 10^3 = 0$, or $T_2 = 130.46\,°\text{C}$.
It follows that $R_2 = 0.34\,\Omega$, $I_2 = 16.3$ A and $I = 36.1$ A. Thus, the current sharing is $I_1 : I_2 = 1.2 : 1$. That is, the MOSFET with the lower resistance carries 20% more current.

6.8.3. Transients

Transients are not such a problem with MOSFETs as they are with BJTs or thyristors. The withstand capability of the MOSFET dv/dt is normally well over 3000 V/ μs. Transient-current effects can be absorbed as long as the temperature at the junction does not exceed 150°C in general. However, the gate is sensitive to voltage transients that are greater than 20 V at which value damage can be done.

Voltage. The main power circuit will have some stray inductance that should be minimized. MOSFET switching is so fast that the interruption of current in the stray inductance can give rise to high induced voltages. The MOSFET can be protected from these by means of a snubber capacitor in parallel with the device. Capacitors tend to oppose a change of voltage and will absorb part of the energy delivered by the stray inductance.

Reverse recovery of the integral diode of the MOSFET after the diode has been on and a forward bias then applied to the drain by the main circuit can lead to high dv/dt. A result of this high dv/dt can be that part of the recovery current that flows laterally can provide a large enough base bias to the parasitic BJT to turn on some cells. This leads to the BJT phenomenon of current crowding, and destruction. The partial short between the *p*-body region and the source tends to protect the MOSFET against this occurring.

Any externally applied dv/dt at the drain while the MOSFET is off can cause capacitive currents in the device and that includes the base of the parasitic BJT. If the dv/dt is high enough the BJT turns on. Once on, the device cannot turn off.

The transient model of the MOSFET has capacitors. See Fig. 6.10. The capacitors C_{GD} and C_{GS} act as a voltage divider. If the drain voltage rises too quickly the rise in gate voltage could turn on the transistor inadvertently.

The dv/dt problem is taken care of mainly in the device fabrication process, so that some devices have a dv/dt withstand as high as 5000 V/μs. Other forms of protection can be taken. For example, an externally connected, fast-recovery diode can be added in parallel with the integral diode and a small, parallel snubber circuit can be used to keep dv/dt low. The component values of the snubber are small because the switching times of MOSFETs are short. Another way to control dv/dt during the transistor recovery at turn-off is by the gate-driver resistance R_G. An increased R_G increases the switching times and decreases dv/dt.

EXAMPLE 6.9

A 250-V, 30-A MOSFET modulates power from a dc supply to a load that has virtually constant current. The frequency of switching is 50 kHz. Data for the transistor are as follows. The transconductance is $G = 10$ S and the gate threshold voltage is $V_{GS(TH)} = 3$ V. The transfer capacitance is $C_{rss} = C_{GD} = 200$ pF. In order to limit the dv/dt during the MOSFET turn-off to 3000V/μs find (**a**) the value of the equivalent gate-drive resistance R_G and (**b**) the parameter values of a polarized snubber circuit.

Solution

The load values are not given so the limits, $V_s = 250$ V and $I_l = 30$ A are chosen.

(**a**) Turn-off in the active region demands that full-load current $I_D = I_l$ exists while the drain voltage rises, because the load diode is reverse biased and blocks and the load current is constant. This is depicted in Fig. 6.15. The gate voltage v_{GS} is constant at V'_{GS} in the active region while $i_D = I_D = I_l$. The transfer characteristic applies. That is, $I_D = G\,(V'_{GS} - V_{GS(TH)})$.

So, from the transient model, shown in Fig. 6.10, the rate of rise of drain voltage at turn-off is given by

$$\frac{dv_{DS}}{dt} = \frac{d}{dt}(v_{DG} + v_{GS}) = \frac{dv_{DG}}{dt} = -\frac{i_G}{C_{GD}} = \frac{V'_{GS}}{R_G C_{rss}}.$$

If dv_{DS}/dt is to be limited to 3000V/μs,

$$R_G = \frac{V'_{GS}}{C_{rss}\, dv_{DS}/dt} = \frac{6}{200\times 10^{-12}\times 3000\times 10^{6}} = 10\,\Omega.$$

(**b**) A polarized snubber circuit is shown in Fig. EX5.2 and comprises R_s C_s and D. If this snubber circuit is connected across the MOSFET it comes into play as soon as the voltage v_{DS} starts to rise during turn-off. Assume that the capacitor is initially uncharged. Its $i-v$ characteristic is,

$i = C_s dv_{DS}/dt$, or $C_s = i/(dv_{DS}/dt)$.

The current I_D is transferred from the MOSFET to the capacitor, so
$C_s = I_D/(dv_{DS}/dt) = 30/(3000 \times 10^6) = 0.01 \times 10^{-6}$ F.
This is a low value.
To find the snubber resistance value R_s, assume the chopper has a duty cycle $m = 0.5$. This gives the MOSFET on-time t_{ON} equal to
$t_{ON} = m/f = 0.5/(50 \times 10^3) = 10^{-5}$ s.
If the snubber capacitor is to discharge completely during the interval t_{ON}, then, $5R_sC_s \le t_{ON}$.
So, $R_s \le t_{ON}/(5C_s) = 10^{-5}/(5 \times 10^{-8}) = 200\,\Omega$.
The snubber would comprise a capacitance 0.01 μF and a resistance 200 Ω.

Current. Transient currents are not a great problem with MOSFETs. The wide SOA indicates that large current pulses can be tolerated with the only limit being created by the junction temperature. Therefore, there is usually no need for protection such as the use of series snubbers (inductance).

In section 3.7 the high current pulse due to diode reverse recovery while the power switch was turned on was discussed. See Fig. 3.7. If this is a problem for a MOSFET, there are two solutions other than using a series snubber. One is to use a fast recovery diode in the load. The other is to slow down the MOSFET turn-on by increasing the gate-drive resistance R_G.

Gate transients. Stray inductance in the power circuit can give rise to high voltages during MOSFET switching. There is transient coupling with the gate capacitance that could give rise to oscillations. These oscillations can be damped out by a resistor of a few ohms connected across the gate-to-source terminals. Zener diodes can be used to prevent forward or reverse bias voltage spikes from damaging the gate region. The gate voltage limit is 20 V. Opto-couplers can also protect the low voltage IC gate drive from any voltages induced in the gate region.

6.9. MOSFET RATINGS AND APPLICATIONS

A single MOSFET switch can be used in any application of power or voltage modulation, where the ratings of voltages lie between 50 V and 500 V and where the ratings of currents lie between 2 A and 50 A.

There are tradeoffs made in design. 500 V devices may have a maximum current rating of 20 A and a 50 A device may be limited to 100 V. Thus, the maximum power that can be controlled by a single device is about 10 kW. However, higher powers at 500 V can be considered if MOSFETs are connected in parallel. The majority of power modulation applications with MOSFETs are more likely to be about 2 kW or less (that is voltages less than 200 V and current less than 10 A).

Of all the power semiconductor switches the MOSFET is about the worst insofar as the magnitude of power handling capability is concerned. However, within its power range of capability it has excellent characteristics that bring it close to the best. The switching times are short, the gate driver is simple and requires very little power, protection is not complex and the voltage and current waveforms are relatively *clean*. The frequency of switching is the highest of all power switches and can be of the order of 100 kHz. (If the switching time $t_{on} + t_{off}$ amounts to 500 ns, the switching frequency will be below 1 MHz to have an on-state conduction time of 500 ns.)

The ratings of an arbitrary MOSFET at 25°C are as follows.

Breakdown voltage, $BV_{DSS} = 500$ V.
Maximum continuous current, $I_D = 15$ A.
Maximum on-state voltage, $V_{DS(ON)} = 6$ V.
Maximum gate voltage, $V_G = 20$ V.
Switching time, $t_{on} = 0.5\,\mu$ s.
Switching time, $t_{off} = 1.0\,\mu$ s.
Maximum average power dissipation, $P_D = 120$ W.
On-state resistance, $R_{DS(ON)} = 0.4\Omega$.

The design and analysis of the circuit incorporating a MOSFET can be accomplished with a knowledge of such data.

In general, the on-state resistance $R_{DS(ON)}$ increases with the value of the breakdown voltage BV_{DSS}. A 50-V, 40-A MOSFET may have an on-state resistance $R_{DS(ON)}$ of 0.03Ω at 25°C ($0.06\,\Omega$ at 150°C) while a 1000-V, 5-A MOSFET may have an on-state resistance of $2\,\Omega$ at 25°C (4Ω at 150°C). This indicates a range of on-state voltage drops of between 1.2 V and 20 V. These are important factors to consider when deciding which semiconductor switch to choose for a particular application.

The applications to which MOSFETs are suited are those whose power level is not much more than 2 kW and whose voltage level is less than 300 V. The other limitation is that the MOSFET cannot block reverse voltage, so that it is best suited to be used in dc-dc or dc-ac converters where a blocking facility is not required. If reverse voltage blocking is required and it is necessary to use MOSFETs, then diodes can be added to the circuit. Series MOSFETs in opposition can be used for ac control.

The other general feature of the power MOSFET is that it switches faster than other semiconductor devices. This makes it suited admirably to pulse-width-modulation (PWM) techniques at switching frequencies of 20 kHz and higher. Within the power limits of the MOSFET, this device can do anything the BJT can do, but it does it better.

The high-speed switching of the MOSFET is an advantage in applications such as switch-mode power supplies, induction heating, high-frequency fluorescent lighting, and welding. High frequency of operation means that components are smaller and cheaper. Motor speed control is more accurate, has a faster response and gives better electrical characteristics if choppers and inverters can operate at a high frequency of switching. In general, the low voltage, low power and high frequency make the MOSFET a common device in automobile applications. Further, the ruggedness, low current, low power and high frequency of switching of MOSFETs mean that they are good candidates for applications in portable equipment.

6.10. SUMMARY

The MOSFET is a low power, very fast acting switch that derives its characteristic speed from the insulating region at the gate that makes it a majority carrier device. It is voltage at the gate that alters the conductivity of a channel for current between the main terminals. Continuous application of a gate voltage is necessary for the device to conduct. Removal of the gate voltage is the cause of turn-off. Toggling the gate voltage between 0 and 10 V will create the on/off conditions for switching. A steady value of gate voltage between these two values is not advised.

The *pnp* structure with the insulated gate makes the MOSFET rugged in comparison with other semiconductor switches in terms of reliability. What is good from a circuit designer's point of view is that the device's characteristics are almost linear.

The breakdown voltage of a MOSFET can be as high as 1200 V with a current rating of 5 A and an on-state voltage drop in excess of 10 V. At the other end of the scale, commonly available MOSFETs have current ratings as high as 60 A with a breakdown voltage of 60 V and an on-state voltage drop of about 1.0 V. In view of the on-state voltage drop being high at high withstand voltages, there is a tendency for power MOSFETs to be used in circuits that have supply voltages less than 400 V. Below this breakdown voltage the on-state voltage would be expected to be less than 4 V. Below 100 V ratings the MOSFET has the lowest on-state voltage drop of all switches. Hence, low voltage applications with MOSFETs are attractive. The MOSFET is asymmetric in that it can block forward voltage but cannot block reverse voltage.

Models that are used to describe the MOSFET are relatively simple. As a switch in the steady state the transistor is represented by a resistor $R_{DS(ON)}$ that varies almost linearly with temperature. The value of $R_{DS(ON)}$ roughly doubles between 25°C and 150°C. Manufacturers usually set this latter temperature value as the upper limit. Doubling of the resistance means that at high temperatures the current rating is much reduced. The transient model, that is used for the study of the turn-on and turn-off behaviour, incorporates the internal capacitance for gate responses. In addition, the almost linear transfer characteristic (drain current versus gate voltage) is used to complete the analysis.

The value of the gate voltage determines the degree of turn-on and turn-off. For example, if the gate voltage v_{GS} is less than the threshold value ($V_{GS(TH)} \approx 2$ V), the device is off, whereas, if $v_{DS} \leq v_{GS} - V_{GS(TH)}$, the device is on. Values of gate voltage between these limits means that, as long as v_{GS} is tending to increase or decrease, the transistor is turning on or off. There is no trigger action. The process of switching involves the transition through the active region of operation. In any voltage range the total switching times ($t_{on} + t_{off}$) are about 250 ns for the low end of the current ratings and about 1000 ns for the high end of the current ratings. Gate drives are simple and the gate power is less than 1 watt. The gate threshold voltage, that is about 2 V, has advantages. There is little risk of spurious switching due to noise, and there is no real need for reverse biasing the gate for turn-off.

Protection of the MOSFET is afforded by sensing voltage, current and temperature and preventing damage by using gate turn-off techniques. In the *smart* MOSFETs, there is integration of protection by using a *chip-on-chip* process. There are gate protection, current limit, temperature limit and overvoltage protection all built in. Snubber circuits and heat sinks are also used. These are to improve the ratings of dv/dt and current respectively. Second breakdown does not occur. A positive temperature coefficient of resistance prevents current focussing.

Applications of MOSFETs are predominantly in the area of high frequency switching (greater than 20 kHz), where no other switches can compete. However, the power levels are limited to a few kilowatts. Speed control of small motors, and switched-mode power supplies are important applications.

6.11. PROBLEMS

Section 6.4

6.1 A power MOSFET modulates power from a dc supply, whose voltage is $V_s = 100\,\text{V}$, to a resistive load, whose value is $R = 5\,\Omega$. For a gate voltage $V_{GS} = 10\,\text{V}$, the on-state resistance between drain and source is $R_{DS(ON)} = 0.06\,\Omega$. If the MOSFET is on for a long time, determine **(a)** the power dissipated in the MOSFET due to conduction and **(b)** the efficiency of operation if the supply is considered ideal.

6.2 A power MOSFET is employed as a chopper to modulate power from a dc supply of 50 V to a resistive load. Figure 6.6a depicts the circuit diagram. From the data sheets of the MOSFET, the on-state resistance is $R_{DS(ON)} = 40\,\text{m}\Omega$, the threshold value of the gate voltage is $V_{GS(TH)} = 3\,\text{V}$ and the device transconductance is $G = 10\,\text{S}$. The gate signal is applied to the MOSFET at 200,000 pulses per second. If the MOSFET is capable of dissipating an average power of 100 W, estimate **(a)** the maximum drain current I_D for a chopper duty cycle $m = 0.5$. Determine **(b)** the value of the load resistance to sustain maximum current and **(c)** the value of the gate voltage V_{GS} for this condition. Neglect switching losses.

Section 6.5

6.3 A power-MOSFET chopper modulates power from a dc supply, whose voltage is $V_s = 500\,\text{V}$, to a resistive load, whose value is $100\,\Omega$. The chopper switches at a frequency of 40 kHz. MOSFET data are delay time $t_{d(on)} = 30\,\text{ns}$, rise time $t_{ri} = 90\,\text{ns}$ for turn-on and a drain leakage current $I_{D\,leak} = 1\,\text{mA}$ for the worst condition (125°C). Determine **(a)** the average power dissipation in the MOSFET due to the delay action of turn-on and **(b)** the average power dissipation in the MOSFET due to the drain current rise over the interval t_{ri} of turn-on.

6.4 A power-MOSFET chopper modulates power from a dc supply of 60 V to an RL load with a freewheeling diode. See Fig. 6.12. The current in the load is 20 A and is virtually constant. The chopper is operating in the steady state at a junction temperature of 100°C. The MOSFET data sheet provides the information that the gate threshold voltage is $V_{GS(TH)} = 3$ V, the device capacitances are $C_{iss} = 2000$ pF, $C_{oss} = 600$ pF and $C_{rss} = 200$ pF, and the transconductance is $G = 12$ S. If the gate supply is $V_G = 15$ V and the resistance is $R_G = 50\,\Omega$, estimate **(a)** the turn-on delay time $t_{d(on)}$, **(b)** the turn-on rise time t_{ri} of the current i_D, **(c)** the turn-on fall time t_{fv} of the voltage v_{DS} and **(d)** the total turn-on time t_{on}.

6.5 A power-MOSFET chopper modulates power from a dc supply, whose voltage is $V_s = 500$ V, to an *RL* load with a freewheeling diode. See Fig. 6.12. The current in the load is 5A and is almost constant. The estimated MOSFET crossover time for turn-on is $t_c = 250$ ns. If the chopper frequency is 40kHz, determine the average power dissipation in the MOSFET due to the turn-on process. Ignore the losses incurred during the delay time $t_{d(on)}$.

Section 6.6

6.6 A power-MOSFET chopper modulates power from a dc supply, whose voltage is $V_s = 500$ V, to a resistive load, whose value is $100\,\Omega$. The chopper switches at a frequency of 40 kHz. MOSFET data are delay time $t_{d(off)} = 200$ ns, fall time $t_{fi} = 80$ ns for turn-off, and the on-state resistance $R_{DS(ON)} = 1\,\Omega$. Determine **(a)** the average power dissipation in the MOSFET due to the delay action of turn-off and **(b)** the average power dissipation in the MOSFET due to the drain voltage rise over the interval t_{fi} of turn-off. Compare the results with the turn-on losses from problem 6.3.

6.7 A power-MOSFET chopper modulates power from a dc supply of 60 V to an *RL* load with a freewheeling diode. See Fig. 6.12. The current in the load is 20 A and is virtually constant. The chopper is operating in the steady state at a junction temperature of 100°C. The MOSFET data sheet provides the information that the gate threshold voltage is $V_{GS(TH)} = 3$ V, the device capacitances are $C_{iss} = 2000$ pF, $C_{oss} = 600$ pF, and $C_{rss} = 200$ pF, and the transconductance is $G = 12$ S. If the gate supply is $V_G = 15$ V for the on-state, and the gate resistance for the off-state is $R_G = 50\,\Omega$, estimate **(a)** the turn-off delay time $t_{d(off)}$, **(b)** the turn-off rise time t_{rv} of the voltage v_{DS}, **(c)** the turn-off fall time t_{fi} of the current i_D and **(d)** the total turn-off time t_{off}.

6.8 A power-MOSFET chopper modulates power from a dc supply, whose voltage is $V_s = 500$ V, to an *RL* load with a freewheeling diode. Refer to Fig. 6.12. The current in the load is 5A and is virtually constant. The estimated MOSFET crossover time t_c for turn-off is 200 ns and the on-state resistance is $R_{DS(ON)} = 1\,\Omega$. If the chopper frequency is 40 kHz, determine the average power dissipation in the MOSFET due to the turn-off process **(a)** for the delay time $t_{d(off)}$ of 250 ns and **(b)** for the crossover interval t_c.

6.9 A 600-V, 5-A, power MOSFET is used in a chopper circuit. The data sheet for the MOSFET provides the information that the delay time is $t_{d(on)} = 30$ ns and the crossover time is $t_{c(on)} = 80$ ns for turn-on, the delay time is $t_{d(off)} = 210$ ns and the crossover time is $t_{c(off)} = 85$ ns for turn-off, and the on-state resistance is $R_{DS(ON)} = 1\,\Omega$. If the chopper modulates power from a

500-V dc supply to a load of resistance $R = 100\,\Omega$ that requires an average power of 1.6 kW, determine **(a)** the on-state voltage drop across the MOSFET **(b)** the maximum frequency of switching and **(c)** the average power dissipation in the MOSFET due to the on-state conduction.

6.10 A 450-V, 50-A power MOSFET acts as a chopper to modulate power from a dc supply of 250 V to a resistive load of 8 ohms. The chopper operates at a switching frequency of 40 kHz and a duty cycle $m = 0.6$. From data sheets the on-state resistance $R_{DS(ON)} = 0.14$ ohm for a gate voltage $V_{GS} = 10$ volts. The turn-on and turn-off times are $t_{d(on)} = 0.1\,\mu$s, $t_{ri} = 0.5\,\mu$s, $t_{d(off)} = 0.5\,\mu$s and $t_{fi} = 0.7\,\mu$s for a gate drive voltage $V_G = 10$ V. Ignore leakage current in the device and calculate **(a)** the energy loss during turn-on, **(b)** the energy loss during turn-off, **(c)** the energy loss during steady conduction of the MOSFET and **(d)** the average power dissipation by the switch. Neglect the gate drive losses.

Section 6.7

6.11 A 250-V, 30-A power MOSFET modulates power from a dc supply of voltage $V_s = 200$ V to a load that absorbs an average power of 2.8 kW. MOSFET data are $C_{iss} = 2500$ pF, $C_{oss} = 600$ pF, $C_{rss} = 200$ pF, $G = 9$ S, $V_{GS(TH)} = 3$ V, $R_{DS(ON)} = 0.12\,\Omega$. The load comprises a resistance in series with a large inductance. A freewheeling diode is connected across the load to allow the current to be virtually constant. The gate drive consists of a series resistance $R_G = 100\,\Omega$ and a rectangular-wave voltage supply with $V_G = \pm 10$ V operating at a frequency of 30 kHz and a nominal duty cycle $m = 0.7$ (time ratio of positive voltage to the period). Confirm that the average power dissipated by the MOSFET for steady-state switching conditions is less than the specified maximum of 150 W.

6.12 Consider the gate-circuit diagram in Fig. 6.16b. A 250-V, 30-A power MOSFET is used as a chopper to modulate power from a 140-V, dc supply to a resistive load whose value is $R = 7\Omega$. The MOSFET data are the input capacitance $C_{iss} = 2500$ pF, the transconductance $G = 9$ S, the gate threshold voltage $V_{GS(TH)} = 3$ V the on-state resistance $R_{DS(ON)} = 0.12\Omega$, and the maximum average power dissipation $P_D = 70$ W. Estimate values of the gate drive parameters, that is, the maximum value of the resistance R_G and minimum value of the source current I_G, if the switching frequency is 100 kHz. Let the crossover times of turn-on and turn-off be the same.

6.13 A 250-V, 30-A power MOSFET is used as a 50-kHz chopper to modulate power from a 140-V, dc supply to an inductive load whose current is virtually constant. Consider the equivalent gate-drive circuit to have a supply $V_G = 10$ V in series with a resistor $R_G = 20\Omega$. If the drive response is that shown in Fig. 6.17 calculate **(a)** the minimum and maximum values of the gate input capacitance C_{iss}, **(b)** the average power delivered by the gate-drive supply and **(c)** the rms value I_G of the gate current.

Section 6.8

6.14 A 250-V, 25-A MOSFET is used in a chopper circuit to modulate power from a 140-V, dc supply to a purely resistive load $R = 6.8\,\Omega$. The device data sheet provides the following information. Maximum power dissipation for $T_J = 150°C$ and $T_A = 25°C$ is $P_D = 90\,W$. The on-state resistance is $R_{DS(ON)} = 0.2\,\Omega$ at 150°C. Switching times are $t_{c(on)} = 180\,ns$ and $t_{c(off)} = 170\,ns$. The device thermal resistance is $R_{\theta JC} = 1°C/W$. **(a)** Find the maximum value of the heatsink thermal resistance $R_{\theta CA}$ so that the junction temperature does not exceed 150°C. **(b)** Find the maximum frequency of the chopper operation so that a drain current $I_D = 20\,A$ does not cause device overheating for any resistive load.

6.15 A 250-V, 15-A MOSFET has a thermal resistance $R_{\theta JC} = 1°C/W$ and an on-state resistance given by $R_{DS(ON)} = 1.6 \times 10^{-3}\,T_J + 0.12$ ohms, where T_J is the value of the junction temperature in °C. If a heatsink has a thermal resistance $R_{\theta CA} = 1\,°C/W$ and the ambient temperature $T_A = 30\,°C$, what is the over-current I_D at which a protection circuit should operate?

6.16 Three 250-V, 20-A, MOSFETs are connected directly in parallel to modulate power from a dc supply to a resistive load. The average load current is 40 A if the chopper duty cycle is $m = 0.6$. The MOSFETs have the same thermal resistance $R_{\theta JC} = 1.2\,°C/W$, and a common heatsink maintains all the case temperatures at $T_C = 50°C$. Measured values of on-state resistance for this load condition are $R_{DS(ON)1} = 0.12\,\Omega$, $R_{DS(ON)2} = 0.125\,\Omega$, and $R_{DS(ON)3} = 0.13\,\Omega$. Estimate the junction temperatures of the three MOSFETs. Neglect switching losses.

6.17 Two unmatched 250-V, 30-A MOSFETs are connected in parallel to control load currents greater than 20 A but less than 40 A. Thermal characteristics are given by

$$R_{DS(ON)1} = 0.2 + 0.95 \times 10^{-3} T_1 \text{ and } R_{DS(ON)2} = 0.14 + 0.86 \times 10^{-3} T_2,$$

where T_1 and T_2 are the junction temperatures of the MOSFETs. Both transistors are connected to the same heatsink, both have the same thermal resistances $R_{\theta JC} = 0.9°C/W$ and both have their case temperatures maintained at 35°C. If the maximum junction temperature is 150°C, **(a)** plot the currents and temperatures of the two devices over the full range of load current and **(b)** what are the values of the limiting currents?

6.18 A 500-V, 30-A MOSFET modulates power at a frequency of 30 kHz from a 400-V dc supply to a load whose resistance is 16 ohms. **(a)** Design a polarized *RC* snubber circuit to limit the turn-off dv/dt to $2000\,V/\mu s$ and the turn-on current to 30 A. **(b)** What is the average power dissipated by the snubber circuit?

6.12. BIBLIOGRAPHY

Baliga, B.J. *Modern Power Devices*. New York: John Wiley & Sons, 1987.

Baliga, B.J., and D.Y.Chen. *Power Transistors: Device Design and Applications*. New York: IEEE Press, 1984.

Blecher, A. *Field Effect and Bipolar Power Transistor Physics*. New York: Academic Press, 1981.

Oxner, E.S. *Power FETs and Their Applications*. Englewood Cliffs, N.J: Prentice-Hall, 1982.

Blicher, A. *Field Effect and Bipolar Power Transistor Physics*. New York: Academic Press, 1981.

Pierret, R.F. *Field-effect Devices*. Modular Series on Solid State Devices, Volume 4., Massachusetts: Addison-Wesley, 1983.

Grant, D.A. and J.Gowar. *Power MOSFETs; Theory and Applications*. New York: John Wiley & Sons, 1989.

-----. *MOSPOWER Applications Handbook*. Siliconix Ltd, 1985.

-----. *Power MOSFET Designer's Manual*. 4th ed. International Rectifier Corp., 1987.

CHAPTER 7

THE IGBT

7.1. INTRODUCTION

The semiconductor switch, that has become called the *insulated gate bipolar transistor* (IGBT), has found a niche in applications of ac and dc motor drive control at power levels well in excess of 50 kW. This transistor literally combines the switching characteristics of the MOSFET with the power handling capabilities of the BJT. Designed in the 1980s, the IGBT has the gating properties of a power MOSFET, fast, voltage driven and low power dissipation. It has a small on-state voltage drop (2 to 3 volts) and low conduction losses, like a BJT. Further, in theory, it can block forward and reverse voltages of equal magnitude, like a thyristor. In practice the reverse voltage withstand is low.

There is a number of circuit symbols to represent the IGBT. Figure 7.1 depicts three symbols. Figure 7.1a illustrates the importance of the MOSFET portion of the structure. The difference from the n-channel MOSFET symbol is the added arrow at the drain D contact. This arrow may represent the emitter of the *pnp* bipolar-transistor portion of the IGBT. If the arrows were in the opposite direction, this would indicate that the IGBT was a *p*-channel device and the drain voltage and current would also be opposite. A circuit symbol that indicates the importance of the MOSFET and illustrates the presence of a bipolar transistor in the device, as shown in Fig. 7.1b, uses a collector and an emitter. In the symbol, the base region of the *pnp* transistor is shown unconnected in order to represent the fact that a base lead is not brought out externally. A variation of this is shown in Fig. 7.1c. This last symbol is simple but it can be misleading, because the bipolar transistor is a *pnp* type, with the drain contact an emitter contact (shown as C) and with the source contact a collector contact (shown as E). Accordingly, this circuit symbol is only symbolic and does not indicate the true structure of the IGBT.

Operation of the IGBT can be described with the aid of the circuit diagram in Fig. 7.2. Consider the switch to be off and blocking ($i_D \approx 0$, $v_{DS} \approx V_s$). In order to turn on the IGBT, the drain terminal D must be positively biased with respect to the source terminal S. The turn-on signal is a positive voltage V_G that is applied to the gate terminal G. This voltage, if applied as a step change of magnitude of about 15 V, can cause the switch to turn on in less than 1 μs, after which the drain current i_D is equal to the load current I_l (assumed constant). Once on, the device has to be maintained on by the voltage signal at the gate. However, by virtue of the voltage control, the gate power dissipation is very low.

The IGBT is turned off simply by removing the voltage signal V_G from the gate terminal. Transition from the conduction state to the blocking state can take as little as 2 μs, so the switching frequency can be in the range of 50 kHz.

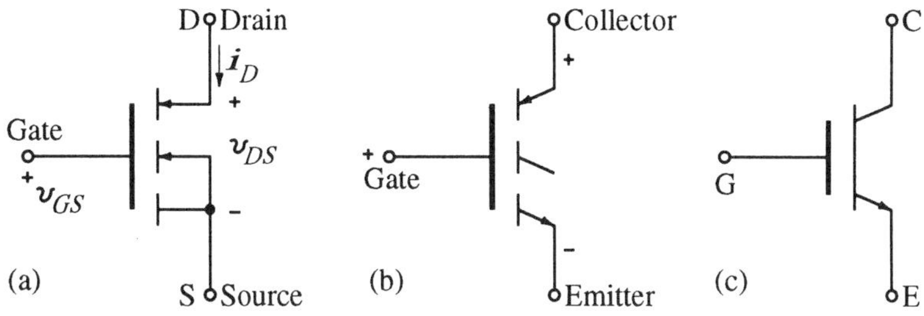

Fig. 7.1 IGBT circuit symbols.

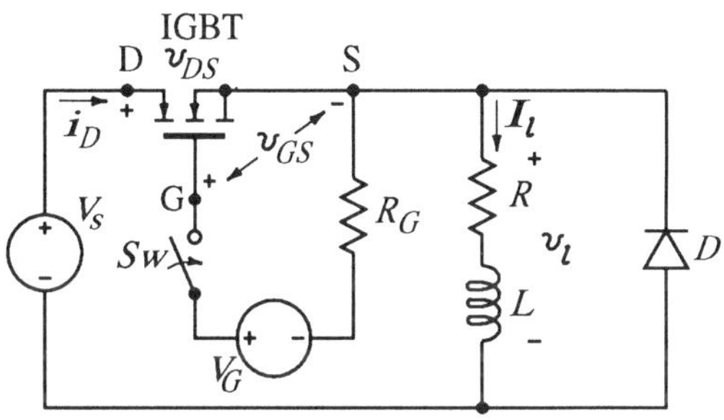

Fig. 7.2 IGBT biasing for controlled switching.

Manufacturing techniques allow the IGBT to have voltage ratings greater than 1500 V and to have current ratings of several hundred amperes. As a fully-controlled on/off switch with forward blocking capabilities and with unidirectional current, the IGBT is an important member of the silicon power-switch family.

7.2. IGBT STRUCTURE

The structure of the insulated gate bipolar transistor is similar to an n-channel MOSFET. There is one major difference and that is the use of a highly doped p^+-type substrate.

Figure 7.3 shows a simplified sketch of the IGBT without the depiction of the normal interdigitation of the source and gate layout. As can be seen by inspection, a portion of the structure is the combination of n^+, p and n^- regions that form the MOSFET between the source S and the gate G with the n^- drift region being the drain D′ of the MOSFET. Another portion is the combination of three layers p^+n^-p, that create a bipolar junction transistor between the drain D and

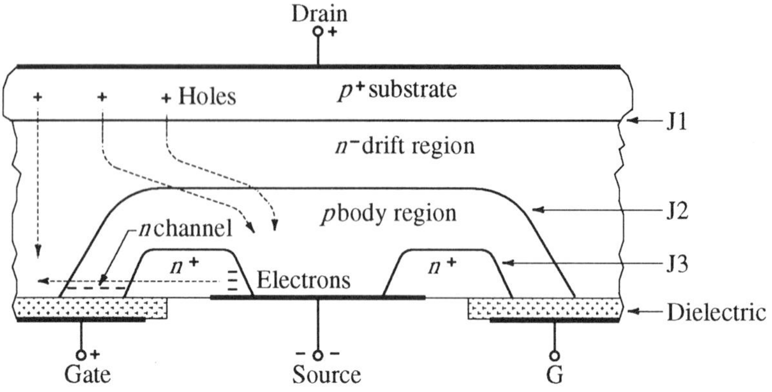

Fig. 7.3 IGBT structure.

source S terminals. The p region acts as a collector C, the n^- region acts as the base B and the p^+ region acts as the emitter E of a *pnp* transistor. Yet another device can be recognized. Between the drain and the source there are the four layers $p^+n^-pn^+$ that form a thyristor. This thyristor is parasitic and its effect is minimized by the designer of the IGBT.

The low-doped n^- drift region of the IGBT is made to be wide. This allows the possibility that the junction J1 can block high reverse-bias voltages across the drain to source terminals. In a like manner, if the gate voltage v_{GS} is below a threshold value (about 4 V), junction J2 will block high forward bias voltages across the drain to source terminals.

The MOSFET at the gate of the IGBT allows the device to be turned on and off. If the drain of the IGBT is positively biased with respect to the source and if the gate voltage V_{GS} is well above the threshold value the IGBT turns on and has a low voltage conduction drop (about 2 V) between drain and source as long as the gate voltage is maintained. Removal of the gate voltage turns off the MOSFET and, in turn, turns off the IGBT.

The turn-on process is explained with the aid of Fig. 7.3. With the gate positive with respect to the source, the electric field across the dielectric at the gate causes negative charge to accumulate in the p body region (just like a capacitor), close to the gate. This is called inversion, and is the creation of an n channel that short-circuits the n^+ and n^- regions near the dielectric. With the n^- drift region positively biased due to the drain bias, MOSFET current is conducted between the drain and source via the short-circuit path. What happens is that electron current (from the source contact) is conducted in the n^+ and n^- regions via the n-channel of the p-body region. This causes holes to be injected from the p^+ substrate into the n^- drift region, if the drain is positively biased, and these holes recombine with the electrons in the drift region. This resultant MOSFET current is a substantial part of the IGBT current. The other component of current, also shown in

Fig. 7.3, is created by the flood of holes in the n^- region drifting and diffusing into the p-body region at junction J2. Electrons from the source contact recombine with these holes. So, the p^+ substrate can be considered to be the emitter, the n^- region can be considered to be the base, and the p-body region can be considered to be the collector of holes in a pn^-p^+ bipolar transistor. Turning on the MOSFET creates a base current that turns on the BJT. The flooding of the n^- drift region with minority carriers from the p^+ substrate during the conduction state causes a much greater reduction of resistance than in the power MOSFET. This is the reason why current densities in the IGBT can be up to 20 times greater than the MOSFET.

The area of junction J1 is large compared with the MOSFET n channel and the conductivity of the n^- drift region is modulated by the gate voltage. A high gate voltage (about 15 V) produces a high conductivity because of the increase in minority carriers. A low gate voltage decreases the conductivity. For these reasons, with positive bias at the drain and gate, the voltage drop across the IGBT between drain and source can be low (about 2 V).

The turn-off process of the IGBT is initiated by reducing its gate voltage below the threshold value. The consequence is the removal of the conducting n-channel in the p-body region. This turns off the MOSFET and curtails the source of base current to the bipolar transistor, so the complete device turns off. The advantage of flooding the n^- region with minority carriers for good on-state characteristics (high current, low voltage drop) becomes a disadvantage at turn-off. The minority carriers cannot be swept out by a bias at the base because the base is unconnected externally. Thus the turn-off time is prolonged and is much greater than for MOSFETs. Minority-carrier lifetime is reduced by adding an n^+ buffer between the n^- drift region and the p^+ substrate. However, this method of speeding up recombination also increases the on-state voltage drop. Design is a compromise. A shorter turn-off time means less losses and a higher frequency of switching but the on-state voltage is higher. For example, an IGBT for operation at 50kHz may have an on-state voltage drop $V_{DS} = 3$ V, whereas an IGBT with an on-state voltage drop of $V_{DS} = 1.8$ V may have a maximum operating frequency of 1kHz.

The *npn* BJT is cheaper to manufacture than the *pnp* type, so it might be questioned why the *npn* bipolar device is not integrated with a p-channel MOSFET onto a single chip. The answer is that the p-channel switch has a much higher resistance than the n-channel type and low on-state voltage drop is a design priority to compete with BJTs.

7.3. IGBT I-V CHARACTERISTICS

Refer to Appendix 5 for data. Figure 7.4a shows a circuit diagram that includes an IGBT switch for the modulation of power from the source of voltage V_s to the load R. The IGBT gating circuit is shown in the simplified form of an ideal voltage V_G in series with a resistance R_G and a generic switch *Sw*.

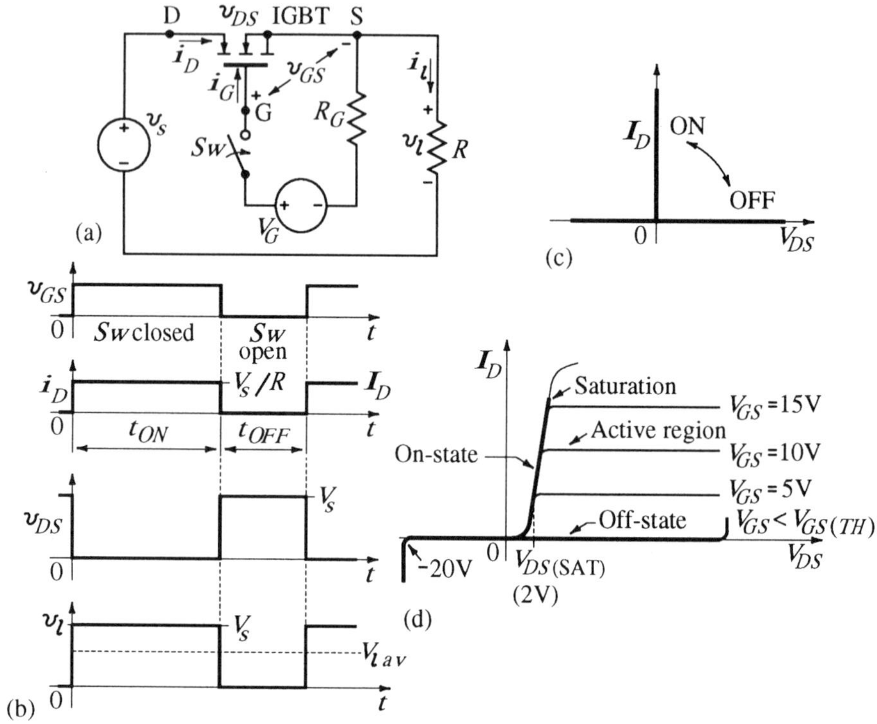

Fig. 7.4 IGBT steady-state I-V characteristics. (a) Circuit, (b) ideal waveforms, (c) ideal characteristics, (d) realistic characteristics.

Of importance are the I-V steady-state characteristics. These are the relations between the drain current I_D, that is also the load current, and the voltage V_{DS} across the IGBT, drain to source. Ideal waveforms of the switching characteristics are sketched in Fig. 7.4b.

Consider the IGBT to be blocking initially. This means that there is no voltage applied to its gate. If a voltage V_{GS} is applied at the gate G by closing the switch *Sw*, the IGBT turns on immediately, current I_D is conducted and the voltage V_{DS} across it drops from the blocking value V_s to zero. The value of the current I_D is determined only by the source voltage V_s and the load resistance R. The current persists for the time t_{ON} that the gate signal is applied. The voltage across the load becomes $v_l = V_s$ during this interval. As soon as the gate voltage V_{GS} is reduced to zero by opening the switch *Sw*, the current I_D is interrupted and the voltage V_{DS} rises to V_s as the IGBT goes into the blocking state. The off-state will persist for a time t_{OFF} until a gate signal is applied again. During the off-state interval the voltage v_l across the load is zero. The power to the load is modulated by adjusting the on-state interval t_{ON} of the IGBT, and the average voltage $V_{l\,av}$ across the load is expressed by

$$V_{l\,av} = V_s \frac{t_{ON}}{t_{ON} + t_{OFF}} = V_s t_{ON} f = mV_s \tag{7.3.1}$$

where f is the switching frequency and m is the duty cycle that is defined by $t_{ON}/(t_{ON} + t_{OFF})$.

An ideal IGBT will block any voltage $\pm V_s$ if there is no gate voltage. Also, if a gate voltage is applied, the device switches to the on-state with no voltage across it and carrying a current $I_D = V_s/R$. For any source voltage V_s and load resistance R, Fig. 7.4c illustrates the ideal I-V characteristics.

No switch is ideal. An IGBT requires a threshold value $V_{GS(TH)}$ for the state to change from on to off or vice versa. This is usually about 4 V. Above this value the voltage V_{DS} drops to a low value of about 2 V. In order that the on-state voltage drop remains low, the gate is driven hard, up to about 15 V. The current I_D is, to a certain extent, self-limiting, as shown in the characteristic of Fig. 7.4d. This is due to the conductance modulation of the n^- drift layer that is controlled by V_{GS}.

With no gate signal, there is a limit to the blocking voltage capability of the IGBT. There is a dependency on the junctions Jl, J2 and the thickness of the lightly doped n^- drift region. Forward-voltage breakdown has a positive temperature coefficient of about 0.7 V/°C. For a particular device that has a specified minimum voltage breakdown of 600 V at 25°C, it would be expected that the breakdown voltages would be 544 V and 687 V at the two operating limits of temperature -55°C and 150°C respectively. This specification can apply to both forward and reverse bias voltages, but the data sheets must be checked. Most manufacturers have a reverse-connected diode built into the package or some have a modified design such that there is avalanche breakdown in the reverse direction at a voltage V_{DS} that is as little as -15 V.

Once the IGBT is in the on-state the voltage drop $V_{DS(\mathrm{SAT})}$ does not vary significantly with temperature or current for a constant drive voltage V_{GS} (about 15 V). However, a 100 V device may have a voltage drop of 1.8 V and a 1000 V device may have a voltage drop of 3 V for the same current density (about 1.7A/mm^2). Typical variations of the voltage drop for the 100 V device for constant current are 1.8 V to 2.4 V over a temperature range from -55°C to 150°C, and for constant temperature the variation can be 1.8 V to 2.2 V for a current range of half full load to full load. The manufacturers' data sheets must be used if preciseness is required.

Although ratings of IGBTs in the neighbourhood of 1400 V and 300 A are available, ratings of 600 V and 50 A are common and inexpensive compared with BJTs.

For the sake of comparison the IGBT cannot block voltages as high as the thyristor, but the transitor has a lower on-state voltage (and lower conduction losses) than the power MOSFET for blocking voltages greater than 200 V.

EXAMPLE 7.1

Power is to be modulated by a single IGBT operating as a chopper. The supply is a dc source of 480 V and the load is a resistor of value 4.8 Ω. The circuit diagram is shown in Fig. 7.4a. If the switching frequency of the chopper is 1kHz and if the average power to be absorbed by the load is 30 kW, determine the IGBT on-time t_{ON} for each period of switching. What are suitable ratings for the IGBT?

Solution

In conjunction with Fig. 7.4a and Fig. 7.4b the average power P absorbed by the load is seen to be

$$P = \frac{1}{T}\int_0^{t_{ON}} \frac{v_l^2}{R}dt = \frac{V_{l\,rms}^2}{R} = I_{Drms}^2 R\ . \quad \text{That is, } P = \frac{V_s^2}{R}\frac{t_{ON}}{T}.$$

Since $T = 1/f = 1$ ms, $t_{ON} = PRT\,/\,V_s^2 = 30\times10^3 \times 4.8\times10^{-3}\,/\,480^2 = 0.625$ ms.
A rule of thumb is that the dc source voltage should be no more than 80% of the voltage rating of the IGBT. This indicates that the IGBT voltage rating must not be less than 600 V. The continuous current of the IGBT in this example would be $V_s\,/\,R$ = 480/4.8 = 100 A, so the current rating of the IGBT for a specified junction temperature would not be less than 100 A. The value of a current pulse would be determined by the transient thermal impedance and the maximum junction temperature (150°C). Some manufacturers specify a current pulse 10 times the continuous current rating without triggering any parasitic effects. That is, the IGBT can be turned off by gate control even at very high load-current pulses.

7.4. IGBT MODEL

The structure of the IGBT is illustrated in Fig. 7.3. It has been described in two parts. The first part involved an n-channel n^-pn^+ MOSFET that initiates the IGBT turn-on. The n^- drift region is the MOSFET drain D′, the p body region provides the gate G and the highly doped n^+ region is the source S. A MOSFET circuit symbol can be used for this portion of an equivalent circuit and is depicted in Fig. 7.5. The MOSFET conducts most of the load current.

The second part of the structure involves the p^+ substrate, the n^- drift region and the p body region. This operates as a pnp bipolar transistor. Its base is the n^- drift region that is internally connected to the drain D′ of the MOSFET. There is no external connection to the base, so that turn-off cannot be accelerated. The emitter is the p^+ substrate and the p-body region collects the holes that cross the base region from the p^+ region. This bipolar transistor is shown with its letter designations B, E and C in Fig. 7.5 together with its regions n^-, p^+ and p. The magnitude of the gate voltage v_{GS} adjusts (or modulates) the resistance (or conductance) of the n^- drift region. This is represented in the equivalent circuit by the resistor R_B. The emitter E has an external connection and the collector C has a common connection with the MOSFET source S as shown. The BJT provides

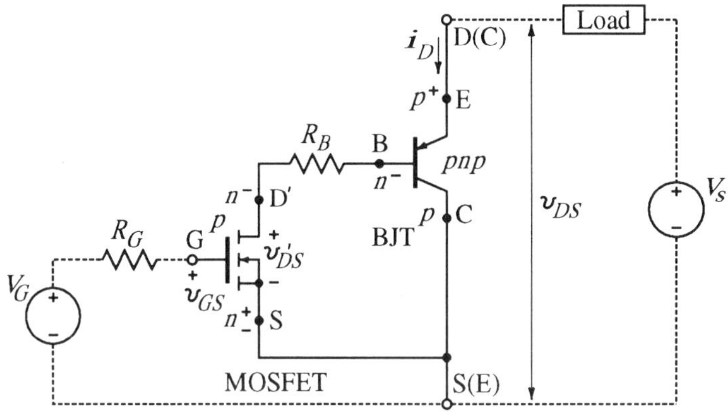

Fig. 7.5 IGBT equivalent circuit model.

further conduction and a low voltage drop.

Since the overall structure of the IGBT is like a MOSFET with an added p^+ substrate at the drain, the external connections are designated D for drain and S for source. However, there is an alternative designation with D replaced by (C) and S replaced by (E), since the equivalent circuit represents a MOSFET-driven BJT. This may tend to be misleading because the collector (C) is the emitter of the transistor and the emitter (E) is really the collector. If the structure is not considered, then the designations G for gate, C for the positive contact and E for the negative connection make the IGBT circuit and symbol more distinctive. See Fig. 7.1c.

An inspection of the structure in Fig. 7.3 and the equivalent circuit diagram in Fig. 7.5 provides a simple explanation of the operation of an IGBT. Without a positive voltage at the gate G the MOSFET has a high resistance in the off-state. So, there can be no base current, and the BJT is also in the off-state. The IGBT blocks the supply voltage V_s and there is no current in the load.

If a gate voltage v_{GS} is applied to MOSFET with G positive with respect to the source S an n channel is formed between the n^+ region of the source and the n^- drift region of the drain D′. This forms a short circuit between D′ and S so the MOSFET is in the on-state. The magnitude of the gate voltage v_{GS} has to be above a threshold value (about 4 V) to turn on the MOSFET. If a voltage $v_{DS} = V_s$ is applied to the contact terminals such that D is positive with respect to S, then the MOSFET drain D′ is positively biased with respect to its source S and there is a path for holes to flow from the p^+ substrate to the n^- drift region. This hole flow constitutes a MOSFET current and a base current injection in the bipolar transistor. Base current causes the BJT to turn on and allow holes to flow from the p^+ region to the p region. That is, there is conduction in the region between the BJT emitter and collector. This is part of the load current i_D .

How much current there is in the load depends on how much current there is in the base (n^-) region. How much base current there is depends on the resistance R_B of the base region (modulation conductance). If the resistance R_B is high, the base current is low and the on-state voltage v_{DS} is high. This limits the current i_D in the load, as shown in the characteristic curves of Fig. 7.4d. A high value of gate voltage v_{GS} (about 15 V) will result in a low value of R_B, a high value of base current and a near saturated BJT to give a low voltage $v_{DS(\text{SAT})}$ (about 2 V) and a current i_D that is only limited by the load. This latter is the required switching characteristic. It is shown by the heavy line in Fig. 7.4d.

Once the IGBT is on, it remains conducting until the gate signal v_{GS} drops below the threshold value, or, as is usual, it is reduced to zero. If the gate voltage is removed, the n channel cannot be sustained so the MOSFET takes on a high impedance mode and turns off. If the MOSFET is off, there is no base current for the BJT and it turns off. With both transistors off the IGBT resumes the off-state. Consequently, there are two required extreme states for IGBT switching. The gate is driven hard, but not too hard, to achieve the on-state with a low voltage drop, and the gate signal is removed to achieve the off-state with a full voltage V_s blocking capability.

7.5. IGBT TURN-ON

In this section we are concerned with the definition of IGBT turn-on and the action of turn-on. Turn-on occurs with the application of a gate signal.

Since the equivalent gate circuit of the IGBT is a MOSFET, then turn-on control is by voltage rather than current, although gate current magnitude affects the speed of turn-on. Voltage control makes the IGBT faster to turn on than the BJT but it is still slower than the power MOSFET. The gate power to enable switching is very small compared with current controlled switches; for the IGBT, the steady gate current is of the order of nanoamperes compared with milliamperes and higher for BJTs and thyristors. Low gating power implies the advantage that computer controlled signals can be applied to integrated driver circuits that include power supplies. Such ICs are on the market.

A term that is used to measure the degree of turn-on of the IGBT in the active region of the I-V characteristics is the transconductance G which is associated with the main terminal current i_D and the gate voltage v_{GS}. This is similar to a measure of the saturation level of BJTs, using the gain h_{fe} of main current to base current and is the same as the transconductance G of the MOSFET. Other terms such as g_{fe}, g_{FE} and g_{FS} are used for transconductance. Once the gate voltage of the IGBT has been driven above the threshold value, the transconductance G is such that the transfer characteristic I_D vs V_{GS} is almost linear. See Fig. 6.7. The IGBT has its current limited by thermal considerations rather than gain. At 40 amperes, 600 volts, a 70-A IGBT has a typical value of 30 siemens for G.

The gate circuits for IGBT turn-on are the same as those for the MOSFET, so reference can be made to section 6.7.

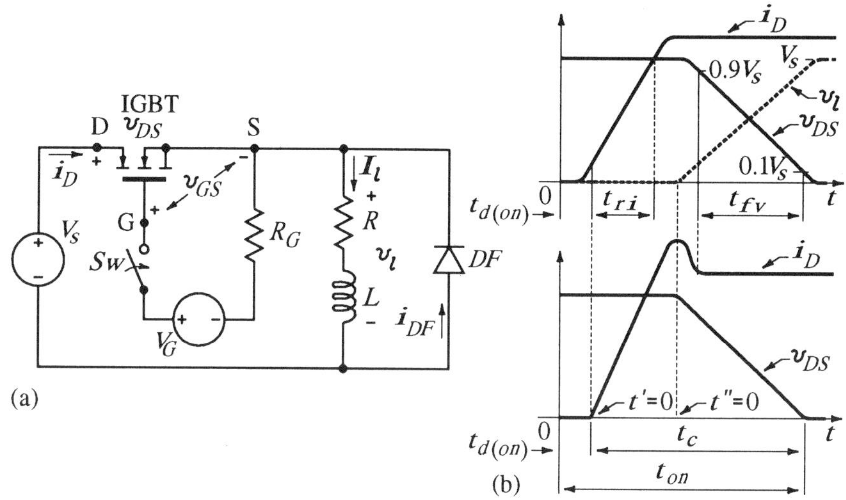

Fig. 7.6 IGBT turn-on with a resistive load.
(a) Circuit diagram, (b) waveforms.

7.5.1. Turn-on Losses

IGBT turn-on times t_{on} are not well characterized. Data sheets provide information about delay times t_d and current rise-times t_{ri} but there is no information about voltage fall-times. This creates a problem if turn-on energy losses are estimated by calculation, because a knowledge of the voltage fall-time and its position relative to current rise is critical for an accurate determination. The type of load influences the fall-time of the device voltage in relation to the current rise.

IGBT circuit with resistive load. Figure 7.6 shows patterns of current and voltage as a function of time during turn-on for the best case (lowest losses). The circuit diagram shows a load which is purely resistive. At time $t=0$ a source of voltage V_G is applied to the gate circuit. The voltage v_{GS} rises according to the values of the gate capacitance and the series resistance R_G. When the gate voltage rises to the threshold value $V_{GS(TH)}$, conduction of the drain current i_D begins. The drain current continues to rise rapidly according to the transfer characteristics of i_D versus v_{GS} that are given by G until the current reaches the steady load value $i_D = I_l \approx V_s / R$. By definition, there is a delay time $t_{d(on)}$ that is the interval from $0.1\,V_G$ to $0.1\,I_l$ and there is a rise-time t_{ri} that is the interval from $0.1\,I_l$ to $0.9\,I_l$.

For an analyst, it is easier if the delay time $t_{d(on)}$ is defined as the interval from the application of a gate drive voltage V_G to the time that the gate voltage rises to the threshold value $v_{GS} = V_{GS(TH)}$. Following this, the current rise-time t_{ri} could then be defined as the interval from the time that the gate voltage $v_{GS} = V_{GS(TH)}$ to the time that $v_{GS} = V'_{GS}$, a value that just sustains the steady on-state current

$I_D = I_l$, where $V'_{GS} = (I_D/G) + V_{GS(TH)}$. This would be in keeping with the MOSFET definitions in section 6.6.

During the time that the drain current rises, the voltage v_{DS} falls. The interval from $0.9\,V_s$ to $0.1\,V_s$ is shown as the fall time t_{fv}. The fall time of the voltage v_{DS} is coincident with the rise time for a resistive load. This is to be expected for a resistive load, since $v_{DS} = V_s - i_D R$.

The energy loss W_{on} in the IGBT during the turn-on process is derived from

$$W_{on} = \int v_{DS}\, i_D\, dt \tag{7.5.1}$$

over the turn-on interval. Turn-on is initiated at $t = 0$. There follows a delay time $t_{d(on)}$, during which there is negligible dissipation. At $t = t_{d(on)}$ the gate voltage has reached a value $v_{GS} = V_{GS(TH)}$. This is the initiation of the drain current rise and the drain voltage fall, where $t_{ri} = t_{fv} = t_c$. Let $t' = t - t_{d(on)}$. We can describe the changes by using the waveforms in Fig. 7.6b.

$$i_D = \frac{I_D}{t_c}t' = \frac{I_l}{t_c}t' \quad \text{and} \quad v_{DS} = -\frac{V_s}{t_c}t' + V_s. \tag{7.5.2}$$

The dissipation is

$$W_{on} = \int_0^{t_c} v_{DS}\, i_D\, dt' = \int_0^{t_c} \left(-\frac{V_s}{t_c}t' + V_s\right)\left(\frac{I_l}{t_c}t'\right) dt' = \frac{V_s I_l}{6} t_c. \tag{7.5.3}$$

The average power dissipated P_{on} due to the turn-on process, if the IGBT is switched on at a frequency f, is

$$P_{on} = W_{on} \times f. \tag{7.5.4}$$

EXAMPLE 7.2

An IGBT chopper modulates power from a 400-V, dc supply to a resistive load, whose value is $20\,\Omega$. The chopper switches at a frequency of 1kHz. IGBT data for a steady junction temperature of 100°C are that the steady on-state voltage is $V_{DS(\mathrm{SAT})} = 1.9\,\mathrm{V}$, the delay time is $t_{d(on)} = 25\,\mathrm{ns}$ and the crossover time $t_c = 35\,\mathrm{ns}$ for turn-on. If the chopper has a maximum duty cycle that is $m = 0.8$, determine **(a)** the IGBT current rating, **(b)** the average power loss due to conduction and **(c)** the average power dissipated due to the turn-on process.

Solution

(a) Steady on-state current $I_l = \dfrac{V_s - V_{DS(\mathrm{SAT})}}{R} = \dfrac{200 - 1.9}{20} = 9.9\,\mathrm{A}$.

$m = 0.8$. So, $I_{D\,av} = I_{l\,av} = 0.8 I_l = 0.8 \times 9.9 = 7.92\,\mathrm{A}$.

(b) Conduction loss P_C is

$$P_C = f \int_0^{t_{ON}} v_{DS(\mathrm{SAT})} I_D\, dt = V_{DS(\mathrm{SAT})} m I_D = 1.9 \times 0.8 \times 9.9 = 15\,\mathrm{W}.$$

(c) From eq. (7.5.3) and eq. (7.5.4) the average power loss P_{on} is

$$P_{on} = W_{on} \times f = \frac{V_s I_l}{6} t_c \times f = \frac{400 \times 9.9}{6} \times 35 \times 10^{-9} \times 10^3 = 23.1 \times 10^{-3}\ \text{W}.$$

At this low frequency the turn-on loss is small compared with the conduction loss.

IGBT circuit with inductive load. Figure 7.7b illustrates the turn-on response if the power-circuit load is inductive with a time constant long in comparison with the switching time of the IGBT. In this case the load current I_l is considered to be unvarying. While the IGBT is off, the freewheeling diode DF provides the path for the load current, and, since the diode is conducting, the voltage across the load terminals is close to zero. Application of a gating signal V_G initiates the IGBT turn-on process. After the delay time $t_{d(on)}$, needed for v_{GS} to reach $V_{GS(TH)}$, the current i_D rises, the current i_{DF} in the diode falls, but it is not until the current $i_D = I_l$ and $i_{DF} = 0$ that the ideal diode recovers its blocking state. At that point the voltage v_l across the load rises and the voltage v_{DS} across the IGBT falls, as shown in the figure. If the rise time t_{ri} and the fall time t_{fv} are known, an approximate calculation of the switching loss can be made.

In real life, ideal diodes are not available. The recovery of a diode from the on-state to the blocking state takes a finite time. So, for a short time after the current i_{DF} drops to zero while the IGBT turns on, the diode is on with a positive voltage applied to the cathode terminal of the diode. This allows conduction of reverse current until the diode recovers its blocking capability and this is reflected as a spike of current in the IGBT as shown in Fig. 7.7b. The extra loss in the switch due to the current spike cannot be calculated, nor does a value appear in data sheets. Data sheets give typical delay times $t_{d(on)}$, rise time t_{ri} and total switching energy losses without the effects of a freewheeling diode. The turn-on energy loss is usually a measure of the dissipation in the IGBT during the interval from $i_D = 0.05\, I_l$ to $v_{DS} = 0.05\, V_s$. This interval is called the crossover time t_c.

Typical parameter values of turn-on for an inductive load with a freewheeling diode are as follows.

(i) 600 V, 70 A, $f = 1$ kHz, $V_G = 15$ V; $t_d = 25$ ns, $t_{ri} = 60$ ns, turn-on loss $W_{on} = 0.35$ mJ.
(ii) 600 V, 50 A, $f = 10$ kHz, $V_G = 15$ V; $t_d = 25$ ns, $t_{ri} = 30$ ns turn-on loss $W_{on} = 0.12$ mJ.

The turn-on times are longer if the gate-circuit resistance R_G is increased because the gate voltage v_{GS} rises more slowly. Typical values of R_G are from 2 to 50 Ω, for which the turn-on switching losses could have values from 0.35 mJ to 0.4 mJ respectively. Every device is different, so these values are stated to give some idea of the order of magnitudes. Another similar device (600 V, 60 A, 10 kHz, $V_{GS} = 15$ V) has the turn-on characteristics $t_{d(on)} = 25$ ns, $t_{ri} = 50$ ns, $W_{on} = 1.5$ mJ, conduction loss 0.03 mJ/A, average power loss $P_D = 30$ W and peak power loss 150 W.

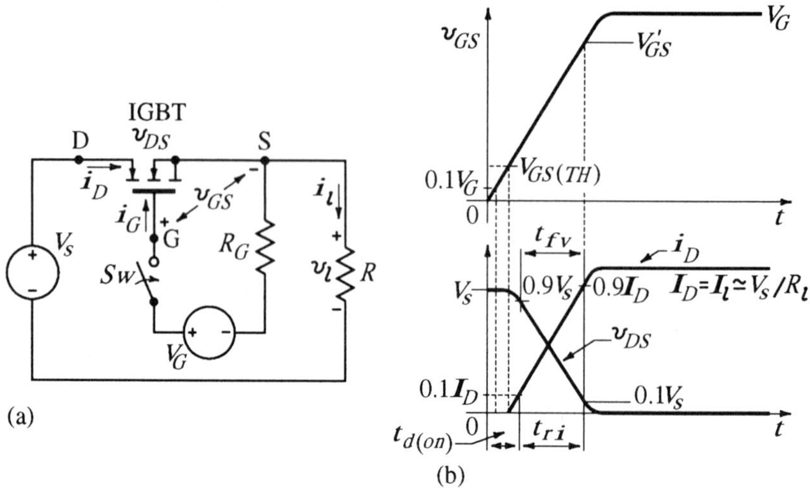

Fig. 7.7 IGBT turn-on for an inductive load.
(a) Circuit diagram, (b) waveforms.

The fall time t_{fv} of the device voltage v_{DS} is influenced by the gate series resistance and the gate-to-drain capacitance. The rate of change of voltage is given by

$$\frac{dv_{DS}}{dt} = \frac{d}{dt}(v_{DG} + v_{GS}). \tag{7.5.5}$$

This is similar to the MOSFET, eq. (6.5.11). However, in an inductive load the voltage v_{GS} is almost constant (V'_{GS}) by the time v_{DS} starts to decrease from the value V_s. So

$$\frac{dv_{DS}}{dt} \approx \frac{dv_{DG}}{dt} = \frac{i_G}{C_{GD}} = \frac{V_G - V'_{GS}}{R_G}\frac{1}{C_{GD}}. \tag{7.5.6}$$

An inspection of this equation shows that if R_G is doubled the voltage fall-time for v_{DS} is also doubled. If, at the time that the current i_D reached its final value I_l, the voltage v_{DS} began to fall from $V_s = 500$ V, and, if $R_G = 20\,\Omega$ and $C_{GD} = 10$ pF, then the fall-time t_f for v_{DS} is approximately 40 ns. This is comparable with the current i_D rise-time during turn-on.

A final note regarding turn-on is that a high dv/dt applied between the drain and source terminals can raise the gate terminal voltage v_{GS} above the threshold value. This would initiate turn-on, but drain-current conduction would clamp the dv/dt, because of the transfer characteristic. It is possible to counteract dv/dt turn-on by means of a negative gate voltage while the IGBT is off. This increased circuit complexity adds to the cost.

An estimation of the energy dissipation W_{on} during the crossover time t_c of turn-on is obtained using the voltage v_{DS} and current i_D waveforms in Fig. 7.7b. The current i_D is linearized and the effect of diode reverse recovery is ignored. Let time $t' = t - t_{d(on)}$ and $t'' = t - (t_{d(on)} + t_{ri})$.

$$W_{on} = \int_0^{t'} p\, dt' = \int_0^{t'} v_{DS}\, i_D\, dt' = \int_0^{t_{ri}} V_s\, i_D\, dt' + \int_0^{t_{fv}} v_{DS}\, I_D\, dt''. \tag{7.5.7}$$

$$W_{on} = V_s \int_0^{t_{ri}} \frac{I_D}{t_{ri}} t'\, dt' + \int_0^{t_{fv}} \left(-\frac{V_s}{t_{fv}} t'' + V_s \right) dt'' = \frac{V_s I_D}{2}(t_{ri} + t_{fv}) = \frac{V_s I_D}{2} t_c. \tag{7.5.8}$$

The energy loss over the delay time of turn-on can be neglected.

EXAMPLE 7.3

An IGBT chopper modulates power from a 400-V dc supply to an RL load with a freewheeling diode connected across it. The chopper switches at a frequency of 1 kHz at a duty cycle $m = 0.8$ such that load current is 25 A and virtually constant. IGBT data for steady operation with the junction temperature at 100°C are that the delay time is $t_{d(on)} = 25$ ns, the drain current rise time is $t_{ri} = 35$ ns and the drain voltage fall time $t_{fv} = 40$ ns for turn-on. Determine the average power dissipation in the IGBT due to the turn-on process.

Solution

Neglect the losses incurred during the delay interval. From the linearized waveforms in Fig. 7.7b the energy loss during turn-on is expressed as

$$W_{on} = \int_0^{t_{on}} p\,dt = \int_0^{t_{on}} v_{DS}\, i_D\, dt = \frac{V_s I_D}{2}(t_{ri} + t_{fv}).$$

$V_s = 400$, $I_D = I_l = 25$ A. Hence, $W_{on} = \dfrac{400 \times 25}{2}(35 + 40) \times 10^{-9} = 375 \times 10^{-6}$ J.

The average power P_{on} dissipated by the turn-on process being repeated 1000 times per second is

$P_{on} = W_{on} \times f = 375 \times 10^{-6} \times 10^3 = 375 \times 10^{-3}$ W.

This power loss is 16 times greater than the case for a purely resistive load.

7.6. IGBT TURN-OFF

In general, removal of the supply voltage V_G from the gate circuit causes the IGBT to turn-off. From Fig. 7.5, if the gate voltage V_{GS} is reduced to zero the MOSFET turns off, so that there is no base current for the bipolar transistor. A bipolar transistor without a base current reverts to the blocking off-state and the main current falls to zero. At this point the IGBT is off until the gate signal is reapplied.

The turn-off procedure takes a finite time that cannot be completely characterized because of the dependency on the load. This interval of turn-off is usually longer than the turn-off time of a power BJT because there in no way to apply a reverse bias to the base of the IGBT in order to sweep out the excess minority carriers in the n^- drift region. Recombination has to take place internally. However, the turn-off time is usually less than $5\,\mu$ s. This is far shorter than the times to turn off thyristors.

7.6.1. Turn-off Losses

An analytical estimate of turn-off loss is prone to larger errors than that obtained for the MOSFET, because of the complication of minority-carrier recombination. For an analysis of turn-off energy dissipation in the IGBT we can consider the two extreme cases of a circuit with a purely resistive load and a circuit with a highly inductive load.

IGBT circuit with a resistive load. Refer to Fig. 7.8 for the turn-off circuit with a resistive load and the simplified waveforms. The initiation of turn-off is to have the gate circuit switch *Sw* closed and the gate-supply voltage reduced to zero, that is, $V_G = 0$. A supply of zero volts is equivalent to a short circuit. At this point the IGBT gate capacitance starts to discharge. There is a delay time $t_{d(off)}$ before the main circuit voltage and current change and this is the time taken for the gate voltage v_{GS} to reach V'_{GS}, the value that just sustains the drain current I_D. As v_{GS} reduces so too does i_D, according to the transfer characteristic $i_D = G(v_{GS} - V_{GS(TH)})$. The voltage v_{DS} increases as i_D falls because of the circuit equation $v_{DS} = V_s - i_D R$. The current fall in time t_{fi}, the voltage rise in time t_{rv} are roughly the same and equal the crossover time t_c from the start of the voltage rise to the end of the current fall. Unlike the MOSFET, the IGBT has a long current tail as the minority carriers recombine in the BJT portion of the device. The estimation of the energy loss W_{off} during the turn-off interval t_c is the same as that for turn-on with a resistive load. Accordingly, if $t' = t - t_{d(off)}$,

$$W_{off} = \int_0^{t_c} p\,dt' = \int_0^{t_c} v_{DS}\, i_D\, dt' = \frac{V_s I_D}{6} t_c . \tag{7.6.1}$$

EXAMPLE 7.4

An IGBT chopper modulates power from a dc supply, whose voltage is 400 V to a resistive load, whose value is $20\,\Omega$. The chopper switches at a frequency of 1 kHz. IGBT data for a steady junction temperature of 100°C are that the steady on-state voltage is $V_{DS(\mathrm{SAT})} = 1.9$ V, the delay time $t_{d(off)} = 1\,\mu$ s and the crossover time $t_c = 1.2\,\mu$ s for turn-off. Estimate the energy losses in the IGBT **(a)** during the delay interval $t_{d(off)}$, **(b)** during the crossover interval t_c, and determine **(c)** the average power dissipation due to the turn-off process.

Solution

The on-state IGBT current is $I_D = I_l = (V_s - V_{DS})/R = (400 - 1.9)/20 = 10.9$ A.

(a) During the delay interval
$t_{d(off)} = 1\,\mu$ s, $v_{DS} = V_{DS(SAT)} = 1.9$ V and $i_D = I_D = 19.9$ A.
So, energy loss $W_{d(off)} = V_{DS(SAT)} I_D\, t_{d(off)} = 1.96\times 19.9\times 10^{-6} = 37.8\times 10^{-6}$ J.

(b) During the crossover interval t_c the energy loss W_c is

$$W_{off} = W_c = \frac{V_s I_D}{6} t_c = \frac{400\times 19.9}{6}\times 1.2\times 10^{-6} = 1592\times 10^{-6}\ \text{J}.$$

This is much higher than the loss during the delay interval by a factor greater than 40.

(c) Average power loss $P_{off} = (W_{d(off)} + W_c)\times f = 1630\times 10^{-6}\times 10^3 = 1.63$ W.
This is 70 times greater than the turn-on loss calculated in EXAMPLE 7.2.

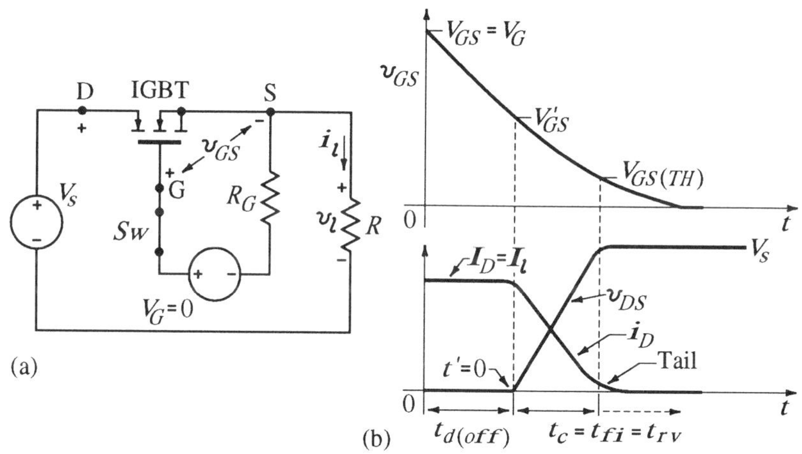

Fig. 7.8 IGBT turn-off for a resistive load.
(a) Circuit diagram, (b) waveforms.

IGBT circuit with inductive load. Figure 7.9 illustrates the sequence of events during turn-off if the load is highly inductive. The voltage source V_G is made zero so the gate voltage v_{GS} falls, as shown in the figure, due to the gate capacitance discharging into the resistance R_G. IGBT on conditions are maintained until the gate voltage $v_{GS} = V'_{GS}$, the voltage that just maintains the operation between the saturated and active regions. Then the drain voltage v_{DS} rises as the capacitor C_{GD} discharges and v_{GD} reduces. During this interval the main current i_D falls a little below $I_D = I_l$ in practice, but, ideally, it remains constant. When the drain voltage v_{DS} reaches the value V_s, the freewheeling diode conducts, the load voltage v_l is zero and the gate capacitance C_{GS} continues to discharge so that v_{GS}

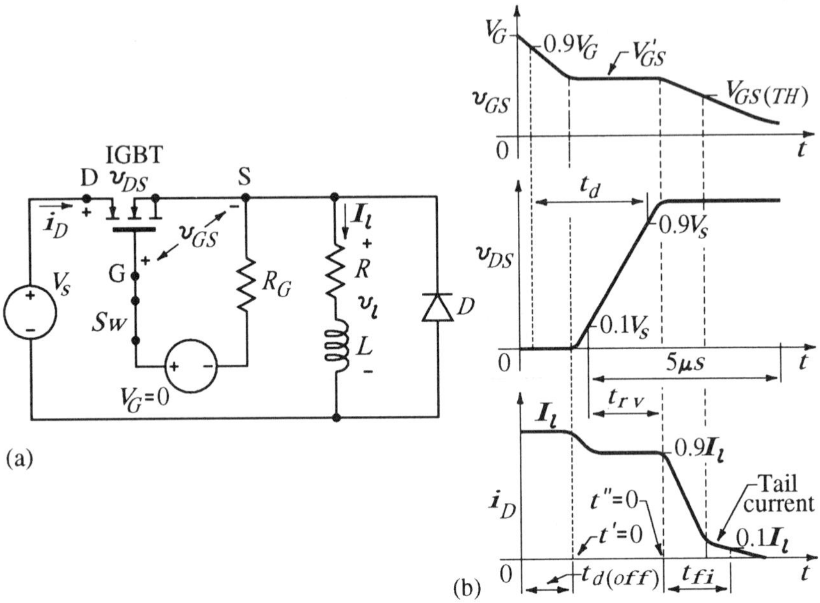

Fig. 7.9 IGBT turn-off. (a) Inductive load, (b) waveforms.

reduces below the value V'_{GS}. The transfer characteristic associated with the transconductance G forces the current i_D to reduce in proportion to the reduction in v_{GS}. At the gate threshold voltage $V_{GS(TH)}$, the n channel is no longer supported and the MOSFET is off. Following this sharp fall of current i_D there is a tail current that is relatively slow to decrease to zero. The tail current is associated with the bipolar-transistor portion of the IGBT turning off. It is a slow process because the decay of current is by recombination of minority carriers in the base (n^-drift region) of the pn^-p^+ transistor. Some manufacturers define the turn-off intervals differently for the IGBT from those of the MOSFET. For example, for the IGBT the time from $v_{GS}=0.9\,V_G$ to $v_{DS}=0.9\,V_s$ is defined as a delay time t_d. The fall time t_{fi} is defined as the interval from $i_D=0.9I_l$ to $i_D=0.1\,I_l$. Since the time from $i_D=0.1\,I_l$ to $i_D=0$ can be relatively considerable, the turn-off time t_{off} is not well defined. Data sheets generally give values of t_d and t_{fi} for given values of V_s, $I_D=I_l$, V_{GS}, R_G and junction temperature. To cover the complete tail current the turn-off energy dissipation W_{off} is given from a measurement over a $5\,\mu$s interval that starts at the time that $v_{DS}=0.05\,V_s$. This time interval includes a period whereby i_D is a leakage current, but the leakage dissipation is deemed negligible.

The tail current, that is the same as the collector current in a BJT with an open base circuit, can have its decay shortened in the IGBT. The addition of a highly doped n^+ buffer region between the p^+ substrate and the n^- drift region allows faster recombination after the MOSFET portion has turned off. Unfortunately this buffer offers two adverse effects. The on-state voltage drop is increased slightly and the reverse blocking voltage[1] is reduced considerably.

Typical values of turn-off for an inductive load are as follows.

(i) 600 V, 70 A, $f=1$ kHz, $V_G=15$ V; $t_d=1.4\,\mu$s, $t_{fi}=0.7\,\mu$s, turn-off loss $W_{off}=15$ mJ.

(ii) 600 V, 50 A, $f=10$ kHz, $V_G=15$ V; $t_d=0.2\,\mu$s, $t_{fi}=15\,\mu$s, turn-off loss $W_{off}=1.6$ mJ.

Compared with the turn-on times and turn-on losses, the values for turn-off are greater by an order of magnitude. For slow speed switching (1 kHz), a long current tail and slow recovery can be tolerated, and there is the advantage of a low on-state voltage (1.8 V) that gives low conduction losses. For high speed switching (50 kHz), fast recovery and a short current tail are imperative, but there is the disadvantage of a high on-state voltage (3 V) that gives high conduction losses.

IGBTs do not have the second breakdown phenomenon at turn-off like BJTs.

A simplified analysis to determine the turn-off losses in the IGBT, if the load is highly inductive, is carried out using a modified waveform of Fig. 7.9b. The delay time $t_{d(off)}$ is the interval from the initiation of turn-off by the removal of the gate supply voltage V_G to the time that the drain voltage begins to rise. While the drain voltage v_{DS} rises linearly from near zero to its value V_s in time t_{rv}, the drain current remains constant at $i_D = I_D = I_l$. Then, the current falls linearly to almost zero in time t_{fi}. The tail current due to recombination will be ignored since the associated losses can be small. The delay time losses are usually lumped with the on-state losses, so, they will not be included here. The crossover time t_c, from zero voltage to zero current is $t_c \approx t_{rv} + t_{fi}$. If $t' = t - t_{d(off)}$ and $t'' = t - (t_{d(off)} + t_{rv})$, the energy loss W_{off} due to the turn-off process is

$$W_{off} = \int_0^{t_c} p\,dt' = \int_0^{t_{rv}} v_{DS} I_D\,dt' + \int_0^{t_{fi}} V_s\,i_D\,dt''. \tag{7.6.2}$$

$$W_{off} = I_D \int_0^{t_{rv}} \frac{V_s}{t_{rv}} t'\,dt' + V_s \int_0^{t_{fi}} \left(-\frac{I_D}{t_{fi}} t'' + I_D \right) dt'' = \frac{V_s I_D}{2}(t_{rv} + t_{fi}) = \frac{V_s I_D}{2} t_c. \tag{7.6.3}$$

The error from this calculation may be substantial, but if other, more specific, loss data are not available, the estimate will give an indication of acceptability.

EXAMPLE 7.5

An IGBT chopper modulates power from a 400-V, dc supply to an *RL* load with a freewheeling diode connected across it. The chopper switches at a frequency of 10 kHz and the duty cycle *m* is such that the load current is 20 A and virtually constant. IGBT data for a steady junction temperature of 100°C and $R_G = 50\,\Omega$, are that the delay time is $t_{d(off)} = 1\,\mu$s, the voltage rise time is $t_{rv} = 0.2\,\mu$s and the current fall time is $t_{fi} = 0.12\,\mu$s for turn-off. Estimate the average power loss due

[1] The reverse blocking voltage can be reduced to as little as 15 V.

to the turn-off process.

Solution

The energy dissipated W_{off} in the IGBT during the crossover $t_c = (t_{rv} + t_{fi})$ is

$$W_{off} = \frac{V_s I_D}{2} t_c = \frac{400 \times 20}{2} \times (0.2 + 0.12) \times 10^{-6} = 1280 \times 10^{-6} \text{ J}.$$

The average power loss P_{off} due to the turn-off process is

$$P_{off} = W_{off} \times f = 1280 \times 10^{-6} \times 10 \times 10^3 = 12.8 \text{ W}.$$

7.7. IGBT PROTECTION

Protection of the IGBT is required, but it appears to be less complex than that for the thyristor, because of the inherent nature of this switch being voltage controlled. However, consideration has to be given to overvoltages, overcurrents and transients, just like other semiconductor switches.

The IGBT is an integrated circuit comprising a MOSFET gating switch and a bipolar transistor. It has the structure of a MOSFET but the operating features of a BJT. Since the gating is voltage controlled, the gate driver can be an integrated circuit. It is also possible to integrate the circuits that are used in the sensing of voltage, current and temperature for protection processing. ICs are fragile and measures must be taken to protect them. For example, the high frequency of IGBT switching produces electromagnetic interference. Protection of the low power control circuits against this noise is afforded by optical links to the gate driver circuits.

7.7.1. Overvoltages

If an IGBT is in the off-state and if no gate voltage is applied, the device will remain off until its rated drain-to-source blocking voltage is exceeded. At this point avalanche breakdown[2] occurs at junction J2, that is, breakdown of the p^+n^-p transistor. The resulting uncontrolled high currents give rise to high power dissipation in the switch, and with overheating there is the possibility of destruction. The rated value of voltage, above which breakdown occurs, is given in the data sheets. Primary protection against voltage breakdown is by limiting the supply voltage to no more that 80% of the IGBT rated value. Auxiliary protection is obtained by connecting across the IGBT a non-linear voltage arrester, which is a metal-oxide device whose resistance decreases rapidly with voltage increase. The arrester diverts current from a blocking IGBT if the applied voltage rises to a value beyond that permitted, either as a transient or as a steady value.

Theoretically, protection against reverse bias voltages across the drain-to-sink terminals of the IGBT is the same as the forward bias protection. However, in those IGBTs, that have an n^+ buffer layer between the p^+ substrate and the n^-

[2] This is called first breakdown also.

drift region to shorten the switches' turn-off times, the junction J1 of the p^+n^-p transistor cannot block more that about 15 V to 25 V of reverse-bias voltage. In this case, a reverse-connected diode is placed across the IGBT to give protection. If this form of protection is used, care must be taken to minimize the stray inductance L_s in series with the diode. As load current i_l is suddenly diverted to the diode, as i_{diode}, due to a switching action elsewhere in the circuit, the IGBT will experience a reverse bias voltage equal to $L_s di_{diode}/dt$. A low value of L_s will not create a problem.

The use of the diode in parallel with the IGBT does limit the application of the device. In order to allow high reverse voltages to be blocked in circuits with IGBTs, that have low reverse-bias voltage withstand, it is possible to add a series-connected diode. This solution decreases the efficiency of power modulation because of the voltage drop across the diode during its conduction interval.

7.7.2. Overcurrent

The IGBT data sheet usually gives the maximum continuous current that is allowed for the temperatures 25°C and 100°C. This is associated with the maximum power that can be safely dissipated without the temperature rising above its limit. In general, thermal limitations impose the restriction of a defined maximum current. Any value of continuous current above this defined value is not permissible, so protection is taken against all overcurrents. Common methods of protection are fuses and gate control to interrupt or reduce currents within a few microseconds. Pulsed currents can be much higher that the continuous current rating because the thermal limit is dictated by average power dissipation.

Current limits are directly associated with thermal considerations so that the modelling of heat flow for IGBTs can follow the same format as that used for thyristors. Refer to section 5.10.

EXAMPLE 7.6

An IGBT is used in a chopper circuit. From data sheets for a 600-V, 16-A IGBT the following information is given.

The maximum on-state voltage $V_{DS(\mathrm{SAT})} = 2.4\,\mathrm{V}$ (for $V_G = 15\,\mathrm{V}$).

The thermal resistance, junction to sink, $R_{\theta JS} = 2.5\,^\circ\mathrm{C/W}$.

Determine **(a)** the maximum continuous current I_D and **(b)** the magnitude of the current pulse for a duty ratio $m = 0.5$, if for both cases the heat sink is maintained at 50°C and the maximum junction temperature is 150°C. Assume that the switching losses are negligible in comparison with the conduction losses.

Solution

For steady operating conditions, the thermal equation relating temperature difference ΔT, average power dissipation P_D and thermal resistance $R_{\theta JC}$ is

$\Delta T = T_J - T_s = P_D R_{\theta JS}$,

where T_J is the junction temperature and T_S is the temperature of the heat sink.

Upon substitution, $P_D = (T_J - T_S)/R_{\theta JS}, = (150 - 50)/2.5 = 40\,\mathrm{W}$.

So, for all conditions, the power dissipation in the IGBT must not exceed 40 W.

(a) For continuous current $P_D = I_D V_{DS(\mathrm{SAT})}$, so
$I_D = P_D / V_{DS(\mathrm{SAT})} = 40/2.4 = 16.7$ A.
This is the maximum continuous current for the given thermal conditions.

(b) For pulsed current ($m = 0.5$, $t_{ON} = 0.5\mathrm{T}$) the average power P_D is

$$P_D = \frac{1}{T}\int_0^T v_{DS} i_D dt = \frac{1}{T}\int_0^{t_{ON}} V_{DS(\mathrm{SAT})} I_D dt = \frac{t_{ON}}{T} V_{DS(\mathrm{SAT})} I_D = m V_{DS(\mathrm{SAT})} I_D .$$

Thus the magnitude of the current pulse is
$I_D = P_D / m\, V_{DS(\mathrm{SAT})} = 40/(0.5 \times 2.4) = 33.3$ A. This assumes that the switching frequency is greater than 1kHz.

There is no second breakdown phenomenon[3] in IGBTs, so that the safe operating area (SOA)[4] is wide. That is, the SOA is rectangular, minus triangles that are associated with increasing intervals of on-time t_{ON}, as shown in Fig. 7.10. The safe operating region takes into account all values of gate voltage V_{GS}, so that the IGBT can be in saturation or in the active region.

As an arbitrary example of an IGBT rated at 600 V, 100 A, let $I_D = 100$ A, $V_{DS} = 600$ V, $t_{ON1} = 100\,\mu$s and $t_{ON2} = 10\,\mu$s. For a continuous main current i_D the device would be controlled to operate at any i_D and v_{DS} within the hatched area. As the duty cycle decreases then the operating area can be increased.

The main current in an IGBT can be self limiting to some extent. An inspection of Fig. 7.4d indicates the limit of overcurrent. If, for a given gate voltage V_{GS}, the switch is on and $v_{DS} = V_{DS(\mathrm{SAT})}$, a fault condition could make the current I_D increase. If the current increase were high enough, the IGBT would leave the on-state and enter the active region, in which the current I_D is virtually unvarying and independent of the voltage V_{DS}. However, the IGBT must be turned off quickly, because it cannot tolerate for long the dissipation that occurs in the active region beyond that indicated by the safe operating area.

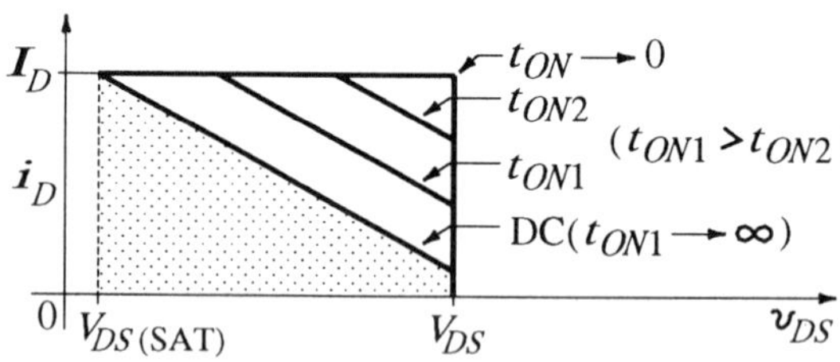

Fig. 7.10 IGBT safe operating region (SOA).

[3] Second breakdown occurs if there is constriction of the current into areas of high density that leads to hot spots and breakdown.

[4] For different intervals of the current pulse I_D the SOA shows the combined I_D and V_{DS} bounds beyond which maximum dissipation is exceeded.

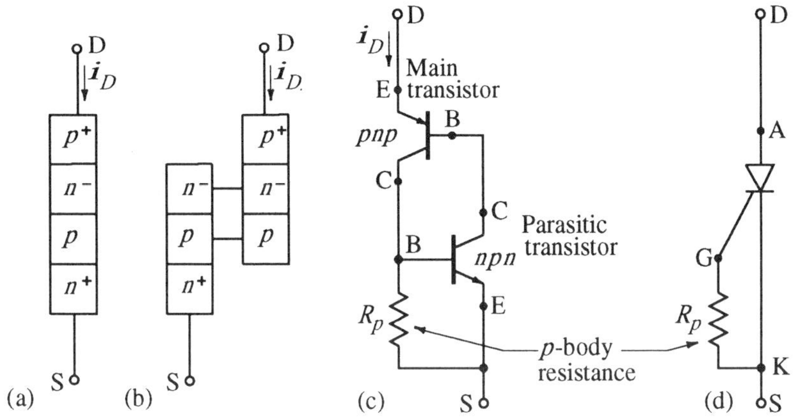

Fig. 7.11 IGBT's parasitic thyristor. (a) Thyristor structure, (b) equivalent structure, (c) two transistor model, (d) thyristor circuit.

Fault currents as severe as short-circuit values must not be allowed to persist in any semiconductor switch because of the thermal considerations. The inherent speed of switching the IGBT allows gate-control circuits to provide protection. Upon sensing an overcurrent, the gate voltage v_{GS} can be reduced (by switching in a zener diode across the gate terminals within a time of about 1 μs). The characteristic gain (the transconductance *G)* causes the current I_D to reduce. The IGBT can tolerate this reduced current for a longer period. If the fault is cleared within this longer period, then the gate voltage can be increased to its normal value again. If the fault is not cleared, a timer can be used so that the gate terminals will be short-circuited. With no gate voltage the main current is reduced to zero by the IGBT reverting to its high-impedance off-state.

If the load current rating is greater than the current rating of an IGBT then sharing of current by a number of IGBTs in parallel has been used as a solution to the problem. Care is taken to match the devices, especially in terms of the on-state voltage drop $V_{DS(\mathrm{SAT})}$. Similar precautions to balance steady and transient currents as for paralleled thyristors are taken, that is, resistors are added for an equitable share of steady current and inductors are added to reduce an imbalance of the rate of change of currents.

A problem that has been recognized and more or less solved in the design stage of the semiconductor switch is the problem of the parasitic thyristor that is inherent in an IGBT. An inspection of Fig. 7.3 shows that there is a $p^+n^-pn^+$ structure between the drain and source terminals. This is a thyristor structure, so that the device may be turned on by dv/dt or by a main current that rises beyond a certain limit. A high main current is the problem. Figure 7.11 illustrates how the main current can turn on the parasitic thyristor. Converting the thyristor (Fig. 7.11a) to a transistor structure (Fig. 7.11b) and then to an equivalent two-transistor model

(Fig. 7.11c) or a thyristor circuit (Fig. 7.11d) depicts the resistance R_p of the p-body region above the n^+ region. If the main current is high enough, the voltage across R_p can be large enough to cause regeneration in the two transistors or cause the thyristor to latch on. Once on, the thyristor remains on until the main current is brought to zero by some external means. The IGBT gate voltage removal is ineffective and control is lost. At turn-off the problem can be exacerbated, even though the main current may only be limited thermally in the on-state. If turn-off at the MOSFET gate is too rapid and the load is inductive, the total hole current is transiently concentrated in R_p. So, the tendency is for thyristor turn-on. By increasing the series resistance R_G in the IGBT gate circuit the MOSFET turn-off is slower and there is some electron current in the n channel. This reduces the hole current share of the main current to a level to prevent unwanted thyristor turn-on.

To protect against parasitic thyristor turn-on within the structure, the resistance R_p is made small and the gain α of the p^+n^-p transistor is reduced[5]. In this way the overall gain of the equivalent two transistors is less than unity so regeneration does not occur. Also, the BJT portion of the main current is reduced by designing the MOSFET to carry a larger portion of current in the n channel.

Due to the transconductance G, as the IGBT gate voltage v_{GS} is increased, the main current i_D increases also. Thyristor turn-on is prevented by putting a limit on the value of v_{GS}. This value is usually about 20 V.

A problem might arise during the IGBT turn-off process or during the off-state. If the rate of rise of voltage dv_{DS}/dt is too high, then the parasitic thyristor could turn on. In this case protection is given by a snubber circuit, or an increased value of the gate-drive resistance R_G.

The IGBT is designed to minimize the parasitic thyristor turning on, so that the normal limit of the main current i_D is dictated by thermal limitations. However, beyond the normal limit, recourse can be made to increase the IGBT gate resistance R_G and to connect a snubber circuit across the device.

7.7.3. Transients

High current transients during the on-state of the IGBT can be sensed and the action is usually to reduce the gate voltage v_{GS}. In turn, this action causes a reduction of the main current. This suggests a means to limit the initial inrush current of an IGBT at turn-on. Inrush currents arise from motor start-up, snubber capacitor discharge and reverse recovery of a diode across a load. The normal limitation of the rate of rise of current di_D/dt is the inductor (series snubber). However, for the cases above, a useful way to limit the rise for a short interval is to replace the conventional stepped voltage at the gate terminal with a ramp waveform or a two-step waveform (the first step being 5 to 7 V and the second step being the normal value of 15 V). Low values of gate voltage mean low values of main current due to the transfer characteristic of the device. Not only is the rate of rise of

[5] The gain is reduced if n^- region is made wider. This increases the voltage drop in the on state.

current reduced, so also is the peak value of the current. The first step of gate voltage may have a duration to 2 to 50 μ s, but care must be taken that operation is within the specified safe operating area.

A high rate of rise of voltage dv_{DS}/dt can turn on the IGBT from the off-state or during the turn-off process. IGBT turn-on by dv/dt has two mechanisms. One mechanism is turn-on of the parasitic thyristor. The second, and less damaging, is by taking the voltage v_{GS} beyond the threshold value by capacitor division. This latter effect is reduced if a low resistance is connected between the gate and source terminals. Protection against dv/dt is afforded by a negative gate-bias voltage $-V_G$ during the turn-off and off-state, and increased series resistance R_G in the gate circuit or a parallel snubber across the main terminals D and S, just the same type that is used for a MOSFET.

The maximum temperature of the junction of an IGBT is specified as 150°C. This gives no margin of safety. A slight increase of current i_D will cause further temperature rise and possible damage. It is better to be conservative, by using a large heat sink and by operating at a lower maximum temperature level at which protective circuitry for temperature and current can act in good time.

The low power gate circuit for driving the IGBT is similar to the gate circuit of the MOSFET so the same considerations are given to the protection of both against overvoltages and noise. The user has less to design if he/she chooses a commercial gate driver chip because protection is usually built into the circuit.

7.8. IGBT RATINGS AND APPLICATIONS

A single IGBT can be used in any application of power or voltage modulation where the specification demands ratings of voltages in the 300 V to 1600 V range, currents in the 10 A to 400 A range and a switching frequency in the 1 kHz to 20 kHz range. These ratings put the use of IGBTs in the mid-power range (less than 0.5 MW) and in the mid-frequency range. That is, the power range lies between that of the MOSFET at the low end and that of the thyristor at the high end.

The above statements signify that there is competition between the BJT and the IGBT. The main difference between these two devices is that the BJT is current controlled for switching on and off, whereas the IGBT is voltage controlled. Gate drives with voltage control are simpler and incur lower power dissipation. Also the IGBT can switch faster that the BJT. So it would seem that, except for cost and conduction voltage drop, the IGBT has advantages. However, the technology for the use of BJTs is, at present, more developed.

In terms of applications of completely controllable, unidirectional switches, the IGBT can do anything the BJT can do and it can do it faster. In chopper applications the period of chopping can be shorter, so that filtering is less of a problem and is cheaper. In ac voltage regulation the IGBT offers more versatility in the switching pattern because of the higher switching frequency.

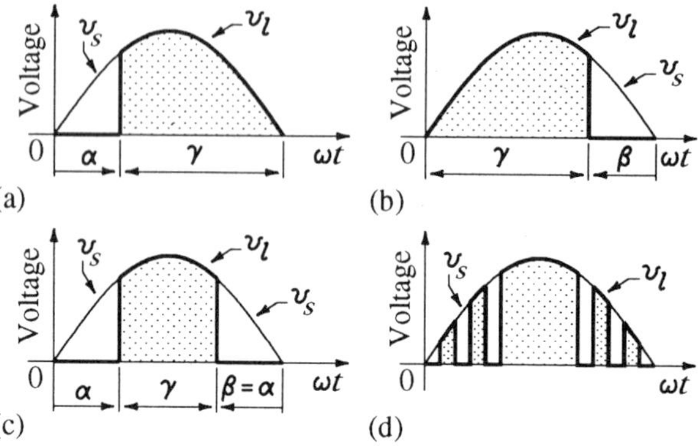

Fig. 7.12 IGBT ac switching patterns. (a) Trigger-angle delay α, (b) trigger-angle advance β, (c) both trigger angles, (d) pulse-width modulation.

Figure 7.12 illustrates some of those patterns. As shown in Fig. 7.12a the IGBT can operate in the same fashion as triacs and thyristors by adjusting the trigger-angle delay α to modulate the ac-source voltage v_s to an adjustable ac or dc load voltage v_l. Whereas the thyristor remains on until the load current reduces to zero, the IGBT turns off at the moment the gate signal is removed.

This IGBT characteristic of turn-off provides the other illustrated switching patterns that are not feasible for thyristors without additional circuitry. These are switching patterns that produce a better power factor or lower harmonics, or both, in the ac source side of the converter. Figure 7.12b depicts trigger angle advance β (also known as extinction angle). It might be seen that this tends to provide a leading power factor, but the same harmonics exist as with trigger angle α control. Symmetry is introduced in the voltage output, shown in Fig. 7.12c, by having both trigger angle α and trigger angle β control with $\alpha = \beta$. Because of the high speed of IGBT switching, any kind of pulse width modulation (PWM) can be introduced, as shown in Fig. 7.12d. Here, voltage regulation is accomplished with a reduced magnitude of lower harmonics but increased magnitude of higher harmonics. If the switching frequency is 20 kHz or higher, then the bothersome acoustic noise is eliminated.

The range of frequency of switching and power handling capability make the IGBT suitable for motor drives, uninterruptible power supplies, transportation applications and possibly induction heating.

7.9. SUMMARY

The IGBT is a fully controllable, unidirectional switch that is suitable for applications in the mid-power and mid-frequency ranges.

The device has the structure of a MOSFET and the operational features of a BJT. That is, it looks like an n-channel MOSFET, except that it has a *p*-type substrate, and it turns on by a voltage applied to the gate. Further, it has a high current density (1 to 3 A/mm^2) and a low voltage drop, like the BJT. In terms of switching frequency, the IGBT is slower that the MOSFET but faster than the BJT. Theoretically, the IGBT could have the attribute of high reverse-bias voltage withstand capability. However, in practice, designers optimize for high frequency or low voltage drop in the on-state. This leads to low voltage avalanche breakdown in the reverse direction. Accordingly, reverse-connected diodes are built into the device, like the MOSFET, or one must be connected externally to give protection. The IGBT has no second breakdown characteristic and there is little tendency to be latched by high currents causing the parasitic thyristor to turn on. The switch is limited by thermal considerations which allow temperatures up to 150°C to be tolerated. The on-state voltage drop across the IGBT is lower than the MOSFET voltage drop but can be greater that the BJT voltage drop.

An IGBT rated at 600 V, 70 A, 10 kHz has characteristics like the following.

Gate voltage for turn on, $V_{GS} = 15$ V, (threshold value $V_{GS(TH)} = 3$ V);
Reverse breakdown voltage, $V_{SD} = 20$ V;
On-state voltage, $V_{DS(\mathrm{SAT})} = 2$ V at $I_D = 70$ A, for $T_J = 25$ °C;
Maximum transconductance $G = 40$ S;
Leakage current $I_{D\,leak}$ is 0.2 to 2 mA, (temperature dependent);
Maximum power dissipation, $P_D = 200$ W;
Turn-on time $t_{on} = 0.1\,\mu$s, for $I_D = 40$ A, $T_J = 25$ °C;
Turn-off time $t_{off} \approx 1\,\mu$s, (load dependent);
Thermal resistance, $R_{\theta JC} = 0.9$ °C/W;
Peak current $I_D = 400$ A for $10\,\mu$s, (interruptible).
At 25°C the current rating is 70 A but at 100°C the current rating is 40 A.

There are devices with voltage ratings of 1500 V and current ratings of 400 A and there are devices that operate at switching frequencies above 50 kHz.

IGBTs can be used in any application where BJTs are used, for choppers, ac to dc converters and dc to ac inverters. IGBTs are admirably suited to the control of electric drives and appear to be the best switches for pulse width modulation in the mid-power range (< 0.5 MW) because of their high frequency of operation.

7.10. PROBLEMS

Section 7.3

7.1 A 600-V, 9-A, low frequency IGBT is rated at 150°C. It modulates power from a 400-V dc supply to a resistive load of $R = 50\,\Omega$. The IGBT data are that for a gate voltage $V_G = 15$ V the on-state voltage is $V_{DS(\text{SAT})} = 2.8$ V. If the switch is on for a long time determine **(a)** the on-state conduction losses of the IGBT and **(b)** the efficiency of the circuit operation, assuming the supply to be ideal.

7.2 An IGBT chopper modulates power from a dc supply of 200 V to a resistive load. Figure 7.4a illustrates the circuit diagram. For steady operation at 100°C at the junction the IGBT data sheet provides the information that the on-state voltage drop is $V_{DS(\text{SAT})} = 2.9$ V, the threshold value of the gate voltage is $V_{GS(TH)} = 2.8$ V, the device transconductance is $G = 1.3$ S and the maximum possible dissipation is $P_D = 30$ W. If the gate signal is applied at a low frequency (< 1 kHz), so that the switching losses can be neglected, determine **(a)** the maximum value of the drain current I_D for a maximum duty cycle $m = 0.8$, **(b)** the current rating of the IGBT at 100°C, **(c)** the load resistance R to sustain the maximum current and **(d)** the gate voltage V_G for this condition.

Section 7.5

7.3 An IGBT chopper modulates power from a 400-V dc supply to a resistive load, whose value is $16\,\Omega$. The chopper switches at a frequency of 30 kHz. IGBT data for a steady junction temperature of 100°C are that the steady on-state voltage is $V_{DS(\text{SAT})} = 2.5$ V, the delay time is $t_{d(on)} = 25$ ns and the crossover time is 30 ns for turn-on. If the chopper has a maximum duty cycle $m = 0.8$, determine **(a)** the IGBT current rating for the given temperature, **(b)** the average power loss due to conduction and **(c)** the average power dissipation due to the turn-on process.

7.4 An IGBT chopper modulates power from a dc supply, whose voltage is $V_s = 400$ V to an *RL* load with a freewheeling diode connected across it. The chopper switches at a frequency of 30 kHz. For a chopper duty cycle $m = 0.7$ the load current is $I_l = 25$ A and is almost constant. IGBT data for a steady junction temperature of 100°C are that the delay time is $t_{d(on)} = 24$ ns, the current rise time is $t_{ri} = 27$ ns and the voltage fall time is $t_{fv} = 43$ ns for turn-on. Estimate the average power dissipation in the IGBT due to the turn-on process.

7.5 A 1500-V, 50-A IGBT is used in a chopper circuit to modulate power from a 1000-V dc supply to an *RL* load with a freewheeling diode connected across it. The current in the load is 40 A and virtually constant. IGBT data available are that the gate threshold voltage is $V_{GS(TH)} = 5$ V, the transconductance is $G = 19$ S, the device input capacitance is $C_{iss} = 9000$ pF, the output capacitance is $C_{oss} = 650$ pF and the transfer capacitance is $C_{rss} = 240$ pF. If the gate-drive voltage is $V_G = 15$ V and if the gate-circuit resistance is

$R_G = 30\,\Omega$, determine **(a)** the delay time $t_{d(on)}$, **(b)** the drain current rise time t_{ri} and **(c)** the drain voltage fall time t_{fv} for turn-on.

Section 7.6

7.6 An IGBT chopper modulates power from a dc supply, whose voltage is $V_s = 400$ V, to a resistive load, whose value is $16\,\Omega$. The chopper switches at a frequency of 30 kHz. IGBT data for a steady junction temperature of 100°C are that the steady on-state voltage is $V_{DS(\text{SAT})} = 2.5$ V, the delay time is $t_{d(off)} = 200$ ns and the crossover time is $t_c = 160$ ns for turn-off. Estimate the average power dissipation due to the turn-off process.

7.7 A 1000-V, 200-A IGBT acts as a chopper to modulate power from a 600-V dc supply to an *RL* load with a freewheeling diode connected across it. The chopper frequency is 10 kHz and for a duty cycle $m = 0.8$ the load current is virtually constant at 120 A. IGBT data for steady operation at 100°C are the on-state voltage drop is $V_{DS(\text{SAT})} = 4$ V the delay time is $t_{d(off)} = 1\,\mu$s and the crossover time is $t_c = 2\,\mu$s for turn-off. Estimate **(a)** the average power dissipated in the IGBT due to conduction, **(b)** the average power dissipated in the IGBT due to the turn-off process and **(c)** the delay interval losses.

7.8 A 1500-V, 50-A IGBT is used in a chopper circuit to modulate power from a dc supply, whose voltage is $V_s = 1000$ V, to an *RL* load with a freewheeling diode connected across it. The current in the load is 40 A and virtually constant for a chopper frequency of 10 kHz and a duty cycle $m = 0.7$. IGBT data available are that the gate threshold voltage is $V_{GS(TH)} = 4.5$ V, the transconductance is $G = 17$ s, the device input capacitance is $C_{iss} = 7500$ pF, the output capacitance is $C_{oss} = 650$ pF and the transfer capacitance is $C_{rss} = 250$ pF. If the gate drive voltage is $V_G = 16$ V and if the gate circuit resistance is $R_G = 25\,\Omega$, determine **(a)** the delay time $t_{d(off)}$, **(b)** the drain voltage rise time t_{rv}, **(c)** the drain current fall time t_{fi}, **(d)** the average power dissipated in the IGBT due to the turn-off process and **(e)** the average power loss in the IGBT due to on-state conduction.

7.9 An IGBT operates as a chopper in a circuit, whose dc supply has a voltage $V_s = 400$ V and whose load is highly inductive and is clamped with a freewheeling diode. The load resistance is $12\,\Omega$ and the maximum duty cycle of the chopper is $m = 0.8$. For a particular gate-circuit configuration and steady operation at a junction temperature of 110°C, the energy loss characteristic for switching is given by $W_{loss} = I_D \times 10^{-4}$ J. If the on-state voltage is $V_{DS(\text{SAT})} = 3$ V and the maximum possible power dissipation of the IGBT is 200 W, determine the maximum frequency of the chopper switching.

7.10 An IGBT, rated at 600 V, 30 A and 1 kHz acts as a chopper to modulate power from a 400 V dc source to a load comprising a resistance of $10\,\Omega$ in series with an inductance of 100 mH. A freewheeling diode is connected across the load. At turn-on the IGBT has a current rise-time $t_{ri} = 50$ ns and a voltage fall-time $t_{fv} = 200$ ns. At turn-off the voltage rise time is $t_{rv} = 100$ ns and the current fall time is $t_{fi} = 1000$ ns. While in the on-state the IGBT has a voltage drop $V_{DS(\text{SAT})} = 1.8\,\text{V}$. The IGBT has a thermal resistance, junction to case, $R_{\theta JC} = 0.7°\text{C/W}$ and the thermal resistance, case to sink is $R_{\theta CS} = 0.5°\text{C/W}$. The heat sink temperature is maintained at 40°C. If the chopper is operated to allow the average power absorbed by the load to be 4 kW, calculate **(a)** the duty cycle m of the chopper, **(b)** the switching and conduction losses, **(c)** the mean junction temperature of the IGBT and **(d)** the maximum average power that the IGBT can dissipate.

7.11 An IGBT has an on-state voltage characteristic $V_{DS(\text{SAT})} = 1.4 + 0.02 I_D^{1.1}$ volts and a switching loss characteristic $W_{loss} = I_D \times 10^{-4}$ joules. The thermal resistance, junction to ambient, is $R_{\theta JA} = 1.67°\text{C/W}$. If the frequency of switching is 10 kHz, estimate a reasonable steady overcurrent setting for the IGBT circuit so that the junction temperature will not exceed 150°C. The ambient temperature can be assumed to be 30°C.

7.12 A 1000-V, 200-A IGBT has a transient thermal impedance characteristic as follows.

$Z_{\theta JC(t)}$ (°C/W)	0.007	0.025	0.05	0.065	0.11	0.115
Time (s)	0.001	0.01	0.05	0.1	1.0	3

The on-state voltage drop is $V_{DS(\text{SAT})} = 4\,\text{V}$. If the case of the IGBT is maintained transiently at 50°C, while the device conducts a current $I_D = 1000\,\text{A}$ for 10 ms, determine **(a)** the junction temperature at the end of the current pulse and **(b)** the junction temperature 10 ms later.

Section 7.7

7.13 A load has a maximum rating of 1000 V, 500 A. Two IGBTs are connected in parallel to share the current. Their on-state characteristics are $V_{DS(\text{SAT})1} = 1.7 + 5.75 \times 10^{-3} I_{D1}$ and $V_{DS(\text{SAT})2} = 1.5 + 11.8 \times 10^{-3} I_{D2}$. Determine **(a)** the natural sharing of the load current, **(b)** the value of the same resistance put in series with each IGBT to force current sharing within 15% of each other, **(c)** the minimum values of each resistance to force equal current sharing and **(d)** the total on-state losses for each case.

7.14 A 1600-V, 50-A IGBT modulates power from a dc supply to a load that has virtually constant current. The frequency of switching is 10 kHz. IGBT data are that the transconductance is $G = 18\,\text{S}$, the gate threshold voltage is $V_{GS(TH)} = 5.5\,\text{V}$, the transfer capacitance is $C_{rss} = C_{GD} = 250\,\text{pF}$. In order to limit the dv/dt at the drain terminal of the IGBT during turn-off to $1000\,V/\mu\text{s}$ find **(a)** the value of the equivalent gate-drive resistance R_G and **(b)** the parameter values of a polarised snubber circuit, $C_s\,R_s$.

Section 7.8

7.15 A single-phase supply is rectified by a full-wave diode bridge. Between the rectifier and a resistive load an IGBT modulates the power delivered to the load. If control is by the delay angle α, show that the power factor on the ac side of the circuit is directly proportional to the rms value of the load voltage.

7.11. BIBLIOGRAPHY

Baliga, J.B. *Modern Power Devices.* New York: John Wiley & Sons, Inc., 1987.

Bose, B.K. (ED). *Modern Power Electronics - Evolution, Technology and Applications.* New York: IEEE Press, 1992.

Ghandi, S.K. *Semiconductor Power Devices.* New York: John Wiley & Sons, Inc., 1987.

International Rectifier IGBT Designer's Manual. 1991.

Mohan, N., Underland, T.M. and Robbins, P. *Power Electronics.* New York: John Wiley & Sons,Inc., 1989.

Rashid, Muhammad Harunur. *Power Electronics.* Englewood Cliffs, N.J.: Prentice Hall,Inc., 1988.

Williams, B.W. *Power Electronics, Devices, Drivers, and Applications.* New York: John Wiley & Sons, Inc., 1987.

CHAPTER 8

THE TRIAC

8.1. INTRODUCTION

The name *triac* is given to a semiconductor switch that can have controlled conduction in either direction and the name is compounded from *tri* from **tri**ode to indicate there are three electrodes or terminals, and the abbreviation *ac* to indicate the control of **a**lternating **c**urrent.

The simplest way to describe the operation of the triac is to say that its behavior is similar to two thyristors connected in antiparallel. It is more convenient to use a single device to regulate ac voltage and power, but if the power to be controlled is in excess of the triac's rating, then two thyristors can be an acceptable substitution.

Triacs, that are common and, therefore inexpensive, are rated at 800 V, 25 A. Accordingly, it would be expected that power of the order of 10 kW or less is well within the limits of a single device. The principal application of the triac is to regulate ac voltage. Examples of this form of modulation are light dimming, temperature control and small ac motor speed adjustment (as for fans).

The triac is sensitive to low values of di/dt and dv/dt so that its switching speed is low. However, there are very few problems if the ac supply frequency is 50 or 60 Hz. Applications are usually limited to below 400 Hz.

Like the thyristor and unlike the transistor, the triac can block voltage of either polarity and unlike any other semiconductor switch it can conduct current in either direction, so it has the unique distinction of being the only true ac switch. As such it is useful as a solid-state relay.

8.2. TRIAC STRUCTURE

There are five layers of silicon in the triac and the structure is similar but more complex than the thyristor. Figure 8.1a depicts a symbolic, two-dimensional representation of the alternate *p* and *n* layers that make up the switch.

The circuit symbol of the triac is shown in Fig. 8.1b and its equivalent representation in terms of thyristor operation is shown in Fig. 8.1c. As it is depicted in the figure, the triac is equivalent to two thyristors connected in antiparallel with their gates connected together. Since there is no preferred direction for conduction the terminal designations of anode and cathode are not used. Instead the main terminals are denoted by T_1 and T_2, as shown. The reference terminal with respect to voltage is T_1. Accordingly, a negative gate bias means that the gate terminal G is negative with respect to the terminal T_1. Also, a positive voltage bias to the main terminals means that terminal T_2 is positive with respect to the terminal T_1.

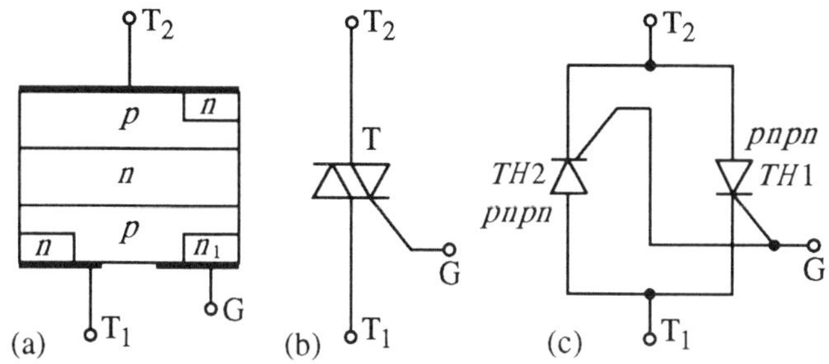

Fig. 8.1 The triac.
(a) Structure, (b) circuit symbol, (c) thyristor equivalent.

An inspection of Fig. 8.1a can indicate why the triac can be represented by a pair of thyristors in inverse parallel. Between terminals T_2 and T_1 a *pnpn* structure *TH*1 can be seen, and, in parallel with this, a *pnpn* structure *TH*2 exists between terminals T_1 and T_2 with two *pn* layers in common. This indicates the possibility of conduction in either direction.

To trigger the device on by means of a gate signal is simple in practice but complex with regard to the physics process. The simple part is that the gate bias can be positive or negative for the triac to turn on whether the terminals T_2T_1 are forward or reverse biased.

There are five modes of operation. If the triac is initially off (high impedance state) and if there is no gate signal, the triac will block current no matter whether terminal T_2 is positive or negative with respect to terminal T_1. That is, the triac remains off. The other four modes comprise the combination of conditions that terminal T_2 is positive or negative with respect to terminal T_1, and a positive or negative gate signal is applied across the terminals G and T_1. For all four conditions the triac turns on (low impedance state) and conducts current. Once the triac is turned on the gate signal can be removed, and, just like the thyristor, the triac remains on until the main current falls to a value below the holding current, which is virtually zero compared with the rated current of the switch. The four conditions for turn-on are explained in section 8.5.

8.3. TRIAC MODEL

The structure of the triac demonstrates that there are *pnpn* layers and *npnp* layers between terminals T_2 and T_1. These two forms are recognizable as thyristors back to back. Each of these forms can be modelled by two transistors. This modelling of the thyristor was discussed in section 5.3.2.

If there is no gate signal applied to the triac, any polarity of voltage will be blocked by the reverse biased *pn* junctions of either group of layers, *pnpn* or *npnp*. No current, other than a leakage current, exists. In the two-transistor model, if there is no gate signal (no base drive), the transistors remain off.

Application of a gate signal at terminal G, and a bias at the main terminals T_2 T_1 , allows the base signal in one transistor to cause base current in the second transistor. This second transistor augments the base current to the first transistor. Regeneration ensues until one of the pair of *pnpn* layers latches on. Section 8.5 gives more details of the turn-on action.

In the on-state, the internal regeneration prevents turn-off if the gate signal is removed. While it is on, the triac offers a nearly constant on-state voltage $V_{T(ON)}$ that is about 1.5 V. The magnitude of the main current I_T does not alter the voltage drop very much. Except for very low supply voltages the value of the current I_T is governed mainly by the supply voltage and the load impedance.

The two transistor model explains the turn-off mechanism also. If the triac is on and the gate signal is removed, the triac remains in the on-state because of the regenerative base currents in the two transistors of the model. Remove the gate signal and allow the main current I_T to fall to near zero due to circumstances in the main circuit. At this point the collector currents of the two transistors are so low that they cannot sustain the base current. A BJT without a base current will turn off naturally. Accordingly, the thyristor turns off. Hence, the triac is off. The triac will remain off until the next gate signal is applied.

With the thyristor modelled by two transistors and with the triac modelled by two thyristors, most of what was written in Chapter 5 about the thyristor applies to the triac. There is one exception. The triac is bidirectional with respect to current conduction.

8.4. TRIAC I-V CHARACTERISTICS

The steady-state, volt-ampere characteristic of the triac can be sketched readily if this ac switch can be likened to two thyristors that are connected back-to-back, as shown in Fig. 8.1c. Using this figure as reference, we can derive the triac's characteristics that are illustrated in Fig.8.2. Figure 8.2a is the test circuit diagram.

Consider there to be no gate signal. An increase in positive voltage at terminal T_2 results in no current other than a small leakage current, until a breakover value V_{BO} occurs. At this point, avalanche breakdown occurs and the triac changes state; that is, it turns on. This form of turn-on is to be avoided in practice, because control by means of the gate signal is lost this way. Curve 1 shows this part of the blocking characteristic in quadrant I. Curve 2 in quadrant III has a similar nature. It is brought about by increasing terminal T_2 voltage negatively while having no gate signal. Only a leakage current exists until the breakover voltage is reached. The broken lines (curves 3 and 4) indicate a transient phase from blocking to turn-on as the plasma spreads dynamically across the wafer of silicon. In the steady-state conduction mode the characteristics are curves 5 and 6.

The controlled form of triac turn-on is to apply a signal to the terminal G. If the signal is of sufficient magnitude and duration, the triac turns on within a few microseconds. The dynamic turn-on is represented by curves 7 and 8, that lead into the steady-state curves 5 or 6. Which particular curve (7 and 8) applies depends on the value of the voltage across the triac at the moment the gate signal

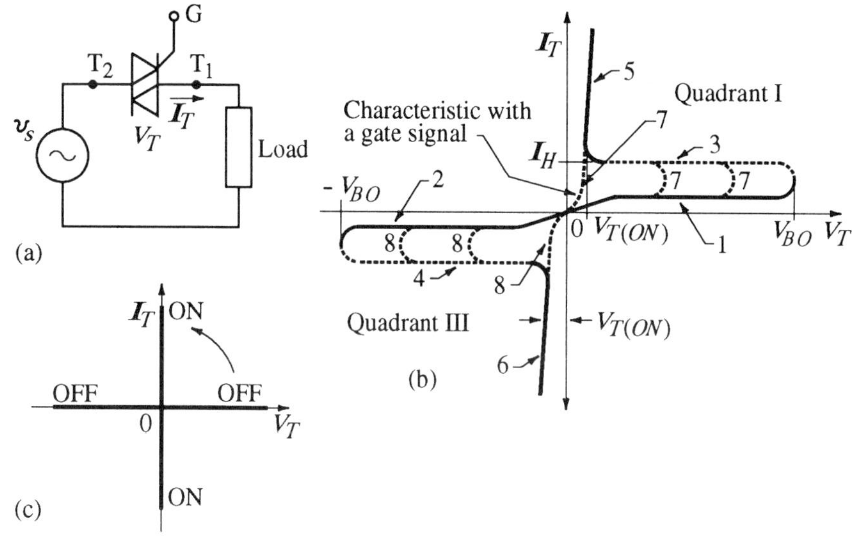

Fig. 8.2 Triac characteristics. (a) Circuit diagram, (b) characteristics: 1,2 blocking; 3,4 avalanche turn-on; 5,6 conduction; 7,8 gate turn-on, (c) ideal characteristics.

is applied. Once on, the magnitude of the current is determined almost completely by the source voltage v_s and the load impedance. As shown in curves 5 and 6, the voltage drop $V_T = V_{T(ON)}$ across the triac, while it is on, is about 1.5 volts and is virtually constant.

To go from the blocking state to the conduction state requires a gate signal. However to go from the conduction state to the blocking state there is a requirement that the triac current I_T be reduced below the value of the holding current I_h that is small compared with the rated current of the triac.

If the voltage drop across a conducting triac is small compared with the amplitude of the source voltage, and if the leakage current of the blocking triac is very small compared with the current I_T in the on-state, then these values can be considered insignificant in the analysis of a circuit that includes triacs. That is, the triac is considered to be an ideal switch. It conducts no current when blocking and it has no voltage drop across it when conducting. These ideal characteristics of the triac are depicted in Fig.8.2c. See Appendix 6 for data.

The characteristics of the triac make the device ideally suited to phase angle α delay control in order to regulate the ac voltage across a load. Fig. 8.3a illustrates a simple, but general circuit that allows power modulation of an ac supply to a resistive load. The triac is switched on at a point on the supply wave at which $\omega t = \alpha$. Load current conduction ensues until the current falls to zero (in this case, this is the time the applied voltage v_s falls to zero). The triac switches off until it receives a gate pulse at $\omega t = \pi + \alpha$, at which point the triac turns on again.

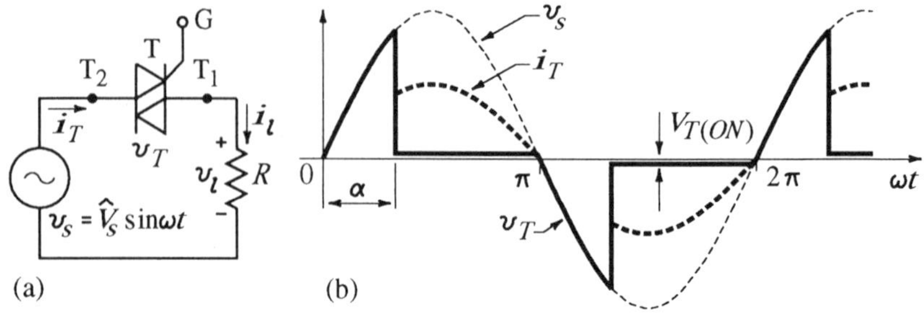

Fig. 8.3 Triac power modulation.
(a) Circuit diagram, (b) waveforms.

Since the applied voltage has reversed polarity, the triac current i_T will have changed direction. When this current is forced to zero by the supply voltage v_s, the triac turns off again, and the cycle repeats. Figure 8.3b shows the waveforms of the supply voltage v_s, the triac voltage drop v_T and the triac current i_T. Changing the value of the delay angle α alters the rms voltage across the load and modulates the power that can be delivered from zero to a maximum value, V_s^2/R.

Since the triac is a low-power-handling device, it is not economic to use any other switching strategy other than delay angle α control. Forced-commutation circuits are expensive compared with the price of a triac.

The quality of the power, that is delivered to the load, is dependent on the switching strategy. The delay angle α control method causes a waveform distortion of both the voltage and current of the load. It is important to know the degradation of the waveform quality. That is, how much deviation is there from an ideal sinewave? Measures of the quality include power factor PF and the total harmonic distortion THD.

EXAMPLE 8.1

Consider the circuit diagram in Fig. 8.3a. The ac supply has a voltage whose value is 120 V at 60 Hz. The load resistance is $R = 10\,\Omega$. A 500-V, 16-A triac is used to modulate power from the supply to the load. The on-state triac voltage is $V_{T(ON)} = 1.6$ V. For the condition of a symmetrical delay trigger angle of $\alpha = \pi/3$ rad for the triac, determine **(a)** the average value of the triac current, **(b)** the average power absorbed by the load, **(c)** the average power dissipation in the triac due to conduction and **(d)** the efficiency of power modulation.

Solution

Figure 8.3b depicts the triac voltage and current waveforms. It is assumed that the switching times (2–4 μs) are short enough compared with the period of the waveform (16.6 ms) that they can be ignored.

(a) The average current in the triac is

$$I_{T\,av} = \frac{1}{\pi}\int_{\alpha}^{\pi} i_T d\omega t = \frac{1}{\pi}\int_{\alpha}^{\pi} \frac{(\hat{V}_s \sin\omega t - V_{T(ON)})}{R} d\omega t\,.$$

So, $I_{T\,av} = \frac{1}{\pi R}[\hat{V}_s(1+\cos\alpha) - V_{T\,(ON)}(\pi - \alpha)]$.

$R = 10\,\Omega$, $\hat{V}_s = \sqrt{2} \times 120 = 169.7$ V, $\alpha = \pi/3$ rad, $V_{T(ON)} = 1.6$ V.

Thus, $I_{T\,av} = \frac{1}{10\pi}[169.7(1+\cos\pi/3) - 1.6(\pi - \pi/3)] = (8.1 - 0.96) = 7.14$ A.

If the voltage drop across the triac is ignored the average current estimate is 8.1 A.

(**b**) The rms value of the triac current is given by

$$I_{T\,rms}^2 = \frac{1}{\pi}\int_{\alpha}^{\pi} i_T^2 d\omega t = \frac{1}{\pi}\int_{\alpha}^{\pi} \frac{(\hat{V}_s \sin\omega t - V_{T(ON)})^2}{R^2} d\omega t\,.$$

So, $I_{T\,rms}^2 = \frac{1}{100\pi}\int_{\pi/3}^{\pi} (169.7 \sin\omega t - 1.6)^2 d\omega t = 113.2\text{A}^2$.

If $V_{T\,(ON)}$ is ignored $I_{T\,rms}^2$ is calculated to be 115.8 A^2.

The average power P absorbed by the load is

$P = I_{l\,rms}^2 R = I_{T\,rms}^2 R = 113.2 \times 10 = 1132$ W.

If $V_{T(ON)}$ is ignored the calculation of power gives 1158 W.

This is a 2% difference.

(**c**) The average power P_C dissipated by the triac due to conduction is

$$P_C = \frac{1}{\pi}\int_{\alpha}^{\pi} p\, d\omega t = \frac{1}{\pi}\int_{\alpha}^{\pi} v_T\, i_T\, d\omega t = \frac{V_{T(ON)}}{\pi}\int_{\alpha}^{\pi} i_T\, d\omega t = V_{T(ON)} I_{T\,av}\,.$$

So, $P_C = 1.6 \times 7.14 = 11.4$ W.

(**d**) The modulation efficiency η is

$$\eta = \frac{\text{output}}{\text{output} + \text{losses}} = \frac{P}{P + P_C} = \frac{1132}{1132 + 11.4} = 0.99 \quad (99\%).$$

The power modulation efficiency is high.

8.5. TRIAC TURN-ON

A positive or negative gate signal to a triac, while the main terminal T_2 is positively or negatively biased, will allow the triac to turn on. This makes the triac a versatile switch in an ac circuit.

A typical triac may be rated at 600 V and 15 A for the modulation of ac power. The gate specifications may be that a voltage $V_{GT} = 2.5$ V and a current $I_G = 100$ mA will turn the triac on within a few microseconds under most conditions.

The sensitivity of the gate circuit to turn on the triac is not the same for all the four possibilities of biasing. The triac is more sensitive to a positive gate current if the terminal T_2 is positive, and it is more sensitive to a negative gate current if the terminal T_2 is negative. In practice it is found quite often that a negative gate signal is used for either main terminal polarity.

Since the triac has a *pnpn* structure like the thyristor, the triac is sensitive to turn-on by overvoltage and dv/dt. These methods of turn-on have to be avoided.

8.5.1. Turn-on Action

The turn-on action by a gate current depends on the polarities of the triac in the off-state. The action of turn-on for the combination of biasing is explained below in terms of the thyristor and the two-transistor models.

Positive gate signal and terminal T_2 positive.
The condition that the gate has a positive signal and terminal T_2 has a positive voltage bias is the way a thyristor is turned on. Figure 8.4a indicates that part of the triac structure that is important for the operation with terminal T_2 positive with respect to T_1 and the gate G positive with respect to T_1. Figure 8.4b shows the equivalent circuit diagram in terms of the two-transistor model.

With respect to this model, that was derived from the *pnpn* layers between T_2 and T_1 and the *pn* junction between G and T_1, current from a source V_G is injected into the base region of transistor *BJT*1. Since the emitter is negatively biased at T_1, the collector of *BJT*1 draws current from the base of transistor *BJT*2. Since the emitter of *BJT*2 is positively biased, then a collector current from *BJT*2 augments the gate signal to the base of transistor *BJT*1, which is now driven harder. This form of positive feedback drives the two transistors into saturation. That is, the composite device is on, and maintained on by the regenerative process, so the gate signal can be removed without turning off the switch.

The only difference between this structure and that of the common thyristor is that the triac has a parallel resistive path between the terminals G and T_1 through a *p* region as well as the *pn* junction (J3). This is modelled as a resistor *R* in Fig. 8.4b and this would indicate that the gate current in a triac would be greater than the gate current in a thyristor of the same rating.

Negative gate signal and terminal T_2 positive.
The important part of the triac structure, if the terminal T_2 is positive and the gate terminal is negative, is illustrated in Fig.8.5a. The main structure between terminals T_2 and T_1 is the thyristor structure *pnpn*. In parallel with this is the structure $pnpn_1$, which is a *pilot* thyristor *TH*1. This composite structure is redrawn in Fig.8.5b to show the two thyristors and the terminal T_1 connection to the lower *p*-region of thyristor *TH*1. This indicates that terminal T_1 is connected to the gate region of thyristor *TH*1 and the terminal G is the cathode of thyristor *TH*1.

Since the terminal G is negative with respect to terminal T_1, the gate of the pilot thyristor *TH*1 is positively biased with respect to its cathode. Because of the current i_G, thyristor *TH*1 turns on and its *p* region (gate) begins to assume the potential of its anode T_2.

The turn-on of pilot thyristor *TH*1 initiates the turn on of the main thyristor *TH*. There now exists a sufficient potential gradient across the lower *p* regions of thyristors *TH*1 and *TH* that the pilot anode current i_a is directed to the *p*-gate region of thyristor *TH* so that the main thyristor *TH* turns on.

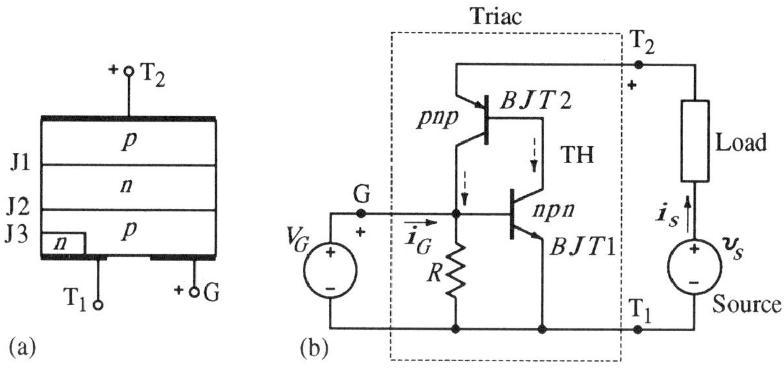

Fig. 8.4 The triac with T_2 and G positive.
(a) Relevant structure, (b) equivalent circuit.

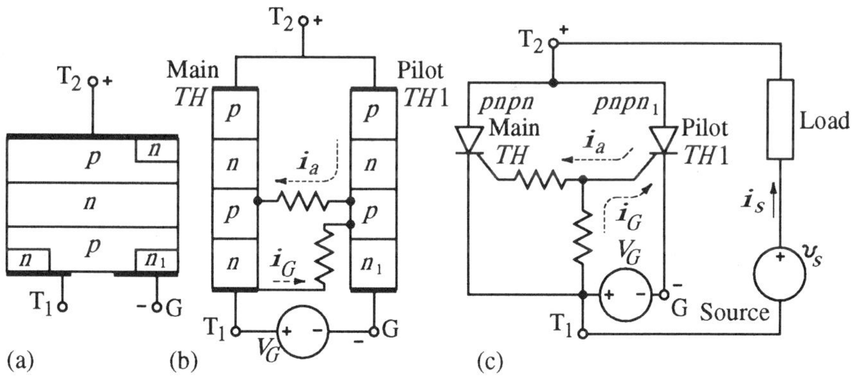

Fig. 8.5 The triac with T_2 positive and G negative.
(a) Relevant structure, (b) two-thyristor model, (c) equivalent circuit.

The equivalent circuit, depicted in Fig. 8.5c, shows the source voltage v_s biasing terminal T_2 positively. The gate voltage V_G biases the terminal G negatively with respect to the gate region of the pilot thyristor *TH*1. Thyristor *TH*1 turns on to allow gate current into the gate region of the main thyristor *TH*. The main thyristor *TH* is forward biased at its anode T_2 so it turns on. Once on, the regenerative process allows the triac to remain on even if the gate signal is removed.

Positive gate signal and terminal T_2 negative.
The main structure for operation with the triac having a positive gate and a negative terminal T_2 is shown in Fig. 8.6a. Between terminals T_1 and T_2 there are the four *pnpn* layers. The two-transistor model can be used for this. In addition we

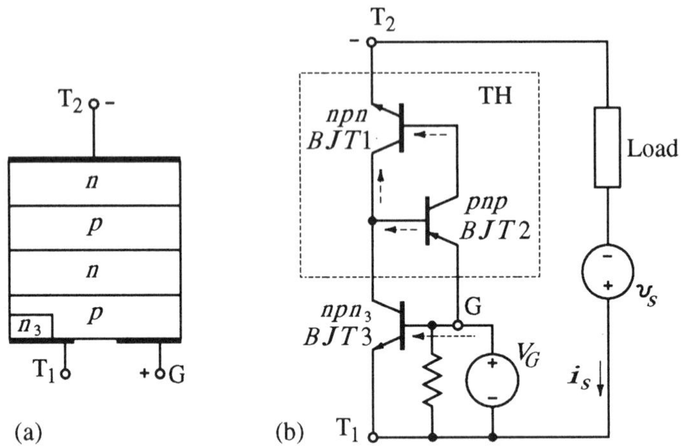

Fig. 8.6 The triac with T_2 negative and G positive. (a) Structure, (b) equivalent circuit.

can create an n_3pn transistor with the n_3 layer connected to terminal T_1, the *p* layer connected to the terminal G, a resistor connected between terminals T_1 and G to account for the conduction between terminals T_1 and G through the semiconductor *p*, and the *n* layer acting as the collector. This leads to the equivalent circuit diagram shown in Fig. 8.6b.

With reference to the equivalent circuit, the gate-source voltage V_G drives the base of transistor *BJT*3, whose collector current drives the base of transistor *BJT*2, whose collector current drives the base of transistor *BJT*1. This sets up the regenerative action between transistors *BJT*1 and *BJT*2 to turn on the thyristor *TH* and hence the triac. The regenerative action maintains the triac in the on-state if the gate signal is removed.

Negative gate signal and terminal T_2 negative.
The main structure for operation with the triac having a negative gate and a negative terminal T_2 is shown in Fig. 8.7a. Between terminals T_1 and T_2 there are the same four *pnpn* layers as in the previous case where terminal T_2 is negative. This is modelled again by two transistors that represent the thyristor TH, as shown in the equivalent circuit diagram of Fig. 8.7b. In addition we can create an npn_1 transistor with the n_1 layer connected to terminal G, the *p* layer connected to terminal T_1, the *n* layer acting as the collector, and a resistor between the terminals G and T_1 to account for the semiconductor path through the *p* region.

The last transistor *BJT*3 initiates the gating of the thyristor TH if a source voltage V_G is connected positive at the base (T_1) and negative at the emitter (G). In this way, the collector current of transistor *BJT*3 provides the base current to transistor *BJT*2. Since the emitter of the transistor *BJT*2 is positively biased (T_1),

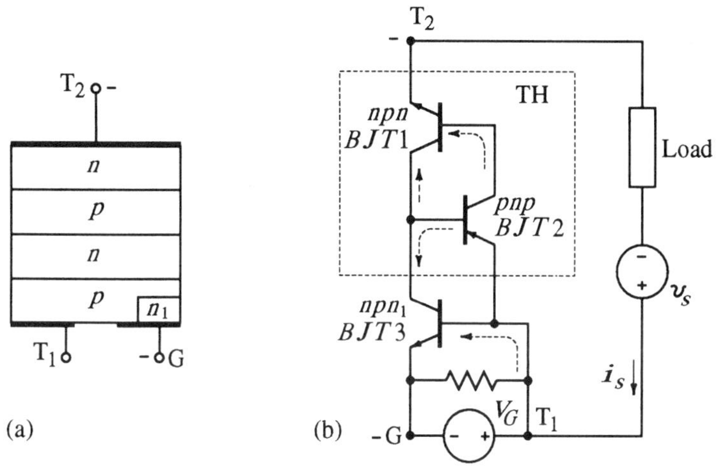

Fig. 8.7 The triac with T_2 negative and G negative.
(a) Structure, (b) equivalent circuit.

then there is collector current to drive the base of transistor *BJT*1. The emitter (T_2) is negatively biased so that the base current of transistor *BJT*1 results in the collector current driving the base of transistor *BJT*2. This base current is in addition to the original gate current and is the initiation of the regenerative action that turns on the switch fully. Once the triac is on it remains on without a gate signal until the main load current is reduced to a value below the holding current.

8.5.2. Turn-on Losses

It might take 2 μs for a triac to turn on after the application of a gate signal. During this interval both the triac voltage v_T and the triac current i_T can have reasonably high values at the same time. Accordingly, the value of the instantaneous power that is dissipated in the switch can be high. How significant the loss is depends on the amount of power dissipated during on-state conduction.

The initiation of turn-on by the application of a gate voltage starts the process of discharging and charging the depletion layer capacitance at the gate. This takes a time t_d (about 0.5 μs), during which time there is no change in the main terminal conditions. There follows a crossover time interval t_c (about 1.5 μs) during which time the current i_T rises and the voltage v_T falls to the steady operating values. The turn-on time t_{on} is

$$t_{on} = t_d + t_c \,. \tag{8.5.1}$$

Precise times of turn-on of the triac are not possible to determine since they are influenced by the supply voltage magnitude, the value of the conduction current

I_T, the value of the gate current and the load impedance. Also, the higher the junction temperature, and the higher the rate of rise of gate current, the shorter is the turn-on time. Any calculations utilizing a value of turn-on time must be expected to be approximate.

EXAMPLE 8.2

A 500-V, 12-A triac modulates power from a supply, whose voltage is 208 V at 60 Hz to a resistive load whose value is $R = 20\,\Omega$, by means of delay angle α control. Data provide the information that the maximum leakage current is $I_{T\,leak} = 0.5$ mA, the turn-on time t_{on} is about 2 μs and the on-state voltage $V_{T\,(ON)}$ across the triac is 1.6 V. If the trigger angle is $\alpha = \pi/2$ radians, determine **(a)** the energy loss W_{on} due to the turn-on process, **(b)** the energy dissipation in the triac in one half cycle due to on-state conduction.

Solution

(a) Without further information, assume that the delay time of turn-on is $t_d = 0.5\,\mu s$. During this time $v_T \approx \hat{V}_s = 208\sqrt{2} = 294$ V, and the current is $i_T = I_{T\,leak} = 0.5 \times 10^{-3}$ A. Consequently, the energy loss W_d during this delay time is

$$W_d = \int_0^{t_d} v_T\, i_T\, dt \approx \hat{V}_s I_{T\,leak} t_d = 294 \times 0.5 \times 10^{-3} \times 0.5 \times 10^{-6} = 73.5 \times 10^{-9} \text{ J.}$$

During the crossover time t_c, if $t' = t - t_d$, and if the changes are linear, the energy loss W_c is

$$W_c = \int_0^{t_c} v_T\, i_T\, dt' \approx \int_0^{t_c} \left(\frac{\hat{V}_s}{t_c} t' + \hat{V}_s \right) \frac{\hat{I}_T}{t_c} t'\, dt' = \frac{\hat{V}_s \hat{I}_T}{6} t_c .$$

Since $\hat{I}_T \approx \hat{V}_s / R = 294/20 = 14.7$ A,

$$W_c = \frac{294 \times 14.7}{6} \times 1.5 \times 10^{-6} = 1080 \times 10^{-6} \text{ J.}$$

The energy loss during the delay interval is negligible.

(b) During conduction the energy loss W_C over one half cycle is

$$W_C = \int_0^{T/2} v_T\, i_T\, dt \approx \int_{T/4}^{T/2} V_{T(ON)} \frac{v_s}{R} dt = \frac{V_{T(ON)}}{\omega R} \int_{\pi/2}^{\pi} \hat{V}_s \sin\omega t\, d\omega t .$$

$$W_C = \frac{V_{T(ON)} \hat{V}_s}{\omega R} [-\cos\omega t]_{\pi/2}^{\pi} = \frac{V_{T(ON)} \hat{V}_s}{\omega R} = \frac{1.6 \times 294}{2\pi \times 60 \times 20} = 0.062 \text{ J.}$$

The conduction loss in the triac is nearly 58 times greater than the turn-on loss. Consequently, the conduction losses dominate all losses in the triac at 50 or 60 Hz operation.

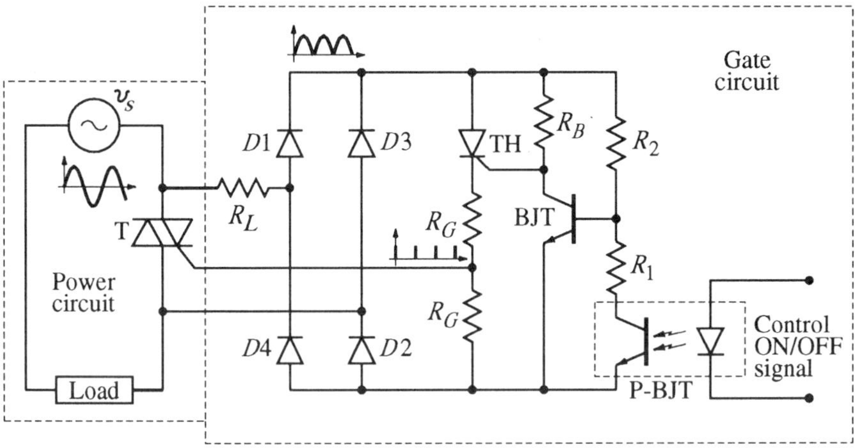

Fig. 8.8 Solid-state relay gating circuit.

8.5.3. Turn-on Circuits

If a triac is used as a solid-state relay in an ac circuit, a constant gate current I_G is injected into the triac for as long as the relay is to be in the on-state. Remove the gate signal and the triac turns off as soon as the main circuit current i_T drops below the holding current, which is close to zero[1].

A dc gate signal to a triac can be implemented by those circuits that are illustrated in section 5.5.2, Figs 5.10 and 5.11. These simple circuits provide a gate signal for as long as the triac is expected to conduct. Pulses at the gate at zero voltage crossing provide a more sophisticated form of triac turn-on. This can be implemented by the gate circuit that is illustrated in Fig. 8.8. Power to the load occurs if the triac T is on. The triac T is on if the control signal is ON. Relay action occurs by means of the state of the control signal. If the control signal is OFF the triac is off. The photo transistor P-BJT provides electrical isolation of the signal source from the power circuit.

Operation of the relay gate circuit in the on-state relies on the photo transistor P-BJT turning on if a signal is applied to the light-emitting diode. For this condition, the BJT will be off as the main voltage v_s passes through zero; otherwise the BJT is always on because of the base bias. Only while the BJT is off can the thyristor TH be turned on with a gate current through the resistor R_B. As soon as the thyristor turns on the voltage across the resistor R_G develops to turn on the triac T. Once the triac is on the gate circuit becomes disabled. Turn-on signals for the triac appear at every voltage crossing as long as the control ON-signal is present. Removal of the control signal means the photo transistor turns off. This

[1] The holding current of a 25-A triac is about 100 mA.

causes the bias of the BJT to rise so that the BJT is on and this disables the gate circuit of the thyristor TH. The off-state of the thyristor means that there are no triac gate signals generated. The resistor R_L is a current limiter.

In triac applications that utilize delay angle α control, a pulse of gate current or a train of pulses is all that is necessary for turn-on. The regenerative nature of the triac, like the thyristor, allows the gate signal to be removed, once the main current i_T has reached the latching value. How long the triac takes to reach the latching value depends on the rate of rise of the gate current and the rate of rise of the main current. The values of latching currents I_l are of the order of 100 mA and are greater than the holding current I_h value by up to 100%.

Any gate signal polarity will trigger on a triac with any polarity at terminal T_2 as long as the gate current is high enough. Some of the gate currents have to be higher than others. For example, the case of negative bias at terminal T_2 provides less sensitivity to turn on, if the gate signal is positive. The gate current may be twice the value than that needed for any other polarity combination. Figure 8.6b shows that this insensitive circuit configuration has a bleeding resistance between the gate terminals. There is a tendency for the use of negative gate signals to maintain triac sensitivity to turn-on, although the latching current can increase by up to 50% if a negative gate signal is applied while terminal T_2 is positive.

If the triac is not being used as a solid-state relay, then it is being used to modulate power by delay angle α control. Basic circuits to achieve this end are similar to the thyristor gating circuits depicted in Figs 5.12 to 5.14 and Fig. 5.24. The difference is that the thyristor is triggered on the positive anode bias whereas the triac is triggered every half cycle.

A very common gate circuit for the triac is the relaxation oscillator. This is a practical circuit for light dimmers. Figure 8.9 shows this type of circuit. The resistance R is adjustable to alter the RC time constant. In this way, the time to reach the breakover voltage of the trigger device (a diac in this case; see section 5.5.2) can be varied. Once the breakover voltage has been reached the capacitor discharges into the triac gate terminal. The triac turns on and the resulting near short circuit disables the gate circuit until the triac regains its blocking condition. This provides a positive gate pulse for a positive terminal T_2 and a negative gate pulse for a negative terminal T_2. These are both sensitive combinations for turn-on, so the asymmetry of trigger angle α for positive and negative half cycles will be minimal.

EXAMPLE 8.3

Consider the circuit diagram of the gating circuit in Fig. 8.9. The power circuit comprises a supply of ± 240 V at 50 Hz, with a rectangular wave pattern, and the load has an equivalent resistance $R_l = 24\,\Omega$. Triac data are that a gate current $I_G = 50$ mA and a gate voltage $V_{GT} = 1.5$ V will turn on the device in 2 μs and the maximum average gate power must not exceed 0.3 W. Determine **(a)** the values of the gate-circuit parameters, if the diac has a breakover voltage of 10 V and an on-state voltage drop of 1V, and **(b)** the value of the resistance R to control the

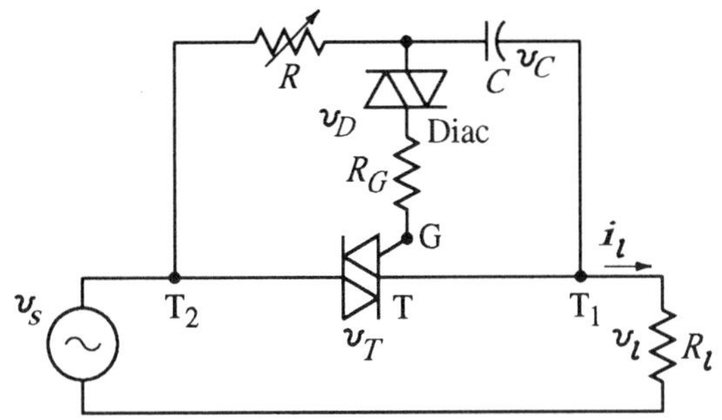

Fig. 8.9 Triac gate circuit.

power absorbed by the load to 50% of the maximum value.

Solution

The time constant of the gate circuit is $(R + R_l)C$ while the triac and the diac are in the off-states.

(a) Let the capacitor discharge its energy into the gate in $10\,\mu$s to ensure turn-on. If the gate current must be greater than 50 mA, we can assume an initial discharge current of 100 mA into the gate of the triac. An initial discharge of 100 mA means the resistance of the gate and R_G is
$(V_C - V_{\text{diac}(ON)})/I_G = (10-1)/10^{-1} = 90\,\Omega$.
A gate voltage of 1.5 V, a gate current of 50 mA and turn-on time of $2\,\mu$s suggest a gate energy W_G requirement of
$W_G = V_{GT}\, I_G\, t_{on} = 1.5 \times 50 \times 10^{-3} \times 2 \times 10^{-6} = 0.15 \times 10^{-6}$ J and an effective resistance R_{GT} of
$R_{GT} = V_{GT}/I_G = 1.5/50 \times 10^{-3} = 30\,\Omega$.
Hence, $R_G \approx 90 - R_{GT} = 90 - 30 = 60\,\Omega$.
If the capacitor discharges in $10\,\mu$s, then $5(R_{GT} + R_G)C \approx 10^{-5}$ s, or

$$C = \frac{10}{5 \times 90} \times 10^{-6} = 0.022\,\mu\text{F}.$$

The energy discharge W_C by the capacitor is

$$W_C = \frac{1}{2} C V_C^2 = \frac{1}{2} \times 0.022 \times 10^{-6} \times 10^2 = 1.1 \times 10^{-6}\ \text{J}.$$

The energy dissipated W_{diac} in the diac is

$$W_{\text{diac}} = V_{\text{diac}(ON)} \int_0^{5\tau} I_G\, e^{-t/\tau} dt = 1 \times 100 \times 10^{-3} \left[-\tau e^{-t/\tau} \right]_0^{5\tau} \approx 0.2 \times 10^{-6}\ \text{J}.$$

This provides $W_C - W_{\text{diac}} = 0.9 \times 10^{-6}$ J for the resistance. That is 0.6×10^{-6} J for R_G and 0.3×10^{-6} J for the gate. This is more than adequate to turn on the triac.

(b) The maximum power $P_{\max}$ absorbed by the load is
$P_{\max} = V_l^2/R = 240^2/24 = 2400$ W.

At half load $V_{l\,rms} = (1200 \times 24)^{1/2} = 169.7$ V.

$$V_{l\,rms} = V_s\left(\frac{\pi - \alpha}{\pi}\right)^{1/2}.$$

$$\text{Hence, } \alpha = \pi\left(1 - \frac{V_{l\,rms}^2}{V_s^2}\right) = \pi\left(1 - \frac{169.7^2}{240^2}\right) = 1.57 \text{ rad.}$$

In terms of time $t_\alpha = \alpha/\omega = 1.57/314 = 1.8 \times 10^{-3}$ s,

t_α is the time for the capacitor to charge up to 10 V.

That is, $V_{\text{diac}(BO)} = 10 = V_s(1 - e^{-t_\alpha/(R+R_l)C})$.

$$R = \frac{t_\alpha}{C \ln V_s/(V_s - V_{\text{diac}(BO)})} - R_l = \frac{1.8 \times 10^{-3}}{0.022 \times 10^{-6} \ln(240/230)} - 24.$$

$R \approx 1.92 \times 10^6\ \Omega$.

8.6. TRIAC TURN-OFF

Consider Fig. 8.10a. If the triac is conducting and if the supply voltage v_s reverses polarity, the triac becomes reverse biased and the device turns off as soon as the current is forced below the value of its holding current I_h. Once off, the triac will block conduction in either direction until an appropriate gate signal is reapplied. This is the action of the ideal triac and the transition between the on-state and off state at near zero current takes no time.

In practice, the value of the holding current I_h may be as low as 50% of the latching current I_l and the turn-off time is of the order of microseconds. Since the triac is a minority carrier device it takes time for the excess charge carriers that are stored in the device to be swept out, or recombine.

The stored charge, as the current comes to zero, makes the triac still look like a virtual short circuit. The rate of change of current is maintained at current zero by the reverse polarity of the supply voltage v_s. Consequently, the current i_T reverses and the triac voltage changes little from $v_T = V_{T(ON)}$. As the triac reverse current rises, the excess charge carriers are swept out of the region until current can no longer be supported. The current falls, the potential barrier increases and a reverse voltage spike appears. The spike is a function of the stored charge, which, in turn, depends on di_T/dt and junction temperature. In most cases the triac recovers its blocking capability. See Fig. 8.10b.

If the triac is sensitive to low values of gate current for turn-on, it is also sensitive to dv/dt during recovery from the on-state to the off-state. In the off-state a typical withstand value of dv/dt is 100 V/μs (this is the static dv/dt) at terminals T_1 and T_2. However, because of the charge already in the device during turn-off, a reapplied dv/dt as low as 10 V/μs can turn the device back on again.

In practice the worst condition for high reapplied dv/dt is for highly inductive loads. This is seen in Fig. 8.10c. At current zero, the triac recovers and the device assumes a reverse voltage equal to a high value of supply voltage in a short time. Some means of protection can be taken to prevent the dv/dt reaching a

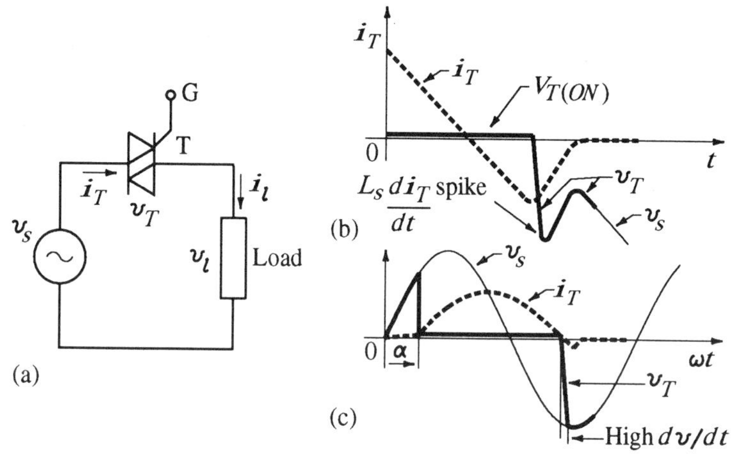

Fig. 8.10 Triac turn-off.
(a) Circuit diagram, (b) reverse recovery, (c) turn-off dv/dt.

limit whereby the switch is turned on.

Since the triac begins to turn off as soon as the main current i_T drops below the holding value ($I_h \approx 25$ mA for a 25-A triac), a low current associated with $V_{T(ON)} \approx 1.5$ V indicates a low power loss at turn-off. The reverse recovery losses can also be considered negligible compared with the conduction losses. Accordingly the turn-off losses in triacs are not considered further.

EXAMPLE 8.4

A 500-V, 8-A triac modulates power from an adjustable frequency 250-V ac supply to a resistive load whose value is 100 Ω. The triac data sheet stipulates that the maximum withstand dv/dt at commutation is 5V/μs. What is the maximum supply frequency that can be tolerated at turn-off?

Solution

At any instant the rate of change of supply voltage is

$$\frac{dv_s}{dt} = \frac{d}{dt}(\hat{V}_s \sin\omega t) = \omega \hat{V}_s \cos\omega t .$$

The maximum value of dv_s/dt occurs at $\omega t = \pi$, so,

$$\text{maximum } \frac{dv_s}{dt} = \omega \hat{V}_s = 2\pi f \hat{V}_s .$$

$$\text{That is, } f = \frac{dv_s}{dt} \times \frac{1}{2\pi \hat{V}_s} = \frac{5 \times 10^6}{2\pi \times \sqrt{2} \times 250} = 2.25 \times 10^3 \text{ Hz.}$$

8.7. TRIAC RATINGS

The triac is a relatively low voltage, low current and low frequency switch that does not control power much above 15 kW. However, it has important applications because it can block voltage of either polarity and allow controlled current in either direction. These characteristics make it unique as a power-semiconductor switch.

Table 5.1 displayed typical ratings of a thyristor. This is repeated in Table 8.1 for a general purpose triac.

Table 8.1 Triac ratings.

$I_{T\,rms}$ A	V_{DRM} V	V_{DWM} V	I_G mA	V_{GT} V	di_T/dt A/μs	I_{TSM} A	dv/dt V/μs	dv_T/dt V/μs
16	1000	700	70	1.5	40	130	10	200

Voltage rating. The forward, repetitive, peak voltage V_{DRM} of the triac whose ratings are shown in Table 8.1, is the same as the reverse, repetitive, peak voltage V_{RRM}, but the working value of the repetitive, peak voltage V_{DWM} is kept to 70% of these. V_{DRM} indicates the very maximum value that must not be exceeded. In the on-state, the voltage drop $V_{T(ON)}$ is usually less than 1.6 V for low voltage devices but for a 1000-V triac the on-state voltage may be as high as 2 V. It is usual that the higher voltage ratings require thicker wafers and these take longer to turn-off. The turn-off times are similar to those of thyristors.

In Table 8.1 there are two dv/dt ratings. One is a static rating that is involved with the junction capacitance and the turn-on value from the off-state. A dv_T/dt equal to 200 V/μs is not high, but the second dv/dt rating of 10 V/μs does limit the frequency of operation and necessitates precautions if loads are inductive. The low value of dv/dt withstand arises from a change of state from on to off. There is storage charge in the device at turn-off, so, if the dv/dt is above the specified limit, the triac can turn on again. This is to be avoided.

Current ratings. The current rating $I_{T\,rms}$ is determined by the maximum temperature at which the triac is allowed to operate. This temperature limit is about 125°C. This is a steady-state rating. There is also a surge rating I_{TSM} which may be up to about ten times the steady value. The surge rating is based on a transient power pulse of about 10 ms duration that raises the junction temperature 45°C while maintaining the case transiently at about 80°C.

Another transient value given in Table 8.1 is the limit of the turn-on di_T/dt. A value above this limit during turn-on can cause local heating and damage.

Other current ratings given in data sheets include the latching current and holding current for each of the four particular gate-drive configurations. The 1000-V, 16-A triac, whose partial ratings are given in Table 8.1 may have a latching current of 200 mA, a holding current of 100 mA and an off-state leakage current of 5 mA. These values are generally the order for all triacs. Actual values may differ only by a factor of about two.

8.7.1. Thermal Ratings

The same thermal considerations must be afforded the triac as are given to other semiconductor switches whose current ratings are controlled by the temperature limit at their junctions.

In terms of the power P generated at the junction, the temperature difference ΔT between the junction and ambient and the thermal resistances between junction, case, sink and ambient, the steady power flow is

$$P = \Delta T/(R_{\theta JC} + R_{\theta CS} + R_{\theta SA}). \tag{8.7.1}$$

In transient terms, transient thermal impedance $Z_{\theta(t)}$ is used in the form

$$P = \Delta T/Z_{\theta(t)}. \tag{8.7.2}$$

See section 5.10 for more details.

EXAMPLE 8.5

A 500-V triac has an on-state voltage drop $V_{T(ON)} = 1.7$ V. Its thermal resistance, junction to sink is $R_{\theta JS} = 6°$C/W. The semiconductor switch is connected to a heatsink whose thermal resistance, sink to ambient, is $R_{\theta SA} = 4.5°$C/W. If the ambient temperature is 35°C and if the junction temperature must not exceed 125°C, determine **(a)** the maximum average power that can be dissipated in the triac and **(b)** the rms current rating of the triac.

Solution

Assume that the triac modulates ac power and that $\alpha = 0$ for maximum power. Neglect switching and gate-circuit losses.

(a) The average power P_D dissipated in the triac is
$P_D = (T_J - T_A)/(R_{\theta JS} + R_{\theta SA}) = (125 - 35)/(6 + 4.6) = 8.49$ W.

(b) Average current $I_{T\,av} = P_D/V_{T(ON)} = 8.49/1.7 \approx 5$ A.

The rms rated current of the triac is $I_{T\,rms} = I_{T\,av} \times \dfrac{\pi}{2\sqrt{2}} = 5.5$ A.

8.8. TRIAC PROTECTION

The triac is a low-power, low-voltage, low-frequency semiconductor switch. It is sensitive to overvoltages, overcurrents and rates of change of voltage and current. Accordingly, some kind of protection must be considered to prevent the device being damaged. The triac is an inexpensive switch, so that the economics of the added circuitry must be kept in mind. Where possible it may be simpler to overate the triac rather than add extra components.

The triac is similar to the thyristor except that the triac conducts current in either direction. Therefore, whatever considerations can be given to the thyristor can be given to the triac. Refer to section 5.9 and Figs 5.32 and 5.33.

Overvoltage.
The voltage ratings of the triac V_{DRM}, V_{RRM} are the peak *(M)* forward *(D)* and reverse *(R)* repetitive *(R)* voltages, above which the triac will have avalanche breakdown and will turn-on. Voltage arresters can give protection against higher voltages. However, if the triac voltage ratings are 50% higher than the peak supply voltage of the circuit then operation should be satisfactory in most cases.

Overcurrent.
If the triac experiences an overvoltage that turns on the switch, a possible overcurrent might be encountered. A ground fault or a partial load short circuit could give rise to overcurrents. Fuses with the appropriate I^2t rating can give protection to the triac. If the current fault is an alternating current and its magnitude does not exceed ten times the current rating for one half cycle, gate-signal suppression is a means of allowing the triac to turn off. A low value of overcurrent can be limited by increasing the trigger angle α delay.

Whatever current protection is applied, it is to limit the junction temperature from rising above the value specified in the data sheets.

For circuit applications that have higher voltages and currents than can be handled by the available switches, thought has been given to devices connected in series and parallel. This is not the case with triacs. If ac control is sought at higher ratings than the available triac, the triac is replaced by two anti-parallel thyristors. The gating circuits may be slightly more complex but the thyristor is more robust to withstand higher transients.

Transients.
The triac is susceptible to relatively low values of transient voltages and currents. A static, off-state dv/dt withstand is limited to about 200 V/μs, because of the depletion capacitance, and the turn-off withstand value may be as low as 10 V/μs. Capacitive snubbers and low value inductive loads ease this problem. However, the ac supplies used in triac applications are low frequency sources.

The turn-on and turn-off times of triacs are similar to thyristors of the same rating. The di/dt ratings for each of these actions are low. At turn-on the di/dt at terminal T_2 has a maximum value of the order of 10 A/μs. A higher value might cause overheating in areas that had not turned on properly. Inductive snubber elements in series with the triac keep the di/dt within limits.

During turn-off the reverse recovery voltage depends on $-di/dt$. A high di/dt leaves a high storage charge and a high reapplied dv/dt that may turn the triac on again. Inductance and low frequency of operation give protection.

A series snubber of value 30 μH may be more than adequate to prevent damaging di/dt occurring. A non-polarized RC snubber with values $R \approx 50\,\Omega$ and $C \approx 0.1\,\mu F$ satisfies most cases. An accurate design calculation of the values of R and C is not straightforward[2].

---

[2] See W.McMurray, *Optimum Snubbers for Power Semiconductors*, IEEE Trans IA, pp503-510, 1972.

EXAMPLE 8.6

A triac regulates the voltage from a 120-V, 60-Hz supply to 10-Ω resistive load. If the stray inductance L_s of the power circuit is measured to be 20 μH specify the required di/dt capability of the triac at turn-on.

Solution

The voltage across an inductance is
$v = L\,di/dt$.
The worst case of di/dt is when the peak value of the supply voltage is across the inductance at the point of triac turn-on with $i_T = 0$. This occurs at $\alpha = \pi/2$ radians and $\hat{V}_s = \sqrt{2}V_s = 169.7$ V.
The highest value of di/dt is
$di/dt = v/L_s = \hat{V}_s/L_s = 169.7/(20\times10^{-6}) \approx 8.5\text{A}/\mu\text{s}$.
This is within the range of specified limits of triac di/dt.

8.9. TRIAC APPLICATIONS

The triac conducts current in both directions. Accordingly, it is best suited to control low power levels in ac circuits. The device turns of naturally when the current falls to zero but it is slow to turn-off. Changes of current di/dt and changes of voltage dv/dt are low compared with other semiconductor switches. These characteristics keep the frequency of switching to below 400 Hz. In general, the triac is well suited to ac-ac conversion in domestic applications where voltage, frequency, inductance values and power levels are all low.

8.9.1. AC-AC Conversion

The method of power modulation in ac circuits is by voltage regulation and voltage regulation is achieved by point-on-wave control using the delay angle α as the variable. Common examples of triac applications are light dimmers, speed controllers for fan motors and temperature controllers.

The basic circuit for ac power control is illustrated in Fig. 8.11. It is usual that the rms value of the ac supply voltage is fixed and that the load is resistive. With delay angle α control the positive and negative half cycles of the load current and voltage are symmetrical. Since the current falls to zero at the zero crossing of the supply voltage each half cycle, full control of the load voltage, from 0 to V_s, can be obtained. The principal variable to be controlled, no matter what the specific application, is average power P. The average power P delivered to the load is

$$P = \frac{1}{\pi}\int_0^{\pi} p\,d\omega t = \frac{1}{\pi}\int_{\alpha}^{\pi} v_l\,i_l\,d\omega t = \frac{1}{\pi}\int_{\alpha}^{\pi}\frac{\hat{V}_s^2}{R}\sin^2\omega t\,d\omega t. \qquad (8.9.1)$$

That is,

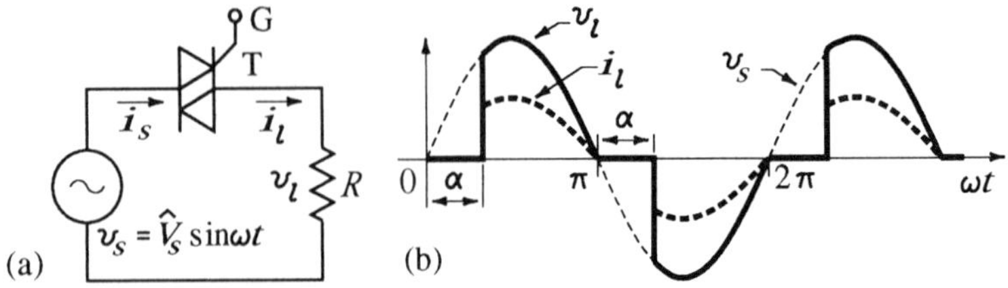

Fig. 8.11 General triac application.
(a) Circuit diagram, (b) waveforms.

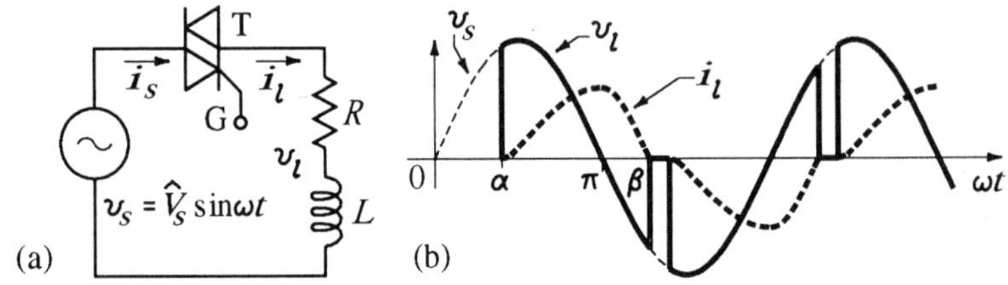

Fig. 8.12 Triac control of inductive load.
(a) Circuit diagram, (b) waveforms.

$$P = \frac{\hat{V}_s^2}{2R}\left[\left(1 - \frac{\alpha}{\pi}\right) + \frac{1}{2\pi}\sin 2\alpha\right]. \tag{8.9.2}$$

Since $P = I_{l\,rms}^2 R$, the rms value of the load current is

$$I_{l\,rms} = \frac{\hat{V}_s}{\sqrt{2}R}\left[\left(1 - \frac{\alpha}{\pi}\right) + \frac{1}{2\pi}\sin 2\alpha\right]^{1/2}. \tag{8.9.3}$$

The power can be adjusted from zero to a maximum value V_s^2/R by adjusting α from π to zero, but the relationship between average power P and delay angle α is nonlinear.

Most circuits have some inductance. If inductance is taken into account, the load current has a different pattern than that for the resistive load and some control of the power is lost. Figure 8.12a shows lumped inductance L that may have components associated with the source, with the load and with the stray inductance. The current waveform is shown in Fig. 8.12b. The functional form of the current is a steady-state component and a transient component. That is,

$$i_l = i_{ss} + i_{trans} = \frac{\hat{V}_s}{Z}\sin(\omega t - \theta) - \frac{\hat{V}_s}{Z}\sin(\alpha - \theta)e^{-t/\tau} \tag{8.9.4}$$

where $Z = (R^2 + \omega^2 L^2)^{1/2}$, $\tan\theta = \omega L/R$ and $\tau = L/R$. It is from this equation, given $\hat{V}_s$, ω, L, R and α that the rms value of the load current and the average power P are calculated, since

$$I_{l\,rms} = \left[\frac{1}{\pi}\int_{\alpha}^{\beta} i_l^2\, d\omega t\right]^{1/2} \quad \text{and } P = I_{l\,rms}^2 R. \tag{8.9.5}$$

Power control is limited to a range of α given by $\theta \le \alpha \le \pi$. For the delay angle $\alpha \le \theta$, half-cycle conduction is always $\beta - \alpha = \pi$ and the triac does not turn off. That is, the steady-state current response is purely sinusoidal in this latter case.

Waveform quality for resistive loads.
The control of more or less power delivered to the load is achieved by chopping the sinusoidal waveform of the supply voltage. The load voltage is no longer sinusoidal so the current response is not sinusoidal. The control aspect has added distortion to the ac waveform. Harmonics are introduced and these have an effect on the system power factor.

Power factor. Consider the triac ac controller illustrated in Fig. 8.11a. The load is resistive. We want to find the power factor as a function of the delay angle α. The power factor PF is defined as

$$\text{PF} = \frac{P}{S} \tag{8.9.6}$$

where P is the average power and S is the apparent power delivered by the source. Average power is

$$P = \frac{1}{\pi}\int_0^{\pi} p\, d\omega t = \frac{1}{\pi}\int_{\alpha}^{\pi} \frac{\hat{V}_s^2}{R}\sin^2\omega t\, d\omega t = \frac{V_s^2}{R}\left[\left(1 - \frac{\alpha}{\pi}\right) + \frac{\sin 2\alpha}{2\pi}\right]. \tag{8.9.7}$$

Apparent power S is

$$S = V_s I_{l\,rms} = V_s\left[\frac{1}{\pi}\int_{\alpha}^{\pi} \frac{\hat{V}_s^2}{R^2}\sin^2\omega t\, d\omega t\right]^{1/2} = \frac{V_s^2}{R}\left[\left(1 - \frac{\alpha}{\pi}\right) + \frac{\sin 2\alpha}{2\pi}\right]^{1/2} \tag{8.9.8}$$

The average reactive power Q (VAr) introduced by α is

$$Q = (S^2 - P^2)^{1/2}. \tag{8.9.9}$$

The power factor PF is

$$\text{PF} = \frac{P}{S} = \left[\left(1 - \frac{\alpha}{\pi}\right) + \frac{\sin 2\alpha}{2\pi}\right]^{1/2}. \tag{8.9.10}$$

The value of power factor ranges from unity for $\alpha = 0$ to virtually zero for $\alpha \rightarrow \pi$ radians.

Harmonics. The input current i_s from the source is the same as load current i_l. In terms of the harmonics we can reiterate eq. (2.3.5), that has to be solved to find the instantaneous value of the supply current i_s in terms of the harmonic components. That is,

$$i_s = I_{dc} + \sum_{n=1,2,3\cdots}^{\infty} \hat{I}_{sn}\sin(n\omega t + \theta_n) \tag{8.9.11}$$

where $\hat{I}_{sn}^2 = a_n^2 + b_n^2$, a_n, b_n are the Fourier coefficients and $\theta_n = \tan^{-1} a_n/b_n$. From this, the total harmonic distortion THD can be calculated.

EXAMPLE 8.7

Consider the circuit diagram shown in Fig. 8.11a. The supply is 250 V at 50 Hz and the load is $353.5\,\Omega$. If the delay angle α is varied from 0 to 180°, plot **(a)** the fundamental, the third and the fifth harmonic components of the supply current as a ratio of $\hat{V}_s/(\sqrt{2}R)$ against the delay angle α, **(b)** the power factor PF versus α, **(c)** the power factor PF versus the average power delivered by the supply and **(d)** the total harmonic distortion THD versus the average power P delivered by the supply.

Solution

(a) From an inspection of Fig. 8.11b, the current response is finite and non-sinusoidal between the limits of α and π, the average value I_{dc} is zero and the symmetry indicates there are no even harmonics. Thus,

$i_s = i_l = \sum_{n=1,3,5\cdots}^{\infty} \hat{I}_{sn}\sin(n\omega t + \theta_n)$ and $\hat{I}_{sn}^2 = a_n^2 + b_n^2$ where

$$a_n = \frac{1}{\pi}\int_0^{2\pi} i_s \cos n\omega t \, d\omega t = \frac{2}{\pi}\int_\alpha^\pi \frac{\hat{V}_s}{R}\sin\omega t \cos n\omega t \, d\omega t \,.$$

For $n \geq 3$, $a_n = \dfrac{2\hat{V}_s}{\pi R}\,\dfrac{1 - n\sin n\alpha\sin\alpha - \cos n\alpha\cos\alpha}{n^2 - 1}$.

For $n = 1$, $a_n = \dfrac{\hat{V}_s}{2\pi R}(\cos 2\alpha - 1)$.

$$b_n = \frac{1}{\pi}\int_0^{2\pi} i_s \sin n\omega t \, d\omega t = \frac{2}{\pi}\int_\alpha^\pi \frac{\hat{V}_s}{R}\sin\omega t \sin n\omega t \, d\omega t \,.$$

For $n \geq 3$, $b_n = \dfrac{2\hat{V}_s}{\pi R}\,\dfrac{n\cos n\alpha\sin\alpha - \sin n\alpha\cos\alpha}{n^2 - 1}$.

For $n = 1$, $b_n = \dfrac{\hat{V}_s}{2\pi R}(2\pi - 2\alpha + \sin 2\alpha)$.

At $\alpha = 0$, the rms value of the fundamental component of current is $I_{s1}(\alpha = 0) = \hat{V}_s/(\sqrt{2}R)$. All values of the rms harmonic components of current are normalized by dividing the actual values by $\hat{V}_s/(\sqrt{2}R)$. The results of I_{sn} for $n = 1,3$ and 5 are plotted in Fig. EX8.7a.

(b) The power factor PF is given by eq. (8.9.10) to be

$$\text{PF} = \left[\left(1 - \frac{\alpha}{\pi}\right) + \frac{\sin 2\alpha}{2\pi}\right]^{1/2}.$$

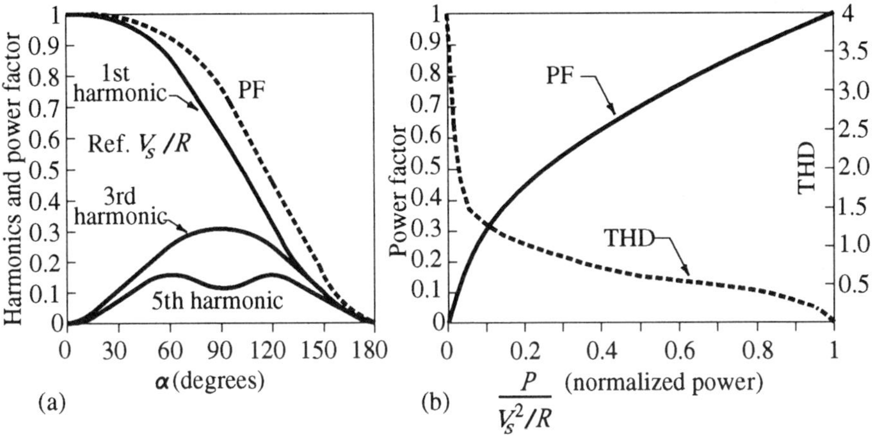

Fig. EX8.7

The plot of PF versus the delay angle α is shown in Fig. EX8.7a. The power factor is lagging and it becomes small at high values of α (low values of average power). The power factor is solely dependent on the angle α.

(c) From eq. (8.9.10), $PF = \left[\left(1-\frac{\alpha}{\pi}\right)+\frac{\sin 2\alpha}{2\pi}\right]^{1/2}$.

From eq. (8.9.2), $P = \frac{\hat{V}_s^2}{2R}\left[\left(1-\frac{\alpha}{\pi}\right)+\frac{\sin 2\alpha}{2\pi}\right]$.

If the maximum possible power, V_s^2/R is used as a reference, the actual value of power can be normalized (per unit value or percentage value divided by 100) by dividing by the reference.

In this case normalized average power $= \frac{P}{V_s^2/R} = PF^2$.

This is plotted in Fig. EX8.7b. The greater is the power to the load, the better is the power factor. That is, the smaller is α, the better is PF.

(d) By definition the total harmonic distortion is

$$THD = \left[\frac{I_s^2}{I_{s1}^2}-1\right]^{1/2}.$$

The rms value of the supply current I_s is, from eq. (8.9.3),

$$I_{l\,rms} = I_s = \frac{\hat{V}_s}{\sqrt{2}R}\left[\left(1-\frac{\alpha}{\pi}\right)+\frac{1}{2\pi}\sin 2\alpha\right]^{1/2}$$

and the rms value of the fundamental component is

$$I_{s1} = \frac{1}{\sqrt{2}}\left(a_1^2+b_1^2\right)^{1/2} = \frac{\hat{V}_s}{2\sqrt{2}\,\pi R}\left[(\cos 2\alpha-1)^2+(2\pi-2\alpha+\sin 2\alpha)^2\right]^{1/2}.$$

For a range of α from 0 to π radians, THD and P can be calculated. The results are plotted in Fig. EX8.7b in the form of THD versus normalized power. The greater is the power delivered to the load, then the lower is the harmonic distortion. However, the THD factor is solely dependent on the angle α. It is independent of the load resistance or the supply voltage and frequency.

Inductive loads. In ac-ac conversion the current response to inductive loads is depicted in Fig. 8.12 and the function is described in eq. (8.9.4). The determination of the circuit power factor and harmonic distortion cannot be carried out by simple analysis like that for the purely resistive load. Computer-aided analysis is required.

Asymmetrical control.
If the load power to be controlled in ac-ac conversion is too great for the triac or if the frequency of the alternating voltage is too high, there is a range of switch configurations that may be suitable. Some give symmetrical switching waveforms and some give asymmetrical output waveforms. Some of the latter may be acceptable for some applications, where a partial range of power modulation is specified.

Figure 8.13 shows some of the circuit diagrams for alternative control of ac power. For those circuits with high frequency and natural turn-off facilities, the delay angle α control can be expanded to α, β control (β is the extinction angle) and even PWM control.

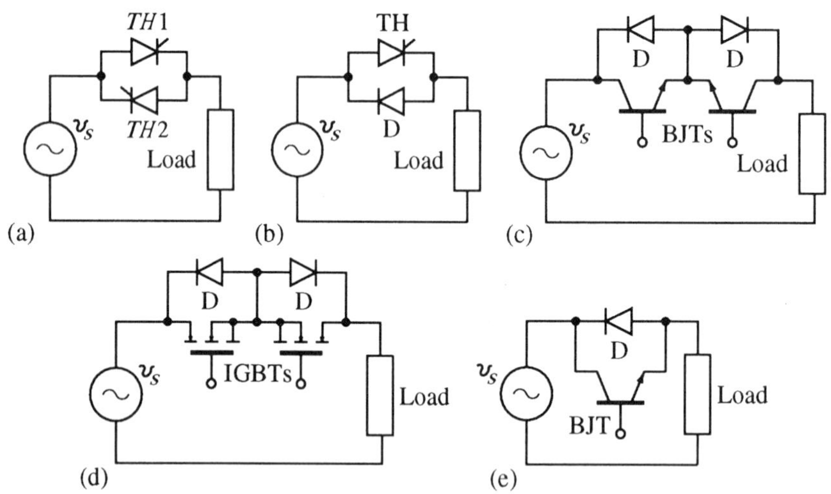

Fig. 8.13 AC voltage regulation. (a) Symmetry, (b) asymmetry, (c) symmetry, (d) symmetry, (e) asymmetry.

8.9.2. AC-DC Conversion

All the applications in this chapter have been associated with ac-ac conversion because this is to what triacs are naturally suited. For special low power applications it is possible to use triacs for modulating power in ac-dc conversion circuits. A simple adaption of any of the circuits given in this chapter (and that includes Fig. 8.13) allows ac-dc conversion. Between the switch, that controls conduction, and the load, a diode bridge can be connected. If the diode bridge is considered ideal then all the calculations for the performance factors that have been made, apply also to this form of ac-dc conversion.

8.10. SUMMARY

One of the important features of the triac semiconductor switch is that it can control the conduction of current in either direction. This implies that the device can also block voltages of either polarity across its two main terminals.

The structure of the triac is a composite form of *pnpn* layers that liken it to the thyristor in form and action. This allows the triac to be modelled as two thyristors connected in reverse parallel with a common gate connection.

All the triac characteristics can be derived from a knowledge of the thyristors behaviour. The triac is a minority carrier device, so that turn-on and turn-off are relatively slow, just like the thyristor. The triac latches on following a gate pulse of sufficient magnitude and duration. However, the gate pulse can be of either polarity no matter what is the polarity of the main terminals. There is a preference for negative gate pulses because of the increased sensitivity to turn-on. Triac turn-off is accomplished by allowing the main current to reduce to zero. This occurs naturally in ac applications. Accordingly, it is to be expected that triacs find application in the control of ac-ac conversion.

The power levels and the frequency of switching that are handled by triacs are both low. Domestic applications such as light dimming, temperature control and speed adjustment of small motors are common and uncomplicated. The power modulation is by voltage regulation, that is controlled by delay angle α triggering. One disadvantage of this means of control is that the quality of power deteriorates as the angle α is increased to obtain a reduction of power. The power factor decreases and the harmonic distortion increases. Nevertheless, the triac can do what no other single switch can do. It can regulate ac voltage.

8.11. PROBLEMS

Section 8.3

8.1 A 1200-V, 15-A(rms) triac requires a minimum signal of 2.5 V and 100 mA at the gate terminals to turn on the switch under all conditions. Determine the maximum **(a)** current gain and **(b)** power gain of the device.

8.2 A 1200-V, 15-A triac acts as a switch in a transformer tap changer whose source voltage is 600 V at 50 Hz and whose load is 10 A. The on-state voltage of the triac is 2.2 V. Determine **(a)** the conduction energy loss per cycle in the triac, (**b**) the maximum average power conduction loss in the triac and (**c**) the efficiency of the switch.

8.3 A triac acts as an ac switch in a circuit whose supply voltage is 400 V at 60 Hz and a resistive load whose value is $R = 40\,\Omega$. Triac data for rated temperature are the on-state voltage is $V_{T(ON)} = 2$ V, the maximum leakage current is $I_{T\,leak} = 5$ mA, the maximum signal at gate terminals are $V_{GT} = 2.5$ V and $I_G = 100$ mA. Compare the on-state losses with the off-state losses and continuous gate signal losses.

Section 8.5

8.4 A 500-V, 25-A triac modulates power from a 300-V, 400-Hz supply to a lamp that can be modelled as an equivalent, constant resistance of $15\,\Omega$ in steady operation. Triac data are that the delay time is $t_d = 0.5\,\mu$s and the crossover time is $t_c = 2\,\mu$s for turn-on, and the on-state voltage drop is $V_{T(ON)} \approx 1.5$ V. If the delay angle for triggering the triac is set at $\alpha = \pi/3$ radians, determine (**a**) the average power dissipated in the triac due to the turn-on process and (**b**) the average power dissipated in the triac due to on-state conduction.

8.5 A 500-V, 8-A triac acts as a solid-state relay in a circuit with a 240-V, 50-Hz supply and a resistive load of $40\,\Omega$. Triac data are that in the worst case a gate signal $V_G = 0.5$ V driving a gate current $I_G = 75$ mA gives a turn-on time $t_{on} = 2\,\mu$s and demands a latching current of 50 mA. If the triac is to allow the supply voltage to be applied to the load for no more than one half cycle, what is the minimum pulse width to ensure turn-on?

8.6 A 600-V, 16-A triac is used to modulate power from a 240-V, 50-Hz supply to an RL load, whose resistance is $R = 20\,\Omega$ and whose inductance is $L = 40$ mH. Triac data are that the latching current is $I_l = 65$ mA for a positive terminal T_2 and is $I_l = 45$ mA for a negative terminal T_2. If the load current is just continuous, determine (**a**) the value of the delay firing angle α, (**b**) the average power absorbed by the load and (**c**) the minimum gate pulse width for satisfactory operation.

8.7 For the given gate-circuit configuration shown in Fig. P8.7 a 500-V, 8-A triac is used. The train of gate pulses are produced by an astable multivibrator (see section 5.5.2, Fig. 5.13). Each pulse has a magnitude of 10 V and a pulse width of $2\,\mu$s. Triac data are that, for successful turn-on, the gate current I_G must be greater than 50 mA and the gate voltage V_{GT} must be greater than 1.5 V. The average gate power must be less than 0.5 W and the peak gate power must be less than 5 W. Determine (**a**) the minimum value of the added resistance R_G, (**b**) the maximum gate-pulse frequency and (**c**) the average gate current I_G.

8.8 Consider the circuit diagram in Fig. P8.7. The gate supply and switch are replaced by a Schmitt trigger circuit for delay angle α control. See section 5.5.2, Fig. 5.24 and EXAMPLE 5.6. If the triac is turned on at angle $\alpha = \pi/3$ radians, determine the gate-pulse width delivered by the one shot to

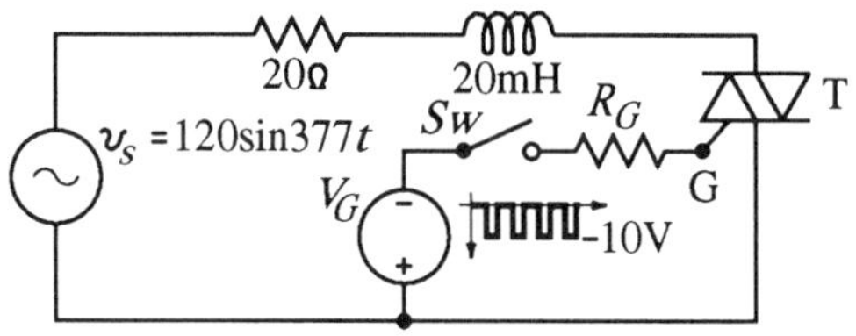

Fig. P8.7

ensure successful turn-on. Triac data are that the device is rated at 500 V, 8 A, the maximum latching current is 65 mA and the minimum gate current is 60 mA.

Section 8.6

8.9 A 500-V, 8-A triac modulates power from a 115-V, 60 Hz supply to a highly inductive load, whose values are $R = 100\,\Omega$ and $L = 100$ mH. If the triac turn-off time is 20 μs, estimate (**a**) the worst possible dv/dt at turn-off, (**b**) the value of di/dt at turn-off. (**c**) What is the di/dt at turn-on if $\alpha = \pi/2$ radians?

Section 8.7

8.10 Consider a triac whose ratings are given in Table 8.1. Estimate the components of maximum power loss in this triac and the efficiency of operation in an ac voltage regulation application. Make any appropriate assumptions.

8.11 A triac modulates power from a 250-V, 50-Hz supply to a magnet coil whose resistance is $R = 10\,\Omega$ and whose inductance is $L = 100$ mH. Determine the voltage and current ratings of the triac.

8.12 A triac has an rms current rating of 4A and the on-state voltage drop is $V_{T(ON)} = 1.7$ V. The maximum value of thermal resistance, junction to sink is $R_{\theta JS} = 3.6$°C/W. If the junction temperature is not to exceed 120°C and if the maximum ambient temperature is 50°C, determine the maximum permitted value of the thermal resistance, sink to ambient, $R_{\theta SA}$.

8.13 A triac has the following data. The maximum on-state voltage drop is $V_{T(ON)} = 1.6$ V and the maximum junction temperature is 120°C. The heat-sink can be maintained transiently at 60°C. If the thermal resistance, junction to sink is $R_{\theta JS} = 2.1$°C/W and if the transient thermal impedance is $Z_{\theta JS(t)} = 0.29$°C/W for a pulse of duration 1ms, determine (**a**) the maximum continuous current that can be conducted by the triac and (**b**) the magnitude of a non-recurrent rectangular pulse of current of duration 1ms.

8.14 A 500-V, 16-A triac modulates power from a ±300-V, 200-Hz, rectangular-waveform supply to a resistive load whose value is $R = 10\,\Omega$. Triac data are that the on-state voltage drop is $V_{T(ON)} = 1.6$ V, the thermal resistance is $R_{\theta JS} = 3.0$°C/W, the transient thermal impedance is $Z_{\theta JS(t)} = 0.24$°C/W for

1.25 ms pulses. If the trigger angle delay is set at $\alpha = \pi/2$ radians and the heatsink is maintained at 40°C determine (**a**) the average value of the junction temperature and (**b**) the maximum value of the junction temperature.

Section 8.8

8.15 A triac regulates the voltage from a rectangular wave, ± 200-V amplitude supply to a 10-Ω resistive load. The stray inductance of the circuit and supply is measured to be 10 μH. What value of inductance must be added to the circuit if the triac has a specified maximum di/dt of 10 A/μs at turn-on?

8.16 A 500-V, 16-A triac modulates power from a 240-V, 50-Hz supply to an *RL* load whose values are $R = 20\,\Omega$ and $L = 20\,\text{mH}$. An unpolarized snubber circuit is employed. The values are $R_s = 120\,\Omega$ and $C_s = 0.1\,\mu\text{F}$. If the delay angle of firing the triac is $\alpha = \pi/3$ radians, (**a**) caculate the extinction angle β of the triac, (**b**) estimate the maximum reverse recovery dv/dt imposed on the triac by the snubber circuit during turn-off, (**c**) compute the peak reverse voltage across the triac and (**d**) determine the order of the power rating of the snubber resistance R_s. Make whatever assumptions are necessary, since no recovery data are given for the triac.

Section 8.9

8.17 A 500-V, 16-A triac modulates power from a 208-V, 60-Hz supply to a purely resistive load, whose value is $R = 16\,\Omega$. If the delay angle of triac triggering is $\alpha = 2\pi/3$ radians, calculate (**a**) the average power P to the load and (**b**) the power factor PF.

8.18 A 500-V, 16-A triac modulates power from a 240-V, 60-Hz supply to a load, whose value is $R = 16\,\Omega$. It is desired to improve the system power factor to unity for the condition that the delay angle α is $2\pi/3$ radians. What is the value of the capacitor that must be placed across the supply to satisfy this requirement?

8.19 A 500-V, 16-A triac modulates power from a 208-V, 50-Hz supply to a purely resistive load, whose value is $R = 16\,\Omega$ and whose inductance is 40 mH. If the delay angle of triac triggering is $\alpha = 2\pi/3$ radians, determine (**a**) the instantaneous value of the load current, (**b**) the conduction interval each half cycle, (**c**) the rms value of the load current, (**d**) the average power delivered by the supply and (**e**) the system power factor.

8.20 A triac modulates the power from a 120-V, 60-Hz supply to a purely resistive load, whose value is $R = 6\,\Omega$. If the delay angle α is set to $\pi/2$ radians, determine (**a**) the total harmonic distortion THD and (**b**) the magnitude of the third harmonic current.

8.21 Consider the circuit diagram in Fig. 8.13b. Let the supply be 440 V at 50 Hz and let the load resistance be $5\,\Omega$. If the thyristor is triggered at $\alpha = \pi/2$ radians, determine (**a**) the dc component of the supply current, (**b**) the average power delivered to the load and (**c**) the system power factor.

8.22 An ac source supplies power to two resistors R_1 and R_2 in series. A triac is connected across the resistor R_1. Determine the input impedance of the circuit as a function of the firing angle α of the triac.

8.23 Figure P8.23 shows a circuit diagram of a triac tap changer. If triac T1 receives gate pulses at the beginning of every half cycle and if triac T2 receives gate pulses at any angle α, determine the average load power for $\alpha = \pi/2$ radians.

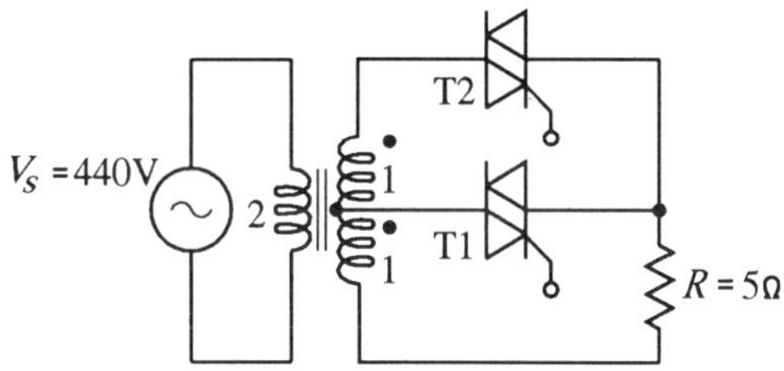

Fig. P8.23

8.12. BIBLIOGRAPHY

Dubey, G.K. *et al. Thyristorised Power Controllers*. Wiley Eastern Ltd., 1986.

Fisher, Marvin J. *Power Electronics*. Boston, Mass.: PWS-Kent Publishing Co., 1991.

Grafham, D.R., F.B.Golden. (EDS). *SCR Manual*. 6th ed. New York: General Electric, 1979.

Humphreys, M.J. *et al.* (EDS). *Philips Power Semiconductor Applications*. Hazel Grove, U.K: Philips.

Laster, Clay. *Thyristor Theory and Application*. Blue Ridge Summit: PA Tab Books Inc., 1986.

CHAPTER 9

THE GATE TURN-OFF THYRISTOR (GTO)

9.1. INTRODUCTION

As its name implies, the gate turn-off thyristor (GTO) is a power semiconductor switch that is like a thyristor except that it has the added advantage that a negative gate-current pulse can turn off the device. In fact, the GTO is the turn-off semiconductor switch with the highest current and voltage ratings presently available.

The electric-circuit symbol for the GTO is similar to that of the thyristor with an addition to the gate electrode to represent the property of gate turn-off. Figure 9.1 illustrates the GTO circuit symbols. The symbol on the left depicts the dual role of the gate terminal. This symbol is used throughout this chapter. The three terminals, anode A, cathode K and gate G have the same representation as the thyristor. If the anode A is positively biased with respect to the cathode K and if a current pulse is injected at the gate by means of a voltage that is positive at G with respect to the cathode K, the GTO latches into the on-state. The GTO remains on until the anode current falls to a value below the holding current I_h (this is a relatively high value). However, the most important characteristic of the GTO is its turn-off ability. If a negative gate signal is injected while the GTO is in the on-state, the device will switch off. It will remain off until the conditions of a positive anode and a positive gate with respect to the cathode are met.

The useful and economic power range of the GTO lies between the bipolar power transistor (or the IGBT) and the thyristor. It does not switch as fast as the transistor but it can handle higher voltage and current levels. It is faster than the phase-controlled thyristor and can be turned off with a gate signal, but it is not capable of modulating such high powers as the thyristor. Typical ratings of a GTO are 1600 V and 350 A(average). However, there are GTOs that have maximum voltage ratings about 4000 V and there are GTOs that have maximum current ratings about 3000 A. In comparison there are thyristors that have maximum voltage ratings of 5000 V and there are thyristors that have maximum current ratings of about 5000 A. Accordingly by cost and ratings thyristors will not be replaced by GTOs at the highest power levels. The turn-on time of the 1600-V, 350-A GTO is less than 8 μs with a dc gate current of 2A, and the on-state voltage drop is about 3.4 V. A negative peak gate current of 24 A will interrupt a load current of 120 A (a gain of 5) in 15 μs. A typical thyristor, that is used in a phase-control application, and that has the same voltage and current ratings as the GTO, will turn on in 2 μs with a gate current of 200 mA, will have an on-state voltage drop of 1.5 V, and will turn off in 150 μs. Here, it is seen that the overall switching speed of the GTO is faster than a thyristor with comparable voltage and current, but the GTO has a higher on-state voltage drop.

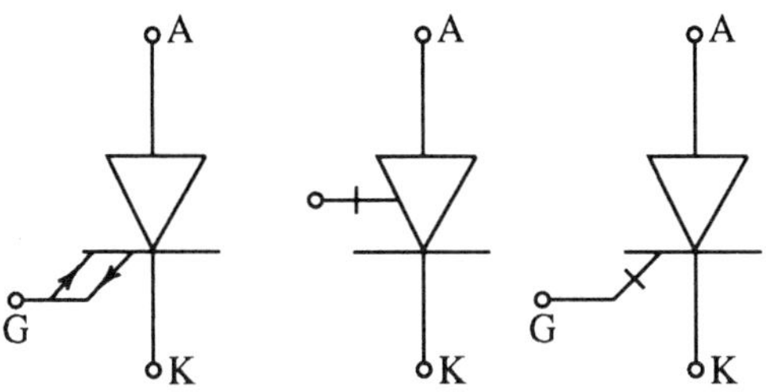

Fig. 9.1 GTO circuit symbols.

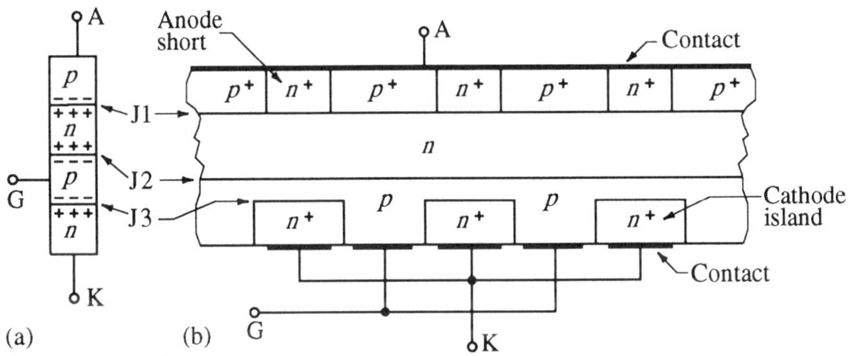

Fig. 9.2 GTO structure.
(a) Basic form, (b) practical form.

Since the GTO has unidirectional current conduction and can be turned off at any time, it finds application mainly in chopper circuits (dc-dc conversion) and in inverter circuits (dc-ac conversion) at power levels above which MOSFETs, BJTs and IGBTs cannot be used. At the low power levels the faster-switching semiconductor devices are preferred. Even in ac-dc conversion GTOs find application because switching strategies other than delay angle α control can be used to improve the quality of the power such as power factor.

9.2. GTO STRUCTURE

The structure of the GTO is essentially the same as the conventional thyristor. As shown diagrammatically in Fig. 9.2a, there are four silicon layers (*pnpn*), three junctions, and three terminals (anode A, cathode K and gate G). As far as the blocking and conduction nodes are concerned, the behavior of the GTO and the thyristor are the same basically. The difference in operation between the two is that a negative gate signal can turn off the GTO, whereas the anode current of the thyristor has to be reduced to zero externally for conduction to cease. In order to create this difference there has to be a design modification that sacrifices the quality of some characteristics of the GTO such as reduced reverse blocking capability and increased conduction losses. Figure 9.2b illustrates the general form of the structure of the GTO. The outer *n*-layer is the cathode connection and the inner *p*-layer is the gate connection, just like the thyristor. However, the GTO is designed to have a reduced *pnp* current gain for the purpose of lower internal regeneration. This allows the switch to turn-off if enough current is drawn from the gate terminal by means of a negative bias, gate to cathode.

In order to accomplish a reduced current gain, highly-doped *n*-spots in the anode *p*-layer create a shorted-emitter effect as shown in Fig. 9.2b. This creates the added advantage of a low on-state voltage, but also reduces the reverse voltage blocking capability to a low value. If a high reverse blocking voltage is a requirement, then the GTO can have its low *pnp* current gain by the diffusion of gold which reduces carrier lifetime. However, attendant with this gold diffusion there is a higher on-state voltage drop across the device. In Fig. 9.2b, cathode islands (or fingers) are illustrated. These are used in high-power GTOs to make the spacing between gate and cathode small. Turn-on is faster with this form of interdigitation, because the distance for dynamic plasma spread at turn-on is kept short.

9.3. GTO I-V CHARACTERISTICS

The steady-state current-voltage characteristics are similar to those of the conventional thyristor. The GTO I-V characteristics are depicted in Fig. 9.3. A simple test circuit is illustrated in Fig. 9.3a to show the voltage and current symbols.

Ideally, with no gate signal and the device off, the GTO remains off no matter what the value or polarity of the voltage V_{AK}. In this state the load voltage is $V_l = 0$ and the GTO voltage is $V_{AK} = V_s$.

For the conditions that the GTO is off and the anode is positively biased by the source voltage V_s, if a small, momentary positive signal I_G is applied to the gate, the GTO changes to the on-state, the GTO voltage drops to $V_{AK} = 0$, the load voltage becomes $V_l = V_s$ and the circuit current is determined by the voltage V_l and the load R. See Figs 9.3a, 9.3b and 9.3c. Once in the on-state, the GTO remains on even if the gate signal is removed.

For the conditions that the GTO is on (with $V_{AK} = 0$, I_s finite and $V_l = V_s$), if a negative signal $-I_G$ is applied to the gate, the GTO changes to the off-state and blocks. This means that the circuit current is $I_A = 0$, the load voltage is $V_l = 0$ and the GTO voltage is $V_{AK} = V_s$. Once off, the GTO remains blocking until a positive gate pulse is reapplied.

If the GTO is on and the anode current I_A is reduced to zero, then the GTO turns off and remains blocking no matter what the anode bias is.

In practical terms the I-V characteristics are not like the ideal case shown in Fig. 9.3c. They are like those shown in Fig.9.3d. While the GTO is off and there is no gate signal, the device blocks with either polarity of anode bias, but a

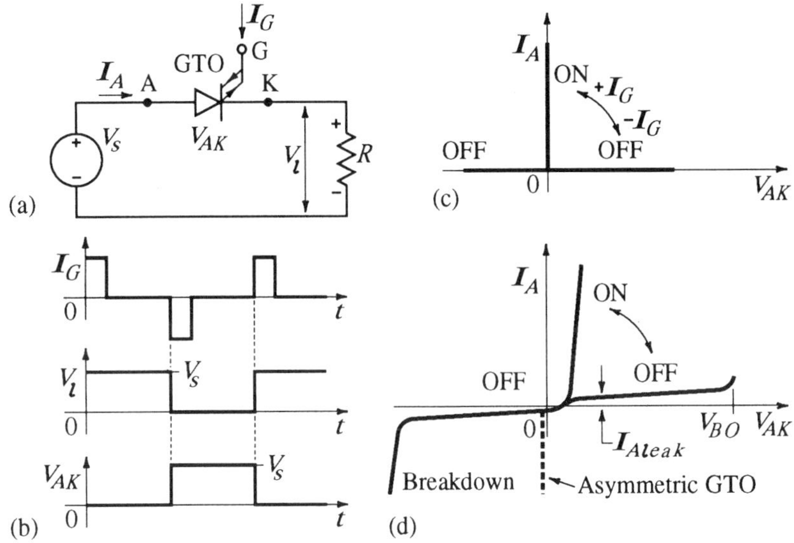

Fig. 9.3 GTO I-V characteristics. (a) GTO circuit diagram, (b) waveforms, (c) ideal characteristics, (d) actual characteristics.

leakage current $I_{A\,leak}$ will exist. With a forward bias (anode A is positive with respect to the cathode K) the GTO blocks until a breakover voltage $V_{AK} = V_{BO}$ is reached. At this point there is a dynamic process of turn-on, $V_{AK} \approx 3$ V and the current I_A is determined by the load. With the GTO off and with the application of reverse bias, only a small leakage current $I_{A\,leak}$ exists. A reverse-bias voltage V_{AK} can be reached at which avalanche breakdown occurs. The value of reverse-breakdown voltage depends on the fabrication method of creating a reduced internal regeneration to facilitate turn-off. It can be as low as -30 V for GTOs with a shorted anode structure, because junction J3 is the only junction blocking. This is a negligible reverse blocking capability. However, this characteristic can be accepted in inverter applications. GTOs with low reverse blocking capabilities are called asymmetric GTOs. The asymmetric GTOs may have a built-in reverse connected diode for protection in inverter circuits.

With a forward-bias voltage applied to the anode and a positive current pulse applied to the gate the GTO turns on and remains on. There is a voltage drop $V_{AK} = V_{GTO(ON)}$ across the device, if it is on, and the value is greater than that across the thyristor. Its value is somewhere between 2.4V and 3.4V. The current I_A is a function of the load once the GTO is on.

For the condition that the GTO is in the on-state, there are two ways to turn it off. One way is to reduce the anode current I_A by external means to a value below the holding current value I_h, at which value the internal regenerative action is no longer effective. The value of I_h in the GTO is much higher than that for the thyristor (of the order of amperes rather than milliamperes), because of the design requirements of the GTO to turn off with the application of a negative current pulse at the gate terminal. Turn-off by means of a gate pulse is the second method of turn-off and is the desired method because it gives the best control.

The characteristics, shown in Fig. 9.3c and 9.3d illustrate that the current in the GTO is unidirectional, that the voltage withstand capability is usually unipolar and that by means of current pulses at the gate there is both controlled turn-on and controlled turn-off. This latter characteristic gives some superiority to the GTO over thyristors of similar ratings. See Appendix 7 for data.

EXAMPLE 9.1

A 1000-V, 25-A GTO controls power from a dc supply of voltage $V_s = 600$ V to a resistive load of value $R = 30\,\Omega$. See Fig. 9.3a. The GTO data sheet provides the information that the on-state voltage drop is $V_{GTO(ON)} \approx 2.2$ V, the average gate power P_G is not to exceed 10 W, gate current I_G for turn-on is 0.5 A and for turn-off is -25 A. Determine (**a**) the device power gain, (**b**) the device turn-on current gain, (**c**) the device turn-off current gain and (**d**) the maximum average power loss in the GTO.

Solution

(**a**) The load current $I_l = (V_s - V_{GTO(ON)})/R = (600-2.2)/30 = 19.93$ A.
The load power $P = V_l^2/R = (V_s - V_{GTO(ON)})^2/R = 597.8^2/30 = 11.91$ kW.
The gate power $P_G = 10$ W.
Thus, the power gain $= P/P_G = 11.91 \times 10^3/10 = 1191$.
This is a reasonable gain for control purposes.

(**b**) For GTO turn-on, $I_G = 0.5$ A.
Thus, the turn-on current gain is $I_A/I_G = I_l/I_G = 19.93/0.5 = 39.9$.
This is a reasonable gain for turn-on control.

(**c**) For GTO turn-off, $I_G = 25$ A.
Thus, the turn-off current gain is $I_A/I_G = I_l/I_G = 19.93/25 = 0.8$.
This is a very low gain. It is the price to pay for controlled turn-off.

(**d**) Conduction power loss P_C in the GTO for continuous current is
$P_C = V_{GTO(ON)} I_A = 2.2 \times 19.93 = 43.8$ W.
The average gate loss over a long period is negligible if gate pulses are applied. Hence, the efficiency η is

$$\eta = \frac{\text{output}}{\text{output} + \text{losses}} = \frac{11.91 \times 10^3}{11.91 \times 10^3 + 43.8} = 0.996. \quad (99.6\%)$$

This is good.

9.4. GTO TWO-TRANSISTOR MODEL

The *pnpn*-structure of the GTO can be divided into *pnp* and *npn* structures just like the thyristor. These two transistors form a model to give an aid to the GTO switching actions. Figure 9.4 shows the arrangement and nomenclature of the two-transistor model.

GTO Turn-on.
Consider that the GTO is in the blocking state, and refer to the model in Fig. 9.4. If the anode A is positively biased with respect to the cathode K and if a positive pulse of current I_G is injected into the gate terminal G, which is the base of the *npn* transistor, transistor *BJT*1 turns on. The base signal I_{B1} is made sufficiently large so that there is no risk of second breakdown. Once *BJT*1 starts to turn on, its collector current I_{C1} is the base current I_{B2} of the *pnp* transistor *BJT*2. So *BJT*2 starts to turn on and its collector current I_{C2} augments the gate current I_G to increase the base current I_{B1} of *BJT*1. This positive feedback increases until the circuit latches on. At this point the GTO stays on, even if the gate signal is removed, because of the internal regeneration that the two-transistor model demonstrates so well. It is to be noted that the regenerative latching effect is not as pronounced in the GTO as in the thyristor. Otherwise the GTO would not switch off by means of a gate signal.

In the on-state the GTO behaves like a thyristor in that it has similar characteristics. That is, both generate their own charge carriers and both remain saturated at overcurrent conditions.

GTO Turn-off.
Consider the GTO to be in the on-state. To turn off the device the gate terminal is biased negatively with respect to the cathode terminal. The resulting negative gate current turns off the GTO and it continues to block after the gate signal is removed. With respect to the two-transistor model, the essence is that a negative gate signal blocks transistor *BJT*1. This removes the base signal from *BJT*2 which then blocks.

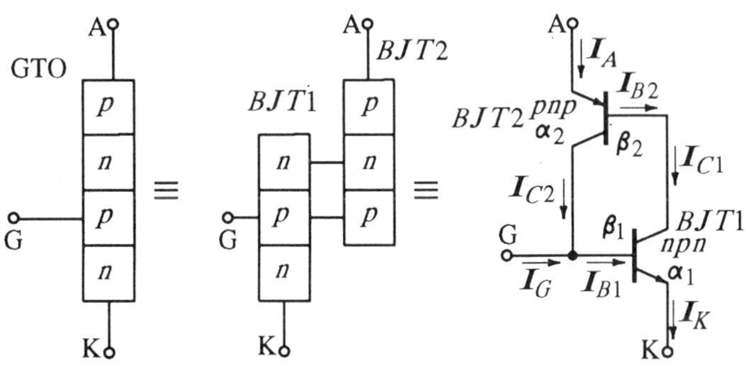

Fig. 9.4 Two-transistor model of the GTO.

From Fig. 9.4 it can be seen that a negative gate bias can draw excess carriers from the base region of the *npn* region of transistor *BJT*1. The collector current I_{C2} of the *pnp* transistor *BJT*2 can be diverted from the base region of *BJT*1 to the external gate circuit. This reduces the internal regeneration because the reduced *BJT*1 base drive reduces *BJT*2 base drive until eventually both are removed. This, in turn, stops conduction and the GTO should turn off and block. This action of reducing the internal regeneration can only be accomplished in the GTO by having a reduced *pnp* current gain for *BJT*2, compared with the thyristor. The lower current gain is achieved by the anode *p* layer having shorting spots (n^+) or by having gold diffusion to reduce carrier life times. A circuit analysis of the two-transistor model demonstrates the effectiveness of a reduced current gain.

Let the GTO be on and consider the two-transistor model, as shown in Fig. 9.4. In saturation the current gain β is the ratio of collector to base current and the current gain α is the ratio of collector current to emitter current. For GTO turn-off the gate G is negatively biased with respect to the cathode K. If the transistor *BJT*1 starts to turn off then we must have the condition that

$$I_{B1} < I_{C1} / \beta_1 . \tag{9.4.1}$$

However, while in saturation

$$\beta_1 \triangleq \frac{I_{C1}}{I_K - I_{C1}} = \frac{I_{C1}/I_K}{1 - I_{C1}/I_K} = \frac{\alpha_1}{1-\alpha_1} . \tag{9.4.2}$$

From Kirchhoff's current law and the current gain definition

$$I_{B1} = I_{C2} + I_G \quad \text{and} \quad \alpha_2 \triangleq I_{C2}/I_A . \tag{9.4.3}$$

Therefore,

$$I_{B1} = \alpha_2 I_A + I_G . \tag{9.4.4}$$

Also from Kirchhoff's current law and the definition of current gain

$$I_K = I_A + I_G \quad \text{and} \quad \alpha_1 \triangleq I_{C1}/I_K . \tag{9.4.5}$$

Therefore,

$$I_{C1} = \alpha_1 (I_A + I_G) . \tag{9.4.6}$$

Upon substitution of I_{B1} and I_{C1} in eq. (9.4.1)

$$-I_G > I_A \frac{\alpha_1 + \alpha_2 - 1}{\alpha_1} \triangleq \frac{I_A}{\beta_{off}} \tag{9.4.7}$$

where β_{off} is defined as the gain for GTO turn-off and is

$$\beta_{off} = \frac{\alpha_1}{\alpha_1 + \alpha_2 - 1} . \tag{9.4.8}$$

From eq. (9.4.7), it is advantageous to have the value of the turn-off gain β_{off} as large as possible. With this in mind, an inspection of eq. (9.4.8) leaves us with two conclusions. The current gain α_1 of transistor *BJT*1 should be as close to unity as possible (this is no different from the thyristor) and the current gain α_2 of transistor *BJT*2 should be as small as possible (this is contrary to the requirement of the thyristor). A thin inner *p*-layer and a heavily doped outer *n*-layer give a high value for α_1 A thick inner *n*-layer with gold diffusion gives rise to a low value of α_2. In practice the value of β_{off} is somewhere between 3 and 5. This is a low value. A controllable anode current I_A of 300 A will require at least −60 A from the gate terminal to turn off the GTO. Fortunately, the negative gate current need exist for only a few microseconds, so the energy requirement is low.

9.5. GTO TURN-ON

The preferred method of GTO turn-on is the application of a small gate current because it is a controllable method. To be avoided is turn-on by a breakover voltage V_{BO} at the anode, or turn-on by current created by a high dv/dt (typically 400 to 1000 V/μs) at the anode. These methods are all similar to thyristor turn-on.

A GTO will turn on if the anode is positive with respect to the cathode and if the gate is positively biased with respect to the cathode such that there is a positive current of sufficient magnitude and duration at the gate. The gate current value of the GTO is about the same as the thyristor of the same rating. For example, a gate voltage V_G of 12 V driving a gate current I_G of 2 A for 10 μs would be normal for turning on a GTO rated at 1000 V, 400 A.

Once the device has been turned on the gate signal can be removed and the device stays on. This assumes that the gate signal is applied long enough for the anode current i_A to exceed the latching value I_{la}, at which value the carrier multiplication, that gives rise to internal regeneration, is self maintaining. This was demonstrated by the two-transistor model. Because the internal regeneration is made lower in the GTO to allow gate turn-off, the latching current is relatively high and the gate pulse duration is relatively long compared with the thyristor (a typical value is 10 μs for the GTO, whereas 3 μs might suffice for the thyristor). A 400-A thyristor might have a latching current whose value is about 1 A, whereas a 400-A GTO might not latch until the anode current reaches 30 A.

For the same reason that the latching current I_{la} is high for GTOs so, too, is the holding current I_h. It is advantageous for a thyristor to turn off once the anode current I_A falls below the holding value, but it may not be convenient to allow a GTO to turn off if the anode current falls below the holding value (30 A for example). This value is too high. Therefore, it is usual to maintain a low-level, dc current at the gate to prevent turn-off if the anode current should fall to a relatively low value momentarily. Further, due to the high concentration of gate-to-cathode interdigitation in the GTO, there is a tendency for some of the cathode islands to turn off, if the anode current falls to a value near the holding-current value and if there is no gate signal. A subsequent rise of anode current might cause hot spots in the remaining inlands that are on. This localized turn-off and possible damage are avoided if a continuous gate current is applied. As a result the on-value of the voltage v_{AK} is kept low.

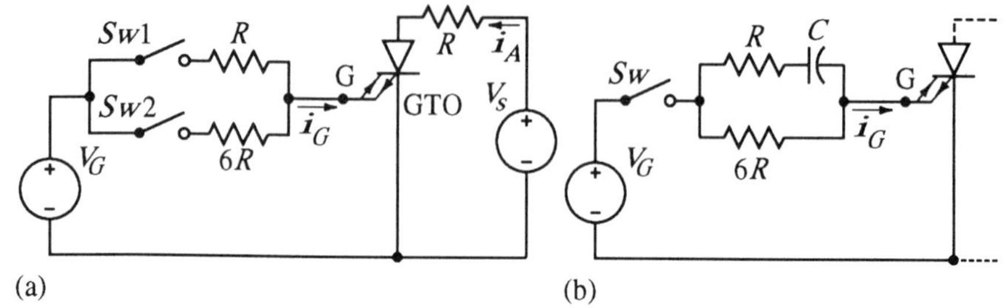

Fig. 9.5 GTO gate circuit diagrams.
(a) Resistive, (b) capacitive.

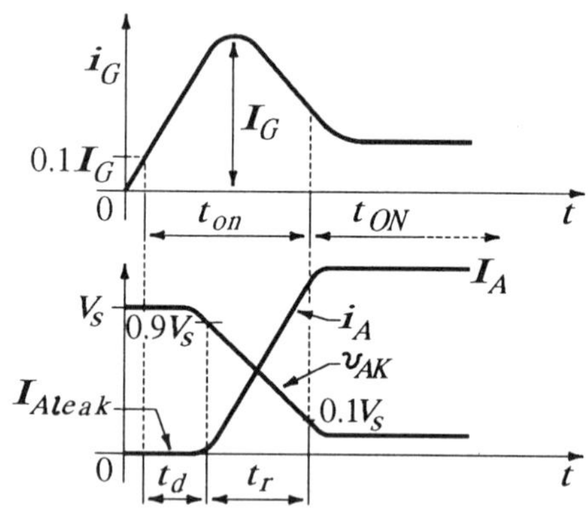

Fig. 9.6 GTO response to a positive gate signal.

Fast turn-on is built into the design of the GTO, because the interdigitation of gate and cathode regions means that the distance for plasma spread is short. However, the speed of turn-on is aided by both the magnitude I_G of the gate pulse and the rate of rise di_G/dt of that pulse. Two simple gating circuits are illustrated in Fig. 9.5. Figure 9.5a shows a resistive gate circuit. Switch *Sw*1 is closed to allow a high pulse of current to be injected at the gate terminal of the GTO. After about a 10-μs interval, switch *Sw*2 is closed and switch *Sw*1 is opened to allow a lower level of continuous gate current to exist as long as the GTO is to conduct load current. Figure 9.5b shows a modified gate circuit with only one switch. Closing switch *Sw* performs the two functions of an initial pulse by means of the capacitor C and then a dc gate current whose value is controlled by the resistor 6*R*.

The general form of gate current is shown in Fig. 9.6 together with the turn-on response of current i_A and voltage v_{AK}. The value of the rate of rise of gate current is typically 5A/µs. After the delay time t_d which is measured from the initiation of the gate signal to the time the voltage v_{AK} falls to $0.9V_s$, the current i_A rises and the voltage v_{AK} falls rapidly to the device's steady on-values. Because of interdigitation the di_A/dt of the GTO can be higher than that for the thyristor. A gate-pulse signal of 10 µs would normally allow uniform conduction throughout the GTO wafer rated at 400 A, so a gate-circuit RC time constant of 2 µs might seem appropriate. However, the turn-on time t_{on}, defined as the interval from $i_G = 0.1I_G$ to $v_{AK} = 0.1V_s$ for a resistive load, may be from 5 to 8 µs. Once on in a steady conduction, the voltage v_{AK} may have a value as high as 3V and the loss incurred during turn-on may be of the order of 200J.

The turn-on time t_{on} may be short, but the time t_{ON} that the GTO must be on is critical and must be taken into account. The time t_{ON} is not of much consequence for the thyristor, but for the GTO, the ON-time prepares the device for safe, successful turn-off. Uniform conduction in the GTO is imperative before a negative gate pulse is applied for turn-off, otherwise hot spots can be created. This requirement tends to limit the switching frequency of GTOs to values below 2 kHz. Choppers and stepped-wave inverters are applications that are not affected by the limitation.

EXAMPLE 9.2

A 1200-V, 400-A GTO modulates power from an 800-V dc supply to a resistive load, whose value is $R = 5\,\Omega$. For steady switching conditions, the GTO data are that the turn-on is $t_{on} = 6$ µs and this includes the delay time $t_d = 1.5$ µs, the leakage current is $I_{A\,leak} = 100$ mA and the on-state voltage drop is $V_{GTO(ON)} = 3.4$ V. The average gate power is 35 W over the turn-on interval. Estimate **(a)** the components of energy loss during the turn-on interval and **(b)** the energy loss during conduction if the minimum on-state time is $t_{ON} = 18$ µs.

Solution

Refer to Fig. 9.5a for the circuit diagram and see Fig. 9.6 for the turn-on waveforms. The waveforms are linearized to allow simple calculations.

(a) The energy loss W_d during the delay interval t_d is

$$W_d = \int_0^{t_d} v_{AK}\, i_A\, dt = V_{AK} I_{A\,leak} t_d = 800 \times 100 \times 10^{-3} \times 1.5 \times 10^{-6} = 120\,\mu\text{J}.$$

The turn-on loss W_r over the current rise and voltage fall interval $t_r = t_{on} - t_d$ is

$$W_r = \int_0^{t_r} v_{AK}\, i_A\, dt \approx \int_0^{t_r} \left(\frac{V_s}{t_r} t + V_s \right) I_A t\, dt = \frac{V_s I_A}{6} t_r \approx \frac{V_s^2}{6R} t_r\,.$$

Thus, $W_r \approx \dfrac{800 \times 800}{6 \times 5.0} \times (6 - 1.5) \times 10^{-6} = 96\,\text{mJ}.$

The delay-time loss can be neglected.

(b) During the minimum time t_{ON} that the GTO is in the on-state for the purpose of obtaining a uniform current distribution in the cathode islands in preparation for turn-off, the energy loss W_C is

$$W_C = \int_0^{t_{ON}} v_{AK}\, i_A\, dt \approx V_{GTO(ON)} \frac{V_s}{R} t_{ON} = 3.4 \times \frac{800}{5} \times 18 \times 10^{-6} = 9.8\,\text{mJ}.$$

The turn-on loss is significant.

9.6. GTO TURN-OFF

Consider the GTO to be in the on-state, conducting an anode current I_A. In simplistic terms, the application of a negative current pulse $-i_G$ at the gate terminal results in the GTO turning off in a few microseconds and blocking further anode current.

In slightly less simplistic terms we can refer to the two-transistor model, shown in Fig. 9.4. If the GTO is on and it is conducting an anode current I_A, the positive gate signal can be removed and the GTO does not change state. Conduction is maintained by the regenerative process that is provided by the collector currents driving the bases of the two transistors to keep them in saturation. Now, a voltage source, that biases the cathode K positive with respect to the gate G, drives a negative current at the base of transistor *BJT*1 so that the collector current I_{C1} reduces to zero. This means that there is no base current to transistor *BJT*2. So, both transistors turn off and conduction ceases. Without regeneration, the device returns to a blocking state. The actual turn-off action is somewhat more complex than this. Otherwise, why does the conventional thyristor not turn off with the injection of a negative pulse of current at the gate?

We can give a description of the turn-off action with the aid of Fig. 9.7. Let the GTO be in the steady-state on-condition, such that all the cathode islands are conducting uniformly and the central region is filled with excess carriers. Let the gate-circuit switch *Sw* be closed to initiate turn-off. The negative gate current sweeps out excess holes from the inner *p*-region, so there is an effective increase in the resistance between the gate fingers and the cathode islands, and an increase in the voltage drop v_{KG} as i_G rises in value. This is shown in the figure. There is a tendency for the anode current to focus into filaments over the cathode islands where the resistance is less. At this point the voltage v_{AK} rises slightly because of the increased anode current density. The reason why the negative gate current starts to interrupt regeneration in a GTO is that, because of the high level of interdigitation, the charges are swept out uniformly over the wafer as opposed to the very local action in the conventional thyristor. Turn-off is made faster by positioning the highly-doped *n*-spots, that form the anode shorts[1], in line with the cathode islands. This tends to squeeze the filaments towards the *n*-spots and further from the gate fingers. As a result, the path for the anode current mainly involves three layers (*npn*), so that the regenerative property is lost. The filaments

---

[1] The anode shorts are resistive in nature.

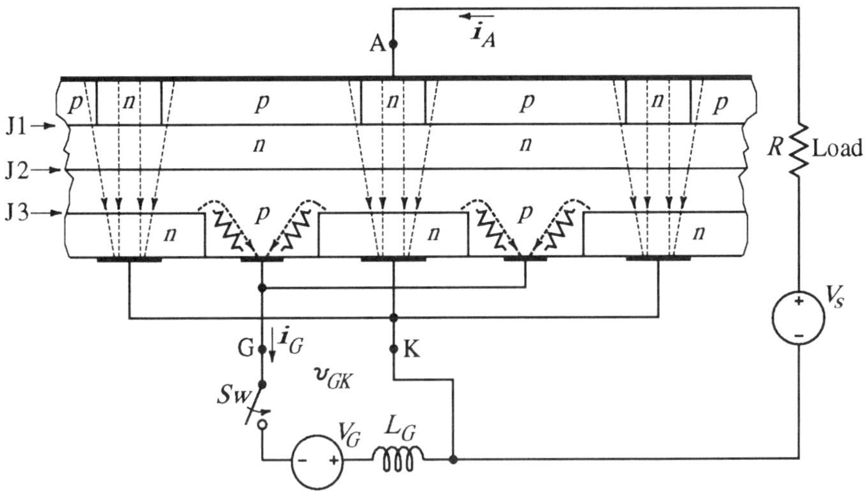

Fig. 9.7 GTO turn-off.

collapse because the current cannot sustain itself without carrier multiplication. Removal of the excess carriers in the inner *n*-region is by recombination and by diffusion in the *n*-spots of the anode region. This makes the current i_A fall quickly to a low value and makes the voltage v_{AK} rise rapidly. The final stage of turn-off is the relatively slow recombination of the minority carriers in the cathode islands; this is manifest by what is called the tail current and final recovery of the device to its blocking state. The faster turn-off with the anode shorts lowers the gain and increases the on-state voltage drop V_{AK}. Also, the off-state reverse blocking voltage is less than 50V.

Some comments should be made about the practical nature of turn-off. This is best done with a description of the dynamic characteristics of the voltages and currents during the change from the on-state to the off-state, as shown in Fig. 9.8. We will consider the turn-off of a 1600-V, 400-A GTO. The gate turn-off pulse is much greater than the turn-on pulse. A pulse length $t_p = 14\,\mu s$, an initial rate of change $di_G / dt = -50A / \mu s$ and peak current $\hat{I}_G = -100A$ would be typical to turn off an anode current $i_A = 400A$(instantaneous). Turn-off tends to be faster if di_G / dt is increased. The controllable I_A is increased, the storage time t_s is shorter and there is less energy dissipation, but there is a limit to I_A because of the need to prevent too high a density of anode current i_A as it focuses into filaments. An inductor L_G is connected in the gate circuit to set the limit of di_G / dt, and one of these (L_G or di_G / dt) has a value that is usually specified in the data sheets for the device.

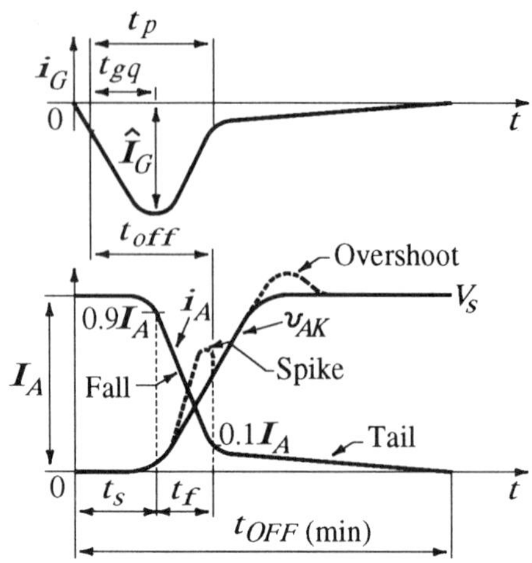

Fig. 9.8 Turn-off characteristics.

At a gate voltage $V_G = -20$ V the peak pulse power to turn off the 400 A device would be 2000 W at the gate. However, the average gate power might be only 10 W and the energy of the gate pulse only 20 mJ. This is small compared with the commutation energy of a 400 A thyristor. The current pulse for commutation in a thyristor can be twice the anode current for twice the time to turn off, and this could require energy that is typically 20 J, a much greater loss.

A reason to limit di_G/dt by an inductance L_G is to prevent avalanche between the cathode and gate. Because there is resistance between the cathode and gate, as depicted in Fig. 9.7, the rising gate current means an increase in the lateral voltage drop V_{GK} between the cathode islands and gate fingers. If the increase is much over 10 V avalanche occurs in this region. The gate current becomes confined to the low resistance avalanche region and does not contribute towards turn-off, because the surrounding stored charge is not removed. This means there is a maximum gate current ($\hat{I}_G = 100$ A in our example) that will turn off a particular anode current ($I_A = \beta_{off} \hat{I}_G$). An anode current above this particular value cannot be extinguished. Consequently high fault currents cannot be turned off by a gate signal, and gate signals must be blocked in such circumstances.

Snubber circuits were discussed in relation to limit forward dv/dt for a blocking thyristor. A snubber circuit is imperative for a GTO, not only for blocking but also to limit the dv/dt during the interval that the GTO recovers its blocking state. Figure 9.9 shows a GTO circuit diagram with an inductive load and a snubber circuit. The elements C_s, D_s and R_s are devices in the snubber circuit, but L_s represents the stray inductance in the loop. The protection afforded by the snubber circuit is explained in conjunction with the turn-off characteristics shown in Fig. 9.8. It can be assumed that commutation of the GTO occurs so quickly that the load current I_l can be considered to be virtually constant over this period. This is valid

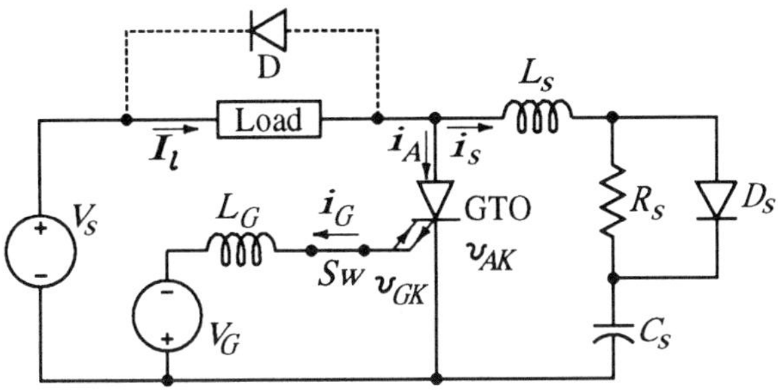

Fig. 9.9 GTO snubber circuit.

if the inductive load has a time constant of the order of milliseconds and the GTO has a turn-off time t_{off} of the the order of microseconds. The turn-off time is measured from $0.1\hat{I}_G$ to $0.1I_A$.

To initiate turn-off a negative gate pulse i_G is applied at $t=0$. The gate current rises at a rate that is limited by the value of the inductance L_G, the anode current remains nearly constant ($i_A = I_A = I_l$) during the storage time[2] and the voltage drop v_{AK} remains low. The increasing gate current i_G removes carriers. Then the anode current i_A decreases sharply (the rate can be 300 A/μs) to a low value over a time called the fall time[3] t_f, which can be as low as 2 μs for a 400-A device. After this, the anode current decreases slowly with a tail characteristic, while excess minority carriers in the cathode islands recombine until there is a uniform off-condition. After this, turn-on would also be uniform. From the time $t=0$ to the time the tail current is zero is the minimum interval t_{OFF} that the GTO can be off. The value of t_{OFF} could be in the region of 70 μs for a 400 A device, since the tail characteristic can persist for up to 50 μs.

While the current i_A is reducing and being focused into filaments during the fall time t_f the voltage v_{AK} across the switch is rising. The rate of rise dv_{AK}/dt must be limited to a value between 500 and 1000V/μs. Otherwise the GTO may turn on again by displacement current. Because the load current I_l is reasonably constant during commutation, the anode current is quickly diverted to the snubber capacitor C_s. Ignoring the stray inductance L_s, the current-voltage characteristic of the capacitor is

$$i_s \approx C_s \frac{dv_{AK}}{dt}. \qquad (9.6.1)$$

For $i_s \approx I_l = 400$ A and dv_{AK}/dt limited to 800V/μs the value of the snubber capacitor would be 0.5 μF. The value of the capacitance used in practice may be

[2] Storage time is the interval that the GTO remains saturated with charge carriers.

[3] The fall time is measured from 90% to 10% of the initial on-state anode current.

larger than this. C_s is made large to keep the voltage v_{AK} (and the value dv_{AK}/dt) low during the fall-time interval t_f. In this manner the power loss in the GTO during turn-off is kept low, although the snubber circuit loss is increased. During the tail current interval the same dv_{AK}/dt would be maintained until the voltage v_{AK} reached the supply voltage V_s.

Stray inductance L_s in the snubber circuit must be kept to a minimum (for example, leads must be kept short) to keep two adverse effects small. One effect is a voltage spike ($L_s di_s/dt$) on the voltage v_{AK} recovery curve. Since the anode current i_A falls quickly, the snubber current i_s rises quickly, so the spike can increase dv_{AK}/dt. The second effect is the voltage v_{AK} overshoot due to the resonance between the stray inductance L_s and the snubber capacitance C_s.

As far as maximum frequency of switching a GTO is concerned for a minimum on-time t_{ON} of 14 μs, and a minimum off-time t_{OFF} of 70 μs, this suggests a maximum frequency $f_{\max} = 1/(t_{ON} + t_{OFF}) = 10^6/84 \approx 12\,\text{kHz}$. However, the switching losses (about 200 mJ for each operation) limit the frequency to less than 2 kHz. Unlike thyristors, GTOs have finite voltages and currents during turn-off so the switching losses of GTOs are greater.

There is an advantage to maintain a negative bias gate voltage v_{GK} while the GTO is off, because it increases the dv_{AK}/dt withstand ability by up to a factor of two and so it reduces the possibility of spurious turn-on.

For most semiconductor switches it has been possible to model the turn-on and turn-off processes mathematically so that some estimate of the losses could be made. We can do this for the GTO turn-on process. However, it can be seen from Fig. 9.8 that the waveshapes of the GTO current and voltages are not only complex by the need to have a snubber circuit at turn-off but the phase relation between voltage and current is far from easy to define. Consequently, it is simpler to rely on the energy loss information given in the data sheets in order to calculate average power dissipation due to the turn-off process.

EXAMPLE 9.3

Consider the turn-off circuit diagram for the GTO in Fig. 9.9 together with the waveforms in Fig. 9.8. A 1600-V, 400-A GTO has the following data. The maximum on-state controllable current is 900 A and the maximum gate-current rise time is $t_{gq} = 10\,\mu\text{s}$. The gate-to-cathode voltage at which avalanche breakdown occurs is $V_{GK} = -25\,\text{V}$ and the turn-off current gain is $\beta_{off} = 5$. If the gate supply voltage for GTO turn-off is 12 V, estimate **(a)** the required value of the inductor L_G in the gate drive circuit, the limit of the rate of rise of gate current and **(b)** the gate pulse width t_p.

Solution

(a) Let the gate pulse, that is shown in Fig. 9.8, be linearized so that the pulse is triangular. If the gate current is to interrupt the anode current I_A, then

$$\hat{I}_G = \frac{I_A}{\beta_{off}} = \frac{900}{5} = 180\,\text{A}.$$

For the gate-drive circuit during the negative gate-current rise from zero to its peak value, Kirchhoff's voltage law gives

$$V_G = L_G \frac{di_G}{dt} - v_{GK} = L_G \frac{\Delta i_G}{\Delta t} - v_{GK}.$$

$V_G = 12\,\text{V}$, $\Delta i_G = \hat{I}_G = 180\,\text{A}$, $\Delta t = t_{gq} = 10\,\mu\text{s}$ and $v_{GK} \approx 0$.

$$\text{So, } L_G = \frac{V_G t_{gq}}{\hat{I}_G} = \frac{12 \times 10 \times 10^{-6}}{180} = 0.67\,\mu\text{H}.$$

This could be the stray inductance of the gate circuit.

$$\frac{di_G}{dt} = \frac{\hat{I}_G}{t_{gq}} = \frac{180}{10 \times 10^{-6}} = 18\,\text{A}/\mu\text{s}.$$

The gate current rise time t_{gq} is approximately equal to the storage time t_s. The interval $t_p - t_{gq}$ is approximately equal to the anode current fall time t_f at turn-off.

(b) For the time interval that $-i_G$ is reducing Kirchhoff's voltage law around the gate-drive loop is

$$V_G + L_G \frac{di_G}{dt} + V_{GK} = 0.$$

Hence, $V_G + L_G \Delta i_G / \Delta t + v_{GK} = 0$.

$\Delta i_G = \hat{I}_G = 180\,\text{A}$. $\Delta t = t_p - t_{gq}$ and $v_{GK} = V_{GK} = -25\,\text{V}$.

$$\text{So, } L_G \frac{\hat{I}_G}{t_p - t_{gq}} = V_{GK} - V_G.$$

$$\text{That is, } t_p = \frac{L_G \hat{I}_G}{V_{GK} - V_G} + t_{gq} = \frac{0.67 \times 180 \times 10^{-6}}{25 - 12} + 10 \times 10^{-6} = 19.2\,\mu\text{s}.$$

This pulse for turn-off can be short compared with the gate pulse for turn-on. For turn-off the anode current has to be reduced to the holding current ($I_h \approx 30\,\text{A}$ in this case), then there will be a long tail of anode current as the carriers around the cathode fingers recombine. During turn-on the gate pulse is applied until the anode latching current ($I_{la} \approx 30\,\text{A}$ in this case) is reached.

EXAMPLE 9.4

A 1000-V, 600-A GTO modulates power from a 600-V dc supply to an *RL* load with a freewheeling diode connected across it. The resistance of the load is $R = 1\,\Omega$ and the inductance of the load is high enough to consider the load current to be virtually constant. The chopper operates at a frequency of 2 kHz with a duty cycle $m = 0.8$. The on-state voltage drop of the GTO is $V_{GTO(ON)} = 3.4\,\text{V}$, and the switching losses for the particular temperature of operation, voltage and snubber protection are given in Fig. EX9.4. Determine the total average power dissipated by the GTO.

Solution

The load current I_l is

$$I_l = m \frac{V_s - V_{GTO(ON)}}{R} \approx m \frac{V_s}{R} = 0.8 \times \frac{600}{1} = 480\,\text{A}.$$

The turn-on loss W_{on} for $I_A = 480\,\text{A}$ from the graph is $W_{on} = 320\,\text{mJ}$, so the

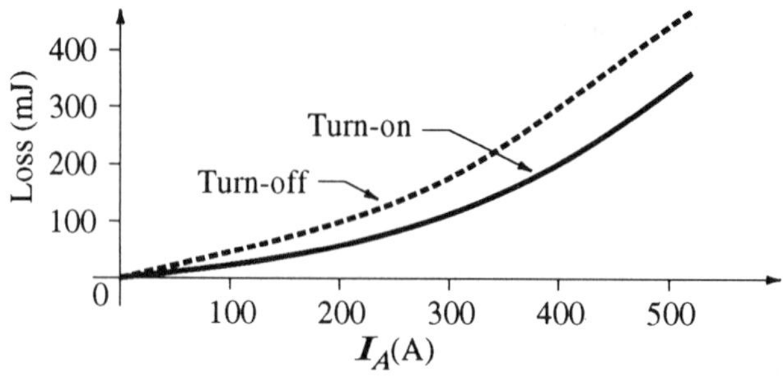

Fig. EX9.4

average power loss is $P_{on} = W_{on}f$.
That is, $P_{on} = 320 \times 10^{-3} \times 2 \times 10^3 = 640$ W.
The turn-off loss W_{off} for $I_A = 480$ A is $W_{off} = 420$ mJ, so the average power loss is $P_{off} = W_{off}f$.
That is, $P_{off} = 420 \times 10^{-3} \times 2 \times 10^3 = 840$ W.
The average conduction loss P_C is

$$P_C = \frac{1}{T}\int_0^{t_{ON}} V_{GTO(ON)} I_A \, dt = mV_{GTO(ON)} I_A = 0.8 \times 3.4 \times 480 = 1306 \text{ W}.$$

The total average power dissipated by the GTO is
$P_D = P_{on} + P_C + P_{off} = 640 + 1306 + 840 = 2786$ W.
This is 1.2% of the average power absorbed by the load.

9.7. GTO GATE CIRCUITS

There are gate drives for GTO turn-on and there are gate drives for GTO turn-off. The turn-on circuits are quite similar to the gate drives for thyristor turn-on. A small, positive current pulse at the gate terminal will turn on the GTO. Any difference between the thyristor and GTO current pulse is in the time of the pulse width. The GTO has a much higher latching current than the thyristor. Consequently, the GTO gate pulse has to be much longer.

The gate drive for GTO turn-off is quite unique. Turn-off will not be accomplished in dc circuits unless a negative current pulse is injected into the gate terminal. The pulse width is short but the pulse magnitude has to be at least a fifth of the magnitude of the anode current.

There are many gate circuits that differ greatly in detail, but in general they have the same form. Figure 9.10 shows a general circuit with gate supplies and switches for GTO turn-on and turn-off. The switches are shown as BJTs but any appropriate semiconductor switch can be used. The base signals from the controller will have been isolated and perhaps amplified and perhaps shaped to satisfy good responses.

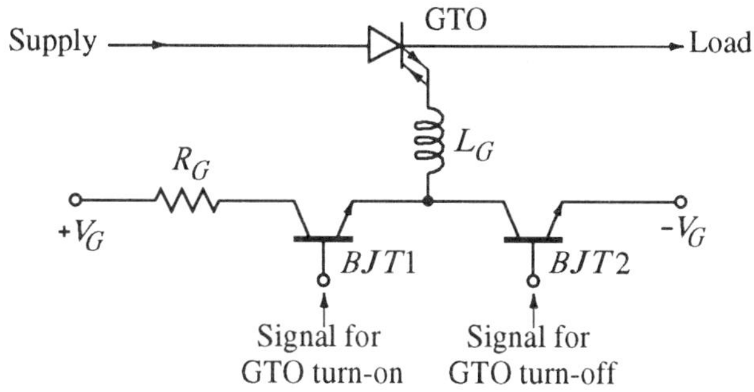

Fig. 9.10 General gate circuit for turn-on and turn-off.

EXAMPLE 9.5

A 1600-V, 300-A GTO controls power from a 1000-V dc supply to a resistive load whose value is $R = 4\,\Omega$. The specifications are that a gate current $I_G = 2\,\text{A}$ will trigger the switch on in time $t_{on} = 10\,\mu\text{s}$ with a delay time $t_d = 2\,\mu\text{s}$, and the current gain for turn-off is $\beta_{off} = 4$ to give a turn-off time $t_{off} = 18\,\mu\text{s}$ that includes a storage time $t_s = 8\,\mu\text{s}$. If the gate supplies are $\pm 12\,\text{V}$ and if the GTO is switched at a frequency of 1 kHz, determine **(a)** the average power delivered by the gate supply for turn-on, **(b)** the peak power delivered by the gate supply for turn-off and **(c)** the average power delivered by the gate supply for turn-off.

Solution

(a) For turn-on the instantaneous power delivered by the gate supply is
$P_G = V_G I_G = 12 \times 2 = 24\,\text{W}$.
The average power delivered for turn-on is

$$P_G = \frac{1}{T}\int_0^{t_{on}} P_G dt = P_G \times t_{on} \times f = 24 \times 10 \times 10^{-6} \times 10^3 = 0.24\,\text{W}.$$

This is low compared with the turn-on losses.

(b) For turn-off the peak gate current is
$I_G = I_A / \beta_{off} \approx V_s / (R\beta_{off}) = 1000/(4 \times 4) = 62.5\,\text{A}$.
The peak power delivered by the gate supply is
$\hat{P}_G = V_G \hat{I}_G = 12 \times 62.5 = 750\,\text{W}$.
This is a high value.

(c) The average power delivered by the gate supply for turn-off is
$P_G = W_G \times f \approx 0.5 \times t_{off} \times \hat{P}_G \times f = 0.5 \times 18 \times 10^{-6} \times 750 \times 10^3 = 6.75\,\text{W}$.
The energy delivered is $W_G = 6.75\,\text{mJ}$.
Average power and energy delivered are small compared with turn-off losses.

9.8. GTO PROTECTION

The GTO is a device much like the thyristor in structure. So, it is not surprising that the protection of the GTO against overcurrents overvoltages and transients is similar to the protection of the thyristor. The protective elements described in section 5.9 of Chapter 5 for the thyristor would work well for the GTO. However, in view of the GTO turn-off action there are some additional contingencies to take into account.

Overcurrent protection is afforded by gate turn-off up to a certain maximum anode current $I_{A\,max}$, given by $I_{A\,max} = \beta_{off} \hat{I}_G$ where β_{off} is the specified turn-off gain (3 to 5) and I_G is the permissible maximum value of the negative gate current. This permissible value of gate current is less than the breakdown value. If the anode overcurrents are greater than I_A max, the gate current must be blocked since it would be ineffective and either the main circuit breaker would have to be opened or a fuse would have to blow. Electronic detection of overcurrent could allow all GTOs in a converter to turn on, short-circuit the supply and ensure interruption at the fuse. Alternatively a fast acting parallel connected thyristor could be triggered to ensure that the fuse blows. See Fig. 9.11.

Some GTOs have a reverse withstand voltage that is less than 50 V. Protection against reverse voltage is provided by connecting a diode in reverse parallel. This is done in converters in any case. The GTO could have a diode in series.

Thermal considerations are similar for the GTO as for the thyristor, only more stringent. The on-state losses are greater because the voltage drop across the GTO is greater. Further, turn-off losses exist for the GTO because the anode current i_A is finite as the device voltage v_{AK} rises, to its blocking value.

At turn-on the rise of current di_A/dt is not such a problem as with the thyristor, because the GTO has a highly interdigitated gate-to-cathode structure. However, at turn-off the rise of voltage dv_{AK}/dt does offer a problem for the GTO, because of the rapid reduction of anode current i_A and the ensuing large $L\,di_A/dt$ voltage. Consequently GTOs utilize snubber circuits for protection during and after turn-off. It is rare that a GTO does not have a capacitive snubber circuit.

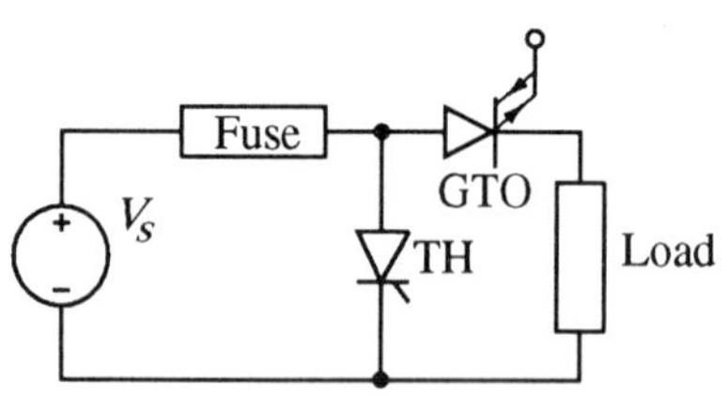

Fig. 9.11 GTO overcurrent protection by turning TH on.

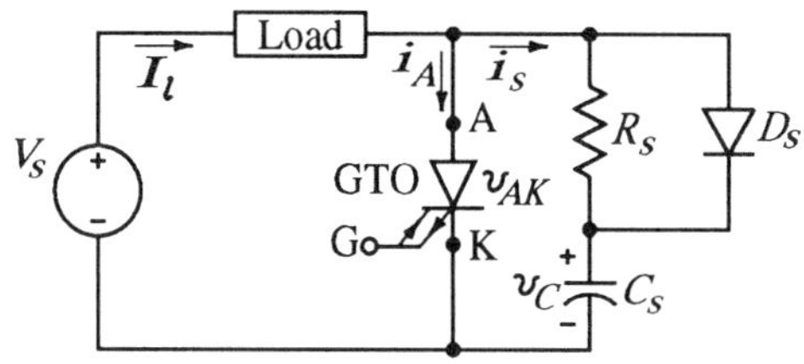

Fig. 9.12 GTO snubber circuit.

The main element of the snubber circuit is the capacitor C_s, that allows diversion of the load current from the GTO to the capacitor during turn-off. This reduces GTO turn-off losses. Also, the capacitor limits the rate of rise of the anode voltage as the GTO recovers its blocking state. It will be seen in the following example that the value of the capacitor C_s in a GTO snubber circuit is several times greater than that for a thyristor circuit. Although a high value of C_s means a large value of stored energy, that is usually dissipated, it also means less turn-off loss, that is due to the voltage v_{AK} rise as the anode current i_A reduces during the fall time t_f. The capacitor C_s helps to reduce the loss by keeping the voltage v_{AK} low during this interval. The other elements in the snubber circuit, shown in Fig. 9.12, are the resistance and the diode. At turn-on the capacitor C_s needs a resistance R_s to limit the discharge rate, and at turn-off the diode D_s allows very rapid charge rates.

EXAMPLE 9.6

A GTO is to act as a chopper to modulate the power from a 1000-V dc supply to a resistive load of 1.0 ohm. Figure 9.12 depicts the circuit diagram that incorporates a snubber circuit for the protection of the semiconductor device. For conditions that might prevail in this circuit a data sheet provides the following information on the GTO. The peak allowable turn-off gate current $\hat{I}_G$ is 100 A, above which avalanche breakdown occurs. The turn-off gain β_{off} is 4.0. The maximum allowable rate of rise of voltage during turn-off is dv_{AK}/dt = 1000V/μs. Turn-on and turn-off times are t_{on} = 12 μs and t_{off} = 14 μs. During turn-off the anode tail current tends to persist for 70 μs. Find **(a)** the maximum power that can be absorbed by the load, **(b)** the values of the snubber circuit elements, **(c)** the apparent maximum frequency of GTO switching and **(d)** the snubber-circuit power dissipation if the frequency of operation is limited to 2 kHz.

Solution

(a) The maximum current that can be drawn from the supply is the maximum current that can be extinguished by the GTO. This is given by

$I_l = I_A = \beta_{off}\hat{I}_G = 4 \times 100 = 400$ A.

If this is an average value, the peak value $I_{l\,max}$ is

$I_{l\max} = V_s / R_l = 1000 / 1.0 = 1000$ A.
For this, the duty cycle ratio is $m = 0.4$ and the maximum rms value of the load current to suit the GTO is $I_{l\,rms} = \sqrt{m}\, I_{l\max} = 632.5$ A.
The maximum permissible average power P_l absorbed by the load is
$P_l = I_{l\,rms}^2 R_l = 632.5^2 \times 1.0 = 400$ kW.

(b) At turn-off the load current is diverted from the GTO to the snubber capacitor C_s and the diode D_s diverts the capacitor current i_s from the resistor R_s, so $i_s = C_s\, dv_C / dt$. The initial value of the current i_s is approximately equal to the load current $I_{l\max}$ and the maximum value of the rate of rise of voltage is $d\,v_C / dt = d\,v_{AK} / dt \,|_{\max}$, so

$$C_s = \frac{I_s}{d\,v_C / dt} = \frac{I_{l\max}}{d\,v_{AK} / dt} = \frac{1000}{1000 \times 10^6} = 1.0\,\mu\text{F}.$$

Capacitors have stray inductance that contributes to a voltage spike across the GTO at turn-off. This is reduced by paralleling a number of capacitors. The resistor R_s is used to protect the GTO against a high discharge current from the capacitor C_s at turn-on. The discharge current should be low, but the discharge time should be relatively short. These are conflicting requirements so a compromise must be made. A reasonable rule of thumb for determining the value of the resistor R_s is to make the snubber-circuit time constant approximately equal to the time of turn-on t_{on}. In this case $R_s C_s \approx t_{on} = 12\,\mu$s, so $R_s \approx 12\,\Omega$.
The values $C_s = 1\,\mu$F and $R_s = 12\,\Omega$ are typical for GTOs.

(c) The maximum frequency of operation of the GTO is obtained from the value of the minimum period of switching. Such a period $T_{\min}$ comprises the minimum on-time and the minimum off-time. The minimum on-time t_{ON} is associated with the complete discharge of the snubber capacitor C_s so that it will perform its duty to allow satisfactory GTO turn-off. Complete discharge of C_s occurs in approximately five time constants, that is $5R_s\, C_s = 60\,\mu$s. The minimum off-time t_{OFF} is associated with the turn-off time t_{off} plus the interval of the tail current, given here as $t_{OFF} = 14 + 70 = 84\,\mu$s.

Thus $f_{\max} = \dfrac{1}{T_{\min}} = \dfrac{10^6}{60 + 84} = 6.9$ kHz.

This frequency of operation does not account for the temperature rise of the GTO due to the switching losses in the device. Taking the losses into account means that a practical limit to the maximum frequency of operation is somewhat less than 2kHz. See problem 9.12.
The diode D_s has to contend with 1000V reverse voltage at the point of capacitor C_s discharge, and the average current, while C_s charged for $5\,\mu$s each period (a frequency of 2kHz) with the load current virtually unchanged, would be about 10 A.

(d) The snubber circuit average power P_s is
$P_s = (1/2)\, C_s\, V_s^2 \times \text{frequency} = (1/2) \times 10^{-6} \times 10^6 \times 2 \times 10^3 = 1$ kW.
It would be worthwhile if some of this power could be delivered to the load.

At turn-off, the GTO has an anode tail current that is a displacement current. If large enough this current could turn the GTO back on again. A continuous negative bias (about -10 V) at the gate while in the off-state can prevent retriggering.

A good practice is to have a positive bias at the GTO gate at all times while in the on-state. This maintains all cathode islands on, even if the anode current falls to a low value (that is below the latching-current value). A subsequent rise of anode current would still be uniformly distributed.

9.9. GTO RATINGS AND APPLICATIONS

GTOs are manufactured in all sizes up to several kilovolts and kiloamperes ratings (higher than transistors but lower than thyristors), but the most economic use of GTOs in converter applications is in the power range greater than 0.5 MW where transistors cannot be used. In spite of turn-on times of about 10 μs and comparable turn-off times the frequency of operation of the GTO does not exceed 2 kHz. This is because of the switching losses. Those GTOs, that have anode shorts for fast turn-off, suffer from a reverse blocking voltage of less than 50 V. This is not a disadvantage in inverters circuits that must have reverse connected diodes.

Applications of GTOs are in choppers and inverters beyond the power range of transistors and up to the power limit of manufacturing ability, as long as the frequency of switching remains below 2 kHz. The GTO is a controlled rectifier, so those that block reverse voltage can be used in ac-dc conversion applications.

EXAMPLE 9.7

A single-phase, full-wave, fully-controlled, ac-dc, GTO converter modulates the power to a load from a 208-V ac source, as shown in Fig. EX9.7a. The load inductance is such that the load current can be considered constant and ripple free. A freewheeling diode is connected across the load. For three different cases it is required to find **(a)** the average load voltage $V_{l\,av}$, **(b)** the Fourier series for the input current i_s, **(c)** the input harmonic distortion THD, **(d)** the displacement power factor DPF and **(e)** the power factor PF of the input current. The three cases are **(i)** for a trigger angle $\alpha = \pi/3$ rad with natural GTO commutation ($\beta = 0$), **(ii)** for a trigger angle $\alpha = 0$ and an extinction angle $\beta = \pi/3$ rad and **(iii)** for the trigger angle α and the extinction angle β to be the same (to give a symmetrical switching arrangement) and of values such that the average load voltage is the same as in case (i).

Solution

The converter has gate signals to $GTO1$ and $GTO2$ during the positive half cycle of the supply voltage v_s and has gate signals to $GTO3$ and $GTO4$ during the negative half cycle.

Case (i) For $\alpha = \pi/3$ rad and natural commutation the GTO converter has the same controlled performance as a thyristor converter. Figure EX9.7b depicts the waveforms of the voltages and current. The voltage v_s is taken as the reference.

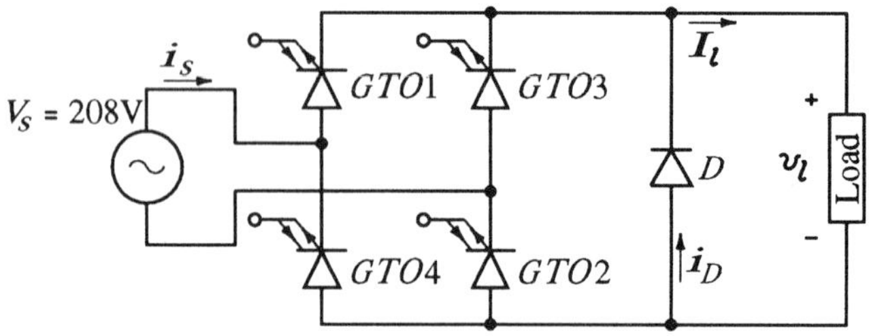

Fig. EX9.7a Converter.

There is symmetry so the input current waveform has no dc component and no even harmonics.

(a) The average load voltage for $\alpha=\pi/3$ is

$$V_{l\,av}=\frac{1}{\pi}\int_{\alpha}^{\pi} v_l\,d\omega t=\frac{1}{\pi}\int_{\pi/3}^{\pi} 208\sqrt{2}\,\sin\omega t\,d\omega t=\frac{208\sqrt{2}}{\pi}(1+\cos\pi/3)=140.4\text{ V.}$$

(b) The Fourier series for the input current is

$$i_s=\sum_{n=1,3,\cdots}^{\infty}(a_n\cos n\omega t+b_n\sin n\omega t)=\sum_{n=1,3,\cdots}^{\infty}c_n\sin(n\omega t+\theta_n)$$

where $\theta_n=\tan^{-1}a_n/b_n$.

$$a_n=\frac{1}{\pi}\int_0^{2\pi} i_s\cos n\omega t\,d\omega t=\frac{1}{\pi}\int_{\alpha}^{\pi} I_l\cos n\omega t\,d\omega t-\frac{1}{\pi}\int_{\pi+\alpha}^{2\pi} I_l\cos n\omega t\,d\omega t\,.$$

So, $a_n=-\dfrac{2I_l}{n\pi}\sin n\alpha$.

$$b_n=\frac{1}{\pi}\int_0^{2\pi} i_s\sin n\omega t\,d\omega t=\frac{1}{\pi}\int_{\alpha}^{\pi} I_l\sin n\omega t\,d\omega t-\frac{1}{\pi}\int_{\pi+\alpha}^{2\pi} I_l\sin n\omega t\,d\omega t\,.$$

So, $b_n=\dfrac{2I_l}{n\pi}(1+\cos n\alpha)$

$$c_n=(a_n^2+b_n^2)^{1/2}=\frac{4I_l}{n\pi}\cos\frac{n\alpha}{2}\quad\text{and}$$

$$\theta_n=\tan^{-1}\frac{a_n}{b_n}=\tan^{-1}-\frac{\sin n\alpha/2\cos n\alpha/2}{\cos^2 n\alpha/2}=-\frac{n\alpha}{2}\,.$$

So, $i_s=\displaystyle\sum_{n=1,3,\cdots}^{\infty}\frac{4I_l}{n\pi}\cos\frac{n\alpha}{2}\sin(n\omega t-n\alpha/2)$.

(c) The input harmonic distortion THD is defined in eq. (1.4.14).

$\text{THD}=(I_s^2-I_{s1}^2)^{1/2}/I_{s1}=(I_s^2/I_{s1}^2-1)^{1/2}$

where the currents are the overall rms value and the fundamental rms value. The rms value of the fundamental component ($n=1$) of the input current is

$$I_{s1}=\frac{1}{\sqrt{2}}\times\frac{4I_l}{\pi}\cdot\cos\alpha/2=0.7798I_l\text{ for }\alpha=\pi/3\text{ rad.}$$

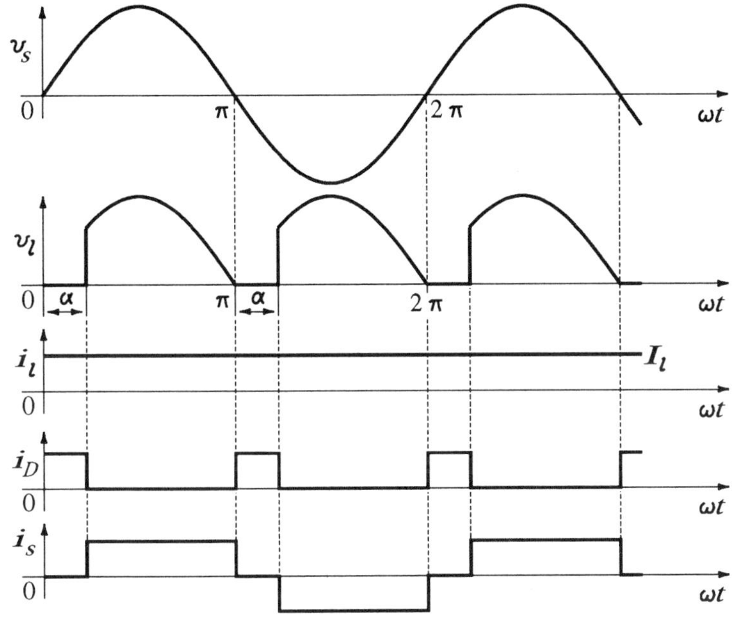

Fig. EX9.7b $\alpha = \pi/3$ rad and natural commutation

The rms value of the input current i_s is

$$I_s = \left[\frac{1}{\pi} \int_{\alpha}^{\pi} I_l^2 \, d\omega t \right]^{1/2} = I_l (1 - \alpha/\pi)^{1/2} = 0.8165 I_l \text{ for } \alpha = \pi/3 .$$

Therefore, the harmonic distortion is $\text{THD} = \left[\frac{0.8165^2}{0.7798^2} - 1 \right]^{1/2} = 0.31$ (or 31%).

(d) The displacement power factor $\text{DPF} = \cos\theta_1$ where θ_1 is the phase angle of the fundamental component of current $\mathbf{I}_{s1}$.
$\text{DPF} = \cos\theta_1 = \cos(-\alpha/2) = 0.866$ for $\alpha = \pi/3$ rad.
The negative sign of θ_1 indicates $\mathbf{I}_{s1}$ lags $\mathbf{V}_s$ by an angle $\alpha/2$.

(e) The overall power factor $\text{PF} = \cos\theta$ is given by the average power input P as $P = V_s I_s \cos\theta = V_s I_{s1} \cos\theta_1$.

So, $\text{PF} = \cos\theta = \frac{I_{s1}}{I_s} \cos\theta_1 = \frac{0.7798}{0.8165} \times 0.866 = 0.827$ lagging for $\alpha = \pi/3$ rad.

Case (ii) For $\alpha = 0$ and $\beta = \pi/3$ the GTO converter has waveforms as shown in Figure EX9.7c.

(a) The average load voltage for $\alpha = 0$ and $\beta = \pi/3$ rad is

$$V_{l\,av} \frac{1}{\pi} \int_{0}^{\pi-\beta} v_l \, d\omega t = \frac{1}{\pi} \int_{0}^{2\pi/3} 208\sqrt{2} \sin\omega t \, d\omega t$$

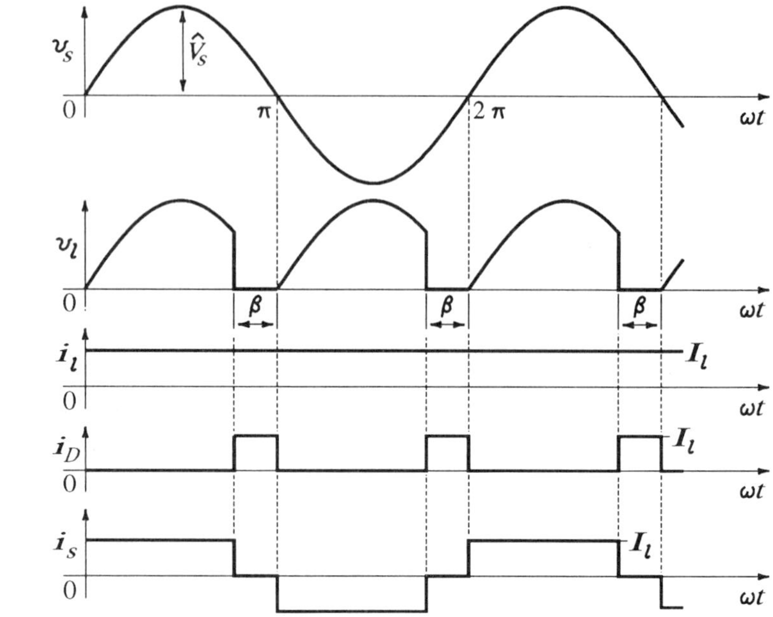

Fig. EX9.7c Waveforms for $\alpha = 0$, $\beta = \pi/3$ rad.

Therefore, $V_{l\,av} = \dfrac{208\sqrt{2}}{\pi}(-\cos 2\pi/3 + 1) = 140.4$ V.

The average load voltage has the same value as in case (i).

(b) The Fourier series for the input current i_s has the coefficients

$$a_n = \frac{1}{\pi}\int_0^{2\pi} i_s \cos n\omega t \, d\omega t = \frac{1}{\pi}\int_0^{\pi-\beta} I_l \cos n\omega t \, d\omega t - \frac{1}{\pi}\int_{\pi}^{2\pi-\beta} I_l \cos n\omega t \, d\omega t = \frac{2I_l}{n\pi}\sin n\beta$$

and $b_n = \dfrac{1}{\pi}\displaystyle\int_0^{2\pi} i_s \sin n\omega t \, d\omega t = \dfrac{2I_l}{n\pi}(1 + \cos n\beta)$.

$$c_n = (a_n^2 + b_n^2)^{1/2} = \frac{4I_l}{n\pi}\cos\frac{n\beta}{2} \quad \text{and} \quad \theta_n = \tan^{-1} a_n/b_n = n\beta/a\,.$$

So, $i_s = \displaystyle\sum_{n=1,3,\cdots}^{\infty} \frac{4I_l}{n\pi}\cos\frac{n\beta}{2}\sin(n\omega t + n\beta/2)$.

(c) The rms value of the fundamental component ($n = 1$) of the input current is

$I_{s1} = \dfrac{1}{\sqrt{2}} \times \dfrac{4I_l}{\pi}\cos\dfrac{\beta}{2} = 0.7798 I_l$ for $\beta = \pi/3$ rad.

The rms value of the input current i_s is

$$I_s = \left[\frac{1}{\pi}\int_0^{\pi-\beta} I_l^2 \, d\omega t\right]^{1/2} = I_l(1 - \beta/\pi)^{1/2} = 0.8165 I_l \text{ for } \beta = \pi/3.$$

Therefore, the total harmonic distortion is

$$\mathrm{THD} = \left[\frac{0.8165^2}{0.7798^2} - 1\right]^{1/2} = 0.31 \text{ or } (31\%).$$

(d) The displacement power factor is

$\mathrm{DPF} = \cos\theta_1 = \cos(+\beta/2) = 0.866$ for $\beta = \pi/3$ rad.

The positive sign of θ_1 indicates $\mathbf{I}_{s1}$ leads $\mathbf{V}_s$ by an angle $\beta/2$. Since most loads are inductive, the extinction of the GTOs by β has a positive influence by introducing a capacitive effect.

(e) The overall power factor PF is

$$\mathrm{PF} = \cos\theta = \frac{I_{s1}}{I_s}\cos\theta_1 = \frac{0.7798}{0.8165} \times 0.866 = 0.827 \text{ leading for } \beta = \pi/3 \text{ rad.}$$

The only difference between cases (i) and (ii) is that phase angle α control provides a lagging power factor, whereas extinction angle β control provides a leading power factor.

Case (iii) This is the case with both phase angle α control and extinction angle β control. We want symmetry, whereby $\alpha = \beta$. Figure EX9.7d depicts the waveforms.

(a) The average load voltage $V_{l\,av}$ is

$$V_{l\,av} = \frac{1}{\pi}\int_0^{\pi} v_s\, d\omega t = \frac{1}{\pi}\int_{\alpha}^{\pi-\beta} \hat{V}_s \sin\omega t\, d\omega t = \frac{\hat{V}_s}{\pi}(\cos\beta + \cos\alpha).$$

For $\alpha = \beta$ and $V_{l\,av} = 140.4$ V and $\hat{V}_s = 208\sqrt{2},\ \alpha = \beta = 0.723$ rad. (41.42°).

(b) The Fourier series for the input current has the coefficients

$$a_n = \frac{1}{\pi}\int_0^{2\pi} is \cos n\omega t\, d\omega t = \frac{1}{\pi}\int_{\alpha}^{\pi-\beta} I_l \cos n\omega t\, d\omega t - \frac{1}{\pi}\int_{\pi+\alpha}^{2\pi-\beta} I_l \cos n\omega t\, d\omega t\,.$$

So, $a_n = \dfrac{2I_l}{n\pi}(\sin n\beta - \sin n\alpha) = 0$ for $\alpha = \beta$.

$$b_n = \frac{1}{\pi}\int_0^{2\pi} i_s \sin n\omega\, d\omega t = \frac{1}{\pi}\int_{\alpha}^{\pi-\beta} I_l \sin n\omega t\, d\omega t - \frac{1}{\pi}\int_{\pi+\alpha}^{2\pi-\beta} I_l \sin n\omega t\, d\omega t\,.$$

So, $b_n = \dfrac{2I_l}{n\pi}(\cos n\alpha + \cos n\beta) = \dfrac{4I_l}{n\pi}\cos n\alpha$ for $\alpha = \beta$.

$c_n = (a_n^2 + b_n^2)^{1/2} = \dfrac{4I_l}{n\pi}\cos n\alpha$ for $\alpha = \beta$ and $\theta_n = \tan^{-1} a_n/b_n = \tan^{-1} 0 = 0$.

Hence, $i_s = \displaystyle\sum_{n=1,3,\cdots}^{\infty} \frac{4I_l}{n\pi}\cos n\alpha \sin n\omega t.$

(c) The rms value of the fundamental component ($n = 1$) of input current i_s is

$$I_{s1} = \frac{1}{\sqrt{2}} \times \frac{4I_l}{\pi}\cos\alpha = \frac{1}{\sqrt{2}} \times \frac{4I_l}{\pi}\cos 0.723 = 0.675 I_l \text{ A.}$$

The rms value of the input current i_s is

$$I_s = \left[\frac{1}{\pi}\int_{\alpha}^{\pi-\beta} I_l^2\, d\omega t\right]^{1/2} = I_l\left[\frac{\pi - \beta - \alpha}{\pi}\right]^{1/2} = I_l\left[1 - \frac{2\alpha}{\pi}\right]^{1/2} \quad \text{for } \alpha = \beta.$$

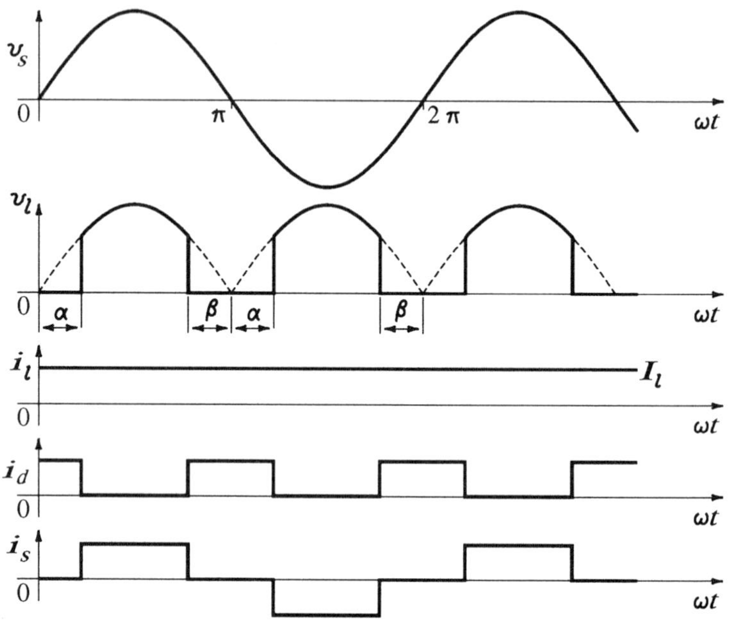

Fig. EX9.7d Waveforms for $\alpha=\beta$.

That is, $I_s = I_l(1-2\times 0.723/\pi)^{1/2} = 0.735 I_l$ A.
Therefore, the harmonic factor THD is

$$\text{THD} = \left[\frac{I_s^2}{I_{s1}^2} - 1\right]^{1/2} = \left(\frac{0.735^2}{0.675^2} - 1\right)^{1/2} = 0.43\ (43\%)$$

(d) The displacement power factor DPF is
DPF $= \cos\theta_1 = \cos 0 = 1.0$ for $\alpha=\beta$.
The current $\mathbf{I}_{s1}$ is in phase with $\mathbf{V}_s$.

(e) The overall power factor PF is

$$\text{PF} = \cos\theta = \frac{I_{s1}}{I_s}\cos\theta_1 = \frac{0.675}{0.735}\times 1.0 = 0.896.$$

Summary: For the same average load voltage of 140.4 V the following table allows a comparison of the input current characteristics.

Control	α	β	THD	DPF	PF
Phase angle	$\pi/3$	0	31%	0.866	0.827
Extinction angle	0	$\pi/3$	31%	0.866	0.827
Symmetrical angle	0.723	0.723	43%	1.0	0.896

For symmetrical angle control the power factor is improved but the harmonic distortion has been increased.

9.10. SUMMARY

The GTO is similar in structure to a thyristor, but it is a fully-controllable, switch. A gate signal can turn the device on and a gate signal can turn it off.

Turn-off is possible due to both high interdigitation of the gate design and anode shorts.

This provides the main advantage over the thyristor. A thyristor must have its anode current reduced to zero before it will turn off. A GTO can have its current extinguished at any time by the application of a negative gate pulse of current. The GTO has higher power handling capabilities that the transistor but not as high as the thyristor. Turn-off losses limit the switching frequency of the GTO to a value below 2 kHz.

Within the power range of a GTO chopper or converter, the GTO circuit is better than the thyristor circuit from the point of view of efficiency, size and cost, mainly because of the commutation circuits that are required for thyristors. However, the GTO has higher switching losses, a higher on-state voltage drop and higher gate driving requirements than the thyristor.

Most loads are inductive. Therefore a snubber circuit is necessary to protect the GTO during turn-off. The snubber capacitor has a value that is higher than that used for the thyristor. Protection must be afforded against both anode overcurrents and negative gate overcurrents if satisfactory turn-off is to be assured.

The GTO is an important member of the silicon, semiconductor-switch family, especially for the relatively low frequency, medium to high power applications. In essence the GTO has controlled turn-on and turn-off characteristics for which gate pulses are required. While in the off-state, the GTO has high voltage-withstand ability in both the forward and reverse biases, but in some cases the withstand is unipolar, where the limit of reverse bias is less than 50 V. Finally, the conduction of current is unidirectional.

9.11. PROBLEMS

Section 9.3

9.1 A 1200-V, 300-A GTO modulates power from an 800-V dc supply to a resistive load whose value is $R = 4\,\Omega$. Refer to Fig. 9.3a. If the switch is on for a long time the junction temperature of the GTO rises to 100°C and the voltage drop is measured to be $V_{GTO(ON)} = 4$ V. Determine the on-state conduction losses.

9.2 Consider the circuit diagram in Fig. 9.3a. A 1000-V, 300-A GTO acts as a chopper to modulate power from a 600-V dc supply to a resistive load whose value is $R = 2.0\,\Omega$. The frequency of chopper switching is 2 kHz and the duty cycle is $m = 0.8$. If the GTO on-state voltage drop is $V_{GTO(ON)} = 3.4$ V, calculate **(a)** the average power loss in the GTO due to conduction and **(b)** the average power absorbed by the load.

9.3 A 1000-V GTO is employed as a chopper to modulate power from a 600-V dc supply to a resistive load. See Fig. 9.3a. From the data sheets of the GTO, the on-state voltage drop is $V_{GTO(ON)} = 3.8$ V, and the maximum average power dissipation is 800 W. If the chopper duty cycle is $m = 0.5$, determine **(a)** the maximum anode current I_A of the GTO, **(b)** the average current rating of the GTO and **(c)** the value of the load resistance to sustain maximum current. Assume that the switching frequency is low enough that the switching losses can be neglected.

Section 9.5

9.4 A 3000-V, 1000-A GTO chopper modulates power from a 2000-V dc supply to an *RL* load that has a freewheeling diode connected across it. The load resistance has a value $R = 2.5\,\Omega$ and the load inductance is high enough to consider that its current is virtually constant. GTO data are that the current rise-time is $t_{ri} = 12\,\mu$s and the voltage fall-time is $t_{fv} = 10\,\mu$s at turn-on, and the on-state voltage drop is $V_{GTO(ON)} = 4.2$ V. If the chopper operates with a duty cycle $m = 0.8$ at a frequency of 1 kHz, estimate **(a)** the average power loss in the GTO due to the turn-on process and **(b)** the average power dissipated in the GTO due to on-state conduction.

9.5 A 600-V, 15-A GTO controls the charging of a superconductive coil whose resistance is $R = 0$ and whose inductance is 0.1 H. The coil has a freewheeling diode connected across it. The dc supply has a voltage $V_s = 100$ V. Calculate the duration of the gate pulse that is necessary to ensure the GTO will turn on, if the device's latching current is $I_{la} = 1.5$ A and its on-state voltage drop is $V_{GTO(ON)} = 2.2$ V. Compare the result with that obtained in EXAMPLE 5.1.

Section 9.6

9.6 A 1000-V, 30-A GTO has the following turn-off specifications in the data sheets. The controllable anode current is $I_A = 130$ A, for which the turn-off current gain is $\beta_{off} = 5$. The storage time $t_s = 1\,\mu$s and the anode current fall-time is $t_f = 0.3\,\mu$s at turn-off. The reverse breakdown voltage, gate to cathode is $V_{GK} = -15$ V. If the gate-supply voltage for turn-off is $V_G = 10$ V, estimate **(a)** the peak value of the gate current for turn-off, **(b)** the minimum gate-pulse width t_p of the current, **(c)** the maximum rate of rise of gate current above which second breakdown may occur, **(d)** the value of the gate-circuit inductance L_G to limit the rate of rise of gate current and determine **(e)** whether the breakdown value of the reverse gate-to-cathode voltage is exceeded. Refer to Figs 9.8 and 9.9.

9.7 A 500-V, 60-A GTO modulates power from a dc source of voltage $V_s = 300$ V to an *RL* load with a freewheeling diode connected across it. The value of the inductance is high enough to consider the load current to be constant at 50 A for steady operation as a chopper circuit, switching at 2 kHz with a duty cycle $m = 0.8$. The GTO data are that the on-state voltage drop is $V_{GTO(ON)} = 2.4$ V, the delay time is $t_d = 0.4\,\mu$s and the total turn-on time is $t_{on} = 5\,\mu$s, and the turn-off losses for these conditions are $W_{off} = 50$ mJ. Determine the total average power dissipation P_D in the GTO. Neglect leakage and gate losses.

9.8 A 500-V, 60-A thyristor modulates power from a 300-V, dc supply to an *RL* load with a freewheeling diode connected across it. The value of the inductance is high enough to consider the load current to be constant at 50 A for steady operation as a chopper circuit, switching at 2 kHz with a duty cycle $m = 0.8$. The thyristor data are that the on-state voltage drop is $V_{TH(ON)} = 1.5$ V, the delay time is $t_d = 0.4\,\mu$s and the total turn-on time is $t_{on} = 5\,\mu$s, and the required time for forced commutation is $t_q = t_{off} = 8\,\mu$s. Determine **(a)** the total average power dissipation P_D in the thyristor and **(b)** the average power loss due to commutation. Compare with problem 9.7.

Section 9.7

9.9 A 1000-V, 400-A GTO controls the rectification from a 600-V, 60-Hz supply to an *RL* load whose resistance is $R = 1.5\,\Omega$ and whose inductance is $L = 2.3$ mH. This single-phase, half-wave rectifier has turn-on by delay angle α control and turn-off by advanced extinction angle β control. The GTO data are that the gate supply voltage is $V_G = 10$ V, the minimum gate current is $I_G = 1$ A for turn-on; the gate voltage is $V_G = -10$ V, the current gain is $\beta_{off} = 5$, the storage time is $t_s = 3\,\mu$s and the anode current fall-time is $t_f = 1.5\,\mu$s at turn-off, and the latching current is $I_{la} = 35$ A. If $\alpha = 45°$ and $\beta = 178°$, determine **(a)** the instantaneous power delivered by the turn-on gate supply, **(b)** the average power delivered by the turn-on gate supply, **(c)** the peak power delivered by the turn-off gate supply and **(d)** the average power delivered by the turn-off gate supply.

Section 9.8

9.10 A 1200-V, 30-A GTO controls power from a 1000-V dc supply to a resistive load of $50\,\Omega$. The GTO must be protected against turn-on di_A/dt greater than 800A/μs and off-state dv_{AK}/dt greater than 8kV/μs. What are the values of snubber elements to give this protection to the GTO?

9.11 Consider the circuit diagram in Fig. 9.9 and let the load be purely resistive with $R = 2\,\Omega$. The 4000-V, 2000-A GTO has a di_A / dt limit of $300\text{A}/\mu\text{s}$ at turn-on and a dv_{AK} / dt limit of $400\text{V}/\mu\text{s}$ at turn-off. The turn-on time is $t_{on} = 10\,\mu\text{s}$ with a minimum on-time $t_{ON} = 30\,\mu\text{s}$. The turn-off time is $t_{off} = 25\,\mu\text{s}$ with a current fall-time $t_f = 2\,\mu\text{s}$ and with a minimum off-time $t_{OFF} = 120\,\mu\text{s}$. The chopper controls the power from a 3000-V dc supply and operates at 1kHz. **(a)** Determine the value of a series snubber element that will give adequate protection to the GTO at turn-on. **(b)** Calculate suitable values of the parallel snubber elements that will give protection to the GTO at turn-off. **(c)** What is the rating of the snubber resistor R_s and what is the initial discharge current in the GTO at turn-on? Comment. **(d)** If the stray inductance in the parallel snubber circuit amounts to as much as $1.0\,\mu\text{H}$, what is the approximate value of the voltage spike appearing across the GTO during turn-off?

9.12 A 1600-V, 400-A GTO acts as a chopper to control the power from a 1200-V dc supply to an *RL* load whose resistance is $R = 3\,\Omega$ and whose inductance is high enough to consider the load current to be virtually constant. The chopper operates at a frequency of 2 kHz with a duty cycle $m = 0.8$. The GTO has an on-state voltage $V_{GTO(ON)} = 3.2$ V and a thermal resistance, junction to sink, $R_{\theta JS} = 0.05°\text{C/W}$. From the data sheets the GTO turn-on loss is 300 mJ and the turn-off loss is 250 mJ. For this operating condition, if the maximum junction temperature of the GTO is 125°C what is the permitted heatsink temperature?

9.13 Two GTOs are connected in parallel[4] to share a 5000-A pulsed load current. The current rises linearly to its full value in 30 ms and the voltages across each GTO in the on-state are 3.2 V and 3.45 V respectively. A centre-tapped reactor helps to force current sharing during the transient interval. The reactor is a 1:1 transformer whose magnetizing inductance is 0.1 mH. If leakage inductance and resistance can be neglected, find the current difference in the two GTOs at the end of the transient interval.

Section 9.9

9.14 A single-phase, full-wave, fully-controlled ac-dc, GTO converter modulates the power so that a resistive load of 1.2 ohms absorbs 12 kW from a 208-V ac source. A freewheeling diode is connected across the load. The load inductance is such that the load current can be considered to be constant and ripple free. For a trigger angle α and natural GTO commutation (extinction angle $\beta = 0$), determine **(a)** the gating angle α, **(b)** the input current harmonic distortion THD and **(c)** the input power factor.

9.15 A single-phase, full-wave, fully-controlled ac-dc, GTO converter modulates the power so that a resistive load of 1.2 ohms absorbs 12 kW from a 208-V ac source. A freewheeling diode is connected across the load. The load

---

[4] Fukui, H., *et al. Paralleling of GTOs*, IEEE IA Record, pp741-746, 1982.

inductance is such that the load current can be considered to be constant and ripple free. For a trigger angle $\alpha = 0$ and an extinction angle β that is finite, determine (**a**) the gating angle β, (**b**) the input current harmonic distortion THD and (**c**) the input power factor.

9.16 A single-phase, full-wave, fully-controlled ac-dc, GTO converter modulates the power so that a resistive load of 1.2 ohms absorbs 12 kW from a 208-V ac source. A freewheeling diode is connected across the load. The load inductance is such that the load current can be considered to be constant and ripple free. For symmetrical switching such that $\beta = \alpha$ determine (**a**) the gating angles α and β, (**b**) the input current harmonic distortion THD and (**c**) the input power factor. See Fig. EX9.7d.

9.17 Consider a GTO ac-dc converter that is single phase, full wave. Refer to the circuit diagram in Fig. EX9.7a. The load is purely resistive. (**a**) Plot the total harmonic distortion THD versus load power for the two cases (i) gating angle α finite and $\beta = 0$ and (ii) gating angles $\alpha = \beta$. (**b**) For half full load, what are the THDs for (i) α finite and $\beta = 0$ and (ii) $\beta = \alpha$.

9.12. BIBLIOGRAPHY

Bose, B.K. (ED) *Modern Power Electronics-Evolution, Technology and Applications*. New York: IEEE Press, 1992.

Fisher, Marvin J. *Power Electronics*. Boston, Mass.: PWS-Kent Publishing Co., 1991.

Grafham,D.R., F.B.Golden (Ed). *SCR Manual*. 6th ed. New York: General Electric, 1979.

Mohan, N., Underland, T.M. and Robbins, P. *Power Electronics*. New York: John Wiley & Sons,Inc., 1989.

Ohno, E. *Introduction to Power Electronics*. Oxford: Clarendon Press, 1988.

Rashid, Muhammad Harunur. *Power Electronics*. Englewood Cliffs, N.J.: Prentice Hall,Inc., 1988.

Taylor, P.D. *Thyristor Design and Realization*. Chichester, U.K: John Wiley & Sons, Inc., 1987.

Williams, B.W. *Power Electronics, Devices, Drivers, and Applications*. New York: John Wiley & Sons, Inc., 1987.

-----. *Semiconductors Data Handbook*. Book S2b. Philips. 1987.

CHAPTER 10

OTHER SWITCHES AND THE MCT

10.1. INTRODUCTION

The seven power semiconductor switches that are discussed in Chapters 3 to 9 are established devices. Each has a range of applications that suits it best. At one end of the scale there is the low-frequency, high power thyristor and at the other end of the scale there is the high-frequency, low-power MOSFET.

No switch is perfect. Due to either technological or theoretical grounds each switch has a less than desirable characteristic. The MOSFET has high on-state losses at high voltage levels, BJTs have second breakdown problems, IGBTs need to switch at higher frequencies and powers with lower voltage drops. Thyristors need to switch even higher powers at higher speeds. In general there is a need to get closer to the ideal characteristics of infinite impedance to any forward and reverse voltages in the blocking state: of zero impedance in the conduction state: and of zero switching times for changing state.

The design of switches seems to require a good deal of compromise. For example, blocking capability has to be sacrificed for high speed of switching. However, development of new switches is continual. If a new switch is not perfect, usually it has one really good parameter characteristic that allows it to be used to some advantage in a particular application.

In some cases the desirable characteristic of one discrete device can be used to overcome an undesirable characteristic of another discrete device if both are used in combination. For example, the BJT can operate at reasonably high voltages and currents compared with the MOSFET, but the BJT requires a high value of base current to turn on and it is slow to turn off. On the other hand, the MOSFET cannot handle high voltages but it requires very little power to switch on and the switching speed is very fast. A combination of the MOSFET and the BJT should give characteristics better than each individual. Figure 10.1a illustrates what is called a *cascade* connection. Little power is needed to turn on the MOSFET. Once it is on, the resulting base current turns on the BJT. Figure 10.1b depicts what is called a *cascode* connection. The BJT handles the high voltage and high current, while the MOSFET handles the high current and the fast turn-off. The use of multiple devices is not favoured. So there is a tendency towards integration of devices on a single chip.

We will review some of the special application switches and some of those that have potential. Those that are described here are the SIT, the SITH, thyristors and the MCT.

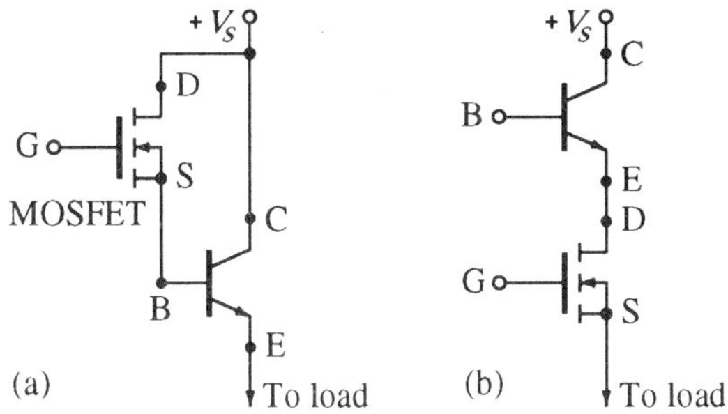

Fig. 10.1 Switch combinations.
(a) Cascade connection, (b) cascode connection.

10.2. THE SIT

The SIT, a **S**tatic **I**nduction **T**ransistor, is a semiconductor switch that is also called the power junction field-effect transistor (power JFET). The name SIT is derived from the electrostatically-induced potential barrier that is used to control the states, on and off. Its characteristics are similar to a MOSFET except that its power level is higher and its maximum frequency of operation is lower. A disadvantage of the SIT is that in the on-state at its maximum current, the voltage drop can be greater than 15 V. However, it finds special use in induction heating applications, that need high frequency switches.

Figure 10.2 depicts the circuit symbols of the SIT. The form is similar to the MOSFET symbol without the gate isolation.

10.2.1. SIT Structure

The structure of the SIT is illustrated in Fig. 10.3. There are three terminals. Two of the terminals, the drain D and the source S, are used for the main current conduction and two of the terminals, the gate G and the source S, are used to apply the control voltage V_{GS} for switching the device.

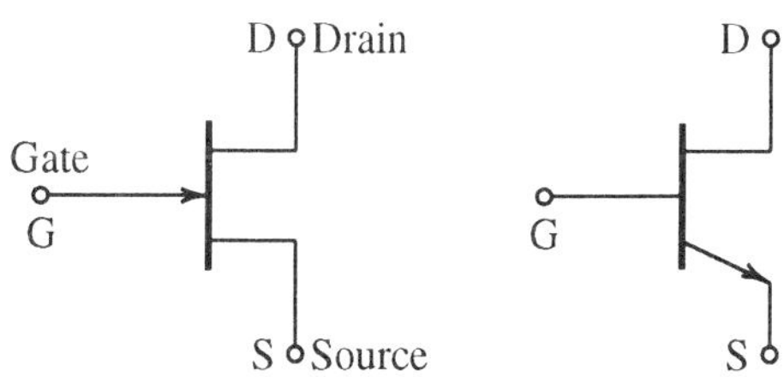

Fig. 10.2 SIT circuit symbols.

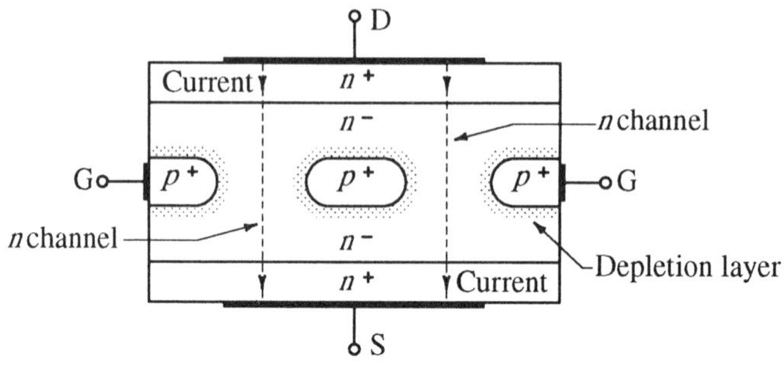

Fig. 10.3 SIT structure.

On the highly doped n^+ region of the drain there is grown a lightly doped n^- drift region to provide a high voltage blocking capability. Buried in the drift region are the many highly doped p^+ regions that form high interdigitation with the source region by means of a grid form. The interdigitation is important because it provides short paths for the electric field of the gate signal. Short paths require low voltages.

Between the drain and source terminals there is an n channel $n^+n^-n^+$ that provides a resistive path for the conduction of majority carriers. The p^+ regions are placed so that the path can be interrupted by a gate signal.

10.2.2. SIT I-V Characteristics

The steady-state current *vs* voltage characteristics of the SIT are not like any of the other transistor characteristics that are discussed in this book. Figure 10.4a illustrates a simple circuit that is used to describe the characteristics. A dc supply has its power to the resistive load modulated by the SIT switch. The dc supply biases the drain D positively with respect to the source S and the gate supply biases the gate terminal negatively with respect to the source S. If there is no gate voltage ($V_{GS} = 0$), there exists a resistive n channel between drain and source. Accordingly, the device is on and conducts current I_D that depends mainly on the supply voltage V_s and the load resistance R. For low values of drain current I_D the voltage drop V_{DS} across the SIT is low. Since the conductive path between drain and source is resistive (with a positive temperature coefficient), an increased drain current produces an increased voltage drop ($V_{DS} = I_D R_{DS}$). This drop is high and can be greater than 15 V at rated current.

If the gate voltage is $V_{GS} = -V_{G\,\max}$, the SIT is off and blocking up to a drain voltage V_{DS3}, as shown in Fig. 10.4b. If the applied voltage V_{DS} exceeds V_{DS3} the SIT begins to conduct again but there is no decrease in V_{DS}. This situation must be avoided because the very large conduction losses in the switch will cause overheating and damage.

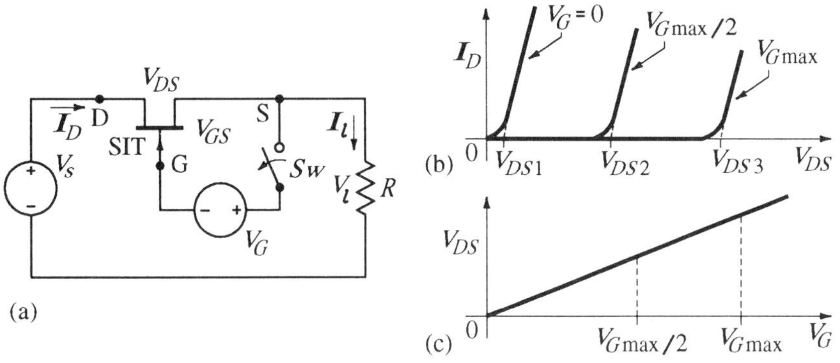

Fig. 10.4 SIT I-V characteristics.
(a) Circuit diagram, (b) $I_D - V_{DS}$ curves, (c) transfer characteristic.

Like all transistors, the gating characteristics of the SIT are linear. If the gate voltage is half the maximum value, the device will block a voltage at the drain up to a value $V_{DS} = V_{DS2} \approx V_{DS3}/2$. A further increase in drain voltage causes conduction without a reduction of drain-to-source voltage. The transfer characteristic is depicted in Fig. 10.4c and it is fairly linear. The transfer coefficient $G = V_{DS}/V_{GS} \approx \text{constant}$. The value of G can be 40 or more.

As a switch the gate voltage must be suppressed by an equivalent short circuit across the gate and source terminals for the on-state. For the off-state, the gate voltage is set at $V_{GS} = -V_{G\max}$ and the voltage V_{DS} must not be allowed to be greater than $GV_{G\max}$. A 1000-V SIT may require a gate voltage as high as -25 V to maintain the off-state.

10.2.3. SIT Turn-off

The SIT is normally on by virtue that there is no signal applied to the gate. In order to switch from the on-state to the off-state a gate voltage V_{GS} is applied. The electric field of the gate voltage widens the depletion layer of the reverse biased p^+n^- junctions of the interdigitated gate-to-source regions. See Fig. 10.3. The widening of the depletion layer pinches the n channel cross section, so that eventually it opens the conduction path and blocks all voltage.

Since the SIT is a majority carrier device and since the establishment of an electric field to pinch the n channel depends only on the rise-time of the gate voltage, the turn-off time for the device is fast. A typical turn-off time is $t_{off} \approx 0.3\ \mu s$.

10.2.4. SIT Turn-on

A voltage V_{GS} maintains the SIT in the off-state. However, an increase in the drain voltage V_{DS} above a critical limit will cause its own electric field to counter the gate field sufficiently to re-establish a conduction path. In this case the device can be said to be on. Since the voltage drop across the SIT remains high, this situation is to be avoided.

The proper way to change states in order to turn on the SIT is to remove the gate voltage. The electric field of the gate voltage collapses very quickly so that the depletion layers shrink and the *n* channel path is available for conduction. The turn-on time is about the same as the turn-off time ($t_{on} \approx 0.3\,\mu s$).

Gate circuits for SITs are simple, because the requirements are the application of a voltage source V_G or the removal of that source. MOSFET gate circuits are similar. The only charge flow at the gate is the discharge of the depletion layer capacitance.

10.2.5. SIT Protection

Care has to be taken against overvoltages at the drain of the SIT, because if $V_{DS} > -GV_{GS}$ then conduction will result without a reduction in the drain voltage drop. The result would be an excessive value of power dissipation in the device.

Overcurrents require protection. Since the SIT switches so quickly, the application of a gate voltage to turn off the switch affords the best protection.

The normal method to protect against internal circuit transients of the form of di/dt or dv/dt is the use of snubber circuits. However, the limits of di/dt and dv/dt are high enough for the SIT that it is possible to operate without snubber circuits.

The on-state voltage drop is relatively high for high drain currents, so that special care must be taken with the design of the heatsink to ensure that the SIT junction temperature does not rise above 150°C.

SITs can be connected in parallel in order to cope with higher load currents than the value of an individual device rating. The positive temperature coefficient of resistance of the SIT facilitates good current sharing.

10.2.6. SIT Ratings and Applications

The SIT is not in common use, but voltage ratings above 1000 V are available and current ratings are above 200 A. With switching speeds greater than 50 kHz, there are application possibilities at power levels the MOSFET cannot reach and switching speeds that the BJT cannot reach. The high on-state voltage drop is a disadvantage that is difficult to cope with. A second disadvantage, that is common to transistors, is that the SIT is asymmetric with respect to voltage blocking. The reverse blocking capability is low.

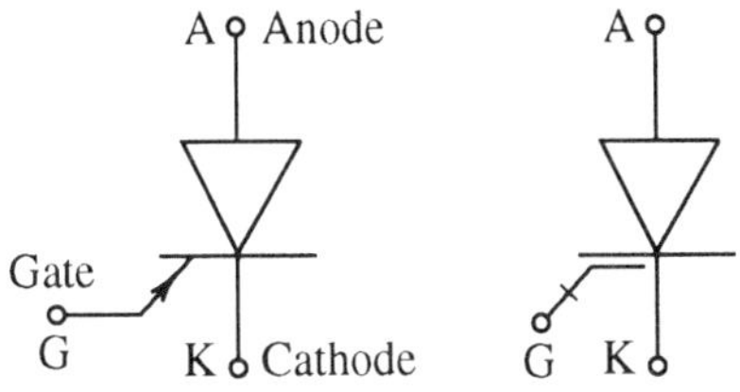

Fig. 10.5 SITH circuit symbols.

10.3. THE SITH

The name SITH given to a particular semiconductor switch is derived from the name **S**tatic **I**nduction **TH**yristor. From the viewpoint of the cross-section of the structure, illustrated in Fig. 10.6, the device may look as though it could be a *pnpn* device and be called a type of field-controlled thyristor (FCT). However, it is really like a *pn* diode with a gate structure control like the SIT.

The SITH has three terminals. The main terminals are called the anode A and cathode K because of the diode nature. The control terminals are called the gate G and the cathode K. These are named on the circuit symbols, illustrated in Fig. 10.5. Both symbols portray the structure of a diode plus a control terminal. The first symbol show a direct electrical contact with the semiconductor region at the gate, in this case a p^+ region. The second symbol gives the impression of electrical isolation of the gate contact and gate region, but it is more probably intended to indicate a grid structure of the regions that form a high level of interdigitation.

The SITH is normally in the on-state. That is, without a gate signal, the device can conduct unidirectionally from anode to cathode, just like a diode. A negative gate signal will turn the device off and the switch will remain off as long as the gate signal is applied. Since the SITH is basically a diode it should block reverse voltage, anode to cathode. However, design constraints, to make the SITH switch off quickly, have reduced the blocking voltage to a low value. Accordingly, the device is classified as an asymmetric blocking switch, like most of the other types of semiconductor switches.

There is not a broad range of applications for the SITH, but a rating of 1000 V and 200 A makes it a possible choice for choppers and inverters. A particular application is induction heating.

10.3.1. SITH Structure

Figure 10.6 shows a symbolic representation of the cross section of a SITH to the extent that two of the interdigitated gate-cathode cells are included.

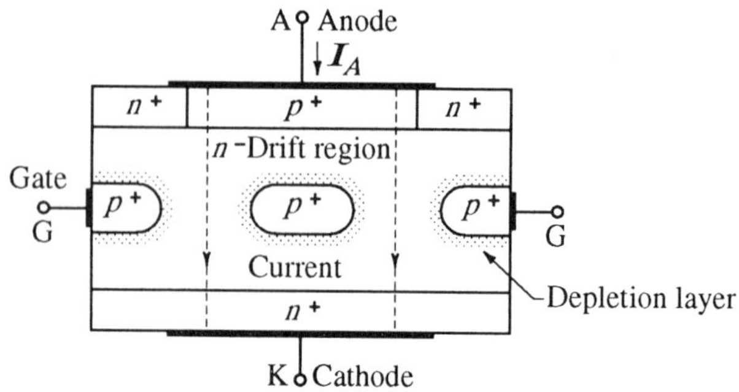

Fig. 10.6 SITH structure.

The basic structure is the p^+n^+ diode between the anode A and the cathode K. In order to increase the blocking voltage at the anode in the off-state a lightly-doped n^- region (drift region) separates the p^+ and n^+ regions. In the first instance, the behavior of the device is expected to be like a diode. A positive voltage at the anode results in the conduction of a current I_A. A reverse voltage would be expected to be blocked so that no reverse current would exist.

The uncontrolled rectification of the diode is changed to controlled rectification by including a buried grid structure of p^+ material that is connected to the gate terminal. A negative gate voltage will cause the voltage at the anode to be blocked so that the anode current I_A will be virtually zero. Removal of the gate signal will allow conduction if the anode is positive.

In addition to the basic structure, there are anode shorts created by the n^+ regions. These provide an aid to faster recombination, but they create the drawback that the reverse blocking capability is greatly reduced.

10.3.2. SITH I-V Characteristics

Figure 10.7a shows a simple circuit diagram of a SITH that can control the power that is delivered by a dc supply V_s to a resistive load R. The control is implemented by the magnitude of the negative gate voltage supply V_G.

If the gate-drive switch *Sw* is open, there is no gate signal and $V_{GK} = 0$. This is the condition that the SITH is on. The value of the anode current I_A depends mainly on the magnitudes of the supply voltage V_s and the load resistance R. As the load current is increased so too does the on-state voltage drop $V_{AK(ON)}$. Its value can be 4V at rated current. This steady-state characteristic is represented by the curve for $V_{GK} = 0$ in Fig. 10.7b.

For the turn-off condition, a negative gate voltage $V_{GK} = -V_G$ is applied. This voltage must be less than the breakover voltage of the reverse biased, gate-to-cathode, *pn* junction. At the maximum value of gate voltage $V_G = -V_{GK\,\max}$, the SITH will block the maximum anode-to-cathode voltage $V_{AK} = V_{AK3}$. This is the maximum voltage rating of the semiconductor switch. It is the electric field of the

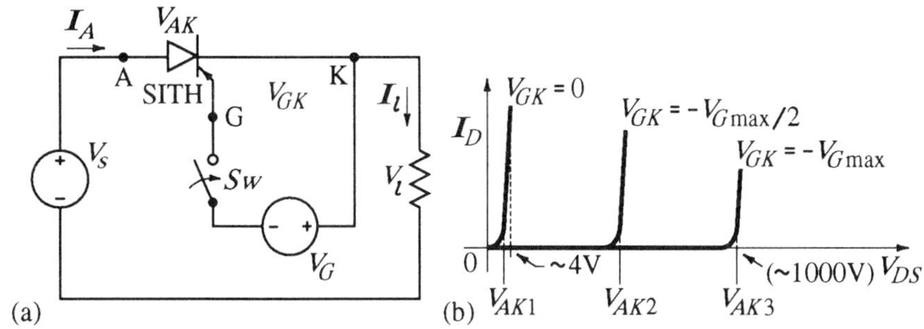

Fig. 10.7 SITH characteristics.
(a) Circuit diagram, (b) I-V characteristics.

gate voltage that causes the anode-to-cathode conduction path to be broken. However, if the anode voltage V_{AK} is increased further, the anode electric field can partially overcome the gate blocking field so that anode current can exist without there being a reduction of the anode voltage. An excessive amount of power would be dissipated in the switch. The curve for $V_{GK\max}$ in Fig. 10.7b represents this condition. For safe blocking operation the anode voltage is kept below V_{AK3}.

The effect of the gating voltage V_{GK} is linear. That is, if the gate voltage were $V_{GK}=-V_{G\max}/2$, the anode voltage that can be safely blocked is $V_{AK2}\approx V_{AK3}/2$. Any anode voltage above this value V_{AK2} would allow some conduction to ensue. The voltage gain $G=V_{AK}/V_{GK}$, where V_{AK} is the blocking voltage, is quite high. It is possible for G to have a value 600. Consequently a 1000-V SITH would require a gate voltage $V_G=1.67\,\text{V}$ for the maintenance of the off-state.

10.3.3. SITH Turn-off

The SITH is in the on-state, if no gate signal is applied to its terminals. The device will remain on until a gate voltage is applied. At this point, a negative gate voltage causes the depletion layers at the grid of p^+ regions to widen due to its electric field. Figure 10.6 shows the depletion layers. An expansion of the depletion layers can completely pinch off the paths of the anode current if the gate voltage is large enough.

If the depletion layers bridge completely the gap between the grid of p^+ regions the SITH turns off. The SITH is a minority carrier device, with excess carriers in the drift region coming from the anode p^+ region. During the turn-off interval, the anode-to-cathode current is diverted to the gate circuit as the excess carriers are swept out. This gate current is high, like a GTO gate current at turn-off, but the current gain $\beta_{off}=I_A/I_G$ is even lower. It may be about 3. Also, just like the GTO, there is a long anode-current tail, while the minority carriers recombine.

Recombination is aided by the anode shorts (the n^+ regions) so that the turn-off time is $t_{off} \approx 10\,\mu s$ or less.

Once off, the SITH remains off only for as long as the gate signal is applied. For a given supply voltage V_s, the value of the gate voltage is given by $V_G = V_s / G$, where G is the transfer voltage gain of the device.

10.3.4. SITH Turn-on

It is assumed that the SITH is in the off-state and this is maintained by a gate voltage that allows the depletion layer to bridge the gaps of the grid of p^+ regions. Removal of the negative gate voltage such that the gate-to-cathode terminals are short-circuited or have a small positive voltage across them, causes the SITH to turn on. The short circuit removes the electric field at the gate, so the depletion layer capacitance discharges, reduces in width and leaves a current path for anode current in the n^- regions between the p^+ regions. The discharge takes time, so that for the SITH the turn-on time is $t_{on} \approx 2\,\mu s$,

Once the SITH is on, the buried gate grid creates the long channel length of the drift region with resistance such that the on-state voltage drop $V_{AK(ON)}$ can be as high as 4V.

10.3.5. SITH Protection

Overvoltage across the anode and cathode terminals of the SITH produces a problem. In the off-state, that is maintained by negative gate voltage, an anode overvoltage will cause anode current to be conducted, as shown in the characteristics of Fig. 10.7b. This must be avoided otherwise there will be excessive power dissipation. An increased gate voltage can counteract this problem up to the extent that the gate voltage reaches an avalanche value itself.

For moderate overcurrents the SITH can be protected by the application of the gate signal in order to interrupt the current. If this method is not satisfactory then fast fuses can be employed in the same way that thyristors and GTOs are protected. Care is taken to match the I^2t properties.

SITHs can exhibit nonuniform turn-on so that the device must be protected against di/dt higher than $1000\,A/\mu s$. A 1000-V SITH would be protected against di/dt damage if the circuit inductance amounted to $1\,\mu H$.

There is no internal regeneration in the SITH like the two-transistor form in the thyristor. That is, there is no latching possibility by displacement currents. Hence, the dv/dt limit for the SITH is reasonably high at about $2000\,V/\mu s$. If values greater than this are expected the usual method of parallel snubber circuits give protection.

10.3.6. SITH Ratings and Applications

The switching frequency of the SITH is about 3 kHz and the power handling capability is about 200 kVA. The SITH can be used in dc-dc, dc-ac and ac-dc conversion applications in competition with other switches at this power and frequency level. It is similar to the GTO performance. Both can be turned on and off, although the SITH is normally on. The power rating of the SITH is lower, but its switching speed is higher, and its transient properties are better.

10.4. THYRISTORS

There is a principal thyristor which is described in Chapter 5 and there are other special thyristors. Some of those special thyristors are described also. The GTO (gate turn-off thyristor) is the subject of Chapter 9, the triac is the subject of Chapter 8 and the SITH (static induction thyristor, which is not thyristor) is discussed in section 10.3. There are other special thyristors that have special purposes. They are the ASCR (asymmetric thyristor), the GATT (gate assisted turn-off thyristor) and the RCT (reverse conduction thyristor).

10.4.1. The ASCR

The ASCR is an **A**symmetric **S**ilicon **C**ontrolled **R**ectifier. The only major difference between this semiconductor switch and the thyristor is that the ASCR does not block any significant voltage in the reverse direction. Other than this the forms of switching and operation are similar.

Voltage-fed inverters use switches that have diodes connected across them in antiparallel to provide a path for inductive load current. Therefore, it is not necessary to have a thyristor switch with reverse blocking capability if a switch can be found with little blocking capability, but with improved characteristics. For such applications the ASCR can replace conventional thyristors up to about 2000-V, 400-A ratings. The only reverse blocking capability that is required is that withstand level of voltage created by the $L_s di / dt$ of stray inductance in the diode branch and this does not exceed 30 V normally.

If the middle n layer of the $pnpn$ device is made thinner the reverse blocking ability of the switch is reduced, but the turn-on speed is increased by faster dynamic plasma spread across the wafer. A reduction of forward blocking capability is prevented by the refinement of adding a buffer of highly doped, low resistivity n^+ material, usually by diffusion, between the anode p layer and the normally adjacent n layer. The combination of n layers gives a lower on-state voltage drop and a faster recombination for shorter turn-off times. See Fig. 10.8.

To give up an unnecessary blocking ability for faster switching times and a lower on-state voltage drop is a real benefit. Since the ASCR cannot block reverse voltage, a current commutation circuit, usually a resonant LC circuit, must be employed.

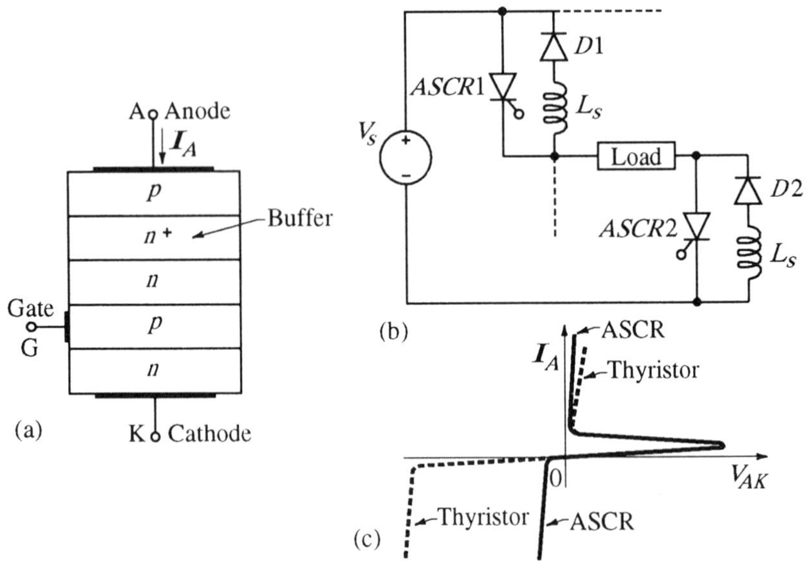

Fig. 10.8 The ASCR
(a) Structure, (b) circuit diagram, (c) characteristics.

10.4.2. The GATT

GATT stands for **G**ate-**A**ssisted **T**urn-off **T**hyristor. Conventional thyristors are minority carrier semiconductor switches and are slow to turn-off. GTOs use a gate signal to turn the switch off and the losses created by the relatively high on-state voltage drop and the switching tend to limit the frequency of operation to about 2 kHz. The GATT uses both a commutation circuit and a negative gate signal without having to sacrifice the low on-state voltage. As a result the frequency of operation is increased to above 10 kHz.

The commutation circuit forces the anode current of the GATT to zero very quickly and the negative gate signal breaks the regenerative gating action, so that a 2000-V, 400-A device can switch off in a time $t_{off} = 4\,\mu s$. This is useful in inverter circuits.

10.4.3. THE RCT

RCT stands for **R**everse **C**onducting **T**hyristor. This device is the integration of the discrete components, the ASCR and the diode. Figure 10.9 shows the structure and the circuit symbol of the RCT. Between the anode and the cathode there is the *pnpn* structure of the thyristor and between the cathode and the anode there is the p^+pnn^+ of the diode in antiparallel There is a region between the two to give isolation. This prevents excess carriers from the conducting diode from

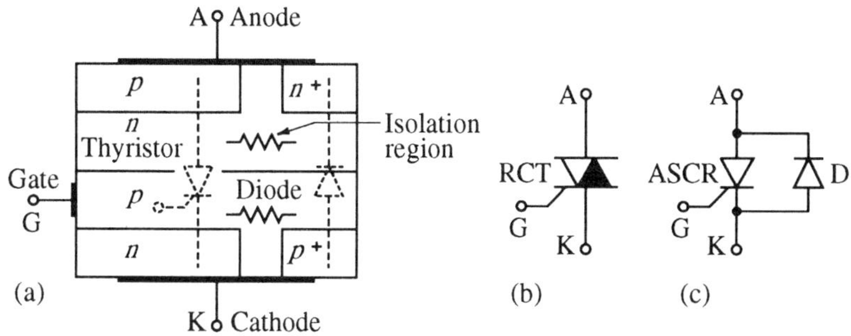

Fig. 10.9 The RCT.
(a) Structure, (b) circuit symbol, (c) circuit representation.

reaching the thyristor. Otherwise the thyristor might fail to turn-off.

The RCT is much like the ASCR and diode in operation, performance and ratings. Because no reverse blocking is required, the wafer can be thinner so that the on-state voltage drop is low and the turn-off is fast.

An advantage of the integrated RCT is that there is no stray inductance in the diode branch that would provide increased reverse voltage during the diode current rise and a forward bias during the diode current fall while the thyristor is turning off.

A disadvantage of the refinement of an RCT is that the matching of the thyristor and the diode is fixed. It is usual that the on-state voltage drops are the same and the current ratings are the same for both.

The applications of the RCTs are the same as those for ASCRs: namely, choppers and inverters.

10.5. THE MCT

MCT stands for **M**OS-**C**ontrolled **T**hyristor. It is a thyristor or a GTO that is integrated with two MOSFETs. One MOSFET turns the thyristor on and one MOSFET turns the thyristor off. In this way low power, fast gating is achieved. The switching frequency of the device may be up to 20 kHz. Consequently, its performance is similar to the IGBT. However the on-state voltage drop of the MCT is lower; it is about 1.1 V.

The MOSFET has a current density about $20\ \mathrm{A/cm^2}$, the BJT and the IGBT have current densities about $50\ \mathrm{A/cm^2}$ and the thyristor has a current density about $200\mathrm{A/cm^2}$. The current density of an MCT is roughly an order of magnitude greater than other power semiconductor switches.

In view of these attributes the MCT has enormous potential to cover a very wide range of applications. The main disadvantage is that it is asymmetric, since the reverse blocking capability has been sacrificed for speed and low on-state voltage.

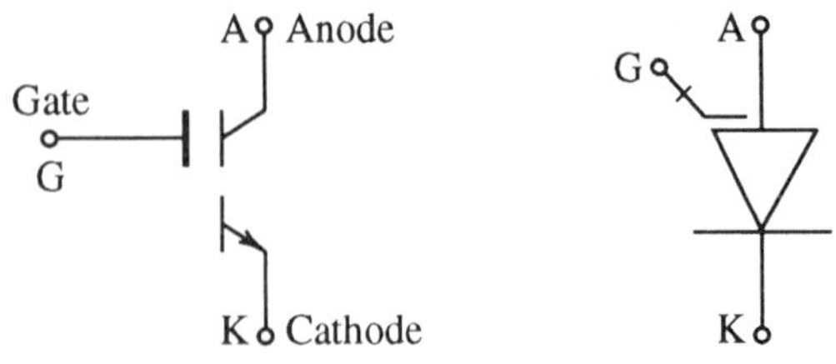

Fig. 10.10 MCT circuit symbols.

Figure 10.10 depicts the circuit symbols of the MCT. Anode and cathode symbolize the *pnpn* structure of the thyristor and the gate signifies the isolation provided by the MOSFET.

10.5.1. MCT Structure

There is a number of different structures, but all have the common features that there is a thyristor *pnpn* structure that determines the conduction (and blocking) properties. Also, all MCTs have two integrated MOS devices for controlling the switching properties.

Figure 10.11 symbolizes the structure of the MCT. Between the anode A and the cathode K there is a recognizable *pnpn* structure that forms the thyristor part of the MCT. The gate-anode region is interdigitated to the extent that there may be as many as 10^5 cells, only two of which are shown in the figure. The large number of cells provide short paths of large cross-sectional area for rapid, uniform switching of current. Within that anode-gate region there are two MOSFETs. One MOSFET is a *p*-channel, *pnp* type that is used for MCT turn-on. The other MOSFET in an *n*-channel, *npn* type that is used for MCT turn-off. Can you discern these two switches which have a common gate G?

The semiconductor layers are numbered in Fig. 10.11 in order to distinguish the connections of the equivalent model components of the MCT.

The *pnpn* device has an n^+ substrate (7), to which a cathode connection is made. On this substrate a thin, *p* layer is epitaxially grown. This *p* region is a buffer layer on which is grown a thick p^- layer (6). The buffer layer allows the value of the forward blocking voltage to be high at the np^- junction (J2). However, the result of the buffer is a higher on-state voltage and a low reverse blocking voltage at junction J3.

The thick p^- layer above the buffer is a multi-purpose region. It forms the base (6) of the n^+p^-n BJT, the collector (6) of the p^-np^+ BJT and the drain (9) of the MOSFET that is used to turn on the MCT.

An *n* region (5) is diffused to the thick p^- layer (6). Together, they form junction J2 of the *pnpn* structure. Further, this *n* region is the base (5) of the p^-np^+ BJT and at the interface (8) with the silicon dioxide insulation it becomes the *p* channel of the MOSFET that is used to turn on the MCT. Also, this *n* region (4)

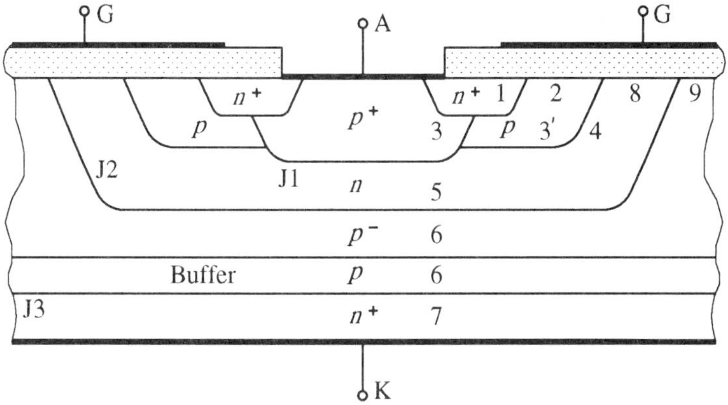

Fig. 10.11 MCT Structure.

is the drain of the *npn* MOSFET that is used to turn off the MCT.

The anode connection is made to the p^+ region (3), that is diffused to the n region (5) to form the p^+n junction (J1). This p^+ region is the emitter (3) of the p^-np^+ BJT and is the source (3′) of the MOSFET that is used to turn on the MCT.

Further diffusion creates the n^+ region (1) that is the source of the *npn* MOSFET that is used for the MCT turn-off. This *npn* MOSFET has its n channel at (2).

The structure described here is general enough to explain the operation of the device. What is not shown is that there are only about 4 percent of the cells that have MOSFETs for turn-on.

10.5.2. MCT I-V Characteristics

The steady-state I-V characteristics of an MCT are similar to an ASCR or an RCT. Figure 10.12a shows a simple circuit diagram that can be used for testing the MCT. The voltage supply V_s is assumed to be adjustable.

If the cathode K is positive with respect to the anode A, no matter what the gate bias is, the MCT will break down at a low voltage. This situation is to be avoided.

If the anode A is positive with respect to the cathode K and if there is no gate voltage, the MCT remains in the off-state until a breakover voltage is reached at which avalanche breakdown occurs. Ideally this never occurs. See Fig. 10.12b. In practice there is a small leakage current $I_{A\ leak}$ in the blocking state until breakdown occurs and the device turns on. See Fig. 10.12c.

If the anode is positive with respect to the cathode and if a negative voltage V_G is applied at the gate, the MCT turns on. The voltage drop $V_{MCT(ON)}$ is small (ideally zero) and may vary from about 0.7 V at no load to about 1.1 V at full load. The anode current I_A is limited only by the value of the load impedance.

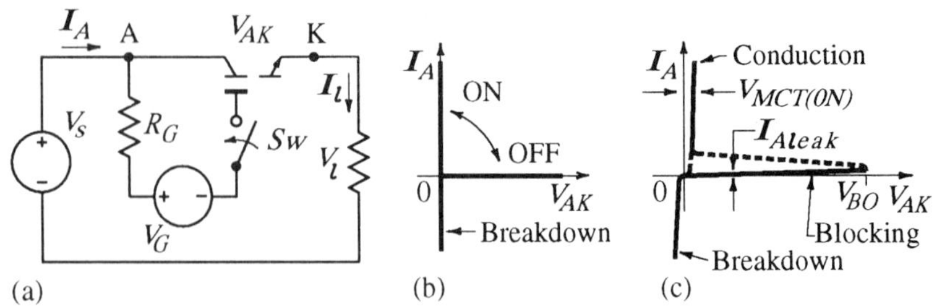

Fig. 10.12 MCT I-V characteristics.
(a) Circuit diagram, (b) ideal characteristics, (c) real characteristics.

If the MCT is on, the application of a positive gate voltage returns the device to the off-state until a negative gate voltage is reapplied.

10.5.3. MCT Models

Figure 10.11 indicates that the MCT has a thyristor structure for conduction and two MOSFETs for switching the thyristor on and off. If these three components are used to model the MCT, then a circuit configuration like that shown in Fig. 10.13a would suffice. However, the switching operation from the on-state to the off-state is not obvious.

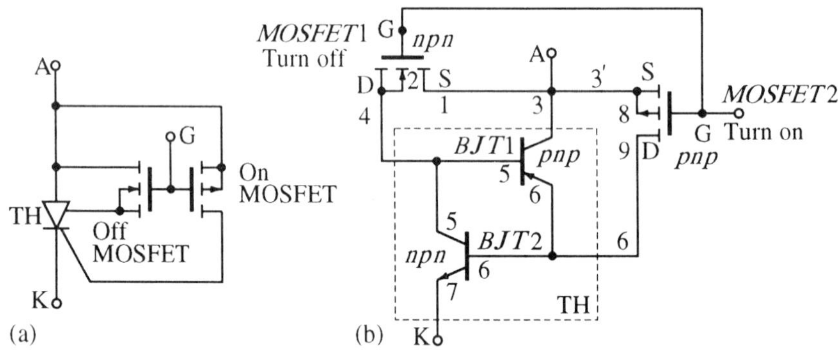

Fig. 10.13 MCT models.
(a) Thyristor model, (b) four-transistor model.

A better model of the MCT is to use the two-transistor model of the thyristor together with the two switching MOSFETs. Consequently, the MCT operation can be described by a four-transistor model. The circuit diagram of this model is shown in Fig. 10.13b. The numbers 1 to 9 are the same as those in the structure diagram of Fig. 10.11 and act as an aid to identify the relevant regions with the circuit connections. If the operation of MOSFETs and BJTs is understood, then the operation of the MCT is straightforward.

10.5.4. MCT Turn-on

Turn-on can be explained with the help of Fig. 10.13b. It is assumed that the MCT is off and that the anode is positive with respect to the cathode. It is usual, although it is not absolutely necessary, that *MOSFET* 1 is maintained on to keep the MCT in the off-state. This prevents spurious turn-on.

A negative voltage V_G can be applied to the gate circuit. This removes the *n* channel from *MOSFET* 1 so that it turns off. At the same time it creates a *p* channel in *MOSFET* 2 so that it turns on. In turn, the *p* channel creates a path for the conduction of current from the anode to the base of *BJT* 2. Since the bias is correct there will be a collector current in *BJT* 2 that provides a base current in *BJT* 1, whose resulting collector current augments *BJT* 2 base current. This is the start of regeneration of the two transistors. Regeneration continues to build to the extent that latching can take place. At this point the negative gate voltage can be removed and the thyristor (and the MCT) stays on.

In practical terms, the threshold value of the gate voltage of *MOSFET* 2 may be about $V_{GA(TH)} = -5$ V. A gate voltage $V_{GA} = -10$ V may cause turn-on with a delay time for plasma spread $t_{d(on)} \approx 400$ ns and an anode current rise-time $t_{ri} \approx 500$ ns. The limit of di_A / dt at turn-on is about 600A / µs or more and depends on the total number of turn-on MOSFETs.

The turn-on losses are not easy to define, except for a resistive load. For an inductive load with a freewheeling diode the losses are not as great as might be expected. For most switches the current i_A rises to its steady value before the voltage v_{AK} falls. In an MCT the voltage v_{AK} drops while the current i_A rises even though the voltage across the load remains virtually zero. This is because the MCT turns on rapidly due to the thyristor regeneration. Any source inductance, stray inductance or series snubber inductance must support the difference between the supply voltage V_s and the MCT voltage v_{AK} during turn-on. Figure 10.14 shows a typical turn-on inductive load. Diode reverse recovery has been ignored. Loss calculations would be very approximate.

10.5.5. MCT Turn-off

Reference can be made to Fig. 10.13 for a clarification of the MCT turn-off process. It is assumed that there is a negative voltage at the gate with respect to the anode so that *MOSFET* 2 and the MCT are on.

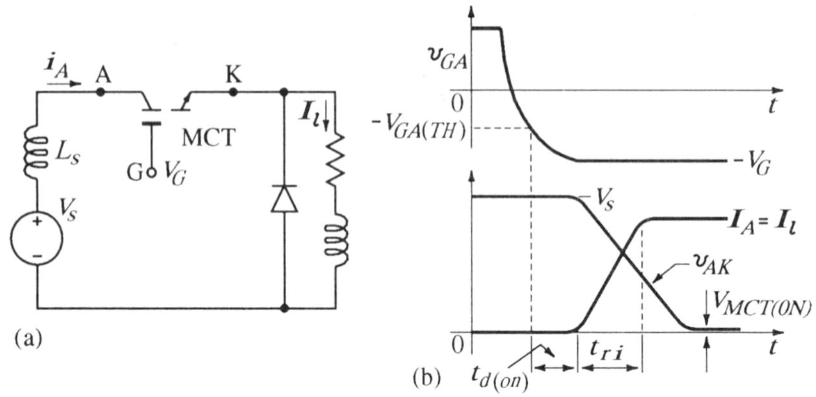

Fig. 10.14 MCT turn-on.
(a) Inductive circuit, (b) waveforms.

A positive voltage of about 15 V at the gate terminal with respect to the anode turns off the p-channel *MOSFET* 2 and turns on the n-channel *MOSFET* 1. Then the thyristor turns off.

The turn-on action of *MOSFET* 1 creates a short circuit across the p^+n junction J1 of the thyristor, that is, across connections 3 and 5, shown in Figs 10.11 and 10.13. In the latter figure it can be seen that the short circuit 3-5 stops the regeneration action of the two-transistor model. Connection 6, the base of *BJT* 2, is deprived of current from the p channel of *MOSFET* 2 and from the collector of *BJT* 1, so *BJT* 2 turns off. Consequently, the thyristor turns off and this means the MCT is off.

The process of turning off is slower than turn-on by a factor of about four, even though there is an n-channel MOSFET in every cell for turn-off. It is much quicker to create the excess charge carriers in the device than it is for their recombination in the two BJT base regions.

The effectiveness of the *BJT* 1 emitter short circuit depends on the value of the on-state resistance $R_{DS(ON)}$ of *MOSFET* 1. As the junction temperature increases so too does $R_{DS(ON)}$. This makes the bypass less effective. Further control is lost with a rise in temperature, because the *BJT* 1 emitter-base voltage decreases. The result is that the peak controllable current I_A of the MCT reduces with an increase in temperature. This sets the current limit and current rating if an external commutation circuit is not to be employed. For example a continuous controllable current might be $I_A = 150\,\text{A}$ at 25°C, whereas control drops to $I_A = 100\,\text{A}$ at 100°C. Circuit design becomes critical for this aspect, because turn-off ability becomes as equally important as the thermal consideration given to the anode current.

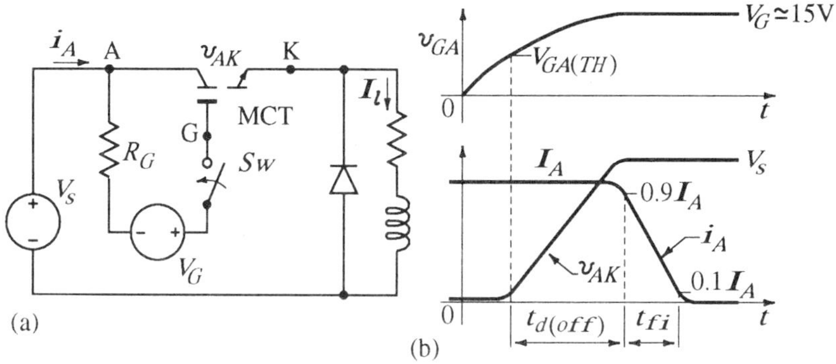

Fig. 10.15 MCT turn-off.
(a) Circuit diagram, (b) waveforms.

It will have been noticed that the components of turn-off times for the semiconductor switches are not standardized. For the MCT with a clamped inductive load, which is the worst case, there is a delay time $t_{d(off)}$ that is the interval from the time that the positive gate voltage reaches its threshold value to the time that the anode current falls to 90% of its initial steady value. Then, there is the standard definition that the current fall-time t_{fi} is the interval from 90% to 10% of the steady value of the anode current. If the energy loss due to the turn-off process is available, then the calculation of average power dissipation is straightforward. Otherwise the estimation of loss is approximate. Fig.10.15 shows the waveforms of the turn-off process.

A 300-A MCT may have a delay time $t_{d(off)} \approx 900$ na, a current fall-time $t_{fi} \approx 1500$ ns and an energy loss $W_{off} \approx 50$ mJ for turn-off. These values are general guidelines since both the turn-off time and the loss can both increase by up to 40% for a junction temperature rise from 25°C to 150°C. The turn-off loss can be four times the turn-on loss.

10.5.6. MCT Protection

Protection of the MCT involves protection of a thyristor and the protection of MOSFETs.

A forward overvoltage at the anode of the MCT is to be avoided, because turn-on would result and the conduction of current is undesirable if the device is supposed to be blocking. A reverse voltage, cathode to anode, cannot be tolerated because of the low avalanche value. Voltage withstand is given to a circuit by connecting a diode in series with the MCT. Alternatively, an antiparallel-connected diode would provide a bypass so that the overvoltage would appear elsewhere and not across the MCT.

Anode current of the MCT is limited by junction temperature. There is another limit placed on the current and that is the value that can be controllably turned off by the MOSFET. This current may have a value as low as 130% of the average current rating. Accordingly, gate turn-off does not afford overcurrent protection. Parallel shorting switches and fast-action fuses must be provided.

The temperature coefficient of resistance of the anode-to-cathode path is negative in the on-state. However, the on-state voltage drop $V_{MCT(ON)}$ is small and does not vary much. Little trouble has been experienced by paralleling matched MCTs in order to handle high values of load current without more than a 10% derating.

Transient currents appear at MCT turn-on and turn-off. At turn-on the di/dt limit to prevent local hot spots depends on the percentage of cells that have turn-on MOSFETs. With the percentage about 4% the di/dt limit has been recorded at about 600A/μs. For protection against a rate of rise of current higher than this a series inductive snubber would be used.

Parallel *RC* snubbers can be used to relieve the stress of turn-off on an MCT but a snubber is not a necessity. The dv/dt withstand is more than 5000V/μs. Typical values of snubber components are $R_s = 25\,\Omega$ and $C_s = 0.1\,\mu F$. If the MCT is in the off-state, a positive voltage V_{GA} at the gate will give protection against almost any likely dv/dt that might otherwise turn on the switch.

Like the GTO during turn-off, the MCT can have current crowding and consequent overheating. Protection is provided by limiting the rate of rise of gate voltage dv_{GA}/dt to a value below about 50V/μs. Since the input capacitance is $C_{iss} \approx 1400\,pF$, a gate resistance $R_G = 10\,\Omega$ would give a reasonable time constant $\tau = 140\,ns$ to limit dv_{GA}/dt to an acceptable level.

The insulation level of the MOSFET gate has a voltage limit of about $\pm 20\,V$. Therefore, the gate leads are connected together when the MCT is not in circuit, so that charge cannot accumulate to build up electrostatic voltages. If the MCT is in circuit, the gate leads are never allowed to be in open circuit. Otherwise, from leakage currents or pickup, dangerous voltage levels can be reached. A simple solution is to give protection by connecting a zener diode across the gate-cathode electrodes.

10.5.7. MCT Ratings and Applications

None of the switches, that are described in this chapter, has been used extensively. The MCT seems like the one that has the most potential for general application. Its frequency of operation is similar to the IGBT, but it has a lower on-state voltage. For the maximum ratings of the MCT, it appears that there is no better switch for choppers or inverters. If MCTs are to be used in ac-dc converters, then it is necessary to use auxiliary diodes because the controlled switch has no reverse blocking capability. Refer to Appendix 8 for typical data.

There is not a great range of MCTs for rating comparison. Arbitrary specifications for a 1000-V, 100-A MCT are as follows.

Withstand $dv_{AK} / dt = 5000\text{V} / \mu\text{s}$;
Limit $di_A / dt = 600\text{A} / \mu\text{s}$ at turn-on;
Turn-on time $t_{on} = t_{d(on)} + t_{ri} = 500$ ns;
Turn-off time $t_{off} = t_{d(off)} + t_{fi} = 2000$ ns;
On-state voltage $V_{MCT(ON)} = 1.1$ V;
Maximum junction temperature $T_{J\max} = 150°\text{C}$;
Thermal resistance $R_{\theta JC} \approx 0.5°\text{C/W}$;
Input capacitance $C_{iss} \approx 14000$ pF;
Turn-on gate voltage $V_{GA} = -5$ V to -10 V;
Turn-off gate voltage $V_{GA} = 10$ V to 15 V;
Maximum frequency of switching is 20 kHz;
Leakage
$I_{A\,leak} = 3$ mA at 150°C.

EXAMPLE 10.1

A 1000-V, 100-A MCT modulates power from a 600-V dc supply to a resistive load. The on-state voltage drop $V_{MCT(ON)} = 1.1$ V. The maximum junction temperature is 150°C. If the case temperature can be transiently maintained at 60°C, determine the magnitude of the current pulse of 10 ms duration that the MCT can tolerate. The transient thermal-impedance characteristic is shown below.

$Z_{\theta JC(t)}$ °C/W	0.17	0.32	0.54	0.82	0.9
Time ms	20	40	80	140	170

Solution

For a time of 10 ms, $Z_{\theta JC(t)} = 0.082°\text{C/W}$.
For a rectangular pulse of current of magnitude I_A, the power dissipated is

$$P_D = V_{MCT(ON)} I_A = \frac{\Delta T}{Z_{\theta JC(t)}} = \frac{150-60}{0.082}.$$

Therefore, $I_A = \dfrac{150-60}{1.1 \times 0.082} = 998$ A.

This is an uncontrolled current pulse, because the MCT cannot turn-off more than about $1.3 \times I_{A\,rated} = 130$ A.

10.6. SUMMARY

This chapter describes uncommon switches, special switches and those switches that have the potential to be used generally.

The SIT is an uncommon switch that has a greater power rating than the MOSFET but its on-state losses are high. Since the efficiency of energy conversion is important, the SIT has its main application in areas where no other switch has its rating and frequency of switching.

The SITH is the only truly controlled rectifier, that is a controlled diode. It is not really a thyristor. Like the SIT, it is normally on without a gate signal and turned off with the application of a negative gate signal. Like the SIT, the SITH is voltage controlled so that changing states can be faster than the GTO. Like the SIT, the SITH is asymmetric regarding voltage blocking capabilities. The SITH is faster than the GTO, but has lower ratings, so there could be special applications for this device.

Thyristors are the high-power semiconductor switches. In order to block forward and reverse biased voltages at the anode the thyristor design creates slow switches with relatively high on-state voltages. The ASCR and the RCT are versions of the thyristor that have no reverse blocking capability. They are asymmetric. They can switch faster and can have lower on-state voltages. Within their power rating they are best used in voltage fed inverters. The GATT is a special thyristor that is designed to be somewhere between the thyristor and the GTO. Both forced commutation and a negative gate signal are used in the turn-off process. This can increase the switching frequency to 10 kHz. This is better than the GTO. There is a niche for the application of the GATT in its power range associated with 2000-V, 400-A ratings.

The MCT has not been proved in general applications, but its low switching losses, low conduction losses, high speed of switching and high current density gives it the potential to be better than the IGBT. A peak power of 1MW can be switched off in 2 μs by a single MCT. These characteristics have been developed at the sacrifice of a reverse blocking capability but that is not required in inverter applications.

Many of the switches that are described in this book are integrated with gating circuits and protection and also with multimodules for matching in parallel or for connection in bridges. There are different forms of packaging to suit different thermal requirements. Descriptions of these are to be found in manufacturers' catalogues. They do not require analysis that is different from what has been done in this text.

10.7. PROBLEMS

Section 10.2

10.1 A 1000-V, 200-A SIT controls the power from a 600-V dc supply to a resistive load whose value is $R = 3\,\Omega$. The chopper operates at 50 kHz with a duty cycle $m = 0.8$. SIT data are that switching times are $t_{on} = 0.3\,\mu s$ for turn-on and $t_{off} = 0.3\,\mu s$ for turn-off, and the on-state resistance is $R_{DS(ON)} = 75\,m\Omega$. If the maximum junction temperature is 150°C and ambient temperature is 30°C determine the maximum thermal resistance $R_{\theta JA}$ between the junction and ambient.

Section 10.3

10.2 A 1000-V, 200-A SITH controls the power from a 600-V dc supply to a resistive load whose value is $R = 3\,\Omega$. The chopper operates at 3 kHz with a duty cycle $m = 0.8$. SITH data are that switching times are $t_{on} = 2\,\mu s$ for turn-on and $t_{off} = 10\,\mu s$ for turn-off, and the on-state voltage drop is 4 V. If the maximum junction temperature is 125°C and ambient temperature is 30°C, determine the maximum thermal resistance $R_{\theta JA}$ between the junction and ambient.

Section 10.4

10.3 Six 2000-V, 400-V thyristors control power in an inverter circuit. The dc supply is 1600 V and the load is purely resistive with a value $R = 4\,\Omega$. Each thyristor is switched at 1kHz with a duty cycle $m = 0.3$. Each thyristor has a turn-on crossover time $t_c = 3\,\mu s$, a turn-off time $t_{off} = 30\,\mu s$ and an on-state voltage $V_{TH(ON)} = 1.5$ V. If the maximum junction temperature is 125°C and the ambient temperature is 30°C, determine the maximum thermal resistance $R_{\theta JA}$ between the junction and ambient.

10.4 Six 2000-V, 400-V ASCRs form an inverter to modulate power from a 1600-V dc supply to a 4-Ω resistive load. Each ASCR is switched at 1kHz with a duty cycle $m = 0.3$. Each ASCR has a turn-on crossover time 3 µs, a turn-off time $t_{off} = 30\,\mu s$ and an on-state voltage $V_{TH(ON)} = 1.23$ V. If the maximum junction temperature is 125°C and the ambient temperature is 30°C, determine the maximum thermal resistance $R_{\theta JA}$ between the junction and ambient.

10.5 Six 2000-V, 400-V ASCRs form an inverter to modulate power from a 1600-V dc supply to a 4-Ω resistive load. Each ASCR is switched at 1kHz with a duty cycle $m = 0.3$. Each ASCR has a turn-on crossover time $t_c = 1.2\,\mu s$, a turn-off time $t_{off} = 12\,\mu s$ and an on-state voltage $V_{TH(ON)} = 1.5$ V. If the maximum junction temperature is 125°C and the ambient temperature is 30°C, determine the maximum thermal resistance $R_{\theta JA}$ between the junction and ambient. Which is better, high speed switching or lower on-state voltage?

Section 10.5

10.6 An MCT acts as a chopper to modulate power from a 600-V dc supply to a 6-Ω resistive load. The chopper switching frequency is 10 kHz. If the power to be absorbed by the load is 40 kW, determine **(a)** the MCT on-time t_{ON} and **(b)** suitable ratings for the switch.

10.7 An MCT chopper modulates power from a 600-V dc supply to a 4-Ω resistive load. The chopper switches at a frequency of 10 kHz. The MCT data for a steady-junction temperature of 120°C are that the on-state voltage is $V_{MCT(ON)} = 1.3$ V, the delay time $t_{d(on)} = 400$ ns and the current rise-time is $t_{ri} = 700$ ns for turn-on. If the chopper has a maximum duty cycle that is $m = 0.7$, determine **(a)** the MCT ratings, **(b)** the average power loss due to conduction **(c)** the average power loss due to the turn-on process and **(d)** the average power absorbed by the load.

10.8 An MCT chopper modulates power from a 600-V dc supply to a 5-Ω resistive load. The chopper switches at a frequency of 10 kHz. The MCT data for a steady junction temperature of 120°C are that the on-state voltage is $V_{MCT(ON)} = 1.2$ V, the delay time $t_{d(off)} = 700$ ns and the current fall-time is $t_{fi} = 1.5\,\mu$s for turn-off. Estimate the energy losses in the MCT **(a)** during the delay interval $t_{d(off)}$, **(b)** during the current fall-time t_{fi} and determine **(c)** the average power dissipation due to the turn-off process.

10.9 An MCT chopper modulates power from a 600-V dc supply to an *RL* load with a diode connected across it. The load resistance is 5 Ω and the load inductance is high enough to consider the load current to be virtually constant. The chopper switches at 10 kHz with a duty cycle $m = 0.8$. MCT data for a steady junction temperature of 100°C are that the steady on-state voltage is $V_{MCT(ON)} = 1.22$ V, the delay time is $t_{d(off)} = 650$ ns and the current fall-time is $t_{fi} = 1.45\,\mu$s for turn-off. Estimate the average power dissipation in the MCT due to the turn-off process.

10.10 A 600-V MCT is used in a chopper circuit that operates at a frequency of 4 kHz. MCT data for a steady junction temperature of 130°C are that the on-state voltage drop is $V_{MCT(ON)} = 1.1$ V, and the thermal resistance, junction to case, is $R_{\theta JC} = 0.6$°C/W. The turn-on loss is 4 mJ and the turn-off loss is 12 mJ. If the case is maintained at 70°C, determine **(a)** the maximum value of continuous current I_A and **(b)** the magnitude of an anode current pulse for a duty cycle $m = 0.4$.

10.11 An MCT has an on-state voltage characteristic $V_{MCT(ON)} = 0.9 + 0.01 I_A^{1.1}$ volts and a switching loss characteristic $W_{loss} = 1.3 I_A \times 10^{-4}$ joules. The thermal resistance, junction to ambient, is $R_{\theta JA} = 0.52$°C/W. The ambient temperature is $T_A = 30$°C. If the frequency of switching is 10 kHz, estimate a reasonable steady overcurrent setting for the MCT control circuit so that the junction temperature will not exceed 150°C.

10.12 A 1000-V, 100-A MCT has a transient thermal impedance characteristic as shown in EXAMPLE 10.1. The on-state voltage drop is $V_{MCT(ON)} \approx 1.1$ V. If the case of the MCT is maintained transiently at 50°C, while the device conducts a current $I_A = 400$ A for 30 ms, determine **(a)** the junction temperature at the end of the current pulse and **(b)** the junction temperature 15 ms later.

10.13 Consider the MCT whose ratings are given in section 10.5.7. Estimate **(a)** the components of power loss in this MCT and **(b)** the efficiency of operation.

10.8. BIBLIOGRAPHY

Baliga, B.J. *Modern Power Devices.* New York: John Wiley & Sons, Inc., 1987.

Bose, B.K. (ED). *Modern Power Electronics - Evolution, Technology and Applications.* New York: IEEE Press, 1992.

Ghandi, S.K. *Semiconductor Power Devices.* New York: John Wiley & Sons, Inc., 1987

Grant, D.A. and J. Gowar. *Power MOSFETs; Theory and Applications.* New York: John Wiley & Sons, Inc., 1989.

Humphreys, M.J. *et al.* (EDS). *Philips Power Semiconductor Applications.* Philips, Hazel Grove, U.K.

Mohan, N., T.M. Underland, and P. Robbins. *Power Electronics.* New York: John Wiley & Sons,Inc., 1989.

Ohno, E. *Introduction to Power Electronics.* Oxford: Clarendon Press, 1988.

Taylor, Paul D. *Thyristor Design and Realization.* Chichester, U.K: John Wiley & Sons, Inc., 1987.

Thorborg, Kjeld. *Power Electronics.* Hemel Hempstead: Prentice-Hall International (UK) Ltd., 1988.

APPENDIX 1

RECTIFIER DIODES

RATINGS

Limiting values in accordance with the Absolute Maximum System (IEC 134).

Voltages			BYV30–300	400	500	
→ Non-repetitive peak reverse voltage	V_{RSM}	max.	350	450	550	V
Repetitive peak reverse voltage	V_{RRM}	max.	300	400	500	V
Crest working reverse voltage	V_{RWM}	max.	200	300	400	V
Continuous reverse voltage*	V_R	max.	200	300	400	V

Currents

Average forward current; switching losses negligible up to 100 kHz				
square wave; δ = 0.5; up to T_{mb} = 113 °C	$I_{F(AV)}$	max.	14	A
up to T_{mb} = 125 °C	$I_{F(AV)}$	max.	10	A
sinusoidal; up to T_{mb} = 118 °C	$I_{F(AV)}$	max.	12.5	A
up to T_{mb} = 125 °C	$I_{F(AV)}$	max.	10	A
R.M.S. forward current	$I_{F(RMS)}$	max.	20	A
Repetitive peak forward current t_p = 20 μs; δ = 0.02	I_{FRM}	max.	320	A
Non-repetitive peak forward current half sine-wave; T_j = 150 °C prior to surge; with reapplied V_{RWMmax};				
t = 10 ms	I_{FSM}	max.	150	A
t = 8.3 ms	I_{FSM}	max.	180	A
I^2t for fusing (t = 10 ms)	I^2t	max.	112	A^2s

Temperatures

Storage temperature	T_{stg}		−65 to +175	°C
Junction temperature	T_j	max.	150	°C

THERMAL RESISTANCE

From junction to mounting base	$R_{th\ j\text{-}mb}$	=	2.0	K/W
From mounting base to heatsink with heatsink compound	$R_{th\ mb\text{-}h}$	=	0.3	K/W
From junction to ambient in free air	$R_{th\ j\text{-}a}$	=	50	K/W

*To ensure thermal stability: $R_{th\ j\text{-}a} \leq 4.6$ K/W.

CHARACTERISTICS

Forward voltage				
I_F = 15 A; T_j = 150 °C	V_F	<	1.05	V*
I_F = 50 A; T_j = 25 °C	V_F	<	1.40	V*
Reverse current				
$V_R = V_{RWM}$ max; T_j = 100 °C	I_R	<	0.8	mA
T_j = 25 °C	I_R	<	50	μA
Reverse recovery when switched from				
I_F = 1 A to $V_R \geq$ 30 V with $-dI_F/dt$ = 100 A/μs; T_j = 25 °C; recovery time	t_{rr}	<	50	ns
I_F = 2 A to $V_R \geq$ 30 V with $-dI_F/dt$ = 20 A/μs; T_j = 25 °C; recovered charge	Q_s	<	50	nC
I_F = 10 A to $V_R \geq$ 30 V with $-dI_F/dt$ = 50 A/μs; T_j = 100 °C; peak recovery current	I_{RRM}	<	5.2	A
Forward recovery when switched to I_F = 10 A with dI_F/dt = 10 A/μs; T_j = 25 °C	V_{fr}	typ.	2.5	V

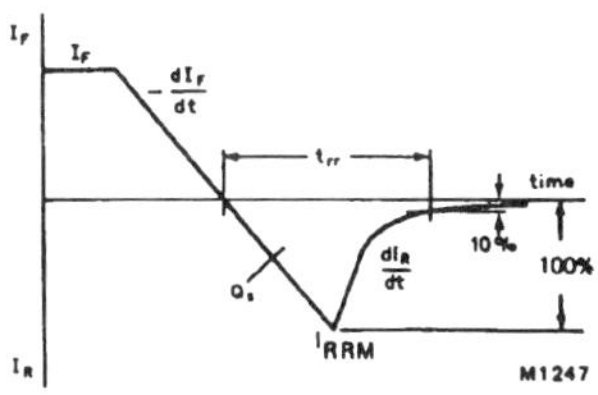

Fig.2 Definition of t_{rr}, Q_s and I_{RRM}.

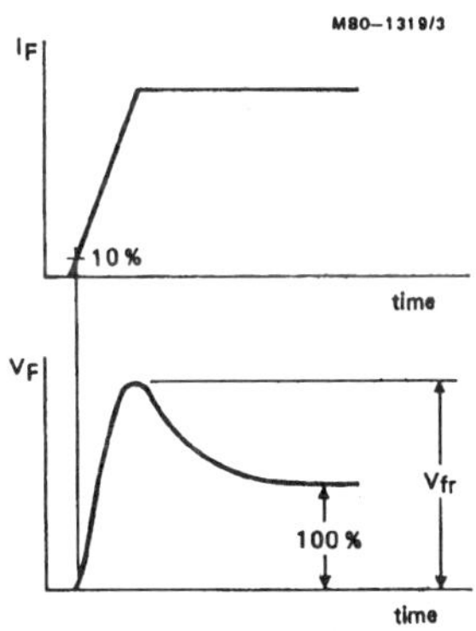

Fig.3 Definition of V_{fr}.

*Measured under pulse conditions to avoid excessive dissipation.

Schottky Barrier rectifier diodes PBYR20100CT series

LIMITING VALUES

Limiting values in accordance with the Absolute Maximum System (IEC 134)

SYMBOL	PARAMETER	CONDITIONS	MIN.	MAX.			UNIT
				2060	2080	20100	
V_{RRM}	Repetitive peak reverse voltage per diode.		-	60	80	100	V
V_{RWM}	Crest working reverse voltage per diode.		-	60	80	100	V
V_R	Continuous reverse voltage per diode.		-	60	80	100	V
	Average forward current[1]	Square wave; $\delta = 0.5$;					
$I_{F(AV)}$	per diode	$T_{mb} = 133\ ^\circ C$	-		10		A
I_O	per device		-		20		A
I_{FRM}	Repetitive forward surge current per diode.	20kHz; $\delta = 0.5$; $T_{mb} = 133\ ^\circ C$	-		20		A
I_{FSM}	Non-repetitive forward surge current per diode.	$T_j = 125\ ^\circ C$; prior to surge; with reapplied V_{RWM}					
		t = 10 ms	-		135		A
		t = 8.3 ms	-		150		A
I^2t	I^2t for fusing per device.		-		93		A^2s
I_{RRM}	Repetitive reverse surge current per diode.	$t_p = 2\ \mu s$; f = 1kHz	-		1.0		A
I_{RSM}	Non-repetitive reverse surge current per diode.	$t_p = 100\ \mu s$	-		1.0		A
T_j	Junction temperature		-		150		°C
T_{stg}	Storage Temperature		-65		175		°C

THERMAL RESISTANCES

SYMBOL	PARAMETER	CONDITIONS	TYP.	MAX.	UNIT
$R_{th j-mb}$	Junction to mounting base				
	per diode		-	2.0	K/W
	both diodes		-	1.0	K/W
$R_{th j-a}$	Junction to Ambient.		-	60	K/W

CHARACTERISTICS

SYMBOL	PARAMETER	CONDITIONS	TYP.	MAX.	UNIT
V_F	Forward voltages[2] per diode.	$T_j = 125\ ^\circ C$; $I_F = 10$ A	-	0.70	V
V_F		$T_j = 125\ ^\circ C$; $I_F = 20$ A	-	0.85	V
V_F		$T_j = 25\ ^\circ C$; $I_F = 20$ A	-	0.95	V
I_R	Reverse leakage current per diode.	$V_R = V_{RWM}$; $T_j = 25\ ^\circ C$	-	150	µA
I_R		$V_R = V_{RWM}$; $T_j = 125\ ^\circ C$	-	150	mA

1 Switching losses negligible up to 500 kHz.

2 Measured under pulsed conditions; pulse width $t_p = 300\ \mu s$

Schottky Barrier rectifier diodes PBYR20100CT series

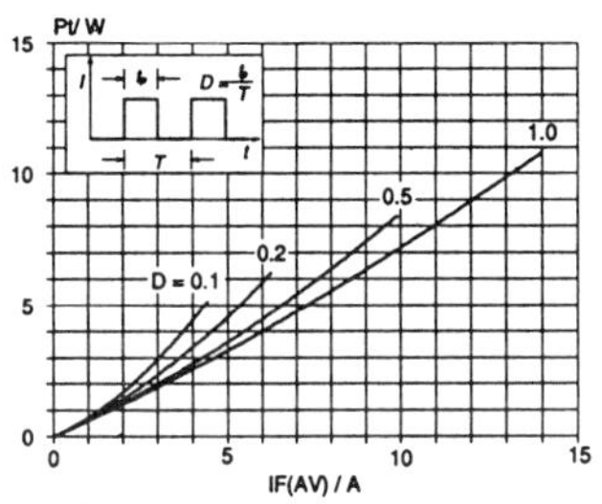

Fig.1. Forward current power rating, square wave operation, where $I_{F(AV)} = I_{F(RMS)} \times \sqrt{D}$.

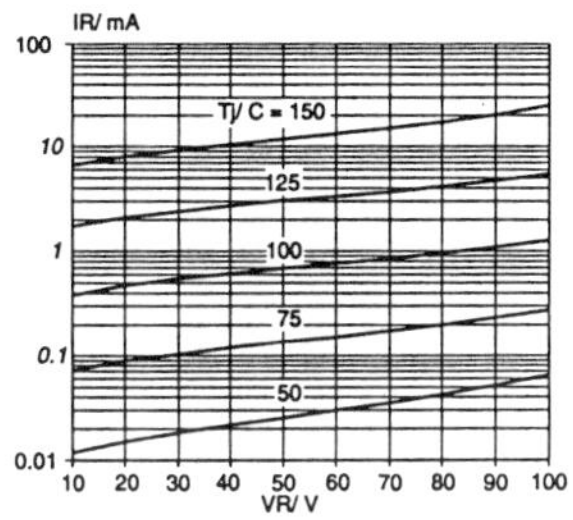

Fig.4. Typical values per diode

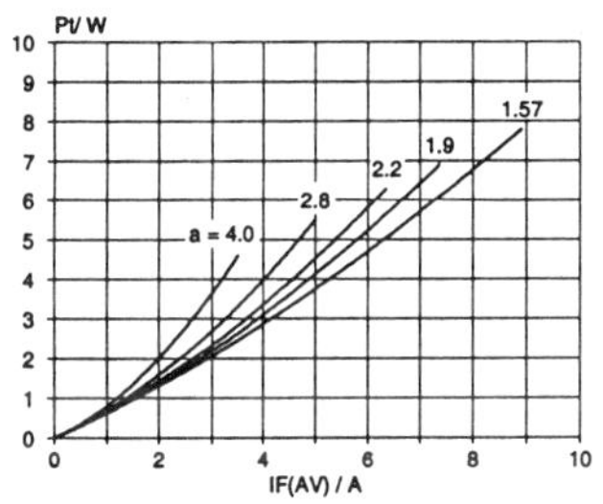

Fig.2. Forward current power rating, sinusoidal operation, where a = form factor = $I_{F(RMS)} / I_{F(AV)}$.

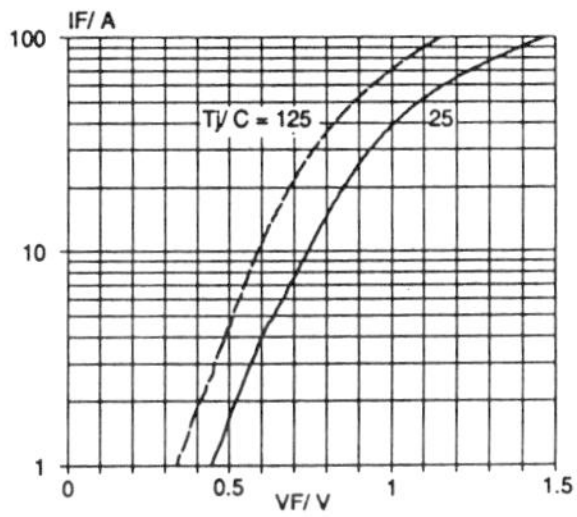

Fig.5. Typical forward voltage.

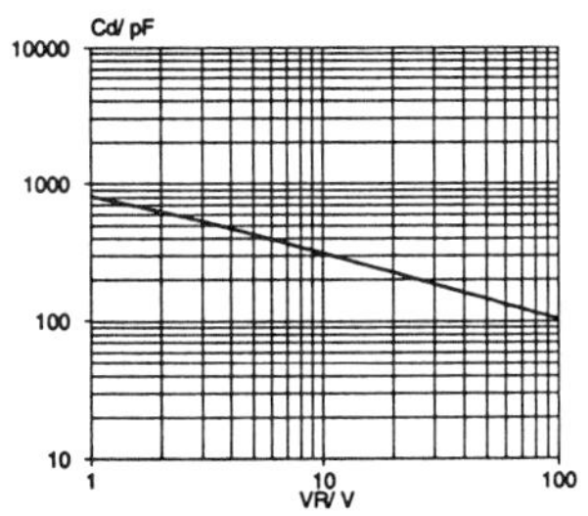

Fig.3. Typical junction capacitance at f = 1 MHz; Tj = 25 to 125 ˚C.

APPENDIX 2

BJT POWER TRANSISTOR

Philips Semiconductors — Product specification

Silicon diffused power transistor — BU1708AX

GENERAL DESCRIPTION

Enhanced performance, new generation, high-voltage, high-speed switching npn transistor in a plastic full-pack envelope specially designed for 277 V high frequency electronic lighting ballast applications.

QUICK REFERENCE DATA

SYMBOL	PARAMETER	CONDITIONS	TYP.	MAX.	UNIT
V_{CESM}	Collector-emitter voltage peak value	$V_{BE} = 0$ V	-	1750	V
V_{CEO}	Collector-emitter voltage (open base)		-	850	V
I_C	Collector current (DC)		-	8	A
I_{CM}	Collector current peak value		-	15	A
P_{tot}	Total power dissipation	$T_{hs} \leq 25$ °C	-	35	W
V_{CEsat}	Collector-emitter saturation voltage	$I_C = 2.5$ A; $I_B = 0.5$ A	-	1.0	V
t_f	Fall time	$I_C = 2.5$ A; $I_{Bon} = 0.5$ A	0.3	0.6	µs

PINNING - SOT186A

PIN	DESCRIPTION
1	base
2	collector
3	emitter
case	isolated

PIN CONFIGURATION

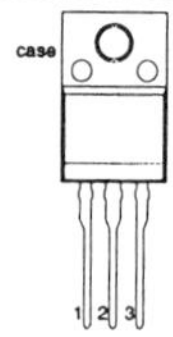

SYMBOL

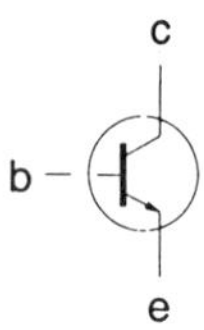

LIMITING VALUES

Limiting values in accordance with the Absolute Maximum System (IEC 134)

SYMBOL	PARAMETER	CONDITIONS	MIN.	MAX.	UNIT
V_{CESM}	Collector-emitter voltage peak value	$V_{BE} = 0$ V	-	1750	V
V_{CEO}	Collector-emitter voltage (open base)		-	850	V
I_C	Collector current (DC)		-	8	A
I_{CM}	Collector current peak value		-	15	A
I_B	Base current (DC)		-	4	A
I_{BM}	Base current peak value		-	6	A
$-I_{B(AV)}$	Reverse base current	average over any 20 ms period	-	100	mA
$-I_{BM}$	Reverse base current peak value [1]		-	5	A
P_{tot}	Total power dissipation	$T_{hs} \leq 25$ °C	-	35	W
T_{stg}	Storage temperature		-65	150	°C
T_j	Junction temperature		-	150	°C

THERMAL RESISTANCES

SYMBOL	PARAMETER	CONDITIONS	TYP.	MAX.	UNIT
$R_{th\,j\text{-}hs}$	Junction to heatsink	with heatsink compound	-	3.6	K/W
$R_{th\,j\text{-}a}$	Junction to ambient	in free air	55	-	K/W

1 Turn-off current.

Silicon diffused power transistor BU1708AX

ISOLATION

T_{hs} = 25 °C unless otherwise specified

SYMBOL	PARAMETER	CONDITIONS	MIN.	TYP.	MAX.	UNIT
$V_{isol(rms)}$	R.M.S. isolation voltage from all three terminals to external heatsink	f = 50-60 Hz; sinusoidal waveform; R.H. ≤ 65% ; clean and dustfree	-	-	2500	V_{RMS}
C_{isol}	Capacitance from T2 to external heatsink	f = 1 MHz	-	10	-	pF

STATIC CHARACTERISTICS

T_{hs} = 25 °C unless otherwise specified

SYMBOL	PARAMETER	CONDITIONS	MIN.	TYP.	MAX.	UNIT
I_{CES}	Collector cut-off current [2]	V_{BE} = 0 V; V_{CE} = 1650 V	-	-	250	µA
I_{CES}		V_{BE} = 0 V; V_{CE} = $V_{CESMmax}$	-	-	1.0	mA
I_{CES}		V_{BE} = 0 V; V_{CE} = $V_{CESMmax}$; T_j = 125 °C	-	-	2.0	mA
I_{EBO}	Emitter cut-off current	V_{EB} = 7.5 V; I_C = 0 A	-	-	1.0	mA
$V_{CEOsust}$	Collector-emitter sustaining voltage. Fig.1 and Fig.2.	I_B = 0 A; I_C = 100 mA; L = 25 mH	750	-	-	V
V_{CEsat}	Collector-emitter saturation voltages	I_C = 2.5 A; I_B = 0.5 A	-	0.25	1	V
V_{BEsat}	Base-emitter saturation voltage	I_C = 2.5 A; I_B = 0.5 A	-	-	1.3	V
h_{FE}	DC current gain	I_C = 10 mA; V_{CE} = 5 V	8	17	-	
h_{FE}		I_C = 300 mA; V_{CE} = 2 V	12	18	35	
h_{FE}		I_C = 2.5 A; V_{CE} = 1 V	5	7	-	

DYNAMIC CHARACTERISTICS

T_{hs} = 25 °C unless otherwise specified

SYMBOL	PARAMETER	CONDITIONS	TYP.	MAX.	UNIT
C_c	Collector capacitance	I_E = 0 A; V_{CB} = 10 V; f = 1 MHz	80	-	pF
	Switching times (resistive load). Fig.3, Fig.4	I_{Con} = 2.5 A; I_{Bon} = $-I_{Boff}$ = 0.5 A			
t_{on}	Turn-on time		1	1.3	µs
t_s	Turn-off storage time		4	7	µs
t_f	Turn-off fall time		0.6	0.9	µs
	Switching times (inductive load). Fig.5 and Fig.6.				
t_s	Turn-off storage time	I_{Con} = 2.5 A; I_{Bon} = 0.5 A; L_B = 1 µH; $-V_{BB}$ = 5 V	2.0	3.0	µs
t_f	Turn-off fall time		0.3	0.6	µs
t_s	Turn-off storage time	I_{Con} = 2.5 A; I_{Bon} = 0.5 A; L_B = 1 µH; $-V_{BB}$ = 5 V; T_j = 100 °C	2.5	3.5	µs
t_f	Turn-off fall time		0.4	0.7	µs

2 Measured with half sine-wave voltage (curve tracer).

Silicon diffused power transistor BU1708AX

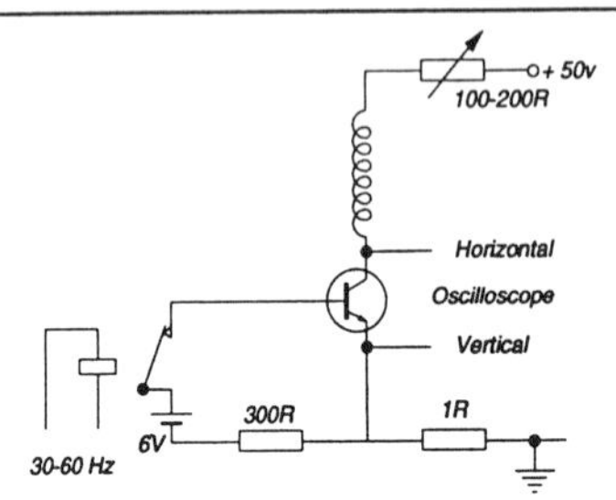

Fig.1. Test circuit for $VCEO_{sust}$.

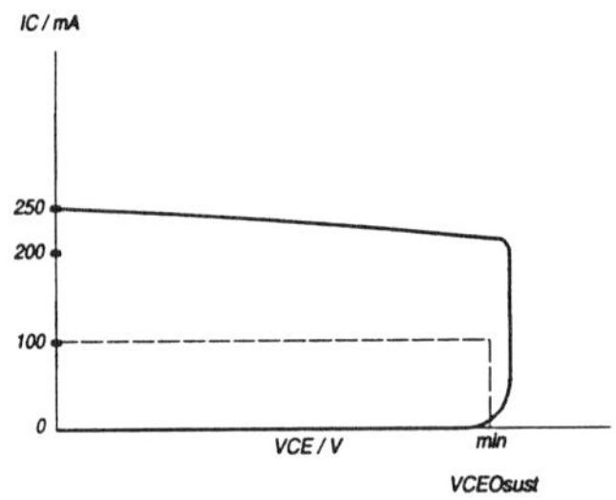

Fig.2. Oscilloscope display for $VCEO_{sust}$.

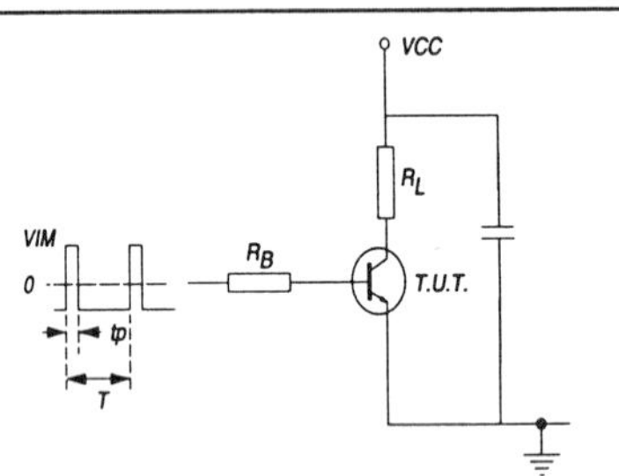

Fig.3. Test circuit resistive load. VIM = -6 to +8 V
V_{CC} = 250 V; tp = 20 µs; δ = tp / T = 0.01.
R_B and R_L calculated from I_{Con} and I_{Bon} requirements.

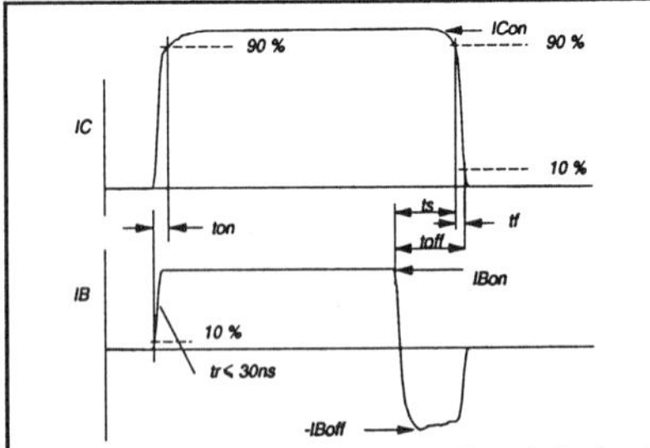

Fig.4. Switching times waveforms with resistive load

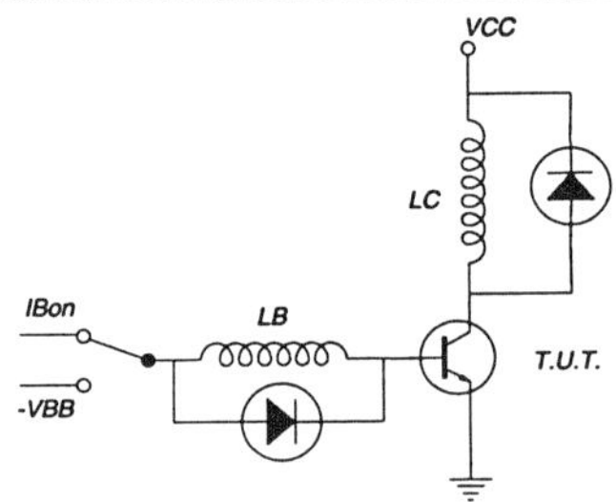

Fig.5. Test circuit inductive load
V_{CC} = 300 V; $-V_{BE}$ = 5 V; L_B = 1 uH

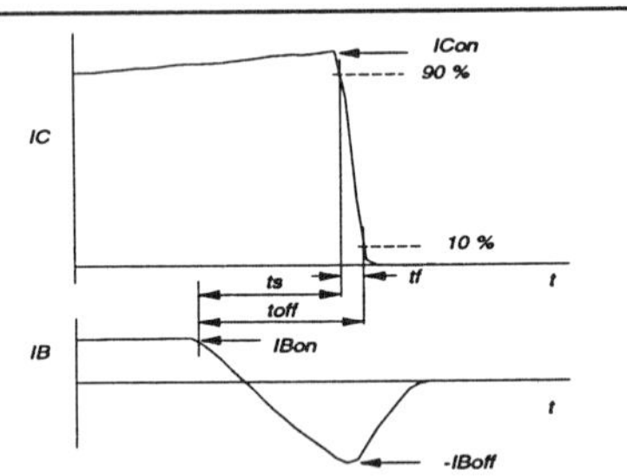

Fig.6. Switching times waveforms with inductive load

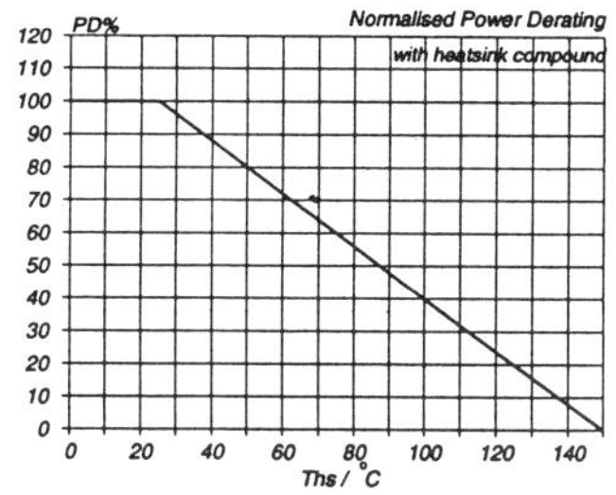

Fig.7. Normalised power dissipation.
PD% = 100·PD/PD 25°C = f (Ths)

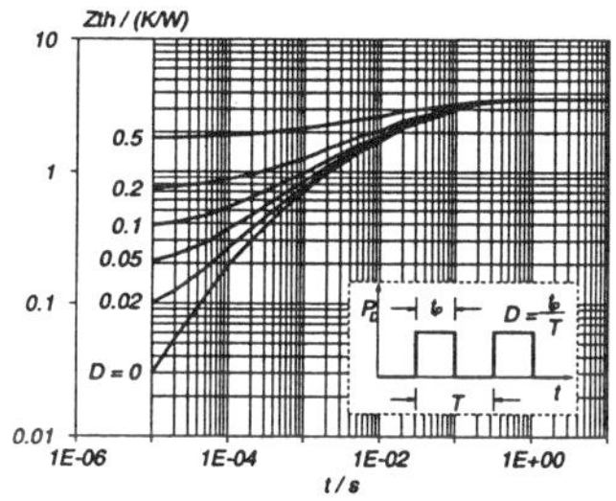

Fig.8. Transient thermal impedance.
$Z_{th\,j-hs}$ = f(t); parameter D = t_p/T

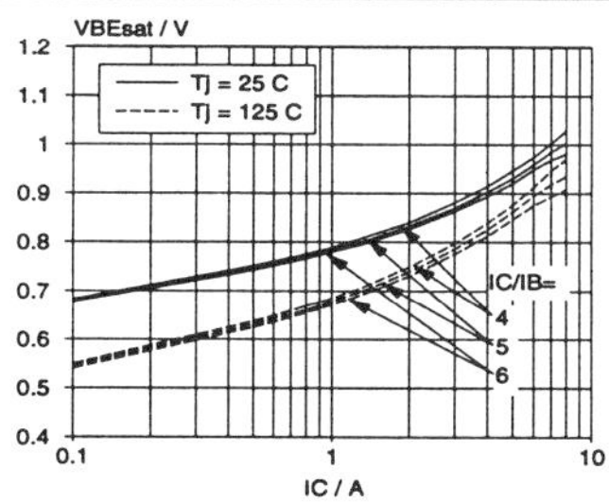

Fig.9. Typical base-emitter saturation voltage.
V_{BEsat} = f(I_C); parameter I_C/I_B

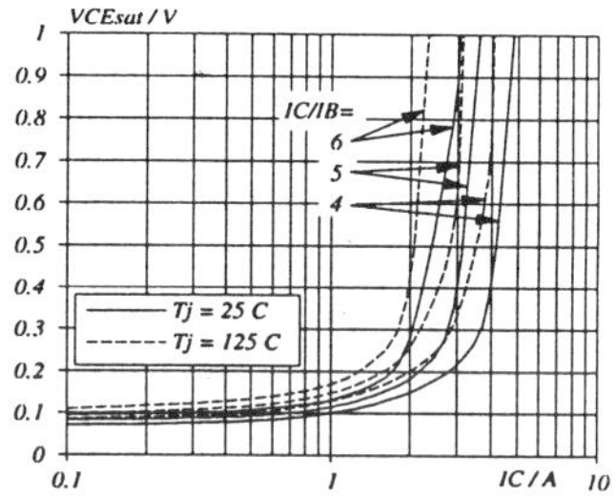

Fig.10. Typical collector-emitter saturation voltage.
V_{CEsat} = f(I_C); parameter I_C/I_B

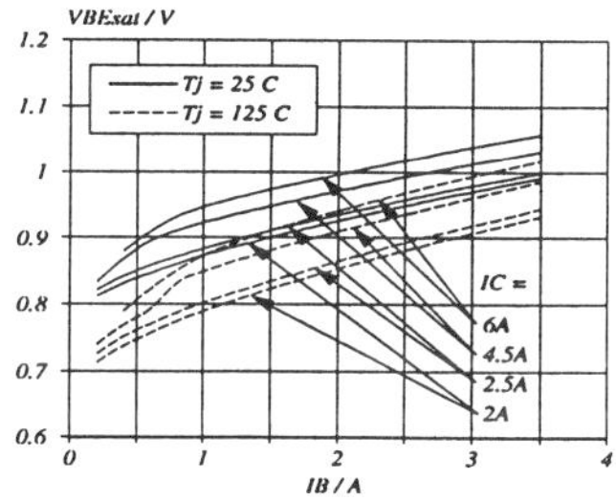

Fig.11. Typical base-emitter saturation voltage.
V_{BEsat} = f(I_B); parameter I_C

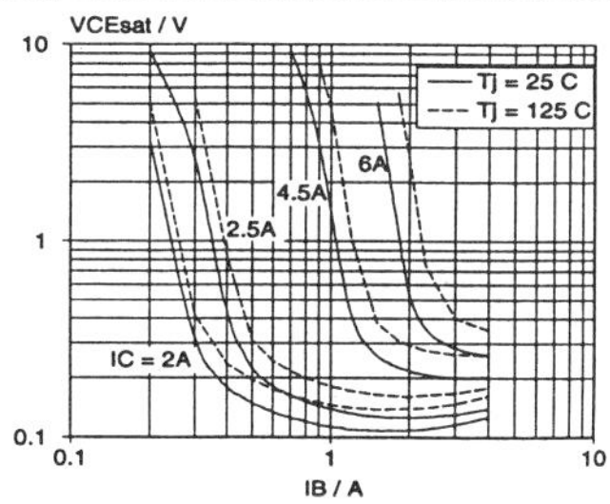

Fig.12. Typical collector-emitter saturation voltage.
V_{CEsat} = f(I_B); parameter I_C

APPENDIX 3

POWERMOS TRANSISTOR

Philips Semiconductors — Product specification

PowerMOS transistor TOPFET — BUK101-50GS

DESCRIPTION

Monolithic temperature and overload protected power MOSFET in a 3 pin plastic envelope, intended as a general purpose switch for automotive systems and other applications.

APPLICATIONS

General controller for driving

- lamps
- motors
- solenoids
- heaters

FEATURES

- Vertical power DMOS output stage
- Low on-state resistance
- Overload protection against over temperature
- Overload protection against short circuit load
- Latched overload protection reset by input
- 10 V input level
- Low threshold voltage also allows 5 V control
- Control of power MOSFET and supply of overload protection circuits derived from input
- ESD protection on input pin
- Overvoltage clamping for turn off of inductive loads

QUICK REFERENCE DATA

SYMBOL	PARAMETER	MAX.	UNIT
V_{DS}	Continuous drain source voltage	50	V
I_D	Continuous drain current	29	A
P_D	Total power dissipation	75	W
T_j	Continuous junction temperature	150	˚C
$R_{DS(ON)}$	Drain-source on-state resistance V_{IS} = 10 V	50	mΩ

FUNCTIONAL BLOCK DIAGRAM

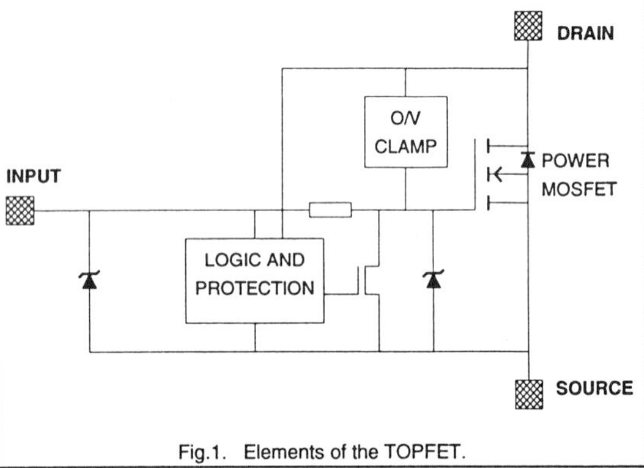

Fig.1. Elements of the TOPFET.

PINNING - TO220AB

PIN	DESCRIPTION
1	input
2	drain
3	source
tab	drain

PIN CONFIGURATION

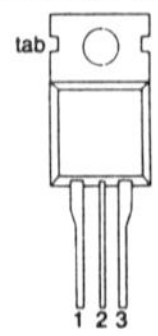

SYMBOL

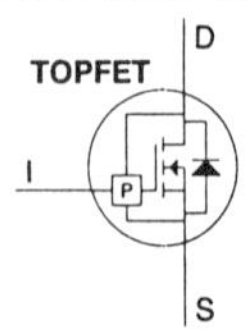

LIMITING VALUES

Limiting values in accordance with the Absolute Maximum System (IEC 134)

SYMBOL	PARAMETER	CONDITIONS	MIN.	MAX.	UNIT
V_{DS}	Continuous drain source voltage[1]	normal operation	-	50	V
V_{IS}	Continuous input voltage	-	0	11	V
I_D	Continuous drain current	$T_{mb} \leq 25$ °C; $V_{IS} = 10$ V	-	29	A
I_D	Continuous drain current	$T_{mb} \leq 100$ °C; $V_{IS} = 10$ V	-	18	A
I_{DRM}	Repetitive peak on-state drain current	$T_{mb} \leq 25$ °C; $V_{IS} = 10$ V	-	120	A
P_D	Total power dissipation	$T_{mb} \leq 25$ °C	-	75	W
T_{stg}	Storage temperature	-	-55	150	°C
T_j	Continuous junction temperature[2]	normal operation	-	150	°C
T_{sold}	Lead temperature	during soldering	-	250	°C

OVERLOAD PROTECTION LIMITING VALUES

With the protection supply provided via the input pin, TOPFET can protect itself from two types of overload.

SYMBOL	PARAMETER	CONDITIONS	MIN.	MAX.	UNIT
V_{ISP}	Protection supply voltage[3]	for valid protection	5	-	V
	Over temperature protection				
$V_{DSP(T)}$	Protected drain source voltage	$V_{IS} = 10$ V	-	50	V
	Short circuit load protection				
$V_{DSP(P)}$	Protected drain source voltage[4]	$V_{IS} = 10$ V	-	20	V
		$V_{IS} = 5$ V	-	35	V
P_{DSM}	Instantaneous overload dissipation	$T_{mb} = 25$ °C	-	1.3	kW

OVERVOLTAGE CLAMPING LIMITING VALUES

At a drain source voltage above 50 V the power MOSFET is actively turned on to clamp overvoltage transients.

SYMBOL	PARAMETER	CONDITIONS	MIN.	MAX.	UNIT
I_{DROM}	Repetitive peak clamping current	$V_{IS} = 0$ V	-	29	A
E_{DSM}	Non-repetitive clamping energy	$T_{mb} \leq 25$ °C; $I_{DM} = 27$ A; $V_{DD} \leq 20$ V; inductive load	-	625	mJ
E_{DRM}	Repetitive clamping energy	$T_{mb} \leq 95$ °C; $I_{DM} = 8$ A; $V_{DD} \leq 20$ V; f = 250 Hz	-	40	mJ

ESD LIMITING VALUE

SYMBOL	PARAMETER	CONDITIONS	MIN.	MAX.	UNIT
V_C	Electrostatic discharge capacitor voltage	Human body model; C = 250 pF; R = 1.5 kΩ	-	1.5	kV

1 Prior to the onset of overvoltage clamping. For voltages above this value, safe operation is limited by the overvoltage clamping energy.

2 A higher T_j is allowed as an overload condition but at the threshold $T_{j(TO)}$ the over temperature trip operates to protect the switch.

3 The input voltage for which the overload protection circuits are functional.

4 The device is able to self-protect against a short circuit load providing the drain-source supply voltage does not exceed V_{DSP} maximum. For further information, refer to OVERLOAD PROTECTION CHARACTERISTICS.

THERMAL CHARACTERISTICS

SYMBOL	PARAMETER	CONDITIONS	MIN.	TYP.	MAX.	UNIT
	Thermal resistance					
$R_{th\,j\text{-}mb}$	Junction to mounting base	-	-	1.3	1.67	K/W
$R_{th\,j\text{-}a}$	Junction to ambient	in free air	-	60	-	K/W

STATIC CHARACTERISTICS

T_{mb} = 25 °C unless otherwise specified

SYMBOL	PARAMETER	CONDITIONS		MIN.	TYP.	MAX.	UNIT
$V_{(CL)DSS}$	Drain-source clamping voltage	V_{IS} = 0 V; I_D = 10 mA		50	-	-	V
$V_{(CL)DSS}$	Drain-source clamping voltage	V_{IS} = 0 V; I_{DM} = 2 A; $t_p \le 300$ µs; $\delta \le 0.01$		-	-	70	V
I_{DSS}	Zero input voltage drain current	V_{DS} = 12 V; V_{IS} = 0 V		-	-	20	µA
I_{DSS}	Zero input voltage drain current	V_{DS} = 50 V; V_{IS} = 0 V		-	-	10	mA
I_{DSS}	Zero input voltage drain current	V_{DS} = 50 V; V_{IS} = 0 V; T_j = 125 °C		-	-	10	mA
$R_{DS(ON)}$	Drain-source on-state resistance	I_{DM} = 13 A; $t_p \le 300$ µs; $\delta \le 0.01$	V_{IS} = 10 V	-	35	50	mΩ
			V_{IS} = 5 V	-	45	60	mΩ

OVERLOAD PROTECTION CHARACTERISTICS

TOPFET switches off when one of the overload thresholds is reached. It remains latched off until reset by the input.

SYMBOL	PARAMETER	CONDITIONS	MIN.	TYP.	MAX.	UNIT
	Short circuit load protection[1]	T_{mb} = 25 °C; L ≤ 10 µH				
$E_{DS(TO)}$	Overload threshold energy	V_{DD} = 13 V; V_{IS} = 10 V	-	0.4	-	J
$t_{d\,sc}$	Response time	V_{DD} = 13 V; V_{IS} = 10 V	-	0.8	-	ms
	Over temperature protection					
$T_{j(TO)}$	Threshold junction temperature	V_{IS} = 10 V; from $I_D \ge 1$ A[2]	150	-	-	°C

INPUT CHARACTERISTICS

T_{mb} = 25 °C unless otherwise specified. The supply for the logic and overload protection is taken from the input.

SYMBOL	PARAMETER	CONDITIONS	MIN.	TYP.	MAX.	UNIT
$V_{IS(TO)}$	Input threshold voltage	V_{DS} = 5 V; I_D = 1 mA	1.0	1.5	2.0	V
I_{IS}	Input supply current	V_{IS} = 10 V; normal operation	-	0.4	1.0	mA
V_{ISR}	Protection reset voltage[3]		2.0	2.6	3.5	V
V_{ISR}	Protection reset voltage	T_j = 150 °C	1.0	-	-	
I_{ISL}	Input supply current	V_{IS} = 10 V; protection latched	1.0	2.5	4.0	mA
$V_{(BR)IS}$	Input clamp voltage	I_I = 10 mA	11	13	-	V
R_{IG}	Input series resistance	to gate of power MOSFET	-	4	-	kΩ

1 The short circuit load protection is able to save the device providing the instantaneous on-state dissipation is less than the limiting value for P_{DSM}, which is always the case when V_{DS} is less than V_{DSP} maximum. Refer to OVERLOAD PROTECTION LIMITING VALUES.

2 The over temperature protection feature requires a minimum on-state drain source voltage for correct operation. The specified minimum I_D ensures this condition.

3 The input voltage below which the overload protection circuits will be reset.

TRANSFER CHARACTERISTICS

T_{mb} = 25 °C

SYMBOL	PARAMETER	CONDITIONS	MIN.	TYP.	MAX.	UNIT
g_{fs}	Forward transconductance	V_{DS} = 10 V; I_{DM} = 13 A $t_p \leq$ 300 µs; $\delta \leq$ 0.01	10	16	-	S
$I_{D(SC)}$	Drain current[1]	V_{DS} = 13 V; V_{IS} = 10 V	-	80	-	A

SWITCHING CHARACTERISTICS

T_{mb} = 25 °C. R_I = 50 Ω. Refer to waveform figures and test circuits.

SYMBOL	PARAMETER	CONDITIONS	MIN.	TYP.	MAX.	UNIT
$t_{d\,on}$	Turn-on delay time	V_{DD} = 13 V; V_{IS} = 10 V	-	1.5	-	µs
t_r	Rise time	resistive load R_L = 2.1 Ω	-	6	-	µs
$t_{d\,off}$	Turn-off delay time	V_{DD} = 13 V; V_{IS} = 0 V	-	18	-	µs
t_f	Fall time	resistive load R_L = 2.1 Ω	-	9	-	µs
$t_{d\,on}$	Turn-on delay time	V_{DD} = 10 V; V_{IS} = 10 V	-	2	-	µs
t_r	Rise time	inductive load I_{DM} = 6 A	-	1	-	µs
$t_{d\,off}$	Turn-off delay time	V_{DD} = 10 V; V_{IS} = 0 V	-	22	-	µs
t_f	Fall time	inductive load I_{DM} = 6 A	-	1	-	µs

REVERSE DIODE LIMITING VALUE

SYMBOL	PARAMETER	CONDITIONS	MIN.	MAX.	UNIT
I_S	Continuous forward current	$T_{mb} \leq$ 25 °C; V_{IS} = 0 V	-	29	A

REVERSE DIODE CHARACTERISTICS

T_{mb} = 25 °C

SYMBOL	PARAMETER	CONDITIONS	MIN.	TYP.	MAX.	UNIT
V_{SDS}	Forward voltage	I_S = 29 A; V_{IS} = 0 V; t_p = 300 µs	-	1.0	1.5	V
t_{rr}	Reverse recovery time	not applicable[2]	-	-	-	-

ENVELOPE CHARACTERISTICS

SYMBOL	PARAMETER	CONDITIONS	MIN.	TYP.	MAX.	UNIT
L_d	Internal drain inductance	Measured from contact screw on tab to centre of die	-	3.5	-	nH
L_d	Internal drain inductance	Measured from drain lead 6 mm from package to centre of die	-	4.5	-	nH
L_s	Internal source inductance	Measured from source lead 6 mm from package to source bond pad	-	7.5	-	nH

1 During overload before short circuit load protection operates.

2 The reverse diode of this type is not intended for applications requiring fast reverse recovery.

PowerMOS transistor
TOPFET

BUK101-50GS

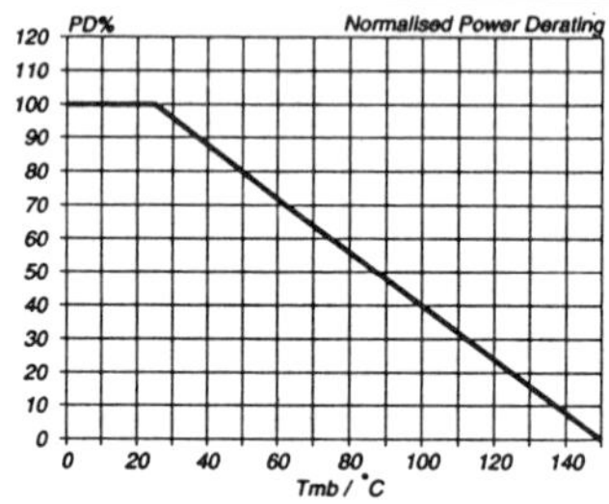

Fig.2. Normalised limiting power dissipation.
$P_D\% = 100 \cdot P_D/P_D(25\ ^\circ C) = f(T_{mb})$

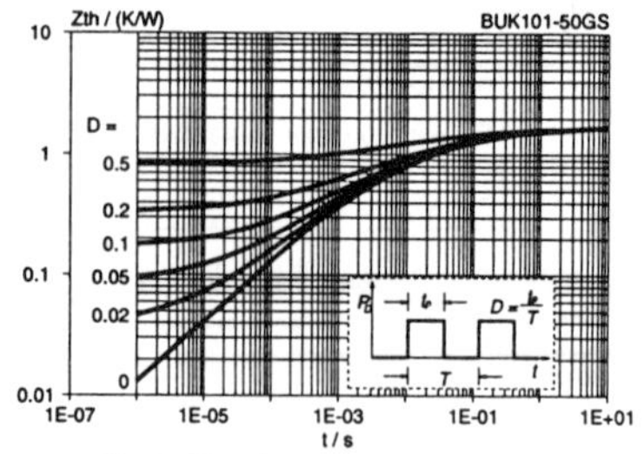

Fig.5. Transient thermal impedance.
$Z_{th\ j\text{-}mb} = f(t)$; parameter $D = t_p/T$

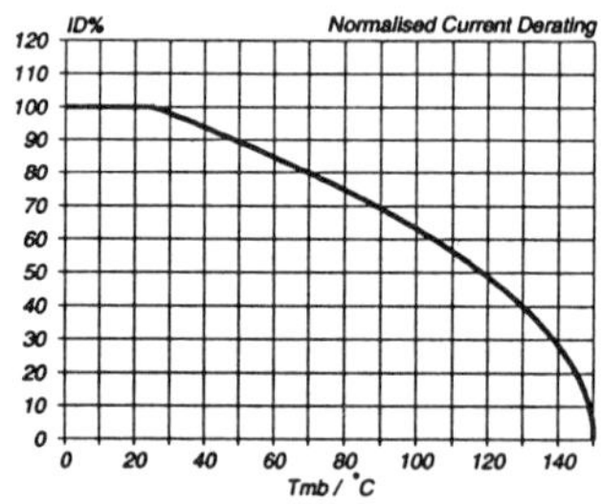

Fig.3. Normalised continuous drain current.
$I_D\% = 100 \cdot I_D/I_D(25\ ^\circ C) = f(T_{mb})$; conditions: $V_{IS} = 5$ V

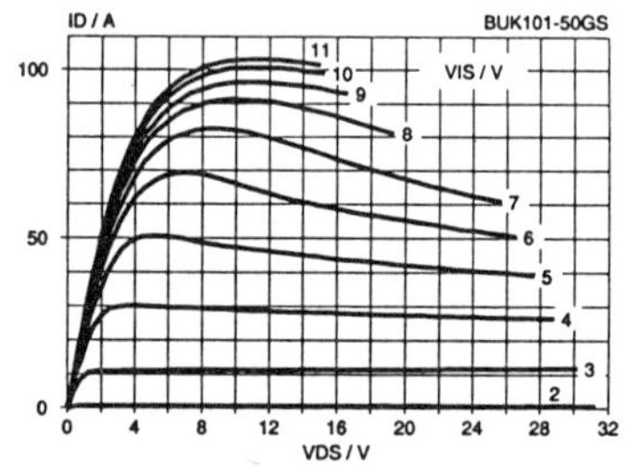

Fig.6. Typical output characteristics, $T_j = 25\ ^\circ C$.
$ID = f(V_{DS})$; parameter V_{IS}; $t_p = 250\ \mu s$ & $t_p < t_{d\,sc}$

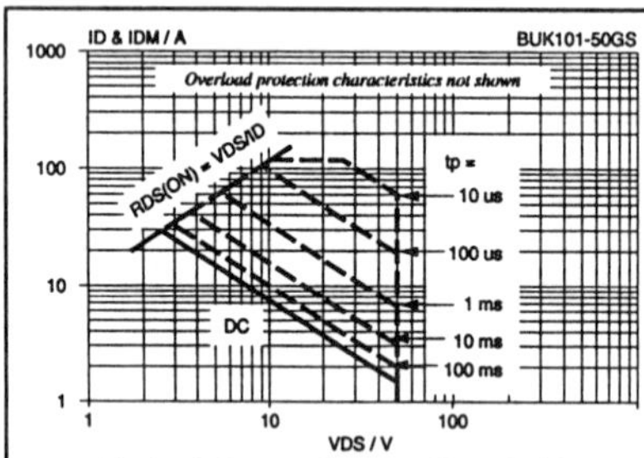

Fig.4. Safe operating area. $T_{mb} = 25\ ^\circ C$
I_D & $I_{DM} = f(V_{DS})$; I_{DM} single pulse; parameter t_p

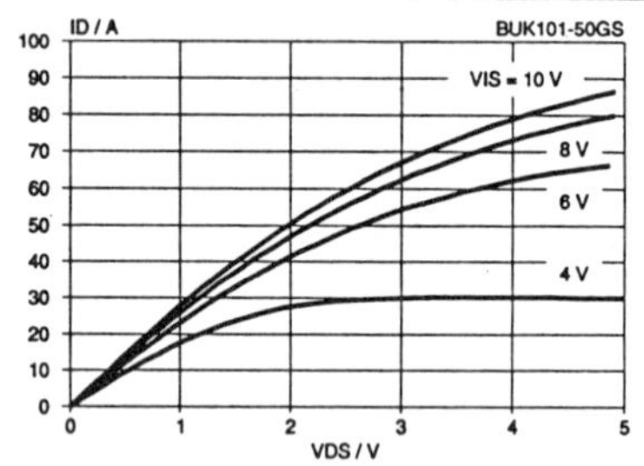

Fig.7. Typical on-state characteristics, $T_j = 25\ ^\circ C$.
$ID = f(V_{DS})$; parameter V_{IS}; $t_p = 250\ \mu s$

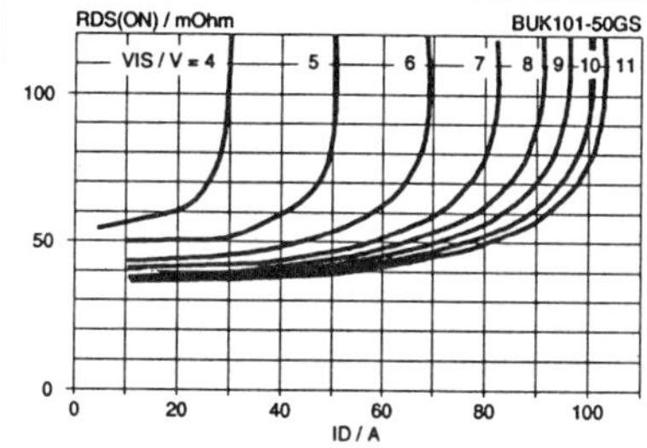

Fig.8. Typical on-state resistance, $T_j = 25$ °C.
$R_{DS(ON)} = f(I_D)$; parameter V_{IS}; $t_p = 250$ µs

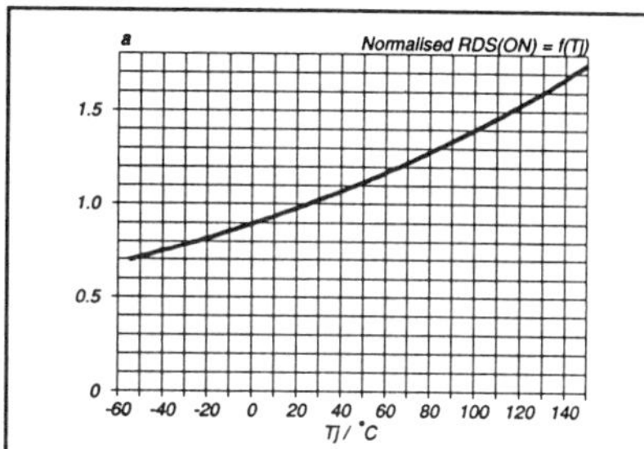

Fig.11. Normalised drain-source on-state resistance.
$a = R_{DS(ON)}/R_{DS(ON)25\ °C} = f(T_j)$; $I_D = 13$ A; $V_{IS} = 10$ V

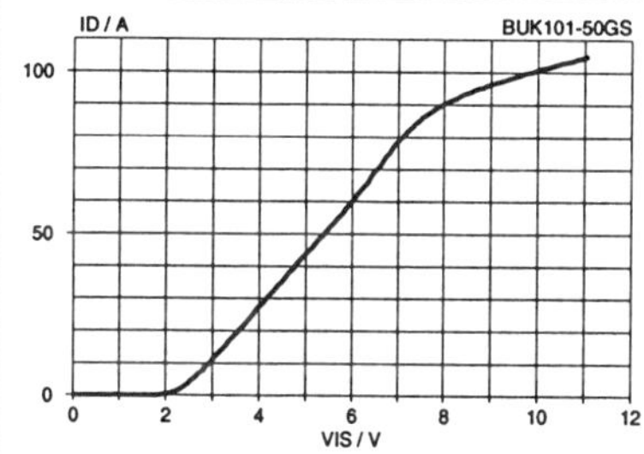

Fig.9. Typical transfer characteristics, $T_j = 25$ °C.
$I_D = f(V_{IS})$; conditions: $V_{DS} = 10$ V; $t_p = 250$ µs

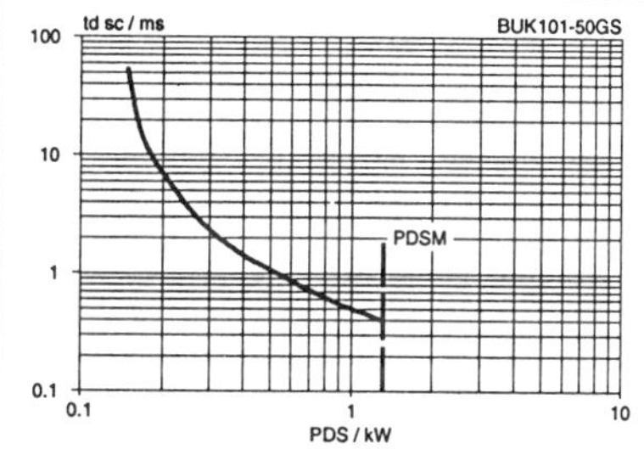

Fig.12. Typical overload protection characteristics.
$t_{d\,sc} = f(P_{DS})$; conditions: $V_{IS} \geq 5$ V; $T_j = 25$ °C.

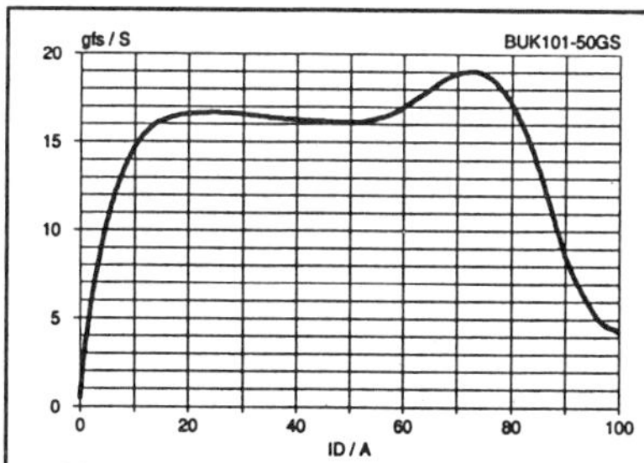

Fig.10. Typical transconductance, $T_j = 25$ °C.
$g_{fs} = f(I_D)$; conditions: $V_{DS} = 10$ V; $t_p = 250$ µs

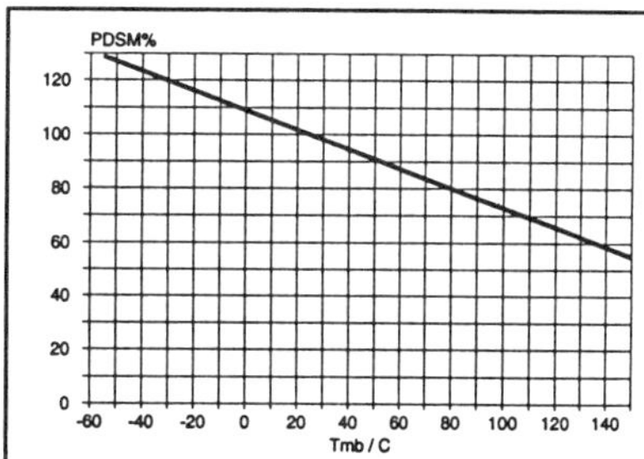

Fig.13. Normalised limiting overload dissipation.
$P_{DSM}\% = 100 \cdot P_{DSM}/P_{DSM}(25\ °C) = f(T_{mb})$

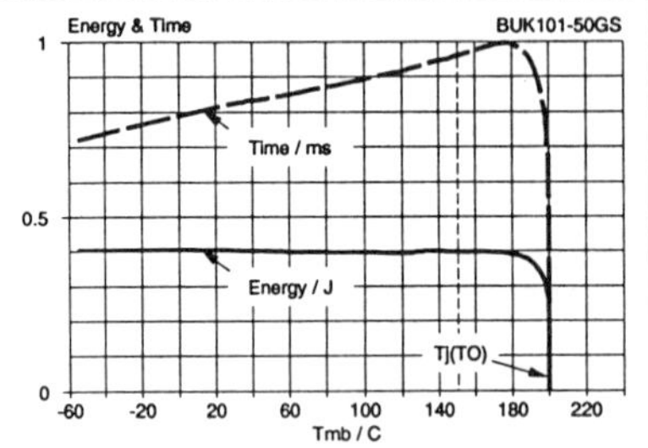

Fig.14. Typical overload protection characteristics. Conditions: V_{DD} = 13 V; V_{IS} = 10 V; SC load = 30 mΩ

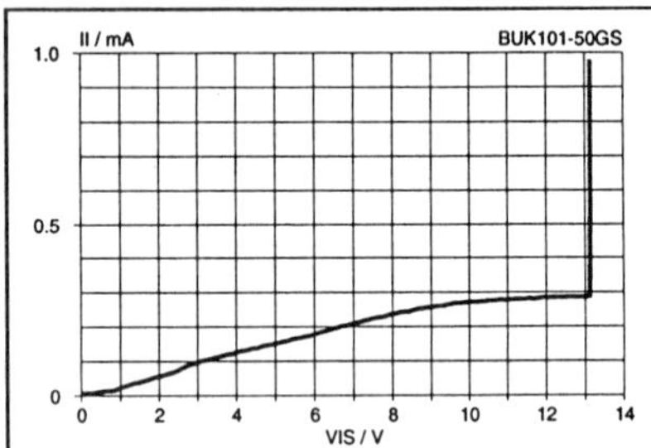

Fig.17. Typical DC input characteristics, T_j = 25 ˚C. $I_{IS} = f(V_{IS})$; normal operation

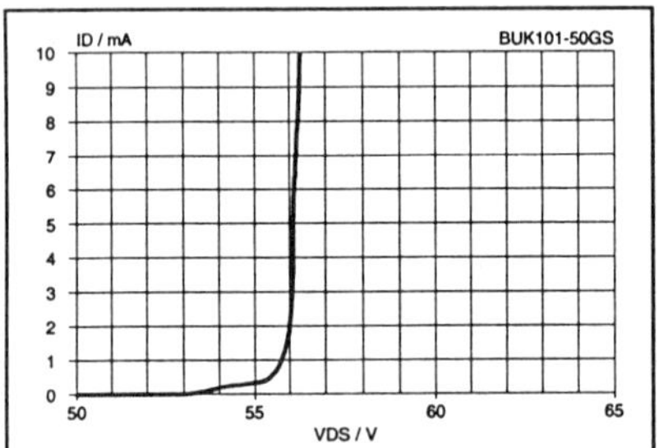

Fig.15. Typical clamping characteristics. $I_{DSS} = f(V_{DS})$; conditions: V_{IS} = 0 V

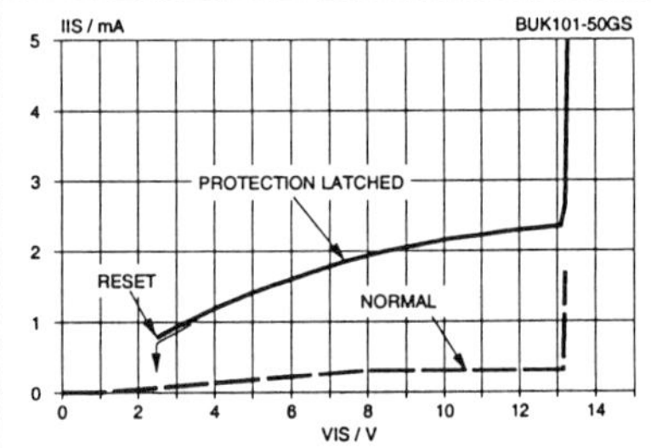

Fig.18. Typical DC input characteristics, T_j = 25 ˚C. $I_{ISL} = f(V_{IS})$; overload protection operated ⇒ I_D = 0 A

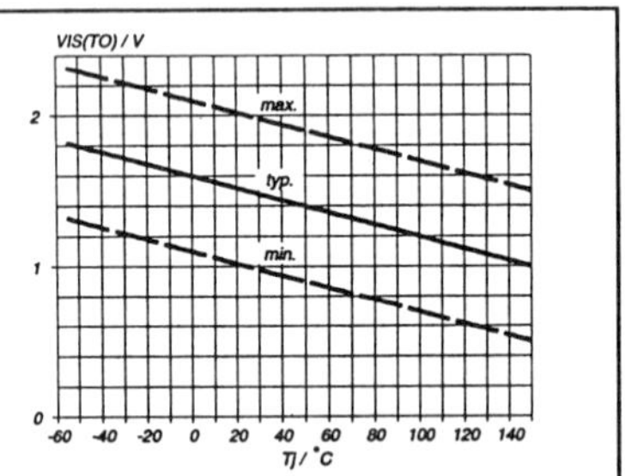

Fig.16. Input threshold voltage. $V_{IS(TO)} = f(T_j)$; conditions: I_D = 1 mA; V_{DS} = 5 V

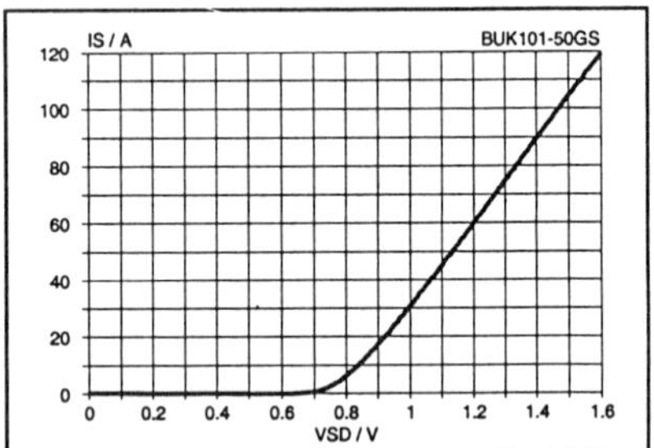

Fig.19. Typical reverse diode current, T_j = 25 ˚C. $I_S = f(V_{SDS})$; conditions: V_{IS} = 0 V; t_p = 250 µs

PowerMOS transistor TOPFET

BUK101-50GS

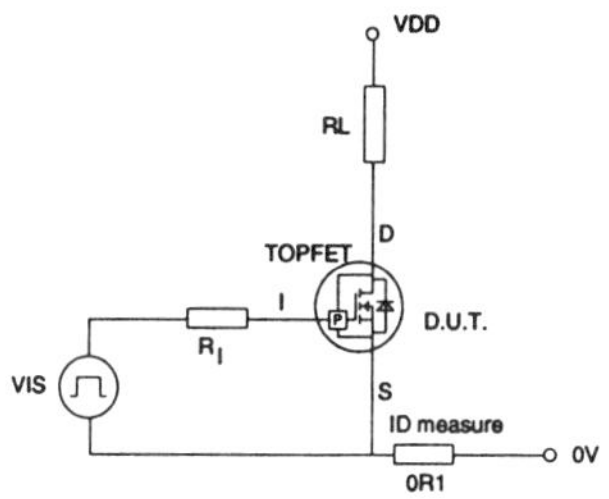

Fig.20. Test circuit for resistive load switching times.

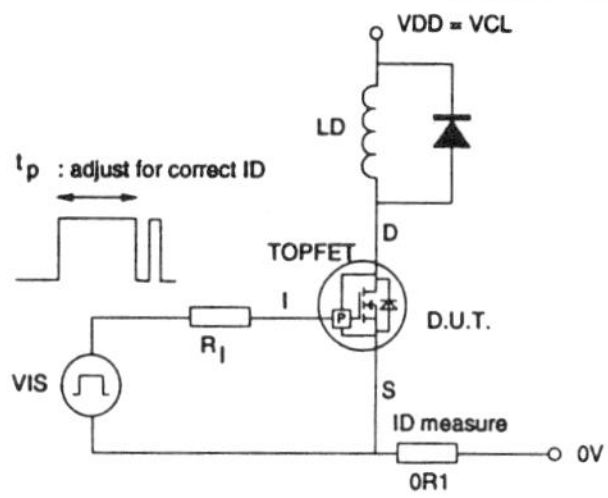

Fig.23. Test circuit for inductive load switching times.

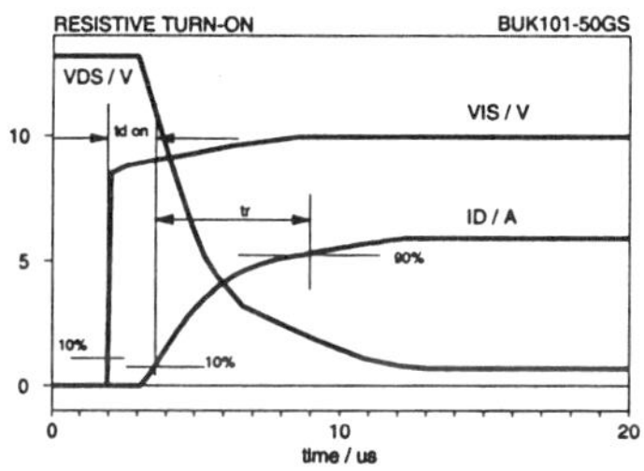

Fig.21. Typical switching waveforms, resistive load. V_{DD} = 13 V; R_L = 2.1 Ω; R_I = 50 Ω, T_j = 25 ˚C.

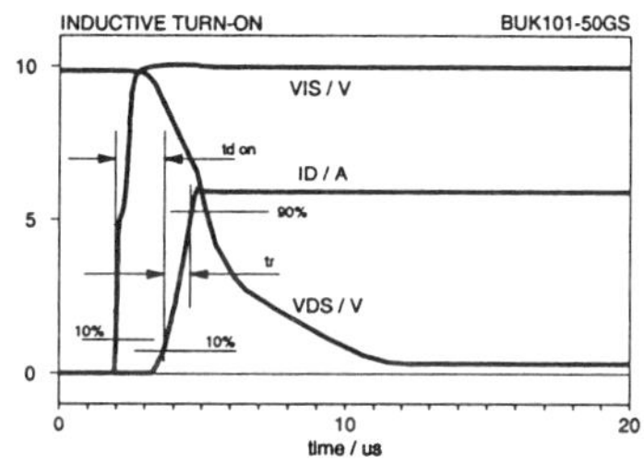

Fig.24. Typical switching waveforms, inductive load. V_{DD} = 10 V; I_D = 6 A; R_I = 50 Ω, T_j = 25 ˚C.

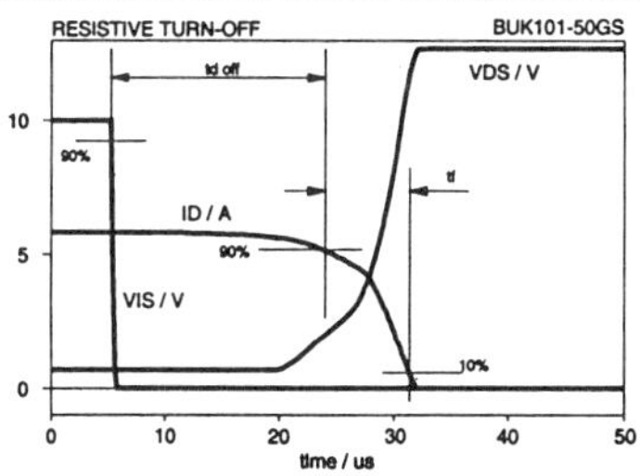

Fig.22. Typical switching waveforms, resistive load. V_{DD} = 13 V; R_L = 2.1 Ω; R_I = 50 Ω, T_j = 25 ˚C.

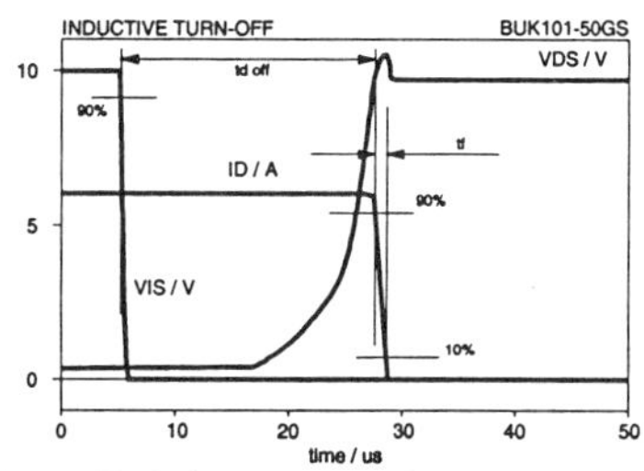

Fig.25. Typical switching waveforms, inductive load. V_{DD} = 10 V; I_D = 6 A; R_I = 50 Ω, T_j = 25 ˚C.

APPENDIX 4
THYRISTOR

BTW62 SERIES

RATINGS

Limiting values in accordance with the Absolute Maximum System (IEC134).

Anode to cathode			BTW62–600R	800R	1000R	
Transient off-state voltage	V_{DSM}	max.	800	1000	1000	V
Repetitive peak off-state voltage	V_{DRM}	max.	600	800	1000	V
Continuous off-state voltage	V_D	max.	500	650	700	V

Transient reverse voltage; $t_p < 5\ \mu s$	V_{RSM}	max.	15	V
Average on-state current (averaged over any 20 ms period) up to T_{mb} = 85 °C	$I_{T(AV)}$	max.	18	A
R.M.S. on-state current	$I_{T(RMS)}$	max.	28	A
Repetitive peak on-state current; $t_p = 50\ \mu s$; $\delta = 0.05$	I_{TRM}	max.	175	A
Non-repetitive peak on-state current T_j = 125 °C prior to surge; t = 10 ms; half sine-wave	I_{TSM}	max.	200	A
I^2t for fusing; t = 10 ms	I^2t	max.	200	A^2s
Rate of rise of on-state current after triggering with I_G = 1.25 A; I_T = 80 A	dI_T/dt	max.	1000	A/μs
Gate to cathode				
Average power dissipation (averaged over any 20 ms period)	$P_{G(AV)}$	max.	1	W
Peak power dissipation; t = 10 μs	P_{GM}	max.	10	W
Temperatures				
Storage temperature	T_{stg}		–40 to +125	°C
Operating junction temperature	T_j	max.	125	°C

ISOLATION*

R.M.S. isolation voltage	V_{isol}	min.	2500	V

THERMAL RESISTANCE

From junction to mounting base	$R_{th\ j\text{-}mb}$	=	1.1	K/W
From mounting base to heatsink with heatsink compound	$R_{th\ mb\text{-}h}$	=	0.2	K/W

*From baseplate to all three terminals connected together.

CHARACTERISTICS

Anode to cathode

On-state voltage				
I_T = 50 A; T_j = 25 °C	V_T	<	2.6	V*
Off-state current				
$V_D = V_{Dmax}$; T_j = 125 °C	I_D	<	6.0	mA
Holding current; T_j = 25 °C	I_H	<	400	mA

Gate to cathode

Voltage that will trigger all devices				
V_D = 12 V; T_j = 25 °C	V_{GT}	>	2.0	V
Current that will trigger all devices				
V_D = 12 V; T_j = 25 °C	I_{GT}	>	250	mA

Switching characteristics (see Fig.2)

Circuit commutated turn-off time				
dV_D/dt = 500 V/μs (linear to V_{DRMmax}); R_{GK} = 10 Ω; V_G = 0; T_j = 125 °C; when switched from I_T = 100 A; t_p = 150 μs				
$-dI_T/dt$ = 50 A/μs				
suffix K	t_q	<	6	μs
suffix N	t_q	<	9	μs
$-dI_T/dt$ = 10 A/μs				
suffix K	t_q	<	4	μs
suffix N	t_q	<	6	μs

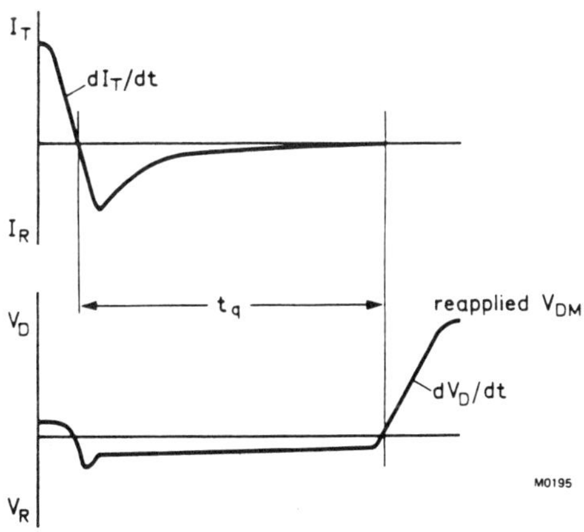

Fig.2 Circuit-commutated turn-off time definition.

*Measured under pulse conditions to avoid excessive dissipation.

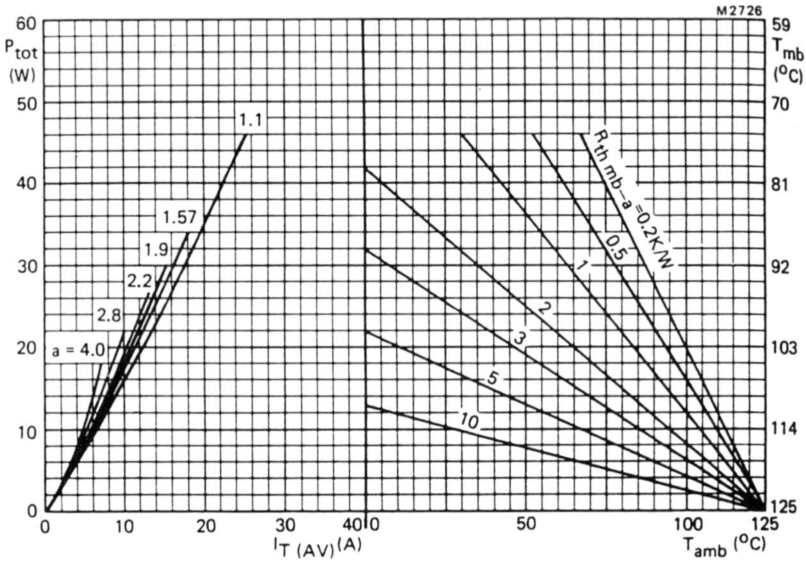

Fig.3 The right-hand part shows the interrelationship between the power (derived from the left-hand part) and the maximum permissible temperatures.

$$a = \text{form factor} = \frac{I_{T(RMS)}}{I_{T(AV)}}$$

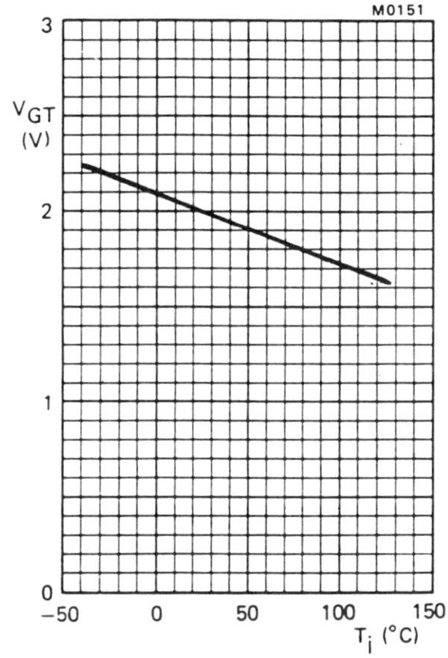

Fig.4 Minimum gate voltage that will trigger all devices plotted against junction temperature.

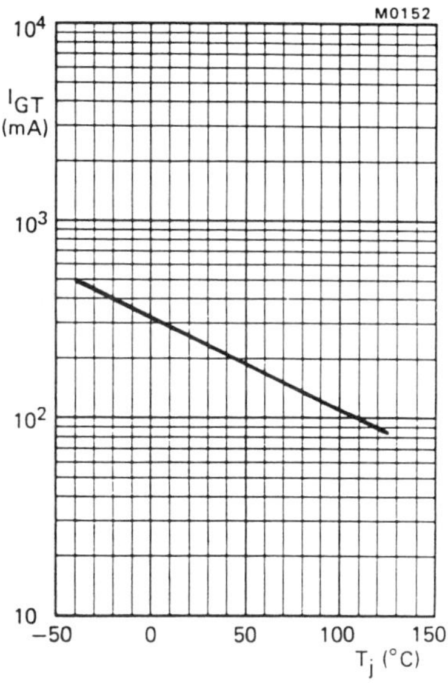

Fig.5 Minimum gate current that will trigger all devices plotted against junction temperature.

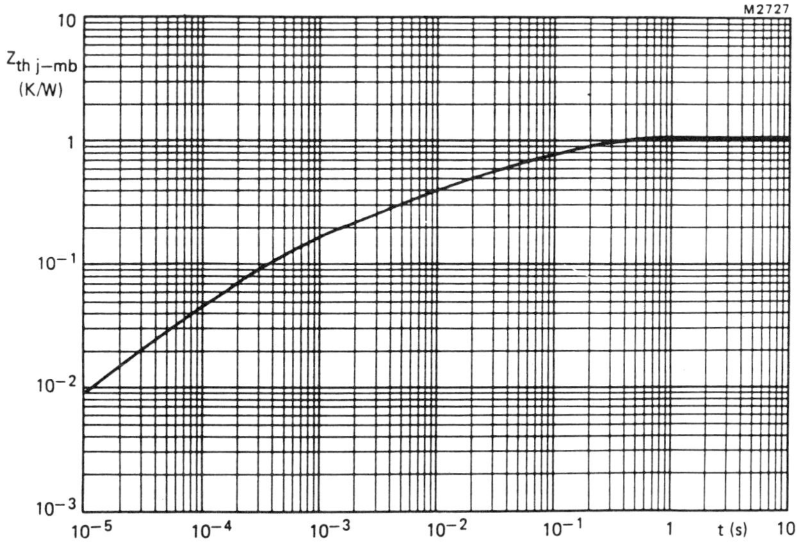

Fig.6 Transient thermal impedance.

APPENDIX 5

IGBT

Philips Semiconductors — Product specification

Insulated Gate Bipolar Transistor (IGBT) — BUK854-500IS

GENERAL DESCRIPTION

N-channel insulated gate bipolar power transistor in a plastic envelope.
The device is intended for use in automotive ignition applications, and other general purpose switching applications requiring low on-state voltage.

QUICK REFERENCE DATA

SYMBOL	PARAMETER	MAX.	UNIT
V_{CE}	Collector-emitter voltage	500	V
V_{EC}	Reverse Collector-Emitter Voltage	25	V
I_C	Collector current (DC)	15	A
P_{tot}	Total power dissipation	85	W
V_{CEsat}	Collector-emitter on-state voltage	2	V

PINNING - TO220AB

PIN	DESCRIPTION
1	gate
2	collector
3	emitter
tab	collector

PIN CONFIGURATION

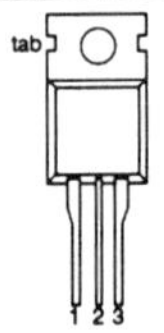

SYMBOL

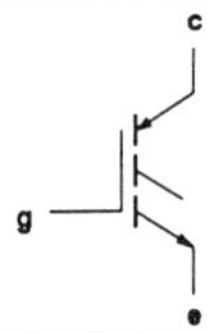

LIMITING VALUES

Limiting values in accordance with the Absolute Maximum System (IEC 134)

SYMBOL	PARAMETER	CONDITIONS	MIN.	MAX.	UNIT
V_{CE}	Collector-emitter voltage	-	-25	500	V
V_{CGR}	Collector-gate voltage	R_{GE} = 20 kΩ	-	500	V
$\pm V_{GE}$	Gate-emitter voltage	-	-	30	V
I_C	Collector current (DC)	T_{mb} = 25 °C	-	15	A
I_C	Collector current (DC)	T_{mb} = 100 °C	-	8.5	A
I_{CM}	Collector current (pulsed peak value, on-state)	-	-	25	A
I_{CLM}	Collector current (clamped inductive load)	V_{CL} = 350 V $R_G \geq$ 1 kΩ	-	25	A
E_{ECR}	Reverse Avalanche Energy (repetitive)	I_E = 2 A	-	5	mJ
P_{tot}	Total power dissipation	T_{mb} = 25 °C	-	85	W
T_{stg}	Storage temperature	-	- 55	150	°C
T_j	Junction Temperature	-	-	150	°C

THERMAL RESISTANCES

SYMBOL	PARAMETER	CONDITIONS	TYP.	MAX.	UNIT
$R_{th\,j\text{-}mb}$	Junction to mounting base	-	-	1.47	K/W
$R_{th\,j\text{-}a}$	Junction to ambient	In free air	60	-	K/W

STATIC CHARACTERISTICS

T_{mb} = 25 °C unless otherwise specified

SYMBOL	PARAMETER	CONDITIONS	MIN.	TYP.	MAX.	UNIT
$V_{(BR)CES}$	Collector-emitter breakdown voltage	V_{GE} = 0 V; I_C = 0.25 mA	500	-	-	V
$V_{(BR)EC}$	Reverse Collector-Emitter breakdown voltage	I_E = 10 mA	25	30	-	V
$V_{GE(TO)}$	Gate threshold voltage	V_{CE} = V_{GE}; I_C = 1 mA	2.5	4	5.5	V
I_{CES}	Zero gate voltage collector current	V_{CE} = 500 V; V_{GE} = 0 V; T_j = 25 °C	-	1	20	µA
I_{CES}	Zero gate voltage collector current	V_{CE} = 500 V; V_{GE} = 0 V; T_j = 125 °C	-	0.1	1	mA
I_{EC}	Reverse collector current	V_{CE} = -25 V	-	0.5	10	mA
I_{EC}	Reverse collector current	V_{CE} = -25 V; T_j = 125 °C	-	5	-	mA
I_{GES}	Gate emitter leakage current	V_{GE} = ±30 V; V_{CE} = 0 V	-	10	100	nA
V_{CEsat}	Collector-emitter saturation voltage	V_{GE} = 15 V; I_C = 8.5 A	-	1.4	2	V
		V_{GE} = 15 V; I_C = 15 A	-	1.7	-	V
		V_{GE} = 15 V; I_C = 8.5 A; T_j = 100°C	-	1.5	2	V
		V_{GE} = 15 V; I_C = 8.5 A; T_j = -40°C	-	1.4	1.9	V

DYNAMIC CHARACTERISTICS

T_{mb} = 25 °C unless otherwise specified

SYMBOL	PARAMETER	CONDITIONS	MIN.	TYP.	MAX.	UNIT
g_{fe}	Forward transconductance	V_{CE} = 15 V; I_C = 3 A	1.5	4.5	-	S
C_{ies}	Input capacitance	V_{GE} = 0 V; V_{CE} = 25 V; f = 1 MHz	-	400	750	pF
C_{oes}	Output capacitance		-	40	80	pF
C_{res}	Feedback capacitance		-	15	40	pF
$t_{d\,off}$	Turn-off delay time	I_C = 8.5 A; V_{CL} = 350 V; R_G = 1 kΩ;	-	3.5	4.5	µs
t_f	Fall time	V_{GE} = 15 V; T_j = 100°C; Inductive	-	4	6	µs
t_c	Crossover Time	Load	-	5.5	7.5	µs
E_{off}	Turn-off Energy loss		-	6.5	9	mJ

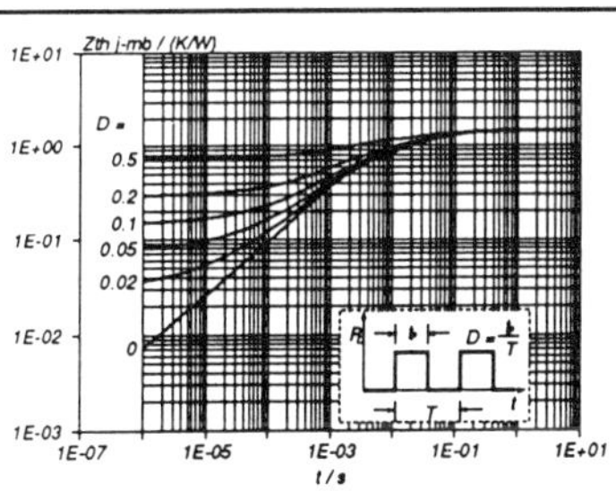

Fig.1. Transient thermal impedance $Z_{th\,j-mb}$ = f(t) ; parameter D = t_p/T

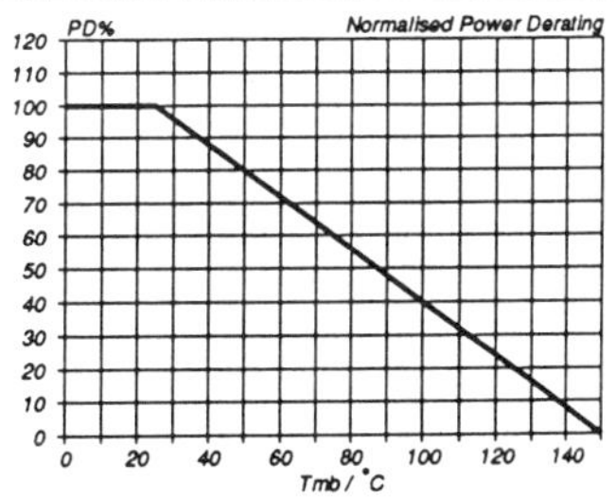

Fig.2. Normalised power dissipation. PD% = $100.P_D/P_{D\,25°C}$ = f(T_{mb})

Insulated Gate Bipolar Transistor (IGBT) BUK854-500IS

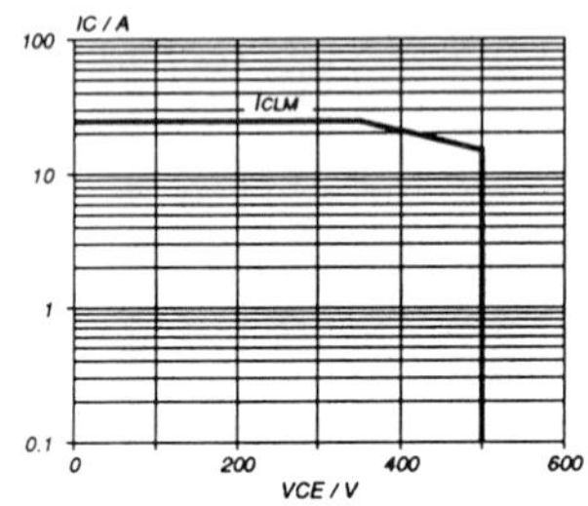

Fig.3. Turn-off Safe Operating Area
conditions: $T_j \leq T_{jmax.}$; $R_G = \geq 1$ kΩ

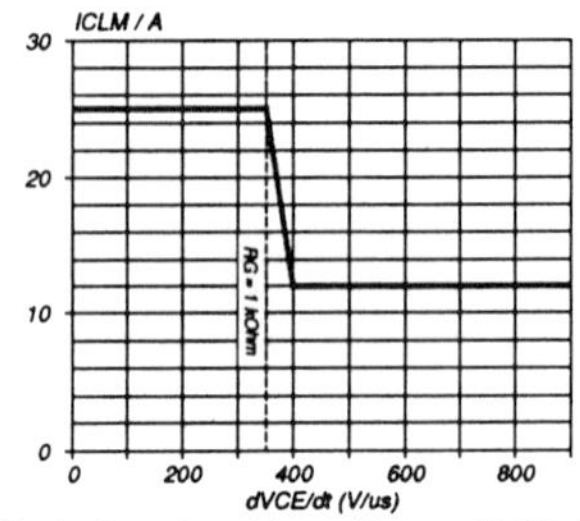

Fig.6. Derating of I_{CLM} with turn-off dV_{CE}/dt
conditions: $V_{CL} \leq 350$ V; $T_j \leq T_{jmax.}$

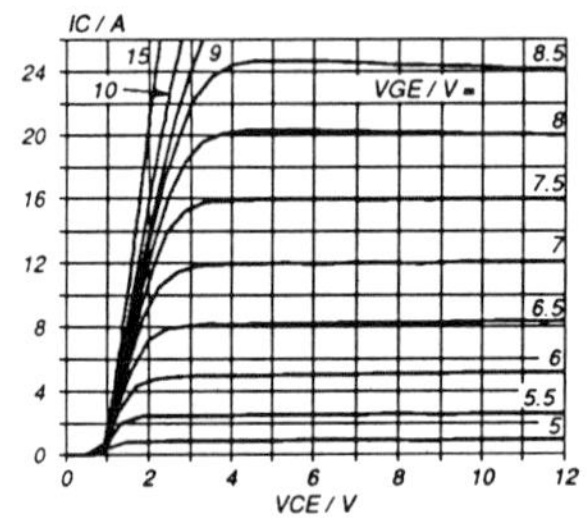

Fig.4. Typical Output Characteristics
$I_C = f(V_{CE})$; parameter V_{GE}; conditions: $T_j = 25$°C

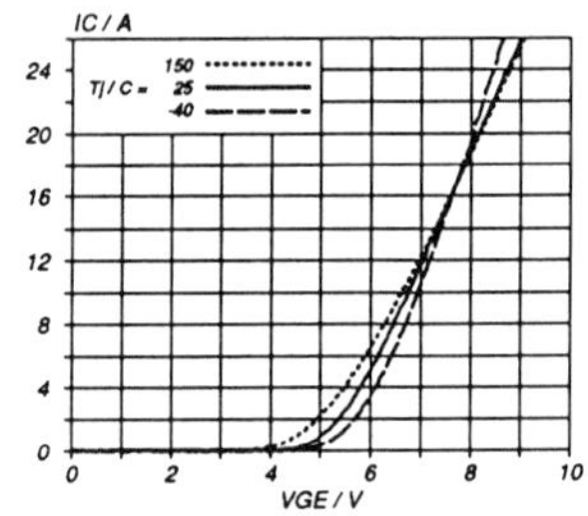

Fig.7. Typical Transfer Characteristics
$I_C = f(V_{GE})$, parameter T_j; conditions: $V_{CE} = 15$ V

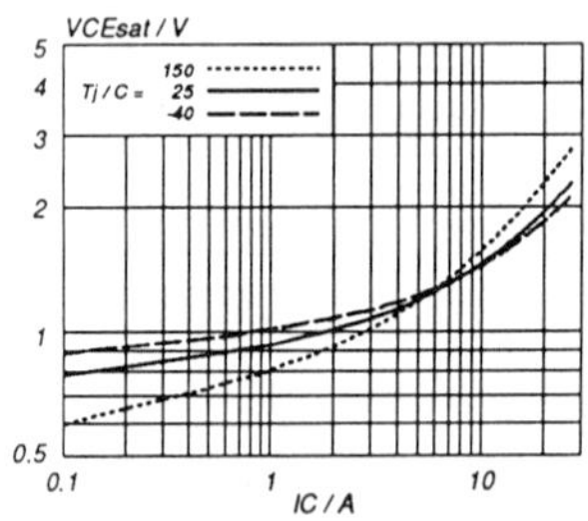

Fig.5. Typical On-state Voltage
$V_{CEsat} = f(I_C)$; parameter T_j; conditions: $V_{GE} = 15$ V

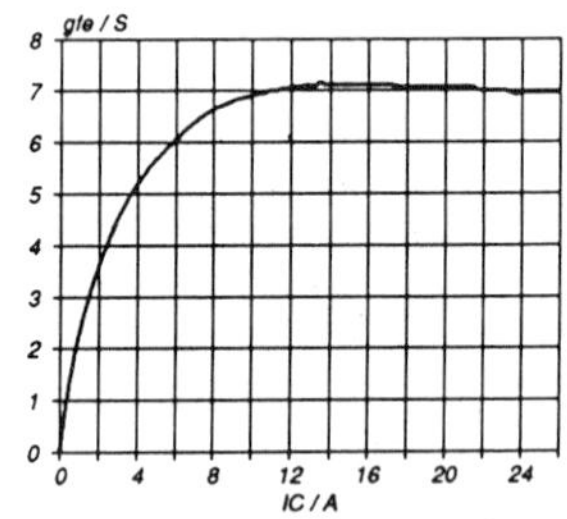

Fig.8. Typical Forward Transconductance
$g_{fe} = f(I_C)$; conditions: $V_{CE} = 15$ V; $T_j = 25$°C

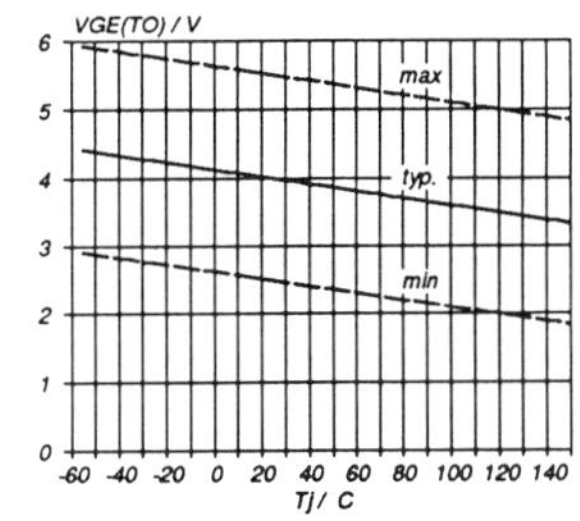

Fig.9. Gate Threshold Voltage
$V_{GE(TO)} = f(T_j)$; conditions: $I_C = 1$ mA; $V_{CE} = V_{GE}$

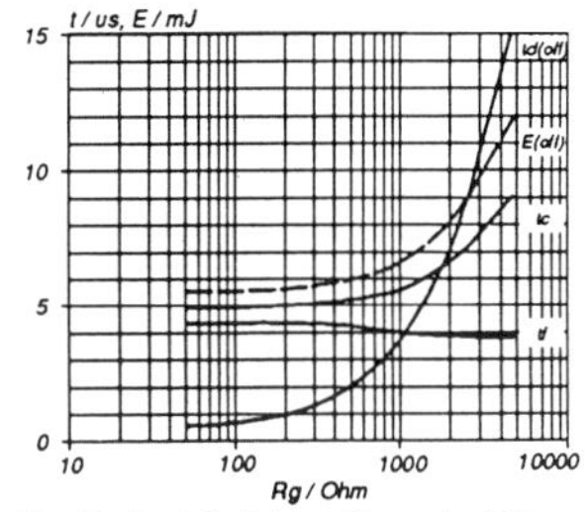

Fig.12. Typical Switching Characteristics vs. R_G
conditions: $T_j = 100$ °C; $I_C = 8.5$ A; $V_{CL} = 350$ V

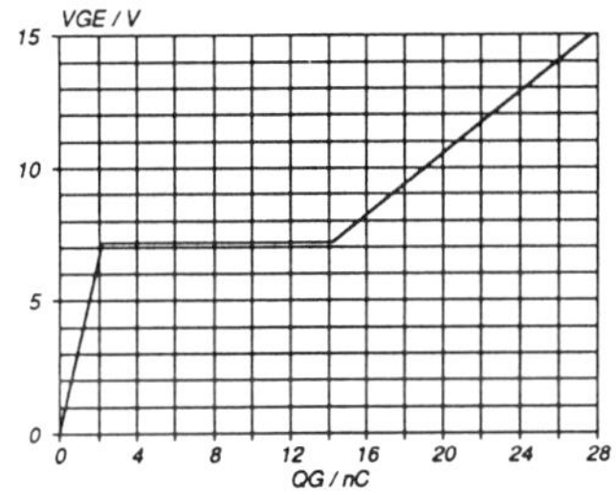

Fig.10. Typical Turn-on Gate Charge Characteristics
$V_{GE} = f(Q_G)$; conditions: $I_C = 8.5$ A; $V_{CE} = 350$ V

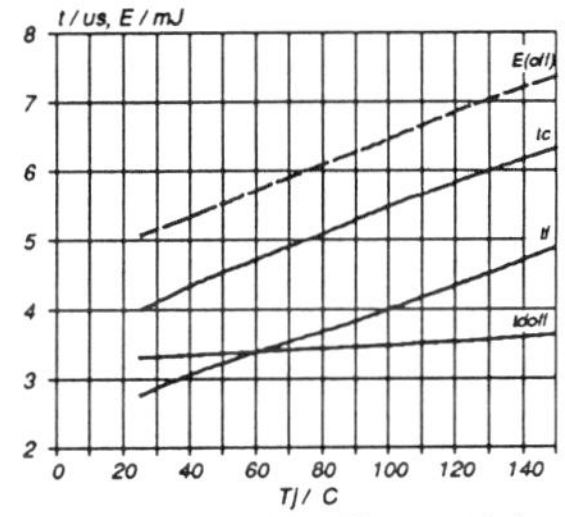

Fig.13. Typical Switching Characteristics vs. T_j
conditions: $I_C = 8.5$ A; $V_{CL} = 350$ V; $R_G = 1$ kΩ

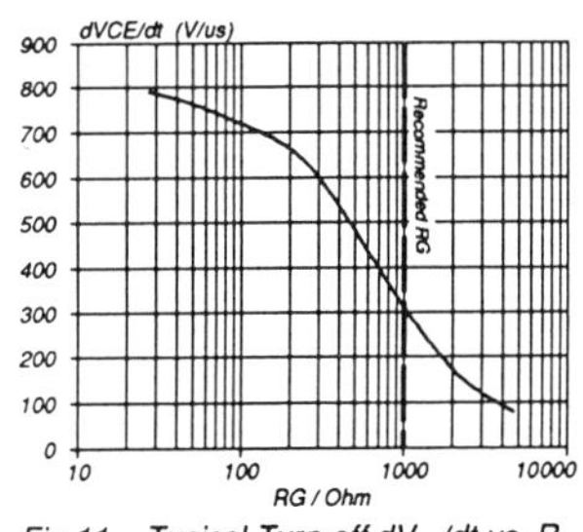

Fig.11. Typical Turn-off dV_{CE}/dt vs. R_G
conditions: $T_j = 100$ °C; $I_C = 8.5$ A; $V_{CL} = 350$ V

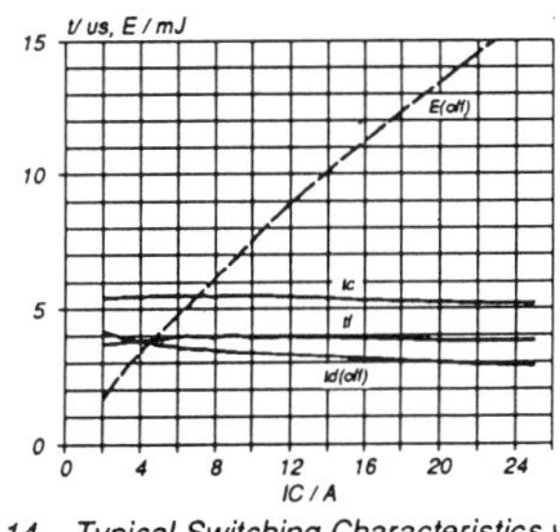

Fig.14. Typical Switching Characteristics vs. I_C
conditions: $T_j = 100$ °C $V_{CL} = 350$ V; $R_G = 1$ kΩ

APPENDIX 6

TRIAC

Triac | BT134W series

GENERAL DESCRIPTION

Glass passivated triacs in SOT223 envelopes suitable for surface mounting. They are intended for general purpose switching and phase control applications.

QUICK REFERENCE DATA

SYMBOL	PARAMETER	MAX.	MAX.	MAX.	MAX.	UNIT
	BT134W-	**500**	**600**	**700**[1]	**800**[1]	
V_{DRM}	Repetitive peak voltages	500	600	700	800	V
$I_{T(RMS)}$	R.M.S. on-state current	1	1	1	1	A
I_{TSM}	Non-repetitive on-state current	10	10	10	10	A

PINNING - SOT223

PIN	DESCRIPTION
1	main terminal 1
2	main terminal 2
3	gate
4	main terminal 2 (tab)

PIN CONFIGURATION

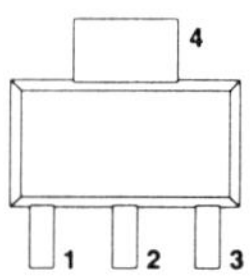

SYMBOL

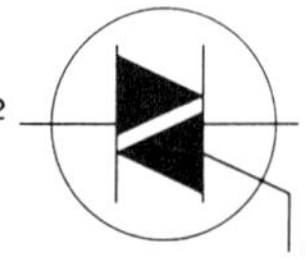

LIMITING VALUES

Limiting values in accordance with the Absolute Maximum System (IEC 134)

SYMBOL	PARAMETER	CONDITIONS	MIN.	MAX.				UNIT
	Voltages (in either direction)			**-500**	**-600**	**-700**[1]	**-800**[1]	
V_{DSM}	Non-repetitive peak off-state voltage.	$t \leq 10$ ms	-	500	600	700	800	V
V_{DRM}	Repetitive peak off-state voltage	$\delta \leq 0.01$	-	500	600	700	800	V
V_{DWM}	Crest working off-state voltage		-	400	400	400	400	V
	Currents (in either direction)							
$I_{T(RMS)}$	R.M.S. on-state current	Conduction angle = 360˚; T_b = 77 ˚C	-	1				A
I_{TRM}	Repetitive peak on-state current		-	10				A
I_{TSM}	Non-repetitive peak on-state current	t = 20 ms; full sine-wave; T_j = *120 ˚C* prior to surge.	-	10				A
I^2t	I^2t for fusing	t = 10 ms	-	0.5				A^2s
dI_T/dt	Rate of rise of on-state current after triggering	I_G = 200 mA to I_T = 1.5 A; dI_G/dt = 0.2 A/μs	-	10				A/μs
	Power dissipation							
$P_{G(AV)}$	Average power dissipation	over any 20 ms period	-	0.13				W
P_{GM}	Peak power dissipation		-	1.3				W
	Temperatures							
T_{stg}	Storage temperature		- 40	125				˚C
T_j	Junction temperature	Full-cycle operation	-	120				˚C
T_j	Junction temperature	Half-cycle operation	-	110				˚C

1 These voltage grades not available for D and E type gate selections. (See next page for details of gate selections)

THERMAL RESISTANCES

From junction to board	P.c.b. mounted (see fig 2), temperature measured 1 - 3 mm from tab.	$R_{th\,j\text{-}b}$ = 30 K/W
From junction to ambient	P.c.b. mounted (see fig 2)	$R_{th\,j\text{-}a}$ = 70 K/W

CHARACTERISTICS

T_{mb} = 25 °C unless otherwise specified

SYMBOL	PARAMETER	CONDITIONS	MIN.	TYP.	MAX.	UNIT
	Voltages and Currents (in either direction)					
V_T	On-state voltage	I_T = 1.5 A	-	1.2	1.7	V
dV_D/dt	Rate of rise of off-state voltage (exponential method)	T_j = 120 °C; $V_D = V_{DWM}$max; gate open circuit				
		BT134W-500 to 800	100	-	-	V/µs
		BT134W-500E/600E	-	30	-	V/µs
		BT134W-500D/600D[2]	-	5	-	V/µs
dV_{com}/dt	Rate of change of commutating voltage	$-dI_{com}/dt$ = 1.8 A/ms; $I_{T(RMS)}$ = 1 A; T_b = 50 °C; gate open circuit; $V_D = V_{DWM}$max				
		BT134W-500 to 800	10	-	-	V/µs
		BT134W-500E/600E		Not applicable		
		BT134W-500D/600D		Not applicable		
I_D	Off-state current	$V_D = V_{DWM}$max; T_j = 120 °C	-	-	0.5	mA
V_{GT}	Gate trigger voltage	V_D = 12 V	-	-	1.5	V
		V_D = 12 V; T_j = -40 °C	-	-	2.3	V
		$V_D = V_{DWM}$max, T_j = 120 °C	0.25	-	-	
I_{GT}	Gate trigger current	G to T1; V_D = 12 V	-	-	Table 1	mA
I_H	Holding current		-	-	Table 1	mA
I_L	Latching current		-	-	Table 1	mA

Table 1: Gate Characteristics - Maximum Values		**T2+ G+**	**T2+ G-**	**T2- G-**	**T2- G+**	**UNIT**
BT134W - 500/600/700/800	I_{GT}	35	35	35	70	mA
	I_H	15	15	15	15	mA
	I_L	20	30	20	30	mA
BT134W - 500E/600E	I_{GT}	10	10	10	25	mA
	I_H	15	15	15	15	mA
	I_L	15	20	15	20	mA
BT134W - 500D/600D	I_{GT}	5	5	5	10	mA
	I_H	10	10	10	10	mA
	I_L	10	15	10	15	mA

2 With $R_{G\text{-}MT1}$ = 1 kΩ

APPENDIX 7

GTO

BTV70 SERIES

RATINGS

Limiting values in accordance with the Absolute Maximum System (IEC134)

Anode to cathode			BTV70-850R	1000R	1200R	
Transient off-state voltage	V_{DSM}	max.	1000	1100	1300	V*
Repetitive peak off-state voltage	V_{DRM}	max.	850	1000	1200	V*
Working off-state voltage	V_{DW}	max.	600	800	1000	V*
Continuous off-state voltage	V_D	max.	500	650	750	V*

Average on-state current (averaged over any 20 ms period) up to T_{mb} = 60 °C	$I_{T(AV)}$	max.	15	A
Controllable anode current	I_{TCRM}	max.	50	A
Non-repetitive peak on-state current t = 10 ms; half-sinewave; T_j = 120 °C prior to surge	I_{TSM}	max.	100	A
I^2t for fusing; t = 10 ms	I^2t	max.	50	A^2s
Total power dissipation up to T_{mb} = 25 °C	P_{tot}	max.	60	W
Gate to cathode				
Repetitive peak current T_j = 120 °C prior to surge				
gate-cathode forward; t = 10 ms; half-sinewave	I_{GFM}	max.	25	A
gate-cathode reverse; t = 20 μs	I_{GRM}	max.	25	A
Average power dissipation (averaged over any 20 ms period)	$P_{G(AV)}$	max.	5.0	W
Temperatures				
Storage temperature	T_{stg}		−40 to +125	°C
Operating junction temperature	T_j	max.	120	°C

ISOLATION**

R.M.S. isolation voltage	V_{isol}	min.	2500	V

THERMAL RESISTANCE

From mounting base to heatsink; with heatsink compound	$R_{th\ mb\text{-}h}$	=	0.3	K/W
From junction to mounting base	$R_{th\ j\text{-}mb}$	=	1.5	K/W

* Measured with gate-cathode connected together.

** From baseplate to all terminals strapped together.

Fast gate turn-off thyristors | **BTV70 SERIES**

CHARACTERISTICS

Anode to cathode

On-state voltage I_T = 10 A; I_G = 0.5 A; T_j = 120 °C	V_T	<	2.3	V*
Rate of rise of off-state voltage that will not trigger any off-state device, exponential method V_D = 2/3 V_{Dmax}; V_{GR} = 5 V; T_j = 120 °C	dV_D/dt	<	10	kV/μs
Rate of rise of off-state voltage that will not trigger any device following conduction, linear method I_T = 20 A; V_D = V_{DRMmax}; V_{GR} = 10 V; T_j 120 °C	dV_D/dt	<	1.0	kV/μs
Off-state current V_D = V_{Dmax}; T_j = 120 °C	I_D	<	5.0	mA
Latching current; T_j = 25 °C	I_L	typ.	1.5	A**

Gate to cathode

Voltage that will trigger all devices V_D = 12 V; T_j = 25 °C	V_{GT}	>	1.5	V
Current that will trigger all devices V_D = 12 V; T_j = 25 °C	I_{GT}	>	300	mA
Minimum reverse breakdown voltage I_{GR} = 1.0 mA	$V_{(BR)GR}$	>	10	V

Switching characteristics (resistive load)

Turn-on when switched to I_T = 10 A from V_D = 250 V with I_{GF} = 1.5 A; T_j = 25 °C				
delay time	t_d	<	0.3	μs
rise time	t_r	<	1.5	μs

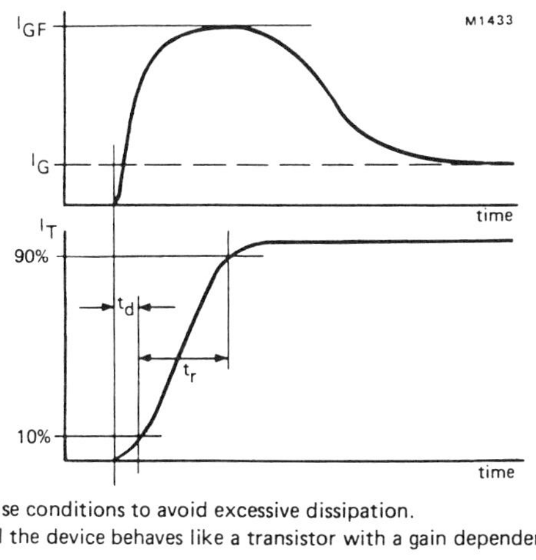

Fig.2 Waveforms.

* Measured under pulse conditions to avoid excessive dissipation.

** Below latching level the device behaves like a transistor with a gain dependent on current.

Switching characteristics (inductive load)

Turn-off when switched from I_T = 10 A to V_D = V_{Dmax};
V_{GR} = 10 V; $L_G \leqslant 0.5\ \mu H$; $L_S \leqslant 0.25\ \mu H$; T_j = 25 °C

storage time	t_s	<	0.60	μs
fall time	t_f	<	0.25	μs
peak reverse gate current	I_{GR}	<	10	A

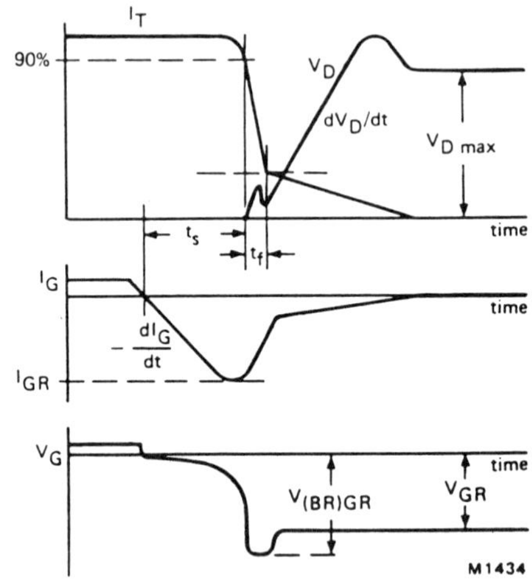

Fig.3 Waveforms.

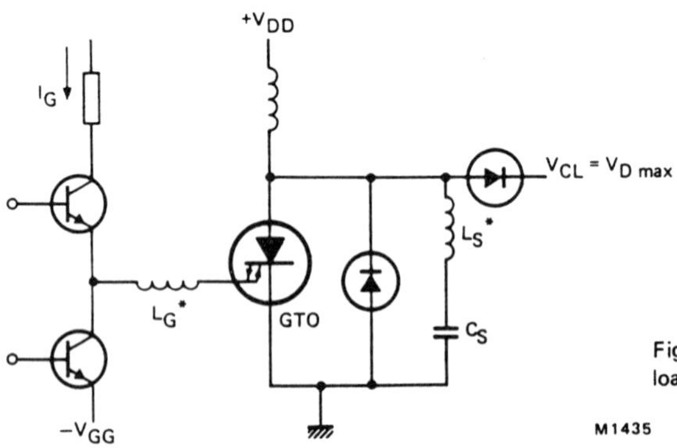

Fig.4 Inductive load test circuit.

* Indicates stray series inductance only

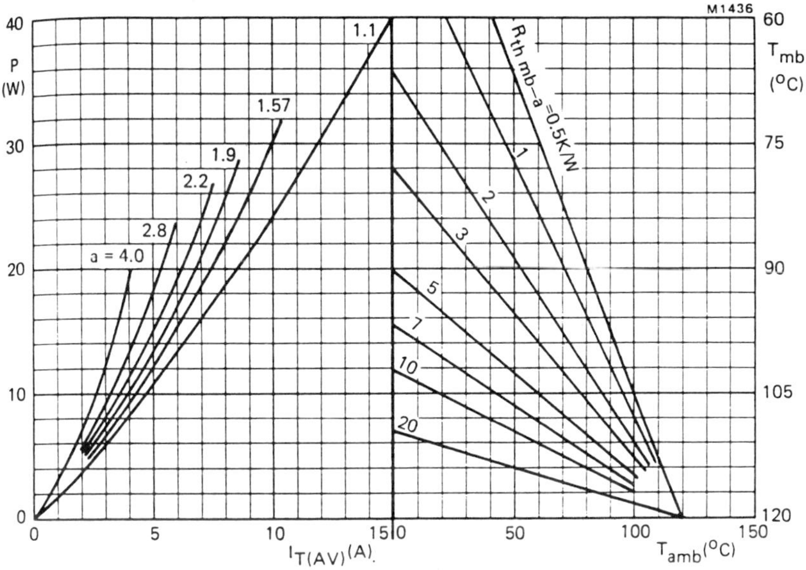

Fig.5 The right hand part shows the interrelationship between the power (derived from the left hand part) and the maximum permissible temperatures.

a = form factor = $\frac{I_{T(RMS)}}{I_{T(AV)}}$

P = power excluding switching losses.

APPENDIX 8

MCT

HARRIS SEMICONDUCTOR

MCTV75P60E1
MCTA75P60E1

75A, 600V
P-Type MOS Controlled Thyristor (MCT)

September 1992

Features

- 75A, -600V
- V_{TM} = -1.3V(Maximum) at I = 75A and +150°C
- 2000A Surge Current Capability
- 2000A/µs d_I/d_T Capability
- MOS Insulated Gate Control
- 120A Gate Turn-Off Capability at +150°C

Description

The MCT is an MOS Controlled Thyristor designed for switching currents on and off by negative and positive pulsed control of an insulated MOS gate. It is designed for use in motor controls, inverters, line switches and other power switching applications.

The MCT is especially suited for resonant (zero voltage or zero current switching) applications. The SCR like forward drop greatly reduces conduction power loss.

MCTs allow the control of high power circuits with very small amounts of input energy. They feature the high peak current capability common to SCR type thyristors, and operate at junction temperatures up to +150°C with active switching.

The MCTV75P60 is supplied in a 5-lead variation of the JEDEC TO-247 plastic package, and the MCTA75P60 in the JEDEC MO-93 plastic package which is a 5-lead variation of the TO-218.

* Formerly TA9836

Package

TO-247 5-LEAD
TOP VIEW

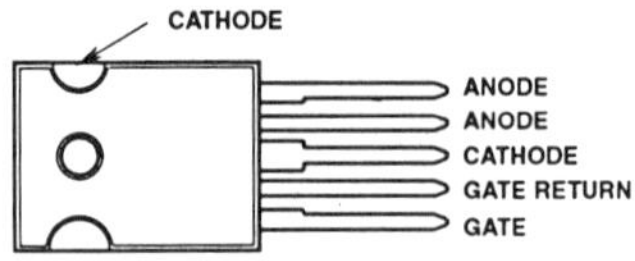

MO-93 (5-LEAD TO-218)
TOP VIEW

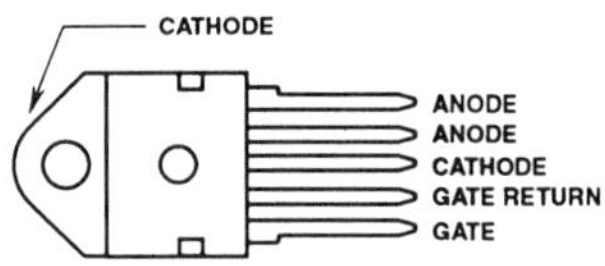

Symbol

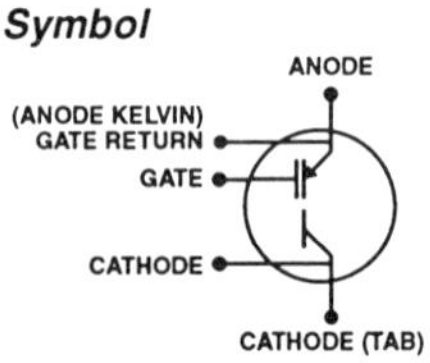

Absolute Maximum Ratings (T_C = +25°C) Unless Otherwise Specified

		MCTV75P60E1 MCTA75P60E1	UNITS
Peak Off-State Voltage (See Figure 11)	V_{DRM}	-600	V
Peak Reverse Voltage	V_{RRM}	+5	V
Continuous Cathode Current (See Figure 2)			
T_C = +25°C (Package Limited)	I_{C25}	85	A
T_C = +90°C	I_{C90}	75	A
Non-repetitive Peak Cathode Current (Note 1)	I_{TSM}	2000	A
Peak Controllable Current (See Figure 10)	I_{TC}	120	A
Gate-Anode Voltage (Continuous)	V_{GA}	±20	V
Gate-Anode Voltage (Peak)	V_{GA}	±25	V
Rate of Change of Voltage		See Figure 11	
Rate of Change of Current	d_I/d_T	2000	A/µs
Maximum Power Dissipation	P_T	208	W
Linear Derating Factor		1.67	W/°C
Operating and Storage Temperature	T_J,T_{STG}	-55 to +150	°C
Maximum Lead Temperature for Soldering (0.063" (1.6mm) from case for 10s)	T_L	260	°C

NOTE:

1. Maximum Pulse Width of 250µs (Half Sine) Assume T_J (Initial) = +90°C and T_J (Final) = T_J (Max) = +150°C

CAUTION: These devices are sensitive to electrostatic discharge. Users should follow proper I.C. Handling Procedures.

File Number **3374**

Specifications MCTV75P60E1, MCTA75P60E1

Electrical Characteristics At Case Temperature (T_C) = +25°C Unless Otherwise Specified

CHARACTERISTICS	SYMBOL	TEST CONDITIONS		LIMITS MIN	TYP	MAX	UNITS
Peak Off-State Blocking Current	I_{DRM}	V_{KA} = -600V,	T_C = +150°C	-	-	3	mA
		V_{GA} = +18V	T_C = +25°C	-	-	100	µA
Peak Reverse Blocking Current	I_{RRM}	V_{KA} = +5V	T_C = +150°C	-	-	4	mA
		V_{GA} = +18V	T_C = +25°C	-	-	100	µA
On-State Voltage	V_{TM}	$I_C = I_{C90}$,	T_C = +150°C	-	-	1.3	V
		V_{GA} = -10V	T_C = +25°C	-	-	1.4	V
Gate-Anode Leakage Current	I_{GAS}	V_{GA} = ±20V		-	-	200	nA
Input capacitance	C_{iss}	V_{KA} = -20V, T_J = +25°C V_{GA} = +18V		-	10	-	nF
Current Turn-on Delay Time	$t_{d(on)i}$	L = 200µH, $I_C = I_{C90}$ $R_g = 1\Omega$, V_{GA} = +18V, -7V T_J = +125°C V_{KA} = -300V		-	300	-	ns
Current Rise Time	t_{ri}			-	200	-	ns
Current Turn-off Delay Time	$t_{d(off)i}$			-	700	-	ns
Current Fall Time	t_{fi}			-	1.15	1.4	µs
Turn-off Energy	E_{off}			-	10	-	mJ
Thermal Resistance	$R_{\theta JC}$			-	.5	.6	°C/W

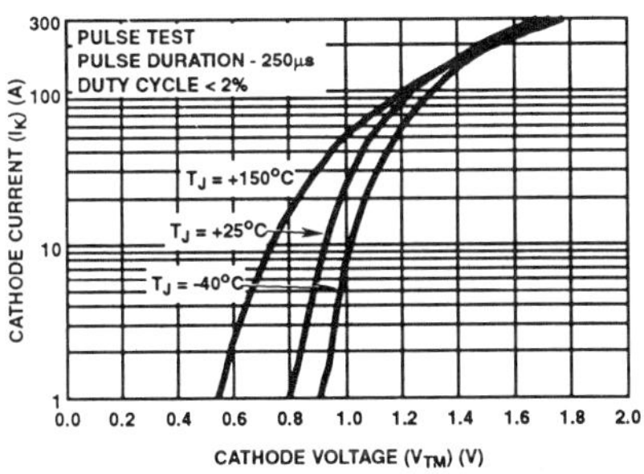

FIGURE 1. CATHODE CURRENT vs SATURATION VOLTAGE (TYPICAL)

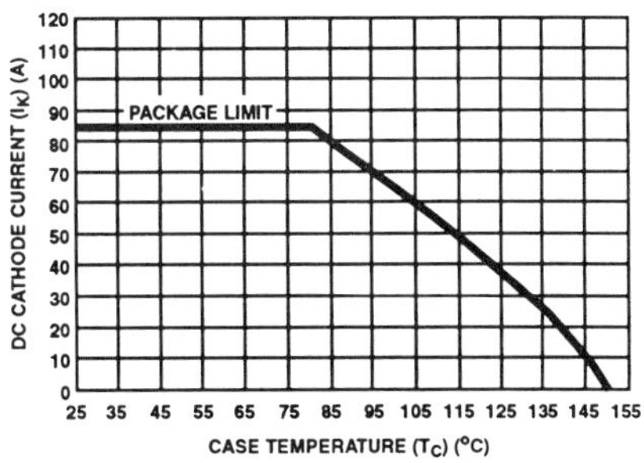

FIGURE 2. MAXIMUM CONTINUOUS CATHODE CURRENT

MCTV75P60E1, MCTA75P60E1

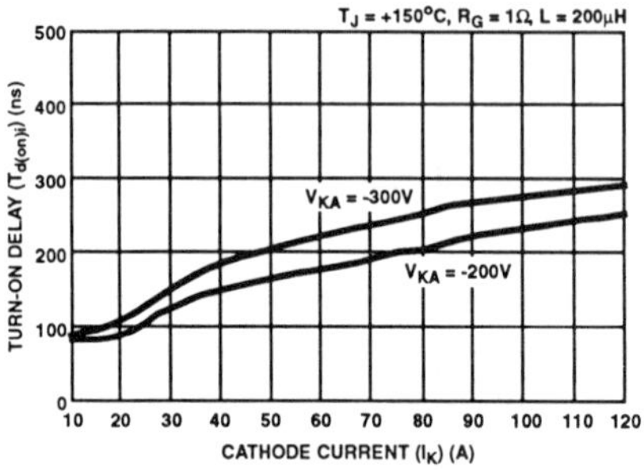

FIGURE 3. TURN-ON DELAY vs CATHODE CURRENT (TYPICAL)

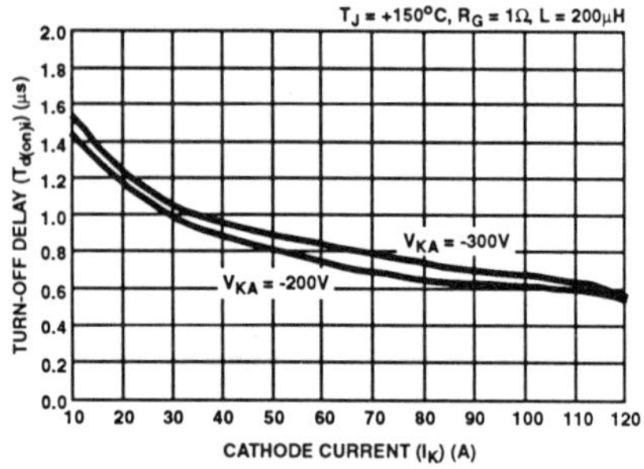

FIGURE 4. TURN-OFF DELAY vs CATHODE CURRENT (TYPICAL)

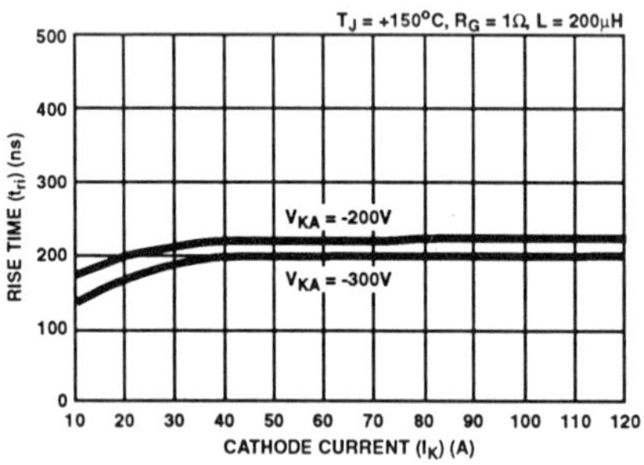

FIGURE 5. TURN-ON RISE TIME vs CATHODE CURRENT (TYPICAL)

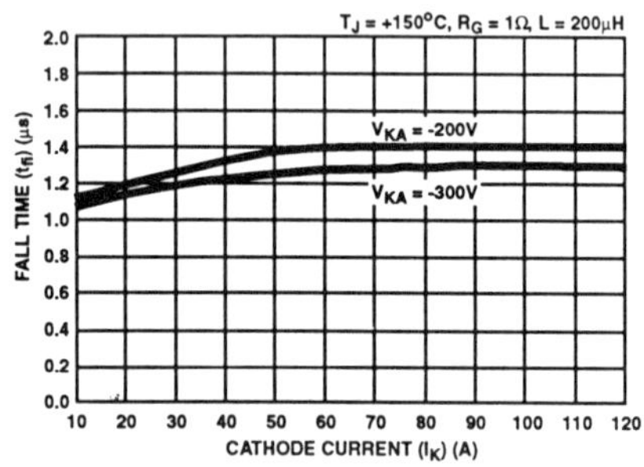

FIGURE 6. TURN-OFF FALL TIME vs CATHODE CURRENT (TYPICAL)

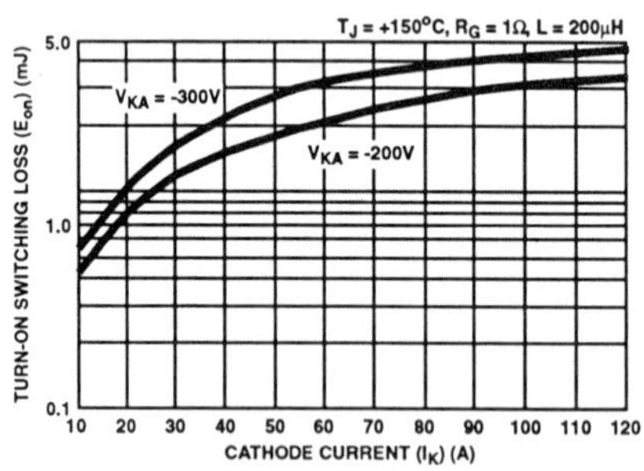

FIGURE 7. TURN-ON ENERGY LOSS vs CATHODE CURRENT (TYPICAL)

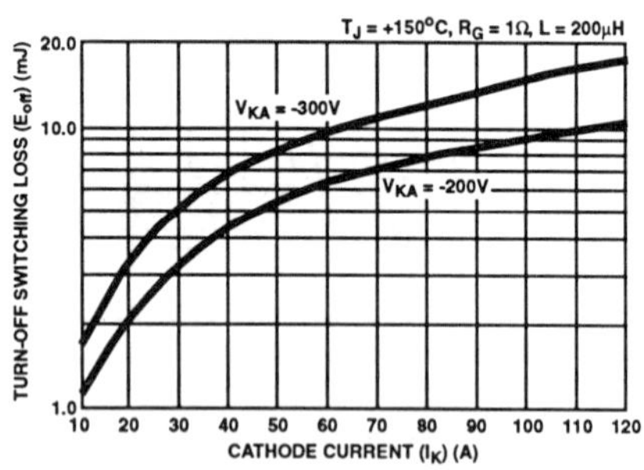

FIGURE 8. TURN-OFF ENERGY LOSS vs CATHODE CURRENT (TYPICAL)

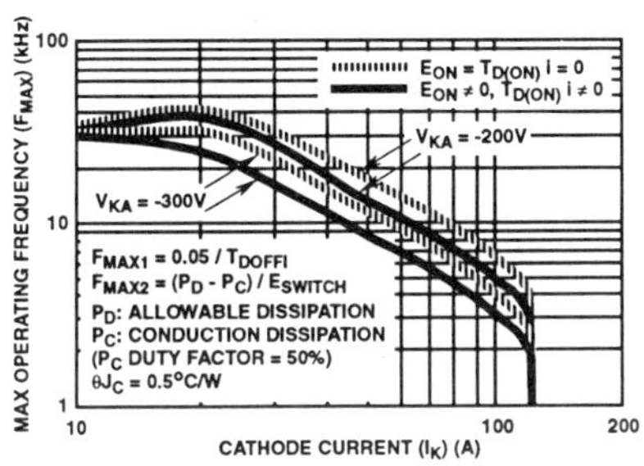

FIGURE 9. OPERATING FREQUENCY vs CATHODE CURRENT (TYPICAL)

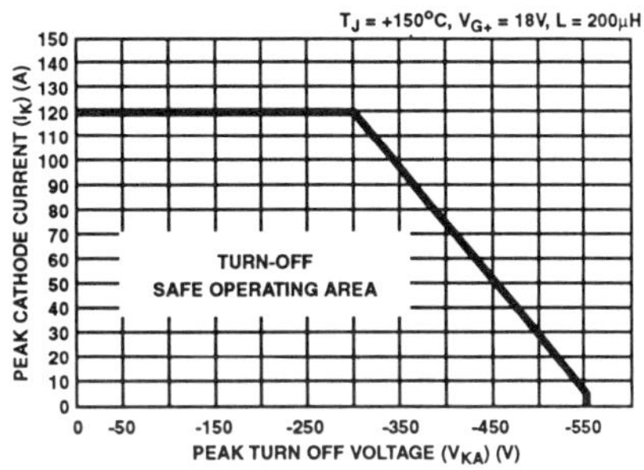

FIGURE 10. TURN-OFF CAPABILITY vs ANODE-CATHODE VOLTAGE

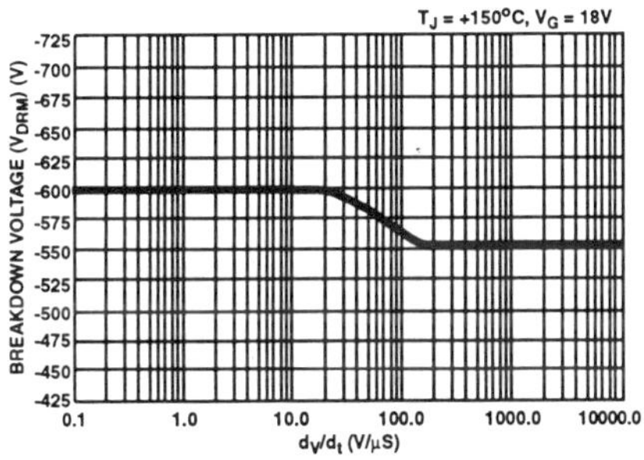

FIGURE 11. BLOCKING VOLTAGE vs D_V/D_T

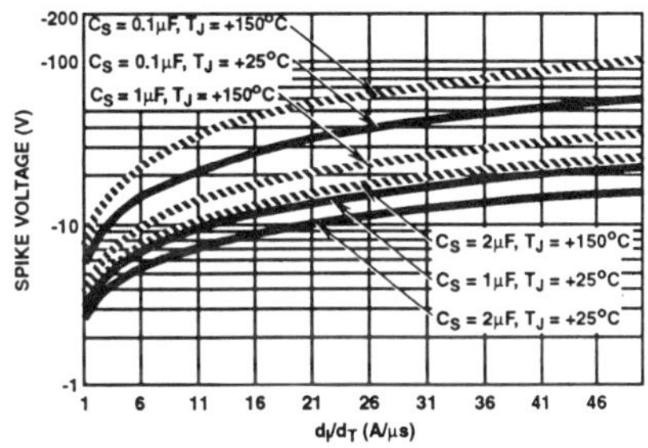

FIGURE 12. SPIKE VOLTAGE vs D_I/D_T (TYPICAL)

Operating Frequency Information

Operating frequency information for a typical device (Figure 9) is presented as a guide for estimating device performance for a specific application. Other typical frequency vs cathode current (I_{AK}) plots are possible using the information shown for a typical unit in Figures 3 to 8. The operating frequency plot (Figure 9) of a typical device shows f_{max1} or f_{max2} whichever is smaller at each point. The information is based on measurements of a typical device and is bounded by the maximum rated junction temperature.

f_{max1} is defined by $f_{max1} = 0.05 / (t_{d(on)i} + t_{d(off)i})$. $t_{d(on)i} + t_{d(off)i}$ deadtime (the denominator) has been arbitrarily held to 10% of the on-state time for a 50% duty factor. Other definitions are possible. $t_{d(on)i}$ is defined as the 10% point of the leading edge of the input pulse and the point where the cathode current rises to 10% of its maximum value. $t_{d(off)i}$ is defined as the 90% point of the trailing edge of the input pulse and the point where the cathode current falls to 90% of its maximum value. Device delay can establish an additional frequency limiting condition for an application other than T_{JMAX}. $t_{d(off)i}$ is important when controlling output ripple under a lightly loaded condition.

f_{max2} is defined by $f_{max2} = (P_D - P_C) / (E_{on}+E_{off})$. The allowable dissipation (P_D) is defined by $P_D = (T_{JMAX} - T_C) / R_{\theta JC}$. The sum of device switching and conduction losses must not exceed P_D. A 50% duty factor was used (Figure 10) and the conduction losses (P_C) are approximated by $P_C = (V_{AK} \bullet I_{AK}) / (\text{duty factor}/100)$. E_{on} is defined as the sum of the instantaneous power loss starting at the leading edge of the input pulse and ending at the point where the anode-cathode voltage equals saturation voltage ($V_{AK} = V_{TM}$). E_{off} is defined as the sum of the instantaneous power loss starting at the trailing edge of the input pulse and ending at the point where the cathode current equals zero ($I_{CE} = 0A$).

The switching power loss (Figure 10) is defined as $f_{max2} \bullet (E_{on}+E_{off})$. Because Turn-on switching losses can be greatly influenced by external circuit conditions and components, f_{max} curves are plotted both including and neglecting turn-on losses.

APPENDIX 9

ANSWERS TO PROBLEMS

CHAPTER 1

1.1 (a) $V_{rms} = 70.7\,\text{V}$, $V_{av} = 0$, $V_{av,\pi} = 63.7\,\text{V}$, form factor = 1.1, (b) $V_{rms} = 70.7\,\text{V}$, $V_{av} = 63.7\,\text{V}$, $V_{av,\pi} = 63.7\,\text{V}$, form factor = 1.11, (c) $V_{rms} = 50\,\text{V}$, $V_{av} = 31.8\,\text{V}$, form factor = 1.57, (d) $V_{rms} = 50\,\text{V}$, $V_{av} = 0$, $V_{av,\pi} = 31.8\,\text{V}$, form factor = 1.57, (e) $V_{rms} = 54.8\,\text{V}$, $V_{av} = 0$, $V_{av,\pi} = 38.22\,\text{V}$, form *factor* = 1.43, (f) $V_{rms} = 100\,\text{V}$, $V_{av} = 0$, $V_{av,\pi} = 100\,\text{V}$, form factor = 1, (g) $V_{rms} = 57.7\,\text{V}$, $V_{av} = 0$, $V_{av,\pi} = 50\,\text{V}$, form factor = 1.15.

1.2 $v(t) = \dfrac{2\hat{V}}{\pi}(1 - \dfrac{2}{3}\cos 2\omega t - \dfrac{2}{15}\cos 4\omega t - \dfrac{2}{35}\cos 6\omega t - \cdots)$, $P = (\dfrac{2\hat{V}}{\pi})^2 + \dfrac{1}{2}(\dfrac{4\hat{V}}{3\pi})^2 + \dfrac{1}{2}(\dfrac{4\hat{V}}{15\pi})^2 + \dfrac{1}{2}(\dfrac{4\hat{V}}{35\pi})^2 + \cdots = \dfrac{\hat{V}^2}{2}\,\text{W}$.

1.3 $v(t) = \dfrac{\hat{V}}{\pi}(1 + \dfrac{\pi}{2}\sin\omega t - \dfrac{2}{3}\cos 2\omega t - \dfrac{2}{15}\cos 4\omega t - \cdots)$.

1.4 $i(t) = \dfrac{\hat{I}}{2\pi} - \dfrac{\hat{I}}{2\pi}\cos\omega t + \sum_{n=2}^{\infty}\dfrac{\hat{I}}{\pi(1-n^2)}(\cos n\pi + n\sin\dfrac{n\pi}{2})\cos n\omega t + \dfrac{\hat{I}}{4}\sin\omega t + \sum_{n=2}^{\infty}\dfrac{-n\hat{I}\cos n\pi/2}{\pi(1-n^2)}\sin n\omega t$.

1.5 0.48.

1.6 6 kW.

1.7 60 kW.

1.8 9.7 kW.

1.9 10 kW.

1.10 12.5 kW.

CHAPTER 2

2.1 (a) 60 A, (b) 6 kW.

2.2 (a) 2450 W, (b) 8.75, 61.27 A, (c) 24.5 A.

2.3 (a) 0.575 mH.

2.4 (a) 0.5, (b) 2.1 kW, (c) 39.5 A, (d) 0.5, (e) 0.447.

2.5 (a) $33.9\sin 314t$ for $0 \le \omega t \le \pi$, (b) 1.11, (c) 0.48, (d) 0.81.

2.6 (b) 54 V, (c) 27 A, (d) 120 V, (e) 7.2 kW, (f) 0.2 per unit, (g) 2.22, (h) 1.98, (i) > 340 V, > 54 A.

2.7 (a) $\pi/3$, (b) > 170 V, > 260 A.

2.8 (a) 1.31 h, (b) 493 W.

2.9 (b) 192.4°, (c) 3.7 kW, (d) > 50 A, > 200 V.

2.10 $i_l = (\hat{V}_s/Z)[\sin(\omega t - \theta) - \sin(\alpha - \theta)e^{(\alpha/\omega - t)/\tau}]$.

2.11 (a) Discontinuous, (b) $192\sin(\omega t - 0.44) - 1010e^{-800t}$ A, (c) 50.6 A, (d) 80.56 A, (e) 194.7 W, (f) 5.49 kW, (g) 82.86 A, (h) > 170 V, > 62 A.

2.12 (a) 1.173, (b) 0.61, (c) 0.727, (d) 6.99 kW, (e) 0.15, (f) 0.985, (g) 0.996, (h) > 170 V, > 25 A.

2.14 (a) Current 1, voltage 1.28, (b) current 0, voltage 0.8, (c) 0.78, (d) 4.38 kW, (e) 0.483,

(f) 0.78, (g) 0.866,
(h) > 170 V, > 27 A.

2.15 (a) 43.3° < α < 180°,
(b) 15.23 kW, (c) 0.73.

2.17 (a) π / 2 rad, (b) 1.57, (c) 1.21,
(d) 0.47, (e) 17 kW, (f) 0.77,
(g) > 4.7 A.

2.18 (b) 1.96, (c) 1.69, (d) 0.32,
(e) 3.56 kW, (f) 0.36.

2.19 (b) For $\pi/3 \le \omega t < \pi$,
$i_l(t) = 169.7\sin(\omega t - 0.304) - 2048e^{-1000T}$ A.
$\pi \le \omega t \le 4\pi/3$ and $t' = t - 0.01$ s,
$i_l(t') = 50.87e^{-1000t'}$ A.
(c) 128.6 A, (d) 0.62,
(e) 1.33, 1.2, (f) 0.88, 0.68,
(g) 23.07 kW, (h) 0.9.

2.20 (a) 27.2 kW, (b) 54 A, (c) 27.4 A,
(d) 0.83, (e) 1.11, 1.08,
(f) 0.48, 0.41, (g) 0.97.

2.21 (b) 0.75, (c) 1.33, (d) 0.88,
(e) 13.12 kW, (f) 0.83, (g) 0.31,
(h) 0.866, (i) > 54 A, > 28 A.

2.22 (b) 0.64, (c) 1.57, (d) 1.21,
(e) 5.8 kW, (f) 0.64, (g) 48%,
(h) 0.707, (i) > 54 A, > 59 A.

2.23 (b) 0.78, (c) 1.28, (d) 0.8,
(e) 13.1 kW, (f) 0.78, (g) 0.69,
(h) 0.95, (i) > 54 A ($TH\,1$, $D\,2$, $D\,4$), > 63 A ($D\,3$).

2.25 (a) 0.3, (b) 0.36, (c) 0.25,
(d) > 600 V, > 8.3 A.

2.26 (b) 17.1 kW, (c) > 50 A, > 8 A.

2.27 (a) 0.075, (b) 2.5 A, (c) 1.55,
(d) 0.41.

2.28 (a) $76.39\sin\omega t + 25.46\sin 3\omega t + 15.28\sin 5\omega t + 10.91\sin 7\omega t + 8.49\sin 9\omega t + \cdots$, (b) 0.48,
(c) 0.11, (d) > 600 V, > 30 A.

2.29 (a) 30 A, (b) > 12.5 V, > 1 A,
(c) 14×10^3 W, (d) 26.5 A.

2.30 (a) 764, 255, 153, 109, 85 volts peak, (b) 37.2, 96.3, 158, 221, 283 ohms, (c) $20.5\sin(\omega t - 1) + 2.6\sin(3\omega t - 1.36) + 0.97\sin(5\omega t - 1.44) + 0.49\sin(7\omega t - 1.48) + 0.3\sin(9\omega t - 1.5)$ amperes,
(d) 0.15, (e) 0.015.

2.31 (a) 90 A, (b) 16.9, 1.9 A,
(c) 18×10^3 W, (d) 32.6%.

2.32 (a) $\dfrac{8V_s}{n\pi}\cos\dfrac{n\pi}{4}\sin n\dfrac{m}{2}\pi$,
(b) 0.12.

2.33 (a) $19.1 sin\,\omega t + 1.28\sin 5\omega t + 2.73\sin 7\omega t + \cdots$,
(b) 0.14, 0, (c) 6.67 A ($I_s/3$).

2.34 (b) 25 A, (c) 5.2 A.

2.36 (a) 150 A, (b) 9×10^3 W,
(c) 245 V, (d) 424 V,
(e) 0, (f) 0.31,
(g) $16.6\cos\omega t - 3.31\cos 5\omega t + 2.37\cos 7\omega t + \cdots$,
(h) 0.14, 0, (i) 5 A.

CHAPTER 3

3.1 (a) 400 W, 0,
(b) 351.6 W, 23.4 W,
(c) 361 W, 19 W.

3.2 (a) 90.8 W, (b) 1.22°C/W.

3.3 (a) 1.34 μs, (b) 14.9 A.

3.4 Yes.

3.5 615 V.

3.6 80 V, No.

3.8 10.8 kW, > 47 A, > 294 V.

3.9 (a) 0.707, (b) 1.0.

3.10 > 100 A.

3.11 (a) 0.6, (b) 1.06.

3.12 100A minimum for both diodes, 23 kW.

3.13 (a) 0.64, (b) 1.03.

3.14 (a) 0 V, (b) 292 J, (c) 270 A.

3.15 (a) 0.64, (b) 0.50.

3.16 (a) 1000 μF, (b) 35 A.

3.17 (a) 6.2 A, (b) 8.96 A.

3.18 (a) 1.556 kW, (b) 53°, (c) 0.509,
(d) 0.64 (25% increase).

3.19 (a) 8025 W, (b) 13,380 W.

CHAPTER 4

4.1 0.009 (≈ 1%).
4.2 (a) 90 kW, (b) 252 W.
4.3 (a) 8 mJ, (b) 24 W.
4.4 (a) 44.2 mJ, (b) 132.5 W.
4.5 (a) Delay, 45 μJ; rise, 15mJ; ON; 23.2mJ; storage, 1080 μJ; fall, 45mJ; OFF, 2745 μJ; base, 3750 μJ, (b) 0.072°C/W.
4.6 (a) 158 W, (b) 179 W, (c) 133 W, (d) 470 W, (e) > 240 V, > 200 A.
4.7 96%.
4.8 (b) 11.08 kW, (c) 96%
4.9 17.1 kHz, 68.68 kW.
4.10 1.25 Ω, 1.25 Ω, <16 μF.
4.11 0.56 Ω, 1.87 Ω, 1.28 Ω, < 8.5 μF.
4.12 3.9 V, 2 V.
4.13 40 mJ, 27 mJ.
4.14 Yes.
4.15 (b) $(\pi - \alpha + (sin 2\alpha)/)^{1/2} / (\sqrt{\pi})$.
4.16 (a) 22°, (b) 0.9.26, (c) 92% of maximum power.
4.17 (a) 0.405 rad, (b) 0.96, (c) 84% of maximum power.
4.18 2.9 kW, 0.48.
4.19 0.80.
4.20 $2\cos\alpha_2 = 1 + \cos\alpha_1$.

CHAPTER 5

5.1 $R = 3.75\,\text{k}\Omega$.
5.2 0.4 μF, 500V / μs.
5.3 (a) > 0.31 μs, (b) 831 W, (c) 680 W.
5.4 5 kW.
5.6 (a) 1Ω, (b) 667 Hz.
5.7 (a) 0.33 Ω, (b) 3.9 A pulse.
5.8 (a) 27.6 μs, (b) 2.5 kHz, (c) 20 kW, (d) > 400, > 50 A.
5.9 (a) 14.5 μF, 1.5 Ω, (b) 1333 A.
5.10 (a) 17 μF, (b) 6.69 kH, (c) 75 μs.
5.11 (a) 141 μs, (b) 5.3 kHz.
5.12 0.4 mH, 64 μF.
5.13 (a) 108 μF, 5.1 mH, (b) $I_{TH1\,av} = 415$ A, $I_{TH2\,av} = I_{D\,av} = 64.8$ A.
5.14 (a) $V_s C / I_l$, (b) $(V_s C / I_l + \pi\sqrt{LC})^{-1}$.
5.15 (a) $31.62 \sin\omega t$, (b) 1ms, (c) 2000 V, (d) no.
5.16 (a) < 110 μs, (b) 18.26 kW, (c) 18.26 A.
5.17 (b) 27 μs, (c) 2.2 kHz, (d) 0.56, (e) 73.4 kW.
5.18 0.25%.
5.19 440 W.
5.20 > 1000 V, > 781 A.
5.21 $L_s = 1\,\mu\text{H}$, $C_s = 0.625\,\mu\text{F}$, $R_s < 64\,\Omega$.
5.22 $L_s = 12\,\mu\text{H}$, $C_s = 2\,\mu\text{F}$, $R_s < 250\,\Omega$.
5.23 (a) 1500 W, (b) 1000 A.
5.24 0.055°C/W.
5.25 (a) 815 A, (b) 0.12 s.
5.26 0.22 s.
5.27 88.8°C, 109°C.
5.28 180ms.
5.29 (a) 92°C, (b)127°C.
5.30 4000 V.
5.31 (a) 1.85m Ω, (b) 1.34 kW.
5.32 (a) 9444 A, 556 A, (b) 1.77×10^{-3}, (c) 75%.

CHAPTER 6

6.1 (a) 23.4 W, (b) 98.8%.
6.2 (a) 100 A, (b) 0.45 Ω, (c) 8 V.
6.3 (a) 60 μ W, (b) 1.5 W.
6.4 (a) 22 ns, (b) 37 ns, (c) 58 ns, (d) 117 ns.
6.5 12.5 W.
6.6 (a) 196 mW, (b) 1.32 W.

6.7 (a) 117 ns, (b) 38.5 ns, (c) 44 ns, (d) 200 ns.
6.8 (a) 0.25 W, (b) 10 W.
6.9 (a) 4 V, (b) 500 kHz, (c) 12.8 W.
6.10 (a) 640 μJ, (b) 895 μJ, (c) 1992 μJ, (d) 141 W.
6.12 $R_G = 170\Omega$, $I_G = 23.6$ mA.
6.13 (a) 766 pF, 2500 pF, (b) 14 mW, (c) 18.7 mA.
6.14 (a) 0.39°C/W, (b) 61 kHz.
6.15 13 A.
6.16 96.2°C, 94.4°C, 92.6°C.
6.18 0.0125 μF, 80 Ω.

CHAPTER 7

7.1 (a) 22.2 W, (b) 99.3%.
7.2 (a) 12.9 A, (b) 10.3 A, (c) 15.2 Ω, (d) 12.7 V.
7.3 (a) 20 A, (b) 40 W, (c) 1.5 W.
7.4 10.5 W.
7.5 (a) 109 ns, (b) 269 ns, (c) 1.95 μs.
7.6 8.3 W.
7.7 (a) 384 W, (b) 720 W, (c) 4.8 W.
7.8 (a) 159 ns, (b) 0.91 μs, (c) 94 ns, (d) 201 W, (e) 126 W.
7.9 51 kHz.
7.10 (a) 0.5, (b) 23.4 W, (c) 68°C, (d) 91.7 W.
7.11 51.1 A.
7.12 (a) 150°C, (b) 78°C.
7.13 (a) 324.8 A, 175.2 A, (b) 24 mΩ, (c) 5.25 mΩ, 0, (d) 1.784 kW, 4.866 kW, 2.225 kW.
7.14 (a) 33.1 Ω, (b) 0.05 μF, 200 Ω.

CHAPTER 8

8.1 (a) ≈150, (b) ≈51,000.
8.2 (a) 0.39 J, (b) 19.8 W, (c) 99.7%.
8.3 Conduction: leakage: gate losses = 36: 2.8: 1.
8.4 (a) 1.2 W, (b) 20.3 W.
8.5 20 μs.
8.6 (a) 0.56 rad, (b) 2068 W, (c) 14.3 μs.
8.7 (a) 5 Ω, (b) 50 kHz, (c) 100 mA.
8.8 12.4 μs.
8.9 (a) 8.13V / μs, (b) 41A / ms, (c) 244 A/s.
8.10 11.3 kW, 99.7%.
8.11 > 350 V, > 6.4 A.
8.12 7.84°C/W.
8.13 (a) 17.9 A, (b) 129.3 A.
8.14 (a) 112°C, (b) 115.4°C.
8.15 10 μH.
8.16 (a) 3.45 rad, (b) 1.8 / μs, (c) 182.9 V, (d) 0.4 W.
8.17 (a) 529 W, (b) 0.442.
8.18 65.8 μF.
8.19 (a) $i_l = 16.68\sin(\omega t - 0.337) - 16.39e^{-1.274\omega t}$, (b) 4.37 ms, (c) 6.9 A, (d) 763 W, (e) 0.46.
8.20 (a) 0.65, (b) ≈6 A.
8.21 (a) 19.8 A, (b) 29 kW, (c) 0.866.
8.22 $a / \sqrt{bc + 2\pi R_2^2}$;
$a = \sqrt{2}\,\pi R_2(R_1 + R_2)$,
$b = R_1(R_1 + 2R_2)$,
$c = 2\pi - 2\alpha + \sin 2\alpha$.
8.23 8.23.

CHAPTER 9

9.1 796 W.
9.2 (a) 813 W, (b) 142.9 kW.
9.3 (a) 410 A, (b) 205 A, (c) 1.45 Ω.
9.4 (a) 1.4 kW, (b) 2.15 kW.
9.5 1.5 ms.
9.6 (a) 26 A, (b) 1.3 μs, (c) 26A / μs, (d) 0.38 μH, (e) yes.
9.7 265 W.
9.8 (a) 129 W, (b) 217 W.
9.9 (a) 10 W, (b) 0.82 W, (c) 514 W, (d) 7 mW.
9.10 1.25 μH, 2500 pF.

9.11 (a) 10 μH, (b) 4 μF, 1.5 Ω, (c) 18 kW, 2000 A, (d) 750 V.
9.12 29°C.
9.13 37.5 A.
9.14 (a) 73.6°C, (b) 36%, (c) 0.75.
9.15 (a) 73.9°, (b) 36%, (c) 0.75.
9.16 (a) 50.1°, (b) 57%, (c) 0.867.
9.17 (b) (i) 0.65, (ii) 1.0.

CHAPTER 10

10.1 0.042°C/W.
10.2 0.07°C/W.
10.3 0.19°C/W.
10.4 0.2°C/W.
10.5 0.31°C/W, high speed switching.
10.6 (a) 66.7 μs, (b) > 600 V, > 100 A.
10.7 (a) > 600 V, > 105 A, (b) 136.2 W, (c) 104.8 W, (d) 62.7 kW.
10.8 (a) 100 μJ, (b) 18 mJ, (c) 181 W.
10.9 604 W.
10.10 (a) 90.9 A, (b) 81.8 A.
10.11 70 A.
10.12 (a) 164°C, (b) 146.8°C.
10.13 (a) 110 W, 106 W, 1600 W, (b) 97.8%.

APPENDIX 10

LIST OF SYMBOLS

Throughout this text there is the convention that a lower-case italicized symbol represents the instantaneous value of the variable. Any upper-case italicized symbol represents a steady or constant value of the variable; it could be a dc, an average or an rms value. Usually the subscript denotes the form of the upper-case variable. If no subscript is employed, the upper-case variable is a constant for a dc supply and is an rms value for an ac supply. A bold variable can denote a phasor.

A	amperes
A	anode terminal
A	natural angular frequency of resonance (rad/s)
ASCR	asymmetric thyristor
B	base terminal
BJT	bipolar junction transistor
BV_{DSS}	MOSFET breakover voltage (V)
C	degree celcius
C	constant
C	collector terminal
C	capacitance (F)
C_{DS}	drain-source capacitance (F)
C_{GD}	gate-drain capacitance (F)
C_{GS}	gate-source capacitance (F)
C_{iss}	input capacitance (F)
C_{oss}	output capacitance (F)
C_{rss}	transfer capacitance (F)
C_s	snubber circuit capacitance (F)
C_{TH}	equivalent thyristor capacitance (F)
D	diode
DC	distortion component
DF	freewheeling for forward biased diode
DPF	displacement power factor
DR	reverse diode model
E	emitter terminal
E	electric field (V/M)
F	force (N)
FCT	field controlled thyristor
F_e	electromagnetic force (N)
FF	form factor
G	gate terminal
G	transconductance (S)
GATT	gate-assisted turn-off thyristor
GTO	gate turn-off thyristor
H	henry
HF	harmonic factor
I	steady current (A)
$\hat{I}$	current amplitude (A)
I_A	anode current (A)
$\hat{I}_A$	peak anode current (A)
I_{av}	average current (A)
$I_{A\,av}$	average anode current (A)
$I_{A\,leak}$	off-state leakage current (A)
$I_{A\,rated}$	rated anode current (A)
$I_{A\,rms}$	rms anode current (A)
I_{ASM}	non-repetitive on-state surge current (A)
I_B	steady base current (A)
IC	integrated circuit
I_C	capacitor current (A)
I_C	steady collector current (A)
$I_{C\,leak}$	collector leakage current (A)
I_D	steady diode current (A)
I_D	steady drain current (A)
$I_{D\,av}$	average diode current (A)
I_{dc}	dc component of a current (A)
$I_{D\,rms}$	rms value of drain current (A)
I_E	steady emitter current (A)
I_e	emitter current (A)

I_F	average forward current (A)
I_G	gate current (A)
IGBT	insulated gate bipolar transistor
I_h	holding current (A)
I_l	steady load current (A)
I_{la}	thyristor latching current (A)
$I_{l\,av}$	average load current (A)
$I_{l\,rms}$	rms value of load current (A)
I_o	initial current (A)
I_R	diode reverse leakage current (A)
I_{rms}	rms value of current
I_{RR}	peak reverse recovery current (A)
I_s	steady source current (A)
$I_{s\,av}$	average supply current (A)
$I_{s\,rms}$	rms supply current (A)
$I_{s\,1}$	rms value of the fundamental, component of current (A)
$I_{s\,1\,rms}$	fundamental, rms value of input current (A)
I_{SC}	short-circuit current (A)
I_{SH}	shunt current (A)
I_{sn}	rms value of the *n*th current harmonic (A)
I_T	triac current (A)
$I_{T\,av}$	average triac current (A)
I_{TH}	thyristor current (A)
$I_{TH\,av}$	average thyristor current (A)
$I_{T\,leak}$	triac leakage current (A)
$I_{T\,rms}$	rms triac current (A)
I_{TSM}	current surge rating (A)
I-V	current voltage
J	joules
J	*pn* junction
JFET	junction field-effect transistor
K	cathode terminal
L	inductance (H)
L_s	snubber, stray, source or leakage inductance (H)
M	mega
MCT	mos-controlled thyristor
MOSFET	metal oxide semiconductor field-effect transistor
P	average power (W)
PAM	pulse amplitude modulation
P_{av}	average power (W)
P_B	base power (W)
P_C	average conduction loss (W)
P_D	average power dissipation (W)
PF	power factor
P_G	average gate power (W)
P_l	average load power (W)
$P_{\max}$	maximum average power (W)
P_{OFF}	off-state steady loss (W)
P_{off}	average loss over turn-off (W)
P_{ON}	average on-state loss (W)
P_{on}	average turn-on loss (W)
P_{ph}	average power per phase (W)
P_{pk}	peak power (W)
PUT	programmable unijunction transistor
PWM	pulse-width modulation
Q	charge (C)
Q	average reactive power (VAr)
Q_{RR}	reverse-recovery charge (C)
R	resistance (Ω)
R_B	base-circuit resistance (Ω)
R_{CE}	collector-emitter resistance (Ω)
RCT	reverse conducting thyristor
R_D	diode equiv. resistance (Ω)
R_{DE}	base-emitter resistance (Ω)
R_{DS}	drain-source resistance (Ω)
$R_{DS(ON)}$	on-state resistance (Ω)
RF	ripple factor
R_G	gate-circuit resistance (Ω)
R_i	input resistance (Ω)
R_l	load resistance (Ω)
RR	rectification ratio
R_R	diode leakage resistance (Ω)
R_s	source or snubber circuit resistance (Ω)
R_{SH}	shunt resistance (Ω)
R_θ	thermal resistance (°C/W)
$R_{\theta\,CS}$	thermal resistance, case to sink (°C/W)
$R_{\theta\,JA}$	thermal resistance, junction to ambient (°C/W)
$R_{\theta\,JC}$	thermal resistance, junction to case (°C/W)
$R_{\theta\,SA}$	thermal resistance, sink to ambient (°C/W)

S	seimens
S	softness, snapiness factor
S	average apparent power (VA)
SBS	silicon bilateral switch
SCS	silicon-controlled switch
SIT	static-induction transistor
SITH	static-induction thyristor
SOA	safe operating area
SUS	silicon unilateral switch
Sw	generic switch
T	triac
T	period of cycle or period of switching (s)
T'	pulse width (s)
T_1, T_2	terminals of triac
T_A	ambient temperature (°C)
T_C	case temperature (°C)
TH	thyristor
THD	total harmonic distortion
T_J	junction temperature (°C)
T_S	sink temperature (°C)
UJT	unijunction transistor
UPS	uninterruptable power supply
V	volts
V	steady voltage (V)
$\hat{V}$	voltage amplitude
$V_{ab\,rms}$	rms line voltage (V)
V_{AK}	anode-cathode voltage (V)
$V_{AK(ON)}$	on-state SITH voltage (V)
$V_{an\,rms}$	rms phase voltage (V)
V_{av}	average value of voltage (V)
V_B	steady base supply voltage (V)
V_{BD}	diode breakdown voltage (V)
V_{BE}	steady base-emitter voltage (V)
V_{BO}	breakover voltage (V)
$V_{(BR)R}$	reverse breakdown voltage (V)
V_{CE}	collector-emitter voltage (V)
$V_{CE(BO)}$	BJT breakover voltage (V)
$V_{CE(SAT)}$	on-state transistor voltage (V)
$V_{CE(SUS)}$	BJT sustaining voltage (V)
V_D	steady voltage across a diode (V)
V_{dc}	dc voltage level (V)
V_{DF}	on-state diode voltage (V)
V_{DRM}	repetitive peak off-state voltage (V)
V_{DS}	drain-source voltage (V)
$V_{D(SAT)}$	IGBT on-state voltage drop (V)
V_{DWM}	repetitive peak working voltage (V)
V_C	capacitor voltage (V)
V_e	emitter voltage (V)
V_G	gate-supply voltage (V)
V_{GA}	MCT gate voltage (V)
$V_{GA(TH)}$	threshold gate voltage (V)
V_{GK}	gate-to-cathode voltage (V)
V_{GS}	gate-source voltage (V)
V'_{GS}	gate sustaining voltage (V)
$V_{GS(TH)}$	gate threshold voltage (V)
V_{GT}	triac gate voltage (V)
$V_{GTO(ON)}$	on-state GTO voltage drop (V)
V_L	steady inductance voltage (V)
V_{L1}	inductance voltage for t_{ON} (V)
V_{L2}	inductance voltage for t_{OFF} (V)
V_l	steady load voltage (V)
V_{l1}	rms value of fundamental load voltage (V)
$V_{l\,av}$	average load voltage (V)
$V_{l\,n}$	rms value of the nth voltage load harmonic (V)
$V_{l\,rms}$	rms load voltage (V)
$V_{MCT(ON)}$	on-state MCT voltage (V)
V_n	rms value of nth voltage harmonic
$\hat{V}_n$	amplitude of nth voltage harmonic
V_p	peak point voltage (V)
$V_{ph\,rms}$	rms phase voltage (V)
V_{rms}	rms value of voltage
V_{RRM}	repetitive peak reverse voltage (V)
V_s	supply voltage (V)
$\hat{V}_s$	supply voltage amplitude (V)
$V_{s\,1\,rms}$	fundamental, rms value of input voltage (V)
$V_{s\,rms}$	rms supply voltage (V)
V_T	triac voltage (V)
V_T	trigger voltage (V)
$V_{T(ON)}$	on-state triac voltage (V)

$V_{TH(ON)}$ on-state thyristor voltage (V)
W watts
W energy loss (J)
W_B base drive loss (J)
W_C conduction loss (J)
W_c loss during crossover time (J)
W_D loss per cycle (J)
W_d loss during delay time (J)
$W_{d(off)}$ loss during turn-off delay (J)
W_f loss over fall interval (J)
W_{fi} loss in fall interval (J)
W_{fv} loss in fall interval (J)
W_{OFF} off-state loss (J)
W_{off} turn-off loss (J)
W_{ON} on-state loss (J)
W_{on} total turn-on loss (J)
W_r rise-time turn-on loss (J)
W_{ri} loss in rise interval (J)
W_{rv} loss in rise interval (J)
W_s loss over storage interval (J)
Z impedance (Ω)
Z_n impedance of *n*th harmonic (Ω)
$Z_{\theta(t)}$ transient thermal impedance (°C/W)
$Z_{\theta JS(t)}$ transient thermal impedance, junction to sink (°C/W)

ac alternating current
a_n Fourier coefficient of *n*th harmonic
b_n Fourier coefficient of *n*th harmonic
dc direct current
f supply frequency (Hz)
f pulses per second (Hz)
g_m BJT transconductance (S)
i instantaneous current (A)
i_A anode current (A)
i_B base current (A)
i_C capacitive or collector current (A)
i_{Cs} snubber capacitor current (A)
i_D instantaneous diode current (A)
i_D instantaneous drain current (A)
i_{DFW} diode freewheeling current (A)
i_G instantaneous gate current (A)
i_l instantaneous load current (A)
i_s instantaneous supply current (A)
i_{ss} instantaneous value of steady-state current (A)
i_T triac current (A)
i_{TH} thyristor current at anode (A)
i_{trans} transient component of current (A)
k kilo
ln natural logarithm
m milli
m duty cycle
m mass (kg)
n nano
n *n*th harmonic
p pica
p instantaneous power (W)
p_C on-state instantaneous power (W)
p_c loss during crossover time (W)
p_d power loss during delay time (W)
p_{fi} instantaneous power over fall time (W)
p_{rv} instantaneous power over rise time (W)
q charge (C)
rad radian
rms root mean square
s seconds
s frequency domain operator
t time (s)
$t(0^-)$ just before zero time
$t(0^+)$ just after zero time
t_1 time for turn-off (s)
t' new time reference (s)
t'' new time reference (s)
t_c cross-over time at switching (s)
t_d switching delay time (s)
$t_{d(off)}$ turn-off delay time (s)
$t_{d(on)}$ turn-on delay time (s)
t_f fall time (s)

t_{fi} current fall time (s)
t_{fr} forward recovery time (s)
t_{fv} voltage fall time (s)
t_{gq} negative gate current interval (s)
t_{OFF} off-state time of blocking (s)
t_{off} turn-off time (s)
t_{ON} on-state conduction time (s)
t_{on} turn-on time (s)
t_q thyristor turn-off time
t_r rise time (s)
t_{ri} current rise time (s)
t_{rr} reverse recovery time (s)
t_{rv} voltage rise time (s)
t_s storage time (s)
u commutation overlap (rad)
u speed (m/s)
v instantaneous voltage (V)
v speed (m/s)
v_{ab} instantaneous line voltage (V)
v_{AK} instantaneous anode-cathode voltage (V)
v_{an} instantaneous phase voltage (V)
v_C capacitor voltage (V)
v_{CE} collector-emitter voltage (V)
v_{Cs} snubber capacitor voltage (V)
v_D instantaneous volt-drop across diode (V)
v_{DG} drain-gate voltage (V)
v_{DS} instantaneous drain-source voltage (V)
v_{GA} MCT gate voltage (V)
v_{GS} gate-source voltage (V)
v_L inductor voltage (V)
v_l instantaneous load voltage (V)
$v_{l\,ac}$ rms value of load ripple voltage (V)
v_R resistor voltage (V)
v_s instantaneous supply voltage (V)
v_T triac voltage (V)
v_{TH} anode-cathode voltage of thyristor (V)

$\triangleq$ defined equal to
ΔI current excursion (A)
ΔQ charge change (C)
ΔT temperature difference (°C)
Δt time interval (s)
ΔV voltage excursion (V)
Δv_{GS} gate voltage change (V)
Ω (omega) ohms

α (alpha) switch firing angle (rad)
α transistor current gain
β (beta) extinction angle (rad)
β transistor current gain
β_F forced current gain
β_{off} turn-off current gain
η (eta) efficiency of operation
η UJT stand-off ratio
γ (gamma) interval of conduction (rad)
μ (mu) micro
ω (omega) angular frequency (rad/s)
ω_o natural angular frequency of resonance (rad/s)
ϕ (phi) flux due to anode current
τ (tau) circuit time constant (s)
θ (theta) phase angle (rad)
θ_n phase angle of nth harmonic (rad)

INDEX